Reviews of Accelerator Science and Technology

Volume 1

Editors

Alexander W. Chao
Stanford Linear Accelerator Center, USA

Weiren Chou
Fermi National Accelerator Laboratory, USA

World Scientific

NEW JERSEY · LONDON · SINGAPORE · BEIJING · SHANGHAI · HONG KUNG · TAIPEI · CHENNAI

Published by

World Scientific Publishing Co. Pte. Ltd.

5 Toh Tuck Link, Singapore 596224

USA office: 27 Warren Street, Suite 401-402, Hackensack, NJ 07601

UK office: 57 Shelton Street, Covent Garden, London WC2H 9HE

British Library Cataloguing-in-Publication Data

A catalogue record for this book is available from the British Library.

REVIEWS OF ACCELERATOR SCIENCE AND TECHNOLOGY
Volume 1

ISBN-13 978-981-283-520-8
ISBN-10 981-283-520-2

Reviews of Accelerator Science and Technology
Vol. 1 (2008) v–vi
© World Scientific Publishing Company

Editorial Preface

This is the inaugural volume of the journal *Reviews of Accelerator Science and Technology.*

Particle accelerators are a major invention of the 20th century. From the first linear accelerator built by Rolf Wideröe in an 88-cm long glass tube in Aachen, Germany, in the 1920s to the latest gigantic 27-km circumference deep-underground *Large Hadron Collider* at CERN in Geneva, Switzerland, particle accelerators in the last nine decades have evolved enormously and fundamentally changed the way we live, think and work.

Accelerators are the most powerful microscopes for viewing the tiniest inner structure of cells, genes, molecules, atoms and their constituent protons, neutrons, electrons, neutrinos, quarks, and possible still undiscovered even more fundamental building blocks of the universe such as dark matter and dark energy. Accelerators are a time machine bringing us back billions of years to the first few milliseconds after the *Big Bang* at the creation of our universe.

They are the brightest light sources man has ever made, allowing us to take crystal clear images at the angstrom scale ($1\,\text{Å} = 10^{-10}$ meters) with femto-second resolution ($1\,\text{fs} = 10^{-15}$ seconds). This opens up a whole new world for material science, chemistry, molecular biology and life sciences.

Spallation neutron sources are another invaluable tool provided by accelerators for advanced science and technology. Accelerators with megawatt beam power may ultimately solve a critical problem facing our society, namely, the treatment of nuclear waste and supply of an alternative source of energy.

In addition to scores of large accelerators, there are also tens of thousands of smaller accelerators throughout the world. They are used every day for medical imaging, cancer treatment, design of new drugs, radioisotope production, fabrication of semiconductors and high-density microchips, mass spectrometry, cargo x-ray and gamma-ray imaging, detection of explosives and illicit drugs, and detection of illegal trafficking of nuclear materials. Commercial industrial accelerators have a global market with annual sales of multi-billions of dollars and rapidly growing.

Furthermore, in the course of building larger and more complex accelerators, technologies have been pushed to an extent never imagined before — large volume ultrahigh vacuum, large scale superfluid helium application to cryogenics, superconducting magnets, superconducting radio frequency systems, high power microwave devices, radiation-hard materials, global control systems, advanced instrumentation and diagnostics, super-fast computing and communication networks, and giant data storage and processing systems.

The goal of this new journal, *Reviews of Accelerator Science and Technology* (RAST), is to give readers a comprehensive review of this driving and fascinating field. RAST will document the tremendous progress made in this field in the past decades and describe its bright future. RAST will cover a wide range of topics, including linear and circular accelerators, high beam power accelerators, high brightness accelerators, synchrotron light sources, free electron lasers, medical accelerators, accelerators for industrial applications, advanced accelerator technologies as well as promising new breakthroughs up to the most recent research frontiers.

In order to produce a journal of the highest quality, an Editorial Board consisting of distinguished scientists has been formed. This Board advises the Editors on the long- and short-term planning of the journal, and selection of topics and prospective authors of each volume.

Articles are by invitation only. Each article will be written by one or more leading scientist with recognized authority in their fields and will be peer reviewed. Beginning in 2008, one volume will be published each year. The first volume is an overview of various types of accelerators across the entire field. Each of the later volumes will focus on one or two particular type of accelerators and present in-depth discussions and analysis.

The first volume consists of fourteen articles. The first article lists the milestones in this nearly century-long journey of particle accelerators. The following seven papers describe various types of accelerators, including electron linear accelerators, high power hadron accelerators, cyclotrons, colliders, synchrotron light sources and free electron lasers, medical and industrial accelerators. Three papers introduce advanced accelerator technologies: superconducting magnets, superconducting radio frequency systems and beam cooling. There is a comprehensive review article on the rise and fall of the Superconducting Super Collider (first part; the second part will be published in Volume 2) and an article on the evolution, growth and future of accelerators and the accelerator community. There is also a book review of a new biography of Pief Panofsky.

In addition to these review articles, this volume also includes a poster titled *A Brief History of Particle Accelerators*. The 30 landmark events from 1919 to 2008 selected highlight the dramatic historical movement in the accelerator world.

The efforts required from the community as a whole to initiate and nurture a new professional journal will not be small. We firmly believe that a high quality journal is a demonstration of the identity of our community and of its importance to society and all of science. We are confident that with an excellent Editorial Board, strong support of the accelerator enterprise and active feedback from the readers, this publication will become an authoritative review and a major vehicle for the dissemination of accelerator science and technology to the global scientific, academic and industrial communities.

Alexander W. Chao
Stanford Linear Accelerator Center, USA
achao@slac.stanford.edu

Weiren Chou
Fermi National Accelerator Laboratory, USA
chou@fnal.gov

Editors

Contents

Reviews of Accelerator Science and Technology
Vol. 1 (2008) 1–5
© World Scientific Publishing Company

World Scientific
www.worldscientific.com

Early Milestones in the Evolution of Accelerators

E. D. Courant

Brookhaven National Laboratory,
Upton, NY 11973, USA
ecourant@msn.com

About 80 years ago Rutherford [1] expressed the hope that particles could be accelerated to energies exceeding those occurring in radioactivity, enabling the study of nuclei and their constituents. Physicists and engineers have more than met this challenge, and today the LHC (Large Hadron Collider) at CERN, Geneva is about to accelerate protons to 7 trillion (7×10^{12}) eV. Here we describe some of the crucial steps that have gotten us there.

Keywords: Accelerator; cyclotron; betatron; phase stability; synchrotron; strong focusing.

1. Resonance Acceleration — The Cyclotron

In 1927, in Aachen, Germany, Norwegian engineering student Rolf Wideröe accelerated sodium ions in a sequence of drift tubes to 50 kV using a radio frequency voltage of only 25 kV, i.e. using the same voltage twice. He published an account of this achievement as his doctoral dissertation [2]; it was clear that with more drift tube electrodes the acceleration could be repeated many times, though Wideröe's first accelerator went only to a factor of 2. A few years later Ernest Lawrence came across Wideröe's paper in the Berkeley library. Not being proficient in German he skipped the text, but the diagrams and the equations were enough to give him the idea that if the particle beam was bent in a circle by a magnetic field, the same accelerating electrode could be used many times rather than requiring a long sequence of many electrodes as implied by Wideröe. He noted the crucial and fortunate fact that the angular frequency of revolution of a particle of mass M and charge e traveling in a circle in a magnetic field B is

$$\omega = \frac{eB}{Mc},\qquad(1)$$

independent of energy. Therefore, with two hollow semicircular D-shaped electrodes excited at this frequency, the beam gets accelerated every time it crosses the gap between the two dees, and thus in

n revolutions it picks up a total energy of 2 neV (V is the voltage between the dees). The maximum energy is determined by the size of the circular magnetic poles.

Lawrence had his student M. S. Livingston build a model demonstrating this principle. Using a magnet with pole faces 10 cm in diameter, and a maximum rf voltage of 2000 V on the dees, Lawrence and Livingston [3] succeeded in accelerating hydrogen molecular ions to an energy of 80 kV. And early in 1932 a larger machine, with 11-inch magnet poles and 4000 V on the dees, accelerated protons to 1.22 MeV [4], the first time in the world that particles were accelerated to over a million volts.

In the following years Lawrence and others built ever-bigger cyclotrons, culminating with a 60-inch machine in 1939, which accelerated deuterons to 20 MeV, and alphas to 40 MeV. Cyclotrons were also built in other US universities and in France and Japan (the Japanese cyclotrons were wantonly destroyed by the US occupying forces in 1945).

It became clear that there was a limit to the energy that could be achieved: the mass M in Eq. (1) is the relativistic mass, and therefore the frequency ω in a uniform magnetic field decreases with increasing particle energy rather than being exactly constant. Furthermore, it had already been found [4] that, to maintain vertical focusing, the magnetic field, rather than being exactly uniform, had to decrease with increasing radius and energy, thus exacerbating the decrease of frequency with energy. Therefore the rf phase at which particles cross the accelerating gap changes until it reaches a decelerating phase, limiting the number of turns on which the particles gain energy. Bethe and Rose [5] estimated that, with magnetic fields as high as 18 kG and rf voltages around 50 kV, the maximum attainable energy would be around 17 MeV for deuterons; higher dee voltages would only increase the attainable energy with the square root of the dee voltage.

Lawrence was not discouraged by this. He built a magnet with poles of 184-inch diameter, intending to reach 100 MeV with dee voltages of around a million volts. But by the time the magnet was finished the wartime Manhattan Project was underway and the giant magnet was diverted to use for electromagnetic isotope separation (the "Calutron"), and cyclotron development was suspended.

2. Electron Acceleration — The Betatron; Orbit Theory

Clearly, the cyclotron principle is not suitable for electrons, which are relativistic at all energies in the MeV range. A different approach is needed. One possibility, already thought of by several people in the 1920's, was to accelerate electrons with the electric field associated with a time-varying magnetic field. Wideröe conceived of a "ray transformer" (*Strahlentransformator*): in a radially symmetric magnetic field $B(r)$, electrons of momentum p can travel in a circular orbit of radius R with

$$RB(R) = \frac{pc}{e}. \qquad (2)$$

If the magnetic flux $\Phi = 2\pi \int_0^R rB(r)dr$ enclosed by this orbit is increased, a tangential electric field $E_\phi = \dot{\Phi}/2\pi Rc$ is produced on the orbit, accelerating the electrons. Wideröe saw that if the field $B(R)$ on the orbit and the flux Φ increase proportionately, *and the flux is just twice what it would be if the field throughout the inside of the circle were equal to the field on the circle*, the electrons will be accelerated just enough to stay on the same circle as they gain energy. So he built a device satisfying this condition, hoping to accelerate electrons to 6 MeV. But he did not succeed in accelerating the electrons. Therefore he then turned to the linear drift tube accelerator described in the previous section; his dissertation

paper [2] describes the ray transformer as well as the linear accelerator.

Others also attempted, unsuccessfully, to use this induction method of acceleration. Finally, in 1940, Kerst succeeded [6, 7], obtaining acceleration to 2.3 MeV. The secret of his success, at least in part, was theoretical analysis of orbit stability. The paper of Kerst and Serber [8] shows that around a circular orbit in a radially symmetric magnetic field the particles undergo radial and vertical oscillations of angular frequency

$$\omega_r = \Omega\sqrt{1-n}, \quad \omega_v = \Omega\sqrt{n}, \qquad (3)$$

where Ω is the angular velocity, and the field in the vicinity of the orbit radius R varies as $(R/r)^n$, defining the field index n. Thus, for stability, we require that

$$0 < n < 1. \qquad (4)$$

Equations (3) and (4) and Ref. 8 constitute the pioneering foundation of accelerator orbit theory.

Following this success Kerst proceeded to build a series of ever-larger induction accelerators. They were named "betatrons," because high energy electrons have been called "beta radiation" ever since radioactivity was discovered.

Following Kerst's publication [6–8], Wideröe succeeded in building a 15 MeV betatron in Germany (described by Kaiser [9]).

A 100 MeV betatron at General Electric in Schenectady (designed largely by Kerst) was the world's highest energy accelerator by 1945, eclipsing cyclotrons.

3. Phase Stability — Synchrotron and Synchrocyclotron

As the Manhattan Project approached success toward the end of the war, the thoughts of people at Berkeley returned to the matter of cyclotrons. McMillan [10] saw a way to get around the limitation of the cyclotron. He noted that (1) can be rewritten in the form

$$E = Mc^2 = \frac{eBc}{\omega} \qquad (5)$$

and saw that if we have a field B, an rf system with frequency ω, and a particle with energy E satisfying (5), and that we then slowly change ω and/or B, the energy E will automatically adjust itself so as to maintain the synchronism condition (1) or (5), as long as the speed at which the field or frequency changes is not too fast, and the rf field has enough voltage to keep up with the rate at which the energy has to change in order to keep (5) satisfied. The mechanism is this, as described by McMillan [10]:

"Consider a particle whose energy is just (5), which we call the equilibrium energy for field B and frequency ω. Suppose that the particle crosses the accelerating gaps just as the electric field passes through zero, changing in such a sense that an earlier arrival of the particle would result in an acceleration. This orbit is obviously stationary. Now suppose that a displacement in phase is made such that the particle arrives at the gap too early. It is then accelerated; the increase in energy causes a decrease in angular velocity, which makes the time of arrival tend to become later. A similar argument shows that a change of energy from the equilibrium value tends to correct itself. These displaced orbits will continue to oscillate, with both phase and energy varying about their equilibrium values."

"In order to accelerate the particles it is now necessary to change the value of the equilibrium energy, which can be done by varying either the magnetic field or the frequency (or both). While the equilibrium energy is changing, the phase of the motion will shift ahead just enough to provide the necessary accelerating force; the similarity of this behavior to that of a synchronous motor suggested the name "synchrotron" of the device."

In fact, the same principle of phase stability and its application to accelerators had already been discovered earlier by V. I. Veksler [11] in Moscow. When McMillan became aware of this work he lost no time in acknowledging Veksler's priority [12].

This principle of phase stability overcame the limitation on the cyclotron: simply modulate the frequency ω downward in a cyclotron magnetic field, and the particle will accelerate. The 184-inch magnet (see Sec. 2) that had been built with the hope of reaching 100 MeV in a cyclotron was now equipped with a pair of dees of moderate voltage and a frequency-modulated rf system, and reached 190 MeV with deuterons (380 MeV with alphas).

In principle the frequency-modulated cyclotron or *synchrocyclotron* can go as high in energy as one wishes if one makes the radius of the machine big enough. The radius for a given energy is determined by the equation

$$pc = eBR, \quad R = \frac{pc}{eB}, \tag{6}$$

where p is the particle momentum. But the whole area encircled by the top energy orbit has to be filled with magnetic field; this can get expensive if the size gets too big. In practice, synchrocyclotrons working this way have been built for energies up to about 600 MeV.

For electron acceleration we keep the orbit radius constant, and change the field — and the frequency — with time so as to keep (3) and (4) satisfied. Now magnetic field is needed only in a small annular region surrounding the orbit, in contrast with betatrons, where one needs a field inside the orbit averaging twice the orbit field.

In the immediate postwar years, the late 1940's, synchrocyclotrons and electron synchrotrons in the 300 MeV range proliferated, because there was exciting new physics to be done: the pion had been discovered in cosmic rays, and these machines led to a thorough exploration of pion physics.

Protons can also be accelerated with constant orbit radius and changing field and frequency. Proton synchrotrons were proposed and built at Birmingham (1 GeV), Brookhaven (3 GeV), Berkeley (6 GeV), and Dubna, USSR (10 GeV).

4. Strong Focusing

Once the 3 GeV proton synchrotron at Brookhaven (named the Cosmotron) was completed in the summer of 1952, a study group was formed to explore possible improvements of the design. In the Cosmotron the magnet was constructed of 288 identical sectors, all with a C shape. The aperture faced the outside, with the back leg on the inside of the orbit. The magnets all had a field index n [see Eq. (3)] equal to 0.6, so as to satisfy the stability condition (4). The magnets of the Cosmotron all faced outward; therefore negative secondary beams were easily obtained, but positive secondaries would tend to hit the inside wall of the machine. In addition, magnet saturation effects tended to reduce the usable "good field" region at the fields corresponding to top energy. Therefore it was proposed to alternate the magnet sectors, with some having the back legs on the inside and others on the outside.

This might lead to a problem: the focusing gradients might easily be different in the inward and outward sectors, especially in the fringing fields. Would this lead to instability of the orbits?

Almost at once analysis [13] showed that the alternating gradients could enhance stability rather than weaken it! With the right parameters the stability could be made much stronger than in the conventional case. The strength of the focusing is in effect determined by the ratio of the horizontal and vertical oscillation frequencies to the angular velocity. In the conventional case these ratios (known as the "tunes") are $\sqrt{1-n}$ and $\sqrt{n}$, necessarily less than 1; with the new scheme the tunes can be much larger, and this corresponds to small oscillation amplitudes. This makes it possible to make the magnet aperture really small. That, in turn, makes the magnets — and other components — much cheaper, and so one can go to higher energies than without "strong" focusing. Reference 13 also proposes quadrupole magnets for focusing in straight sections without bending.

Again, it turned out that the discovery of the alternating gradient principle had been anticipated — by N. C. Christofilos, an engineer working in Greece [14]. Our Brookhaven group acknowledged his priority [15].

Alternating gradient (AG) proton synchrotrons were started almost immediately at Brookhaven and at CERN, Geneva. And practically all large accelerators in the world built since 1952, up to and including the LHC, use AG focusing.

In the process of planning and constructing these synchrotrons, it became evident immediately that detailed theoretical analysis (and of course a great deal of engineering innovation) was indicated. The study of the behavior of particle beams in AG focusing accelerators and storage rings has grown into an active discipline and numerous papers and books have been published on the topic; we list a few [16–20].

It soon became apparent that the reduction in oscillation amplitude promised in the initial paper [13] was somewhat exaggerated. Adams, Hine and Lawson [21] pointed out that when the ratios of oscillation frequencies to the frequency of revolution (soon named the "tunes," ν_x and ν_y) are integral, any

small field errors will lead to large orbit excursions, and when the tunes are half-integral the oscillations will be exponentially unstable. Both these effects constitute *resonances* between the oscillations and Fourier components of field errors. The cure is to construct the magnets to very precise tolerances, and to design the fields and focusing gradients so as to ensure that the tunes are well away from resonant (i.e. integral or half-integral) values.

The art and science of devising layouts ("lattices") of the magnets in AG synchrotrons has continued to advance over the years, starting with the simple alternation of horizontally and vertically focusing magnets in the first AG synchrotrons. Among the first innovations was the introduction by Collins [22] of structures incorporating long straight sections, which make it possible to incorporate all sorts of features in accelerators. And Milton White (unpublished) and Danby et al. [23] proposed that it might be useful to separate the bending and focusing functions, i.e. having pure bending magnets (with uniform fields) interspersed with quadrupole focusing magnets, rather than the "combined function" bending magnets with focusing gradients proposed in Ref. 13 and employed in the first AG synchrotrons at Brookhaven and CERN. Almost all AG accelerators built since the first ones employ this "separated function" configuration.

Many developments in the subsequent evolution of accelerators are described in the following articles in this issue.

References

[1] E. Rutherford, *Proc. R. Soc. A* **117**, 300 (1927).
[2] R. Wideröe, *Arch. F. Elektrot.* **21**, 387 (1928).
[3] E. O. Lawrence and M. S. Livingston, *Phys. Rev.* **37**, 1707 (1931).
[4] E. O. Lawrence and M. S. Livingston, *Phys. Rev.* **40**, 19 (1932).
[5] H. A. Bethe and M. E. Rose, *Phys. Rev.* **52**, 1254 (1937).
[6] D. W. Kerst, *Phys. Rev.* **58**, 841 (1940).
[7] D. W. Kerst, *Phys. Rev.* **60**, 47 (1941).
[8] D. W. Kerst and R. Serber, *Phys. Rev.* **60**, 53 (1941).
[9] H. F. Kaiser, *J. App. Phys.* **18**, 1 (1947).
[10] E. M. McMillan, *Phys. Rev.* **68**, 143 (1945).
[11] V. I. Veksler, *J. Phys. (USSR)* **9**, 153 (1945).
[12] E. M. McMillan, *Phys. Rev.* **69**, 534 (1946).
[13] E. D. Courant, M. S. Livingston and H. S. Snyder, *Phys. Rev.* **88**, 1190 (1952).

[14] N. C. Christofilos, Focussing system for ions and electrons. US Patent No. 2736299 (application 1950, patent issued 1956).

[15] E. D. Courant, M. S. Livingston, H. S. Snyder and J. P. Blewett, *Phys. Rev.* **91**, 202 (1953).

[16] E. D. Courant and H. S. Snyder, *Ann. Phys.* **3**, 1 (1958).

[17] H. Bruck, *Accélérateurs Circulaires de Particules* (Institut National des Sciences et Techniques Nucléaires, Paris, 1966).

[18] M. Conte and W. W. MacKay, *An Introduction to the Physics of Particle Accelerators* (World Scientific, Singapore, 1991).

[19] D. A. Edwards and M. J. Syphers, *An Introduction to the Physics of High Energy Accelerators* (John Wiley, New York, 1993).

[20] S. Y. Lee, *Accelerator Physics* (World Scientific, Singapore, 1999), 2nd edn. (2004).

[21] J. B. Adams, M. G. N. Hine and J. D. Lawson, *Nature* **171**, 926 (1953).

[22] T. L. Collins, Cambridge Accelerator Report CEA-86 (July 1961).

[23] G. T. Danby, J. E. Allinger and J. W. Jackson, *IEEE Trans. Nucl. Sci.* **NS-14**(3), 431 (1967).

Ernest D. Courant is a physicist at Brookhaven National Laboratory since 1948, working primarily on particle dynamics in accelerators. Together with M.S. Livingston and H. Snyder, he originated alternating-gradient dynamics ("strong focusing"). He has taught at Princeton, Yale and Stony Brook. He was elected to the National Academy of Sciences 1976, received the Fermi Award 1986, and the Wilson Prize (American Physical Society) 1987. Retired as Distinguished Physicist Emeritus, 1990, he continues to participate in the RHIC project — primarily concerning polarized beams in accelerators.

Reviews of Accelerator Science and Technology
Vol. 1 (2008) 7–41
© World Scientific Publishing Company

Electron Linacs for High Energy Physics

Perry B. Wilson

Stanford Linear Accelerator Center,
2575 Sand Hill Road, Menlo Park, CA 94025, USA
pwilson@slac.stanford.edu

The purpose of this article is to introduce some of the basic physical principles underlying the operation of electron linear accelerators (electron linacs). Electron linacs have applications ranging from linacs with an energy of a few MeV, such that the electrons are approximately relativistic, to future electron–positron linear colliders having a collision energy in the several-TeV energy range. For the most part, only the main accelerating linac is treated in this article.

Keywords: Linear electron accelerators; accelerating structures; beam loading; rf breakdown; wake-fields; beam breakup; linear colliders.

1. Introduction

1.1. *History of electron linacs*

We will first present a short summary of the history of electron linacs. Such a history is covered in several articles. For example, a report by Gregory A. Loew [1] gives some milestones (to about 1982) in the history of linear accelerators for both electrons and protons. Also useful is a report by P. J. Bryant [2]. A recent book by Sessler and Wilson [3] reviews the field of particle accelerators. They also give short biographies and pictures of many of the people who played pivotal roles in accelerator research.

In his review, Bryant breaks the history of both linear and circular machines into three parts. In part 1, from 1895 to 1932, all accelerators were essentially linear dc machines. In 1932 a generator designed by Cockcroft and Walton reached a voltage of 700 kV, while in 1932 R. J. Van de Graaff published a paper on the design for a 1.5 MV electrostatic generator [2, 3]. Although such dc machines usually accelerated protons because physicists were interested in the interaction of protons with matter, they could in principle have accelerated electrons. In Bryant's second "history line," linear and circular accelerators parted company. In 1924 Ising proposed a timed dc field applied across a series of gaps between drift tubes of ever-larger length such that the arrival of the field at successive gaps matched the particle velocity [3]. Quoting from Bryant, "His invention was to become the underlying principle of all today's ultrahigh-energy accelerators. This is known as *synchronous acceleration*." Figure 1, from Ising's 1924 paper (see reference in [3]), illustrates this principle.

Although Ising used a series of timed dc pulses, he could as well have used an rf oscillator to provide the voltage across successive gaps. Wideröe realized this in 1928 and built a more practical model for a linac powered by a 100 MHz oscillator [3].

In 1929 Lawrence [3] added a magnetic field to bend the beam back to the input of the rf accelerating cavity, thereby inventing the cyclotron [3]. The age of circular machines was born. Bryant's third history line concerns the invention and development of the betatron [3]. According to him, "By the 1940's three acceleration mechanisms had been demonstrated: dc acceleration, resonant acceleration and the betatron mechanism. In fact, there were no new ideas for acceleration mechanisms until the mid-1960's, when collective acceleration was proposed."

Returning again to rf linacs, the frequency of the driving oscillator (or rf power source) is an important parameter. Assume that a particle spends one rf cycle traveling between gaps in the drift tubes, and that the gap length is small compared to the length of each drift tube. The distance between gaps, L, is then $L = \beta\lambda$, where $\beta = v/c$, c is the velocity of light, and λ is the rf wavelength. Note that the rf frequency is proportional to β. For a reasonable distance between gaps — say, 10 cm the frequency of the rf source is 3 GHz. High energy electron linacs were, therefore, not possible until high power rf sources in the microwave range were developed.

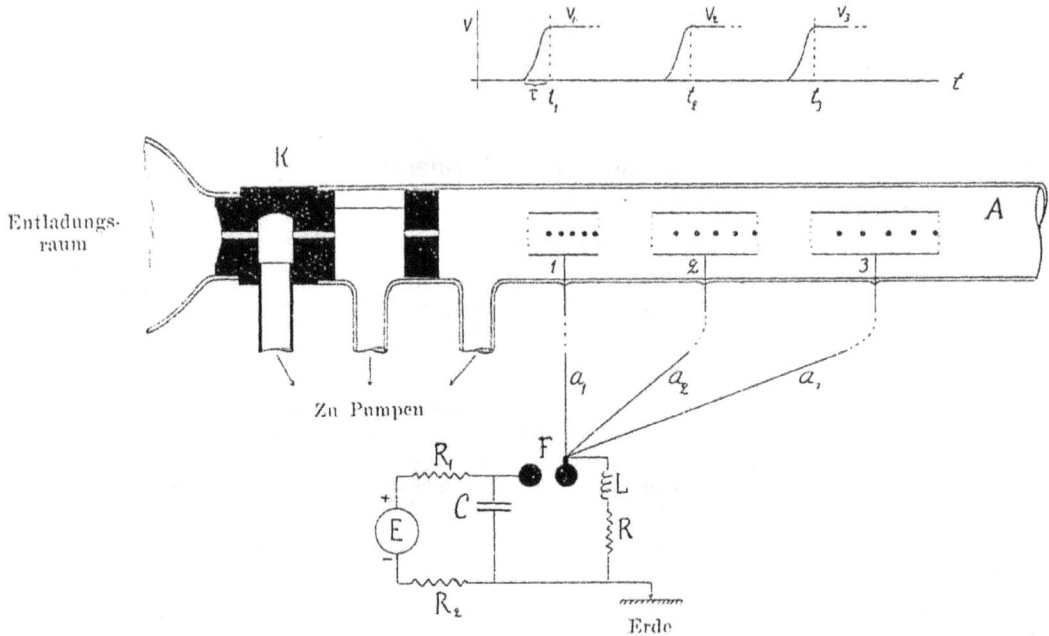

Fig. 1. Figure from Ising's 1924 paper illustrating the principle of resonant, or synchronous, acceleration.

The relativistic relation between particle kinetic energy in electron volts, V_K, the rest energy V_0 and the normalized velocity is

$$V_K = V_0\{(1 - \beta^2)^{-1/2} - 1\}.$$

Here $V_0 = m_0 c^2/e$, where e is the charge on the electron and m_0 its rest mass. The difference between the relativistic and nonrelativistic regimes is approximately marked by a particle having an energy equal to its rest energy, where $\beta = 0.866$. For an electron this is an energy of 511 keV, while for a proton it is about 1 GeV. After an injector of 10 MeV or so, we can assume that for the rest of the linac the electrons are traveling so close to the velocity of light that the difference is negligible for design purposes.

1.2. General references

Much of the material to be presented in the following sections was introduced in Ref. 4. The material presented in the following two sections is also covered in [5], Chapter 1. The material covered in Secs. 2.4 and 2.5 is also covered in [5], Chapter 3.

2. Accelerating Structures

2.1. Waveguides

Let us first consider a structure that is characterized by having any cross-section, but such that the cross-section and electrical properties do not vary with displacement along the axis (z axis). Such structures are referred to as waveguides. It has been proven in a number of books on microwaves that the electric and magnetic fields for a waveguide are solutions to the homogeneous vector Helmholtz equation; see, for example, Ref. 6, Subsec. 3.7. We will not be interested in the general case but will consider only a waveguide of circular cross-section. First, we need to introduce the concept of modes in such a waveguide; for example, see Ref. 6, Subsec. 3.18. For TM (transverse magnetic) modes, all the magnetic field components are transverse, and all the field components are obtained from the Helmholtz equation for E_z. It is shown that the radial variation of E_z follows a Bessel function behavior $J_n(k_c r)$. At the waveguide wall E_z must vanish, giving $k_{c,nm} b = p_{nm}$, where b is the waveguide radius and p_{nm} is the mth root of J_n. We now redefine β as the propagation constant in the waveguide, or $\beta = 2\pi/\lambda_g$ where λ_g is wavelength in the waveguide. The propagation constant in free space (a vacuum with no boundaries) is $k_0 = 2\pi/\lambda = \omega/c$. The Helmholz equation then reduces to $\beta^2 = k_0^2 - (p_{nm}/b)^2$. If the two terms on the right hand side are equal, λ_g approaches infinity and the waveguide becomes "cut off." The propagation constant becomes imaginary (attenuation without propagation) for still smaller values of b. The special case of $n = 0$ and $m = 1$ is of interest for particle acceleration. The TM_{01} mode is the lowest mode that has an axial electric field. For this case the preceding expression for the propagation constant reduces to $\beta = [\omega^2/c^2 - k_c^2]^{1/2}$. Here the cutoff propagation constant is $k_c = p_{01}/b$ and $p_{01} = 2.405$, the lowest root of J_0. The field does not vary azimuthally. Note that k_0 (or ω/c) plotted against β is the equation for a hyperbola.

In any system capable of supporting propagating waves, a number of wave velocities occur that pertain to signal propagation, energy propagation, and wavefront propagation (Ref. 6, Subsec. 3.19). For a propagating wave it is well known that, as a function of time and distance, the wave propagation can be expressed, in complex notation, as $\exp[j(\omega t - \beta z)]$. We will say more about this expression later. But for now we can see that the wave propagates at a velocity given by $z/t = \omega/\beta$. We define ω/β as the *phase velocity* of the wave, v_p. The phase velocity is simply the velocity at which the wave pattern moves along the waveguide. It can also be written as $v_p = c(\lambda_g/\lambda_0)$. In a rather complex discussion, Collin (Subsec. 3.11) shows that, if the wave is modulated to produce a signal composed of a narrow range of frequencies, the signal propagates at a *group velocity* defined by $v_g = d\omega/d\beta$. Using the expression above for β, the group velocity can also be written as $v_g = c(\lambda_0/\lambda_g)$. Thus, for a waveguide $v_p v_g = c^2$.

Collin also derives the important relation that the velocity at which energy propagates in a waveguide is the same as the group velocity defined by $d\omega/d\beta$. For an accelerating structure we want to match the phase velocity of the wave in the structure with the velocity of the electron. Since the electron is moving at a velocity that is slightly less than the velocity of light, the above expression shows that the group velocity (or energy velocity) must be slightly more than c. Matching the electron velocity to the phase velocity of the wave is physically impossible; longitudinally uniform waveguides will not work as accelerating structures.

Feynman [7] nicely illustrates the difference between the phase and group velocities for waves on water. He shows that, for the wake produced by a moving boat or for the ripples produced by a large rock dropped into the water, the phase velocity is exactly twice the group velocity. As the front of the wave pattern spreads out at the group velocity, the

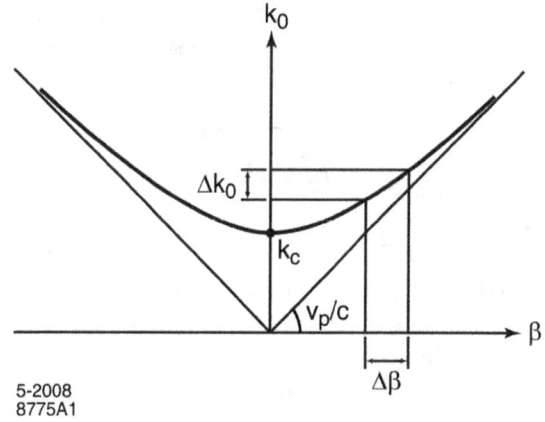

Fig. 2. Plot of k_0 versus β for a waveguide.

waves themselves move outward at twice the wave front velocity and mysteriously die away as they approach the front.

At this point it is useful to introduce a plot of k_0 versus β, as shown in Fig. 2, which illustrates some of these points. The straight lines have a slope v_p/c and the tangent to the hyperbolic curve has a slope v_g/c.

2.2. The pillbox cavity

In this subsection we follow the development in Ref. 4, Subsec. 3.4. A cylindrical cavity, looking something like a pillbox, is formed from a length of circular waveguide between shorts placed a half a guide wavelength apart. This simple cavity is useful because it forms a reasonable model for one cell of a periodic traveling-wave or standing-wave accelerator structure. Structure parameters such as the shunt impedance per unit length r, the quality factor Q, and the parameter r/Q are readily calculated. The expressions also show how these quantities scale with frequency. An end-to-end chain of pillbox cavities could in principle be used as a structure to accelerate a narrow beam, although small axial holes in the end plates would be needed.

Consider a pillbox cavity with radius b and axial length L between metal end-plates. The cavity operates in the so-called TM_{010} mode. Here the first and second subscripts indicate that E_z follows a J_0 variation out to its first root at the cavity wall. The first zero also implies that the field is constant in the θ direction. The third subscript indicates that the field is also constant in the z direction. The full dependence on the radius and time of the electric and magnetic fields for the TM_{010} mode are

$$E_z = E_0 J_0(k_c r) \cos \omega t,$$

$$H_\theta = -\left(\frac{E_z}{Z_0}\right) J_1(k_c r) \sin \omega t,$$

where $Z_0 = 377$ ohms. In the previous section we obtained $k_c = p_{01}/b = \omega/c$, giving $b/\lambda = p_{01}/2\pi$. The stored energy in the cavity, W, can be calculated from $\varepsilon_0/2$ times the integral of E_z^2 over the cavity volume. The power lost in the walls of the cavity is obtained from $R_s/2$ times the integral of H_θ^2 over the cavity surface, where R_s is the surface resistance given by $R_s = (\omega \mu_0/2\sigma)^{1/2} \sim \omega^{1/2}$. Here σ is the conductivity of the wall material. The results for the stored energy, the dissipated power and the cavity Q are

$$W = \left(\frac{\pi \varepsilon_0}{2}\right) E_0^2 b^2 L [J_1(p_{01})]^2,$$

$$P = \left(\frac{\pi R_s}{Z_0^2}\right) E_0^2 b(b + L) J_1^2(p_{01}),$$

$$Q = \frac{\omega W}{P} = \frac{C_1}{R_s} \sim \omega^{-1/2},$$

where $C_1 = (p_{01} Z_0/2)[L/(L + b)] = 453[L/(L + b)]$ ohms and p_{01} is 2.405.

There are two definitions for the cavity shunt impedance per unit length. The major cavity parameters can be obtained from a parallel resonant equivalent circuit (more on this later) with an inductor, a capacitor, and a resistor R in parallel. The resistor represents the losses in the cavity walls; hence the term "shunt impedance." In such a circuit, peak values of voltage and current are used. Hence the power input is related to the peak voltage by $P = V^2/2R$ and $R/L = r$ defines a shunt impedance per unit length. However, accelerator scientists have long used an accelerator shunt impedance defined by $P/L = G^2/r_A$, where G is the accelerating gradient in volts per meter. For a pillbox cavity with resonant wavelength λ, the accelerator shunt impedance per unit length is

$$r_A = G^2 \left(\frac{P}{L}\right)^{-1} = \left(\frac{C_1 C_2 T^2}{R_s \lambda}\right) \sim \omega^{1/2},$$

$$C_2 = \left(\frac{4Z_0}{(p_{01})^2}\right) [J_1(p_{01})]^2 = 967 \text{ ohms},$$

and T, called the transit time factor, is another parameter of interest. The transit time factor needs

some explanation. For the general case, the rf voltage gained by an electron moving at velocity c across a gap is calculated by integrating the field seen in a reference frame moving with the particle (the comoving frame). A full derivation using complex notation is given in Ref. 4, Subsec. 3.1. If the gap is very short, the voltage (energy) gained is simply $E_0 L$. For a longer gap with a field amplitude that is uniform with length (as it is for a pillbox cavity), we adjust the particle entrance time so that the field as a function time reaches a maximum as the particle reaches the center of the gap. The field is lower in amplitude as the particle enters and exits the gap. The transit time factor is a measure of the reduction in voltage compared to that seen by a particle moving with an infinite velocity. The transit time factor is

$$T = \frac{\sin(k_0 L/2)}{k_0 L/2},$$

where $k_0 = 2\pi/\lambda$. Another parameter of interest is the quantity r_A/Q, given by

$$\frac{r_A}{Q} = \frac{G^2}{W_1} = 967 \left(\frac{T^2}{\lambda}\right) \Omega/m \sim \omega.$$

Here $W_1 = W/L$ is the stored energy per unit length. Thus r_A/Q is a measure of how well a structure converts stored energy per unit length into an accelerating gradient. If $r_A L \equiv R_A$ and $V \equiv GL$, then the preceding expression becomes

$$\frac{R_A}{Q} = \frac{V^2}{\omega W}.$$

Let us put some values into the above equations. A traveling wave linac, such as the SLAC accelerating structure, can be modeled roughly as a chain of pillbox cavities with a phase shift between successive cavities. The SLAC structure operates with a phase shift of $2\pi/3$ radians per cavity (three disks per wavelength). At the SLAC frequency of 2856 MHz ($\lambda = 10.5$ cm), and using the cavity length corrected for the disk thickness, Q is about 20,000. This can be compared to a measured value of 13,500, which is 32% lower. The shunt impedance, taking into account the fact that part of the structure length is occupied by disks, is 41 MΩ/m. The measured value is about 56 MΩ/m, which is 39% higher. The effect of the beam apertures between cavities explains these differences between the measured and calculated values. The accelerating field has "shoulders," such that it falls toward zero on either side of the aperture in an end-plate. This produces a reduction in the stored energy and hence a reduction in Q. However, the accelerating field is now more concentrated toward the center of the cavity. This reduces the transit angle and increases the shunt impedance.

If we optimize the shunt impedance per unit length as a function of cavity length, we will find that r_A, is optimum for $L/b = 0.75$. The optimum value of r_A is about $100\,\text{M}\Omega/\text{m}$ at $2856\,\text{MHz}$.

2.3. *Standing-wave structures for high energy copper linacs*

Standing wave (SW) structures are in routine use for superconducting linacs, but not as yet for high energy electron linacs using copper structures. Because of their superior beam loading properties and possibly higher breakdown fields, short SW structures are a reasonable choice for multi-TeV, high gradient linacs of the future. There is, however, a 5.5-cell SW structure that was in operation for about three years in the ASTA (Accelerator Structure Test Area) bunker at SLAC [8]. It operated at a gradient of about $100\,\text{MV/m}$ and a pulse length of $200\,\text{ns}$. The gradient might have been pushed higher, but the experimenters were cautious about possible breakdown damage to the gun. The calculated shunt impedance for a chain of cylindrical cavities that model this structure, including an allowance for

the length occupied by disks, is $115\,\text{M}\Omega/\text{m}$, slightly higher than the measured value of about $100\,\text{M}\Omega/\text{m}$. Figure 3 shows a cut-away view of the structure. Laser illumination for the photocathode enters at the left end of the structure and the accelerated beam also exits there.

The SW structure shown in Fig. 3 was never meant to be a model for an SW structure for use in a multi-GeV linac. The NLC (Next Linear Collider) project at SLAC did, however, test several 15-cell SW structures intended to meet the design criteria for a $500\,\text{GeV}$ linear collider. This work came to an end when, in 2004, the International Technology Recommendation Panel chose superconducting technology rather than copper technology for a $500\,\text{GeV}$ linear collider (now renamed the International Linear Collider, or ILC). However, SW copper accelerating structures using X-band technology are still a possibility for use in a 2 or 3 TeV post-ILC collider.

The question then arises as to how long such an SW structure can be. A longer structure means that fewer rf feeds will be needed and the cost should be less. This is a somewhat complex question. Recall that in the π mode adjacent cavities are $180°$ out of phase. The azimuthal magnetic fields from adjacent cavities exactly cancel in the coupling hole region and there can be no Poynting vector to transmit power. In order to answer the question of structure length, the nature of the π mode resonance must be

RF Gun

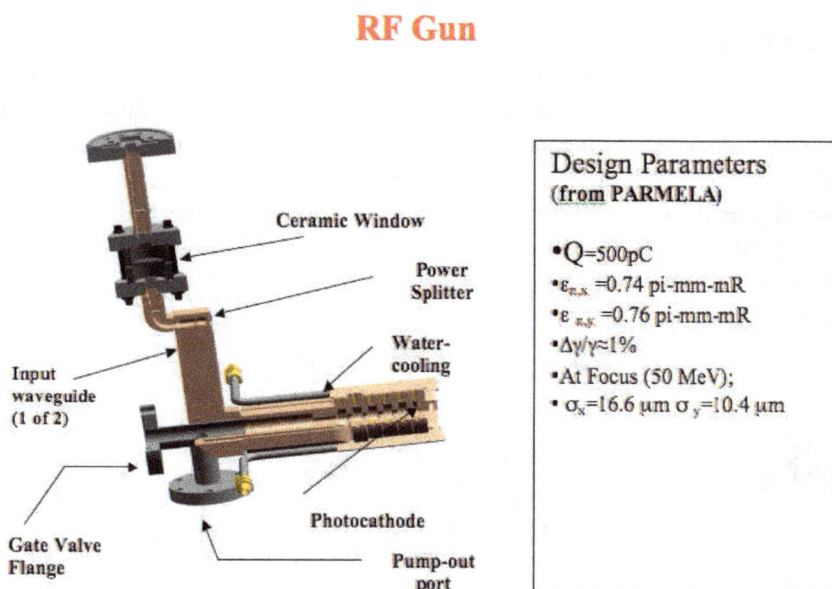

Fig. 3.   Cut-away view of the standing wave photoinjector gun structure (*courtesy of A. Vlieks, SLAC*).

carefully studied (see Ref. 9). To propagate the rf power needed to feed cells further from the input coupler, the peak of the π mode resonance must be shifted slightly from cavity to cavity. Assume that are $2N$ cells in a structure with an input feed in the middle cell. The total phase shift from the feed cell to the end cell is calculated to be $\theta = N^2/kQ$, where k is the half bandwidth of the structure (the frequency of the π mode minus the frequency of the 0 mode divided by the frequency of the $\pi/2$ mode). The phase shift from cell to cell decreases linearly as you move away from the feed cell, because less power is required to feed the remaining cells. Since an electron bunch will be off-crest by a varying amount along the structure, the allowable number for N depends on what energy loss is tolerable.

2.4. *Periodic structures*

Assume that a TM_{010} wave travels through a waveguide. Such a wave has a maximum longitudinal electric field along the axis of the guide, with lines of magnetic field in the azimuthal direction encircling the E field. This ensures that the Poynting vector, proportional to $\mathbf{E} \times \mathbf{H}$, and hence the wave power flow, is also directed along the axis. However, the phase velocity of an electromagnetic wave in a smooth round waveguide is always greater than c. Therefore the wave in such a guide cannot be synchronous with an electron and cannot continuously accelerate it. In order to slow down the phase velocity of the wave to match the electron velocity, yet also have a reasonably low group velocity (we will see the reason for this later), we must add some periodic metallic obstacles along the waveguide. We will assume that these obstacles are metallic disks with circular iris apertures at the center to allow the wave energy to couple from one accelerating cell to the next. If these apertures are relatively small compared to the waveguide diameter, we have a chain of coupled pillbox resonators. Such a periodic resonator chain is well modeled by the equivalent circuit model shown below.

In Ref. 10, Nantista shows in more detail how this circuit is derived and how to obtain a dispersion diagram ($\omega - \beta$ diagram). This equivalent circuit can also yield the transient time response to an arbitrary input pulse. The above circuit is valid for either capacitive coupling (through the electric field on the axis) or inductive coupling (through apertures in a magnetic field region), depending on the sign of k. The coupling constant k is simply one-half the relative bandwidth. This is given by the difference in the frequencies of the π mode and the 0 mode, divided by the frequency of the $\pi/2$ mode. Analogous to Fig. 2, an $\omega - \beta$ diagram (sometimes called a Brillouin diagram) can be drawn for the case of a periodic structure, as is shown in Fig. 5. The fundamental mode and the next higher mode are shown. In Subsec. 2.1 it was shown that the phase velocity of a wave is given by ω/β. Dividing by c normalizes the phase velocity to that of light. Thus the condition $v_p = c$ gives a line of unit slope. This line intersects the curves of allowed operating points at the small circles, fixing both the operating frequency and the phase shift per period d. Note the changes for the case with periodic irises from that of the unloaded waveguide.

The dashed portions of the curves indicate a negative group velocity. Assuming that the rf energy travels in the positive direction, the dashed portions cannot be used in a space harmonic decomposition of the traveling wave. To the left of the y axis the energy velocity is still positive in the solid portions of the curve, but the phase velocity is in the negative direction (a backward wave). Certain microwave oscillators work on the backward wave principle.

Certain features of this diagram are important for periodic structures in general. If there is a wave propagating along the guide, there will be a reflection at each iris. For a band of frequencies

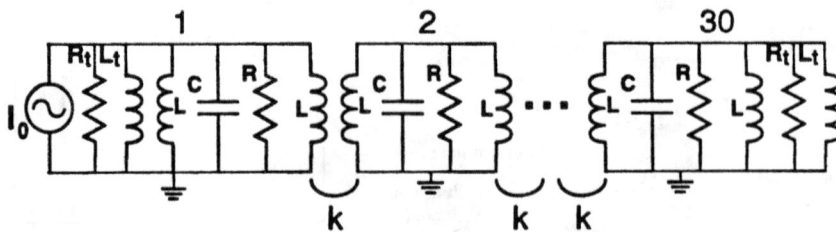

Fig. 4.   Equivalent circuit model for a periodic chain of resonators (from Ref. 10).

these reflections interfere destructively and there will be nearly perfect transmission along the guide, unaffected by the irises. However, for certain frequencies for which $\beta = n\pi/d$ the reflected waves from successive irises are exactly in phase, the group velocity falls to zero, and propagation is impossible; the wave is completely reflected. This creates a "stop band" between the fundamental mode and the next higher mode, as shown in Fig. 5. Next we need to consider space harmonics. We will follow the development by Loew and Talman in Ref. 1, Subsec. 4.7. Figure 6 shows a diagram of a periodic traveling wave structure. In a real structure, the iris tips and the outer cell walls may be curved for a higher shunt impedance. The fields in a single traveling wave mode in the hole region $a < b$ cannot at the same time meet the boundary conditions for $b > a$. Instead, we need a *superposition* of space harmonic modes. We now need to invoke Floquet's theorem for a periodic structure. This theorem states that, for a given mode of propagation at a steady state frequency, the fields at one cross-section differ from those one period away only by a complex constant. Thus, if there is no attenuation and the structure is axially symmetric,

$$E(r, z, t) = F(r, z)\exp[j(\omega t - \beta_0 z)].$$

Since $F(r, z)$ is periodic, we can expand it in a special Fourier series of the form

$$G(r, z, t) = C \sum_{n=-\infty}^{n=+\infty} a_n J_0(k_n r)e^{j(\omega t - \beta_n z)},$$

where a_n is the amplitude of the nth space harmonic. The phase shift per period and propagation constants are

$$\beta_n = \beta_0 + \left(\frac{2\pi n}{d}\right),$$

$$k_n^2(r) = k^2 - \beta_n^2.$$

Once β_0 is known, all the β_n's are known. The possible phase shifts per period are shown by the small circles on the horizontal line in Fig. 5 (this line in principle extends infinitely to the left and right). Each space harmonic has a different phase velocity given by $v_{pn} = \omega(\beta_0 + 2\pi n/d)^{-1}$. The propagation constants are in general a function of the radius. There is one notable exception, which is extremely important for the operation of linear accelerators. If $n = 0$ and $v_p = c$, then $k^2 = \beta_0^2$ and $k_0 = 0$. In the equation above, $J_0 = 1$ and a_0 is constant in region I.

Because an electron is only synchronous with β_0, all the power going into higher space harmonic components (for example $n = 1$ and $n = -1$ in Fig. 5) is lost and does not contribute to the shunt impedance. It is essential therefore that the shunt impedance of only the fundamental space harmonic component is known. Since powerful computer simulations were not available when SLAC was built in the early 1960's, the shunt impedance had to be measured. The job was difficult because of the constant gradient nature of the design chosen for the SLAC structure. In a constant gradient structure, the group velocity and the dimensions for each cell vary along the structure. About ten different resonant test cells had to be built [11] and, using standard microwave measurement techniques, the phase velocity, group velocity, r/Q, Q, and the shunt impedance for the fundamental space harmonic component were measured. Today the shunt impedance and any other desired structure property are calculated using a

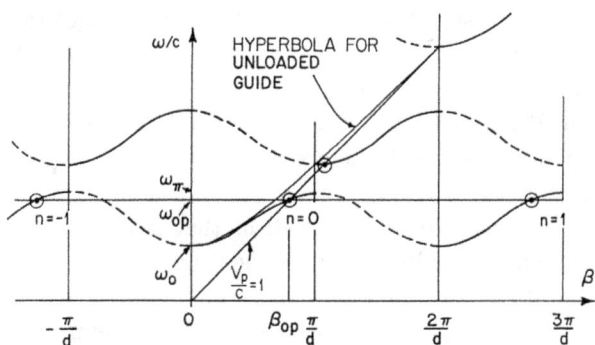

Fig. 5. Brillouin diagram for a chain of coupled cells, each having two modes (from Ref. 1 Fig. 23).

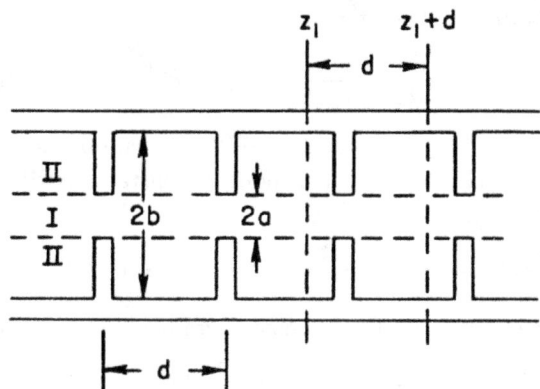

Fig. 6. Diagram of a traveling wave structure with a periodicity d.

computer simulation. Ordinarily, a 2D computer code such as SUPERFISH is used for the typical case of a structure that is cylindrically symmetric. This code is readily available for no charge and also has the advantage of being fast and accurate for calculating the above parameters. Many other codes can, of course, do the same problem but they are either not readily available, expensive, slow, or less accurate.

2.5. Traveling wave structures

The simplest kind of traveling wave (TW) structure is the *constant impedance* (CZ) structure. For such a structure the individual cells and the coupling apertures between them are all identical. If G is the accelerating gradient for the synchronous space harmonic, and w is the total stored energy per unit length for a power flow P, then the shunt impedance, Q, and r/Q are

$$r \equiv \frac{G^2}{\frac{-dP}{dz}},$$

$$Q \equiv \frac{\omega w}{\frac{-dP}{dz}},$$

$$r/Q = \frac{G^2}{\omega w}.$$

The energy flow velocity is $v_g = P/w$. Substituting for w in the second of the above expressions, we obtain

$$\frac{dP}{dz} = -\frac{\omega P}{v_g Q} = -2\alpha P.$$

Since $P \sim G^2$ we can write

$$\frac{dG}{dz} = -\frac{\omega G}{2v_g Q} = -\alpha G,$$

where α is the attenuation parameter per unit length, given by $\alpha = \omega/2v_g Q$.

For a constant impedance structure, α is constant with z. The above two equations can then be integrated to give

$$G = G_0 e^{-\alpha z}, \quad P = P_0 e^{-2\alpha z},$$

where G_0 and P_0 are the accelerating gradient and power flow at the input to the structure. The gradient and power flow at the end of a structure of length L are then $G_L = G_0 e^{-\tau}$ and $P_L = P_0 e^{-2\tau}$, where τ is the total attenuation parameter for the structure,

$\tau = \alpha L = \omega L/2v_g Q$. From the definition above for r and the first expression above for dP/dz, we obtain

$$G = (2\alpha r P)^{1/2} = (2\alpha r P_0)^{1/2} e^{-\alpha z}.$$

The structure energy V is the integral of $G\,dz$. Integrating the above expression from $z = 0$ to $z = L$, we obtain

$$V(CZ) = (rLP_0)^{1/2} \left(\frac{2}{\tau}\right)^{1/2} (1 - e^{-\tau}).$$

This expression points out one disadvantage of a constant impedance structure. If $\tau \approx 0.6$ (a typical value for τ), then the gradient at L is 55% of the gradient at the beginning of the structure. Later we will compare this to the energy gain for a constant gradient structure. Also, one disadvantage of all TW structures is that some fraction of the input power is wasted in a load attached to the output coupler and is not useful for acceleration. For $\tau = 0.6$, 30% of the input power goes into the load for a constant impedance structure. Also, a bad feature of a constant impedance structure is the fact that the gradient at the beginning of the structure is greater than the average gradient. For $\tau = 0.6$, the gradient is 33% greater at the input of the structure than it is at the output end. If rf breakdown is an important consideration, this imposes a performance limitation on a constant impedance structure.

A structure with a gradient that is reasonably constant along its length is, then, highly desirable. The SLAC linac and all modern linacs are designed using structures that have a gradient that is nearly constant with length. The expression above for G is still valid but the group velocity, and hence α, is now a function of z. If G is constant, so is αP (we assume that r is roughly constant), and, since G is constant, so is dP/dz. Therefore

$$-\frac{dP}{dz} = \frac{\omega P}{v_g Q} = \frac{P_0}{L}(1 - e^{-2\tau}).$$

Since $P/P_0 = 1 - (z/L)(1 - e^{-2\tau})$, we obtain the group velocity variation along a constant gradient structure:

$$v_g(z) = \frac{\omega L}{Q}\left[1 - \frac{z}{L}(1 - e^{-2\tau})\right](1 - e^{-2\tau})^{-1}.$$

The energy gain for a constant gradient structure is

$$V = G_0 L = (rLP_0)^{1/2}(2\alpha_0 L)^{1/2}.$$

Since $\alpha_0 = (-dP/dz)(2P_0)^{-1}$ and $-(dP/dz) = (P_0/L)(1 - e^{-2\tau})$, this gives $2\alpha_0 L = (1 - e^{-2\tau})$. The

above expression becomes

$$V(CG) = (rLP_0)^{1/2}(1 - e^{-2\tau})^{1/2}.$$

At $\tau = 0.6$, a CG structure has 1.5% more energy gain than a CZ structure for the same input power.

An important parameter for TW structures is the filling time T_F. The filling time is simply the integral of dz/v_g from $z = 0$ to $z = L$. If the group velocity is constant, then $T_F = L/v_g$. The relation between the filling time and the attenuation parameter is then $T_F = (2Q/\omega)\tau$. Even if the group velocity is a function of z, the above relation between T_F and τ still holds because the two quantities involve the same integral of $1/v_g$.

About 50 TW accelerating structures were built for the NLC project (in collaboration with KEK and FNAL) over a period of about 10 years [12]. Early NLC structures were 1.8 m long and had a v_g/c ranging from 0.12 at the beginning of the structure down to 0.03 at the output end. The filling time was 100 ns and the attenuation parameter was $\tau = 0.6$. A typical 1.8 m NLC structure is shown in Fig. 7.

A number of problems had to be solved before these structures could meet the NLC design parameters. The most difficult one was to meet the allowable

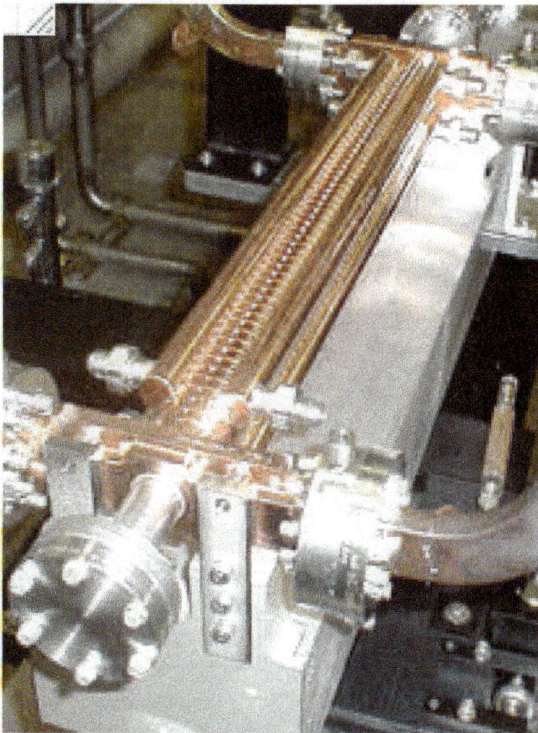

Fig. 7. An early 1.8 m traveling wave accelerating structure for the NLC (*courtesy of Juwen Wang, SLAC*).

breakdown rate of 10^{-6} per pulse per 60 cm of structure length. It was found experimentally that the breakdown rate was improved by using a larger iris aperture in the first half-dozen cells and by designing the shape of both the irises between cells and the input coupler iris.

2.6. Beam loading in TW structures

2.6.1. Two powerful concepts

We begin by introducing two very basic concepts that underlie all calculations of the interaction between the rf field produced in an accelerating structure by an external source and the field induced by an electron bunch. Note that an electron beam is simply a train of individual bunches, normally spaced one rf period apart. The first basic concept is *superposition*. Since Maxwell's equations are linear in both the electric and magnetic fields, a superposition of fields is also a solution. A superposition of rf powers, of course, is never allowable. We first calculate the gradient induced by the beam alone without the gradient produced by an external generator. We then add to this the gradient produced by the input rf. In the general case there may be a phase angle between these two gradients and the addition must be done using vectors. We treat the case in which the electron bunches are on the crest of the rf wave and are maximally accelerated (synchronous acceleration). Also, we assume steady state fields; that is, both the beam and the external rf source have been on for at least one filling time. The case of transient beam loading is analyzed in Ref. 4 Subsec. 7.3.

A second, and better-known, concept is *conservation of energy*. To conserve energy, the field induced by the beam must exactly oppose the accelerating field when the beam is on the crest of the accelerating wave.

2.6.2. Constant impedance structures

We next compute the beam-induced field in a TW structure assuming that there is no input power from the rf generator. From conservation of energy at any point along the structure dP/dz is given by

$$\frac{dP_b}{dz} = I_0 G_b - 2\alpha P_b.$$

Here G_b is the beam-induced gradient and the current is I_0. The beam current induces a gradient that opposes the beam motion and extracts energy from it. The second term gives the wall loss due to the flowing beam power. Since $G_b^2 = 2\alpha r P_b$, we differentiate to obtain $dP_b/dz = (G_b/\alpha r)dG_b/dz$. Equating this to the above expression for dP_b/dz and using $P_b = G_b^2/2\alpha$, we have

$$\frac{dG_b}{dz} = \alpha(rI_0 - G_b).$$

Integrating this expression, the beam-loading gradient becomes

$$G_b = I_0 r(1 - e^{-\alpha z}).$$

Note that the gradient vanishes at $z = 0$, as it must. Now integrate again to obtain the beam-induced voltage for a constant impedance structure:

$$V_b = I_0 r L \left[1 - \frac{1 - e^{-\tau}}{\tau}\right].$$

2.6.3. *Constant gradient structures*

The derivation of the beam-induced gradient and energy is more complex for constant gradient structures. Here we give the final expressions only and refer the reader to Ref. 4, Subsec. 5, for details:

$$G_b = \frac{I_0 r}{2} \ln[1 - \frac{z}{L}(1 - e^{-2\tau})],$$

$$V_b = \frac{I_0 r L}{2}\left[1 - \frac{2\tau e^{-2\tau}}{1 - e^{-2\tau}}\right].$$

As was pointed out previously, practical TW structures, such as those fabricated for the NLC, have a complicated variation in cell parameters along the structure. Overall parameters, such as τ and energy gain, are calculated by a computer simulation. However, the above expressions are useful for physical understanding and for scaling. Figure 8 shows the net energy gain, $V = V_0 - V_b$, as a function of beam current for both constant impedance and constant gradient structures at various values of τ. Note that at high beam currents a low value of τ gives the highest beam energy. The maximum efficiency for conversion of input power into beam power is also of interest. Writing the beam-loaded voltage for a structure as $V = V_0 - mI_0$ and the beam power as $P_b = V_0 I_0 - mI_0^2$, it is easy to show that the maximum beam power occurs at a beam current $I_0 = V_0/2m$ and a beam energy $V = V_0/2$. The maximum conversion efficiency is then $\eta(\text{max}) = (V_0/I_0)/4m$.

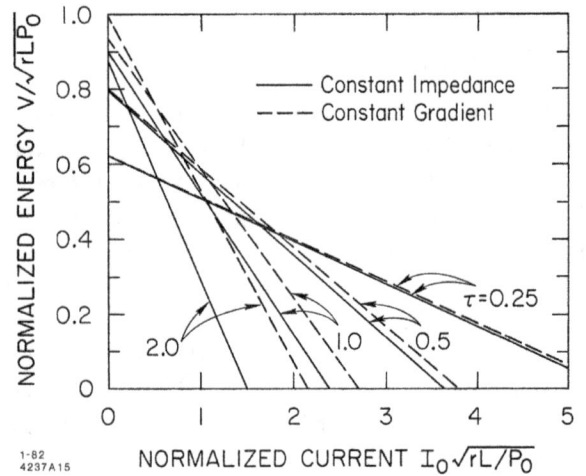

Fig. 8. Beam-loaded energy as a function of beam current for constant impedance and constant gradient structures at several values of the attenuation parameter.

2.6.4. *Vector addition of an rf generator voltage and a beam loading voltage*

For off-crest operation we invoke superposition and add the beam-induced and rf-generator-produced energies as vectors: $V_a = V_0 \cos\theta - V_b = V_0 \,(\text{real}) - V_b$. The net accelerating energy is shown in Fig. 9.

Note that in the above diagram we have chosen the positive real axis to be the negative of the beam-induced voltage. It is only the real component of the generator voltage that contributes to acceleration for synchronous operation. But there is also an imaginary component of V_0. The imaginary component is useful too. In off-crest operation the tail of a bunch will have an energy gain that is different from that of the head. A periodic system of quadrupole magnets, such as a FODO lattice (see, for example, Ref. 5,

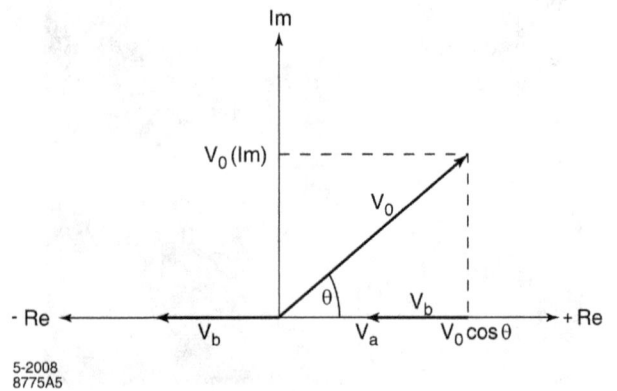

Fig. 9. Vector diagram illustrating the net energy gain in a structure with off-crest bunches.

Subsec. 7.10), can exploit this energy difference to damp transverse beam oscillations (BNS damping) caused by the dipole wake field (see Sec. 6).

3. Equivalent Circuit for a Cavity with Beam Loading

The material for this section is based on Ref. 4, Subsec. 3.5. It is covered in somewhat more detail in Ref. 5, Subsec. 5.4.

Figure 10 shows in (a) the equivalent circuit for an rf source (usually a klystron) connected to a linac cavity by a transmission line. First, note that both the rf cavity and the klystron output cavity are represented by resonant LRC circuits. While this may be intuitively obvious, a rigorous justification is possible for the use of a lumped-element circuit to model a resonant mode in a metal cavity (see for example Ref. 13). Second, note that the beam in an rf cavity is represented by a current generator. This is an excellent approximation for a relativistic electron beam because the beam current is independent of the cavity voltage. The situation is different for a klystron output cavity; a beam loading admittance is needed in the equivalent circuit. Without going into the complications arising from the necessity of matching TW and SW voltages at terminals A and B, we assume there is an isolator just before the cavity; any power reflected from the cavity will be absorbed in the isolator. The simplified circuit in (b) of Fig. 10 can be used, in which the transmission line admittance G_0 and the current generator representing the power source are transformed to the cavity

side of the coupling network. Here β is called the cavity coupling coefficient. If the rf generator (I_{gk}) is turned off and the cavity is then excited by the beam current only, β is the ratio of the power radiated out of the cavity coupling aperture to the power dissipated in the cavity walls.

In using this equivalent circuit, the available power, P_g, from the generator is identified with the incident klystron power. Also, watch out for factors of 2. In terms of the accelerator definitions of shunt impedance and the dc current I_0 that were used earlier, we have $G_c = 2/r_A$, $P_c = G_c V_c^2/2 = V_c^2/r_a$, and $i_b = 2I_0 \exp(-\omega_0^2 \sigma_t^2/2) \approx 2I_0$ for short bunches, where the bunches are assumed to be Gaussian with a time width σ_t. The voltages at resonance produced by the rf source current alone and the beam current alone are

$$V_{gr} = \frac{i_g}{G_c(1+\beta)} = \left[\frac{2\sqrt{\beta}}{1+\beta}\right](r_A P_g)^{1/2},$$

$$V_{br} = \frac{i_b}{G_c(1+\beta)} = \frac{I_0 r_A}{(1+\beta)}.$$

At resonance the net accelerating voltage is $V_a = V_{gr} - V_{br}$. The net accelerating voltage decreases linearly with current along a load line, similar to the behavior of the accelerating voltage for a TW structure with current, as shown in Fig. 8. However, here β is the parameter instead of τ. Roughly, $1/\beta$ plays the same role for an SW structure as τ does for a TW structure. The conversion efficiency is again maximum at a beam current that reduces the loaded accelerator voltage to one-half the unloaded voltage.

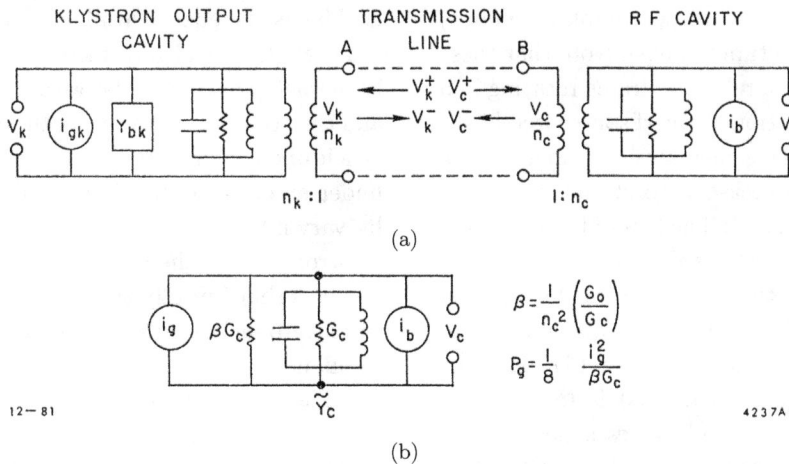

$$\beta = \frac{1}{n_c^2}\left(\frac{G_0}{G_c}\right)$$

$$P_g = \frac{1}{8}\,\frac{i_g^2}{\beta G_c}$$

Fig. 10. (a) Equivalent circuit for a beam-loaded cavity coupled to a klystron; (b) simplified circuit assuming an isolator is used.

We next consider beam loading in an SW structure tuned off-resonance. The admittance of the parallel resonant circuit in Fig. 10(b) that represents the cavity alone is

$$\mathbf{Y}_c = G_c \left[1 + jQ_0 \left(\frac{\omega}{\omega_0} - \frac{\omega_0}{\omega} \right) \right].$$

Note that letters in boldface are complex quantities. Here $\omega_0 = 1/\sqrt{LC}$ is the resonant frequency, $W = (1/2)CV_c^2$ is the stored energy and $Q_0 = \omega_0 C/G_c$. We limit the following discussion to the case of a high-Q cavity such that

$$\delta \equiv \frac{\omega_0 - \omega}{\omega_0}$$

has an absolute value that is small compared to 1. Then

$$\mathbf{Y}_c = G_c(1 + j2Q_0\delta).$$

The total admittance seen by the beam must include the coupled admittance of the input transmission line, βG_c, giving

$$\mathbf{Y_L} = \frac{1 + j2Q_L\delta}{R_0} = \frac{1}{\mathbf{Z_L}}.$$

Here $R_0 = G_c/(1 + \beta)$ is the loaded impedance at resonance and $Q_L = Q_0/(1 + \beta)$ is the loaded Q. We now define a tuning angle ψ by $\tan\psi = 2Q_L\delta$. Using this in the previous equation we obtain

$$\mathbf{Z_L} = R_0(\cos^2\psi)(1 + j\tan\psi) = R_0(\cos\psi e^{j\psi}).$$

In terms of the generator and beam loading voltages at resonance,

$$\mathbf{V_g} = V_{gr}\cos\psi e^{j\psi},$$
$$\mathbf{V_b} = V_{br}\cos\psi e^{j\psi}.$$

Figure 11 shows how these complex (phasor) voltages vary with the tuning angle. Note that these voltages are viewed in a reference frame rotating in a counterclockwise direction at the rf source frequency.

As the tuning angle increases, the magnitudes of both $\mathbf{V_g}$ and $\mathbf{V_b}$ decrease as $\cos\psi$ and the phases rotate through the angle ψ. The tips of these phasors therefore trace out a circle as ψ is varied.

Next, we consider the superposition of the generator voltage and the beam-induced voltage to obtain the net cavity voltage, as shown in Fig. 12. The reference phase (positive real axis) is taken to be the phase of $-\mathbf{i_b}$; this usually gives a net accelerating voltage, V_a, that is positive. Note that only a positive real voltage component exactly opposes the

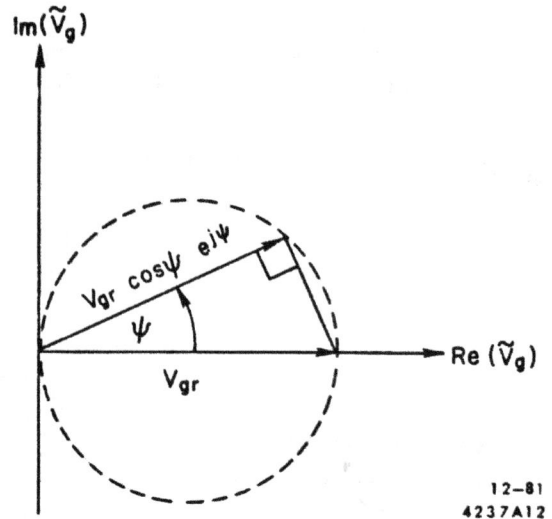

Fig. 11. Diagram showing how the generator voltage varies in the complex plane for the case of a cavity resonant frequency that is *greater* than the generator frequency.

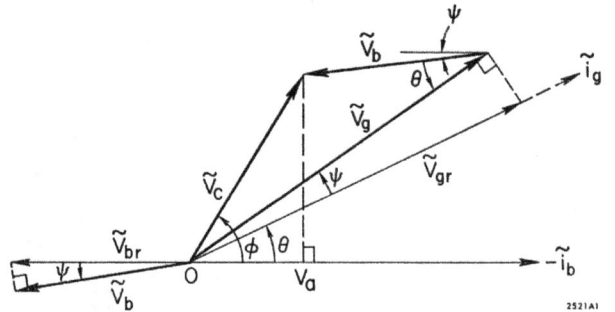

Fig. 12. Diagram showing the vector addition of the generator-produced and the beam-induced voltages in an rf cavity.

beam and takes the maximum possible energy out of it. This is the energy available for acceleration. Two other angles are also defined in the Fig. 12. Angle ϕ is the net phase angle between the current bunches and the crest of the rf wave in the cavity. It is also the synchronous phase angle in a storage ring. Angle θ is under external control in a linac; it can be adjusted by varying a phase shifter in the input drive to the klystron feeding the cavity or group of cavities. This assumes that initially the phase shifter was adjusted to place bunches on the crest of the wave in an on-resonance cavity.

There is one assumption that underlies this entire analysis. Let us assume the rf is turned on nearly instantaneously. A small portion of the rf wave enters the cavity coupling aperture (having a diameter that is usually small compared to the input waveguide transverse dimensions) and spreads through the cavity as a hemispherical wave front moving at the speed of light. The wave front is reflected at the cavity walls and then bounces back and forth many times inside the cavity before the field settles down into the pattern of a single mode. Suppose that it takes N back and forth trips before the field settles down into a single mode. The time needed for this is $t_N = 2\pi N/\omega_0$. But the cavity filling time is $2Q_L/\omega_0$. The filling time should be much larger than t_N, giving the condition

$$Q_L \gg \pi N.$$

If N is about 10, then Q_L must be greater than about 30. We have also assumed that the beam current bunches are spaced apart by a time which is very short compared to the filling time. The physics of the coupling of a waveguide to a cavity, and the transient buildup of the energy stored in the cavity, are discussed in Ref. 5, Subsec. 5.7.

4. RF Breakdown

4.1. *Introductory remarks*

If the reader is interested in digging more deeply into the subject of rf breakdown, the author would recommend that he or she study what has been learned in the past several decades about dc breakdown. An understanding of this subject, often called vacuum arcs, is important for industrial applications such as the use of vacuum switches in transmission systems for high electrical power. A good review of dc vacuum arcs is found in Ref. 14. We also note that some important physical effects, such as field emission, are nearly identical at both dc and rf frequencies. Many of these effects were first extensively studied at dc.

There are two major kinds of rf breakdown. The first kind, destructive breakdown, imposes an absolute limit on the surface electric field that can be reached. At this limit the surface becomes molten over a macroscopic area, on the order of 100×100 microns, in the high field region near the hole in a loading disk. The destructive breakdown process can be roughly divided into four stages: (1) the formation of "plasma spots" at field emission sites, each spot leaving behind a crater-like footprint; (2) crater clustering and the formation of areas with hundreds of overlapping craters; (3) surface melting in the region

of a crater cluster; (4) formation of a plasma covering the molten region. Such a plasma is separated from the surface below by a "Debye sheath," which is essentially a thin (less than a nanometer) space-charge-limited diode that obeys Child's law. The surface below the plasma is bombarded by an intense current ($\sim 10^{12}\,\text{A/m}^2$) of copper ions that were accelerated in the diode toward the metal surface. These ions push on the molten surface with a pressure of about a thousand atmospheres. This pressure in turn causes molten copper to migrate away from the high E-field region near the iris tip, resulting in a measurable change in the shape of the iris tip.

In an operating accelerator one wants to stay well below the surface field at which macroscopic damage can occur. Assuming that extensive surface conditioning (processing) has already taken place at gradients somewhat higher than the operating gradient, the important factor is then the acceptable breakdown rate at the operating gradient. For the NLC design this was 10^{-8} per rf pulse per 60 cm of structure at a gradient of 65 MV/m. The surface field at the iris tip is about twice the on-axis accelerating field. At a surface field of 130 MV/m, breakdowns do not result in surface damage; only single craters or small crater clusters are produced near a field emission site. A theory for the breakdown rate during processing is still under development by the author and others.

The development in the following subsections is based closely on Ref. 15. References to some of the details concerning the topics mentioned in the following subsections will be found there.

4.2. *Plasma spots*

Plasma spots were first studied by Xu and Latham (see Ref. 16) using a small dc diode. They found that, in a high electric field, small bright spots (viewed through a glass port) appeared at locations that were previously field emission current sites. Later examination showed small craters, with a diameter on the order of 10 microns, at the spot sites. Some pictures of dc plasma spots (called cathode spots) are found in Ref. 14, Chap. 3.

As the gradient is increased in an accelerating structure, sharp geometric features in a high field region will begin to field-emit. At some field level the tip of the projection literally explodes, injecting a jet or spray of liquid metal droplets into the field above the projection. The field emission current still flowing quickly vaporizes and ionizes the metal spray, forming a plasma at the emitter site. The plasma melts the metal surface below on a sub-nanosecond time scale. The molten area associated with the plasma will expand until the plasma quenches after some tens of nanoseconds. A small crater-like feature having a diameter $\sim$5–20 μm is left behind. Plasma spots form in both dc and rf fields and the craters produced are indistinguishable, assuming that the dc voltage comes from a high impedance source.

Two models for plasma spot initiation have been proposed. In the mechanical breakup model, the intense force due to the surface field at the tip of a projection exceeds the tensile strength of the metal, causing a fragment of the tip to break loose. While the fragment is still close to the remaining tip, the field emission current impacts the fragment and rapidly vaporizes it. The field emission current then ionizes the copper vapor, forming a plasma. In a second model, trapped gas close to the metal surface in the immediate area of the projection is released by field-emitted electrons (at the correct rf phase) that return to impact the surface. The gas moves out to the emitter tip region and is then ionized by the field emission current. The ions produced return to the emitter and bombard the tip surface, rapidly melting it. Plasma formation and breakdown rapidly follow.

A single plasma spot emits a current of several amperes. Depending on the rf phase at emission, a portion of this current moves away from the metal surface. However, a considerable fraction of the current will first move into the rf field, pick up energy from the field and then return to impact the surface near the emitter over an area about 100 microns in radius (at the X band). The energy spectrum of the impacting electrons ranges from zero to a few hundred keV. However, the energy deposited by a single plasma spot is not enough to melt the metal on a time scale that is short compared to the rf pulse length. In a crater field, however, the impact areas from multiple plasma spots, all alive at once, overlap to provide sufficient power density for surface melting over an area on the order of 0.1 mm² or more.

4.3. *Crater clustering*

The creation of a crater field is possible through the mechanism of crater clustering. The crater left behind by the destruction of a field emitter looks something like a volcano crater. Material that was thrown or pushed out of the central depression forms a jagged rim surrounding it. The sharply pointed features on the crater rim can themselves become field emitters. As the sharpest features burned away during processing, it becomes more likely that a new plasma spot will form on the rim of an existing crater. The resulting double crater will have even sharper rim features than a single crater and be even more likely to attract a new plasma spot. In this way craters tend to cluster together, eventually forming a field of hundreds of overlapping craters. The process of crater clustering is illustrated in Fig. 13.

The surface has been processed to a field of about 400 MV/m at 150 ns. Far from the iris tip we see single craters. Closer to the tip clusters of craters begin to form, and there is some evidence of surface melting in the larger clusters. Finally, in the tip region itself we see smooth, puddle-like areas that are indicative of melting.

4.4. *Calculation of the destructive breakdown threshold*

4.4.1. *Penetration of electrons into a metal*

An electron incident on a metal surface with an energy in the keV range will produce a shower of scattered electrons that will penetrate a considerable distance into the metal. This distance will depend on the energy of the incident electron and on the density of the metal. For a wide range of metals the product of the penetration depth, $X_P(\mu\text{m})$, and the density, ρ (g/cm³), is roughly 5.0. We will need this to simplify the final result of the temperature rise calculation.

4.4.2. *Calculation of the temperature rise*

We will first solve the one-dimensional heat equation using several approximations: the electron trajectories are normal to the surface and the reflection coefficient is independent of surface material; the energy is deposited uniformly over the penetration depth; all incident electrons have the same energy. Details of the calculation are given in Ref. 15. The result is

$$T = \frac{P_A}{X_0 C_s \rho} \int_0^{t_0} erf\left\{ \frac{X_0}{[4D(t_0 - t)]^{1/2}} \right\} dt$$

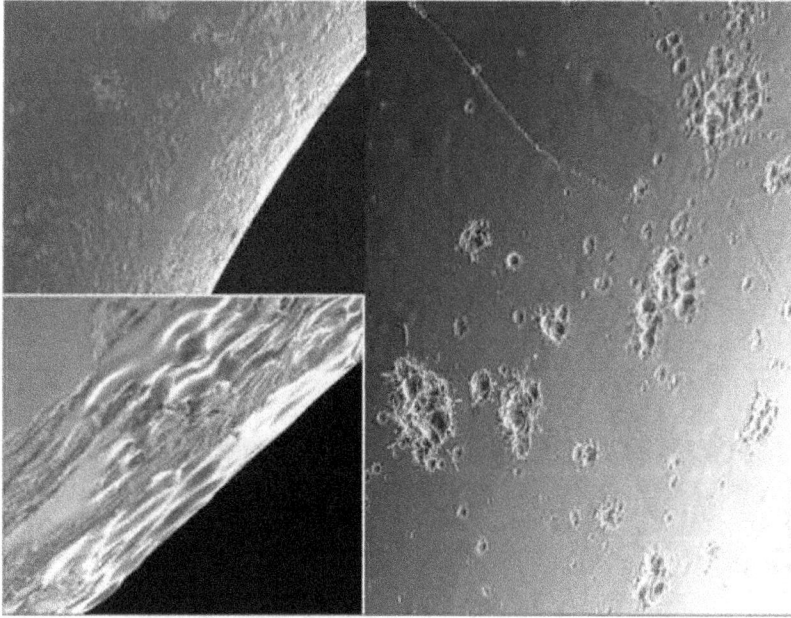

Fig. 13. SEM images of the iris tip region of a traveling wave structure after having been processed to the breakdown limit (*courtesy of C. Adolphsen, SLAC*).

Here P_A (W/cm^2) is the incident electron power per unit area, X_0 (cm) is the penetration depth, $C_s(J/g$-$^\circ$C) is the specific heat, K(W/cm-$^\circ$C) is the thermal conductivity, $D(\text{cm}^2/s) = K/C_s\rho$ is the diffusivity and t_0 is the time to melting. Recalling that $X_0\rho$ is approximately constant, the surface electric field (proportional to $\sqrt{P_A}$) at the melting point $T = T_M$ is proportional to

$$E \sim \left[\frac{T_M C_s}{I(t_0)}\right]^{1/2}.$$

Assuming a total X band pulse width of about 60 ns or greater, we pick a melting time of 30 ns so that the majority of the pulse length will be available to mechanically produce the Taylor cones mentioned later.

After the temperature is raised to the melting point, however, additional energy (the heat of fusion H) must be added to liquefy the metal. The breakdown field scaling then becomes

$$E_b \sim \left\{\left[\frac{T_M C_s}{(I/t_0)}\right] + H\right\}^{1/2}.$$

A small correction must be applied to this expression to account for the fact that the reflection coefficient at the surface of the metal depends on the atomic number of the metal nuclei (see reference in [15]).

Including this correction, we obtain the breakdown fields relative to copper that are shown in Table 1.

Among the conventional metals, beryllium is by far the best. However, it is a difficult material to work with for safety reasons. Note that several nonmetals have also been included. Carbon is a real possibility for an iris tip. We need to note again that these are *ultimate* breakdown fields, not the fields limited by an acceptable breakdown rate with no appreciable iris tip damage.

Several approximations have been made in the process of obtaining Table 1. The most important approximation was to assume a constant impact energy for the electrons that heat the metal surface. The actual energy spectrum of the impacting electrons can easily be taken into account in a simulation.

4.5. *From surface melting to destructive breakdown*

4.5.1. *Action of a dc field on a molten metal surface*

Figure 14 shows the surface of a thin metal layer that had first been melted in a container on a hot plate. An intense dc electric field was then applied to

Table 1.   Breakdown fields relative to copper for various metals.

< 1.00		1.00–1.09		1.00–1.09		1.20–1.29		1.30–1.39		> 1.4	
Zn	0.64	Cu	1.00	Zr	1.10	SS	1.20	Re	1.30	Be	2.0
Au	0.80	Ca	1.03	Mn	1.11	Co	1.22	Se	1.31		
Ag	0.83	Tc	1.07	Y	1.13	Os	1.26	Mo	1.34	Nonmetals	
Pt	0.89			Rh	1.13	Nb	1.27	Cr	1.36	Si	1.2
Hf	0.96			Ta	1.14	Ru	1.27	W	1.37	B	2.3
Pd	0.97			Al	1.14	Ti	1.29	V	1.39	C	3.7
				Mg	1.15						
				Ir	1.17						
				Fe	1.18						
				Ni	1.19						

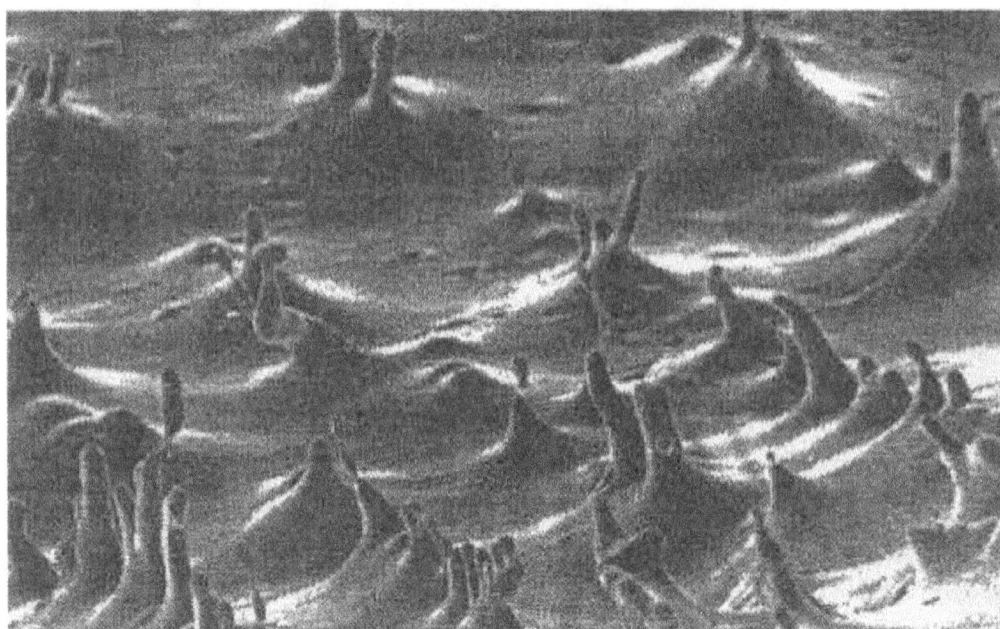

Fig. 14.   Image showing the growth of formations on a molten metal surface after being subjected to an intense dc electric field [17].

the surface and the hot plate turned off. After cooling, the field could be switched off and the surface features readily viewed. Note the conelike features with a base angle of about 45°, with fingerlike projections emerging from the apex of many of the cones. To form these features starting from a flat molten surface, the projections had to be pulled up by mechanical forces. A quick calculation shows that this has to be a slow process compared to the microsecond time scale of a single high power rf pulse. In a pulsed rf field it will therefore take hundreds or more pulses to build up these features because the molten region will cool down after every pulse and must then remelt on the next pulse. A model for the growth of these cones is discussed in

the next subsection. A rough mechanical model for this growth is given in Ref. 15.

4.5.2.   *Theory for the growth of the cones shown in Fig. 14*

We first model the growth of these cones before the projections emerge. Assume that the apex of a cone with a base angle ϕ is capped by a spherical segment of radius r, as shown in Fig. 15. Assume next that the growth starts from a shallow rounded perturbation, with a radius r_1, on the molten surface. It is not clear what sets the scale for this initial perturbation. In any case, the surface field is enhanced at the top of a perturbation compared to the field at the base

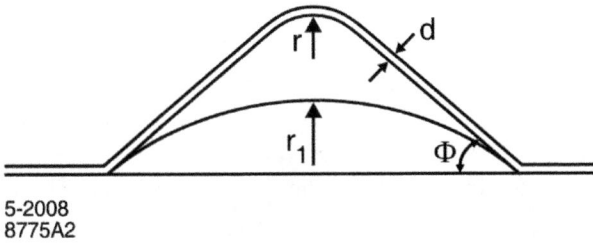

5-2008
8775A2

Fig. 15. Model for the cones shown in Fig. 14.

of the cone. A surface pressure gradient between the apex and the base of the perturbation exists that causes the liquid metal to migrate toward the apex. This further enhances the field at the tip, causing still further growth, etc. The apex must retain a spherical shape since it is in hydrostatic equilibrium between the outward pull of the electric field (proportional to $E_s^2 r^2$) and the restraining force of surface tension (proportional to r). Therefore, $E_s \sim r^{-1/2}$. However, as the cone grows, the enhancement of the field at the apex to that of unperturbed surface, defined as $\beta \equiv E_s/E_0$, also grows. Using the preceding expression for E_s, $\beta \sim r^{-1/2}$. From the geometry of Fig. 15, simulations show that β can be modeled as $\beta \sim r^{-n}$. For n to be exactly $1/2$, the base angle must be about $41°$. This is in agreement with the work by Taylor in 1964 [18]. Figure 16 shows what happens when a water droplet at the tip of a glass capillary tube is subjected to an electric field of increasing strength.

The cross hairs are set at an angle of $40.7°$. In his paper, Taylor shows that this is the base angle that gives a solution to the electrostatic field problem for a cone in a uniform electric field. Note that, as the cone tip radius decreases, it eventually explodes, emitting a spray of water into the vacuum. There is no good theory the author knows of to calculate this critical radius.

4.6. Concluding remarks

We have assumed that some area in the high field region near a structure iris tip is brought to the melting point early in an rf pulse by the electron bombardment mechanism. The E^2 force, acting mechanically on the molten metal surface, then pulls up a large number of Taylor cones over many rf pulses. The cone tips become unstable and spray jets of copper vapor into the electric field. The vapor ionizes in the strong field, forming a plasma over the molten region. The impacting ions from the plasma push down on the molten liquid metal with a force of hundreds of atmospheres. This ponderomotive force pushes the molten liquid toward a region of lower electric field. In turn, this produces macroscopic damage and a deformation of the shape of the iris tip, resulting in a measurable change in the phase shift of the cell.

Although there may be other uses for a very high gradient accelerator, the main motivation for studying rf breakdown is the application to a future multi-TeV linear collider for high energy physics research. Several groups around the world are working on the breakdown problem. In addition to the author, some of the other groups currently active are at Argonne National Laboratory, the University of Maryland, and the CLIC group at CERN. The CLIC group is especially active, because they have proposed a 2 TeV linear collider based on the two-beam technology that has been under experimental investigation at CERN for some time. Most experimental and theoretical results are presented at conferences being held in the US and at CERN on high gradient structures. The reader is referred to the websites of these conferences for current results.

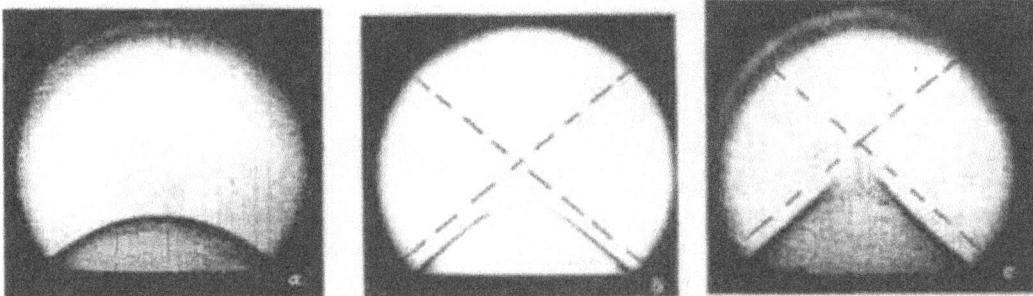

Fig. 16. Formation of a Taylor cone on a water surface at the tip of a glass capillary tube in a strong electric field. The field increases from left to right [18].

First of all, it must be noted that this is a single surface theory; that is, the loading disks are far apart enough so that emission of electrons or ions from a spot on or near one iris does not produce a regenerative interaction with a region near an adjacent iris. We must also point out that some of the material in the preceding two subsections is somewhat speculative. However, the theory does make some specific predictions. First, it ranks various metals according to their surface electric field at the ultimate breakdown limit with surface damage. This has been experimentally verified for gold and copper. Second, at this limit the theory predicts that the pulse width varies with the surface electric field as $t_p \sim E_s^{-1/4}$. This variation has been experimentally verified. Work is in progress to extend the theory to a calculation of the breakdown rate as a function of the surface field, assuming a field that is well below the ultimate field limit.

5. Single Bunch Beam Loading

5.1. *Differential superposition*

Our goal is to answer this question: How does a point charge induce a field as it passes through a resonant cavity? By "how" we mean developing a simple physical model for the process that is readily visualized without too much mathematics. We have already introduced two powerful concepts in Subsec. 2.6: conservation of energy and superposition. Of the two, superposition of fields is somewhat subtler. We will need them both in the following development.

Assume a point charge moving at velocity c along an arbitrary path through a cavity resonant in a single mode. We will call the path the z axis. The stored energy in the cavity, U, is related to the z component of the E field by

$$E_z^2(z) = \alpha U_c,$$
$$dU_c = [2E_z(z)/\alpha]dE_z.$$

The question arises as to why, when we differentiated, there was no $d\alpha/dz$ term. The answer is that $\alpha(z)$ is assumed to be slowly varying (adiabatic) as we move along the z axis. Imagine the axis split up into very many short regions of length Δz. Even though Δz is so short that α is essentially constant over this length, it is still infinitely large compared to the differential element dz.

We have assumed for simplicity that the cavity is empty of energy as the charge enters the cavity at $z = 0$. The charge must put energy into the cavity, so dU_c is positive. The induced field opposes the motion of the charge, so both E_z and dE_z are in the $-z$ direction. In distance dz the charge must lose energy:

$$dU_q = -qE_z dz.$$

By conservation of energy the charge must lose as much energy as the cavity gains. Equating the previous two expressions for dU we obtain

$$dE_z = \left[\frac{-q\alpha(z)}{2}\right] dz.$$

Note that E_z has dropped out and the differential induced field element is now independent of E_z; we can remove our assumption that there was some initial energy stored in the cavity. The above expression gives the differential induced field for a cavity with or without stored energy.

As a charge moves across the cavity, the differential field element induced at $z = 0$ will have rotated by an amount $\exp(jk_0 z)$. The net induced mode field when the charge is at position z is given by

$$E_z(z) = -\frac{1}{2}q \int_0^z \alpha(z)e^{jk_0 z}dz.$$

For the case of constant α the result is shown graphically in Fig. 17.

From the geometry of Fig. 17, the tip of the beam-induced phasor **E** follows a circular path with

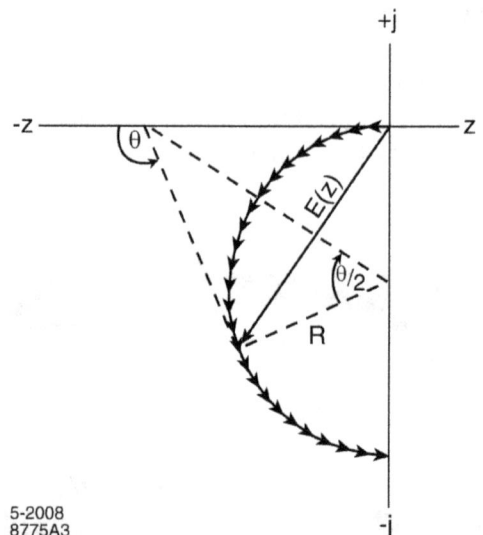

Fig. 17. Differential induced field elements and their summation to obtain the net induced field when the charge is at position z.

radius R and arc length $R\theta$, where θ is the total angle through which the differential field elements have rotated. If the cavity field were not changing with time, this would also be the net magnitude of the induced field, which in this case would be a straight line in the $-z$ direction. The actual magnitude of the phasor is the chord of the arc, which has length $2R\sin(\theta/2)$. Thus the ratio of the length of the chord to that of the arc is $[\sin(\theta/2)]/(\theta/2)$. Note that this is just the transit angle factor obtained earlier for the pillbox cavity.

According to this picture, as a charge induces a differential of energy it puts energy into the entire mode field, even ahead of the charge. This seems to violate the principle of causality, which states there can be no effect moving at a velocity greater than that of light. *Causality*, our third and even more subtle basic principle, must be preserved. It can be preserved while the charge is still in the cavity because the charge also induces a field in an infinite number of orthogonal, higher frequency modes. When the fields of all these modes are summed up, the fields ahead of the charge exactly cancel to zero. The terms in the summation will continually vary as the charge moves along the z axis. This ensures that causality is always obeyed. At the instant the charge leaves the cavity, however, only a single mode remains behind. Such is the magic of Maxwell's equations.

5.2. *The loss parameter*

We will now take the macroscopic value of dE_z over the length Δz. From the previous expression for dE_z we have $\Delta \underline{E}_z = -q\alpha \Delta z/2$. The potential over this region is $V = (\Delta z)(\Delta E_z)$, giving

$$V = \frac{q\alpha(\Delta z)^2}{2}.$$

From the first expression in Subsec. 5.1 and the definition of V,

$$V^2 = \alpha(\Delta z)^2 U_c.$$

We will now define a loss parameter as

$$K_L \equiv \frac{\alpha(\Delta z)^2}{4}.$$

Manipulating the previous three expressions, we have

$$U_c = K_L q^2,$$
$$V = -2K_L q,$$
$$K_L = \frac{V^2}{4U_c}.$$

The loss factor is closely related to the R/Q for a cavity, as defined in Subsec. 2.2:

$$K_L = \left(\frac{\omega}{4}\right)\left(\frac{R_A}{Q}\right) = \left(\frac{\omega}{2}\right)\left(\frac{R}{Q}\right).$$

Here V is the potential behind the charge itself, or the "wake potential" V_w. After the charge leaves the cavity, the cavity fields continue to ring (oscillate) at the mode resonant frequency and decay with a time constant $t_F \equiv 2Q_L/\omega_0$. If another charge comes along during the cavity ringing time, it will interact with the remaining field,

$$V_w(\cos\ \omega_0 t)\exp\left(\frac{-t}{t_F}\right).$$

If the cavity is initially empty of energy, the expression above for U_c gives the stored energy lost by the beam, $U_c = K_L q^2$. This is the reason K_L is sometimes called the loss factor. The energy lost by the beam can also be written as $U_b = qV_b$. From previous expressions we can write $U_c = qV_w/2$, or

$$V_w = 2V_b.$$

This is sometimes called the fundamental theorem of beam loading: a point charge "sees" exactly one-half of the induced potential immediately behind the charge.

5.3. *The fundamental theorem of beam loading*

There are at least two methods of proving this theorem. In Ref. 19 Chao proves the theorem in the time domain using only causality. In Ref. 4, Subsec. 6.1, it is proved in the frequency domain. In the proof presented there it is assumed that the fraction of the induced voltage seen by the beam is an arbitrary fraction, fV_b. Moreover, it is assumed that fV_b lies at an arbitrary phase angle θ with respect to $-V_b$. After a charge passes through the cavity and inducing a voltage in it, the charge is brought back around by a system of bending magnets so that it passes through the cavity a second time. It is assumed that the cavity Q is high enough so that no energy is lost in the process. In this second pass the beam gains some energy (voltage) from the voltage induced on the first pass, but loses the voltage induced on the second pass. A vector addition gives equations that can be true only if $f = 1/2$ and $\theta = 0$.

5.4. *Beam loading by a train of bunches with an arbitrary spacing*

Previously we have assumed a train of bunches with a spacing that is very small compared to the filling time. This allows us to treat an electron beam as a continuous beam current. Now that we know the cavity voltage induced by a single bunch, the net voltage induced by a train of bunches with an arbitrary charge and spacing can be obtained by superposition. A more realistic case is the net beam-induced voltage for a bunch train with bunches having the same charge and spacing. In Ref. 4, Subsec. 6.4, the case where the bunch spacing is a substantial fraction of the filling time is treated. This case has a practical application to electron storage rings with a large diameter. An example is the LEP electron storage ring at CERN (the ring has been dismantled and replaced by the LHC proton collider in the same tunnel). The ring had a diameter of 22 km and a bunch spacing 1.25 times the filling time. As a result, 5% more rf power was required.

6. Impedances and Wakes

6.1. *Introduction*

The literature on impedances and wakes in high energy accelerators and storage rings is both extensive and quite mathematical. Reference 19 devotes Chap. 2 to this topic, while, Ref. 20 is an entire book on the subject of impedances and wakes. References 21 and 22 are also useful introductions to this topic. Reference 4, Subsec. 9, is a less detailed introduction. In the limited space available here we will try to present a few of the most important concepts and refer the reader to the references for more details.

Ideally the vacuum chamber of a linear accelerator would be a smooth, possibly round, pipe with no change in cross-section and having infinite conductivity. All real metals, such as copper, do of course have a finite conductivity, and this already produces a wake field as the result of the resistive wall effect (we will use "wake," "wake field" and "wake potential" somewhat interchangeably). The physics of the resistive wake field is worked out in Ref. 19, Subsec. 2.1. Basically, the wall current in the pipe opposite the bunch diffuses into the metal during the bunch passage, inducing a z-directed electric field. This field in turn diffuses back to the surface

behind the bunch as a wake field. In general, however, we ignore the resistive wall wake field because it is completely dominated by wake fields produced by the many discontinuities in the walls of the vacuum chamber. We will often assume that the vacuum chamber has a center of symmetry, defined as the z axis, and that the beam moves along it at velocity c.

6.2. *The catch-up problem*

Suppose that we have a smooth, highly conductive vacuum chamber with a single charge $-q$ moving along the axis at nearly the speed of light. In the walls of the chamber (assumed round) directly opposite to it, there will also be a ring of charge, with a total charge value of $+q$, that moves with it — the wall current. Now suppose that there is a single small discontinuity at $z = 0$ in the wall of an otherwise smooth vacuum chamber with a radius b. At the discontinuity the wall charge must make sharp turns that produce radiation. These radiated fields have a locally spherical wave front that moves toward the axis at a velocity $v = c$. Although this wave front will never catch up with the point charge q itself, it will catch up with a test charge moving along behind it at a distance $s > 0$. The geometry of the situation is shown in Fig. 18. From the geometry of the figure the catch-up distance, z_c, is $z_c = (b^2 - s^2)/2s$. Note that for $s = b$ the catch-up distance is $z_c = 0$. The figure as drawn is for $s < b$. For $s > b$ the catch-up value of z is negative. It is not shown in the figure, but a toroidal radiated wave from the obstacle also spreads out in the $-z$ direction and the longitudinal impulse given to the test charge also changes sign.

An integration to obtain the total impulse therefore depends on the value of s. In addition, a vacuum chamber will in practice extend over some finite length. To catch all the impulses produced by the

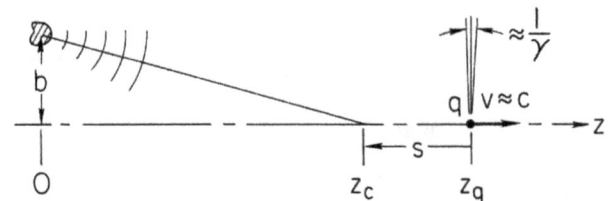

3-88   4998A21

Fig. 18. Radiated field produced by a small metallic object at radius b when charge q passes $z = 0$.

wake fields from a vacuum chamber component, we will need to integrate these wake fields along the z axis from minus infinity to plus infinity.

6.3. *Wake fields*

The electromagnetic effect in a vacuum chamber, left behind by a bunch with a δ function charge distribution and moving with velocity $v \approx c$, can be described in the time domain as a wake field. The effect due to a wall discontinuity is easier to visualize in the time domain but is often easier to calculate analytically in the frequency domain. Either case provides the same physical information. The wake fields excited by some simple wall discontinuities are shown in Fig. 19. In Fig. 19(a) a point charge enters and leaves a length of pipe with thin walls; no wake fields are produced. In Figs. 19(a)–19(e) the dashed curves show only the limits of the wake fields, which are contained in a toroidal region expanding at velocity c. Wake fields fill the entire interior of the toroidal region. Figure 19(f) shows the δ function wake fields between two thin parallel plates after the charge has left the second plate. The solid lines represent wave fronts with a δ function $\mathbf{E} \times \mathbf{H}$ field intensity. The area behind the wave front is filled in with finite, weaker fields. The fields for this geometry were first obtained analytically by A. Chao and P. Morton (see Ref. 19, Subsec. 2.4).

The wake fields excited by a Gaussian bunch moving through a section of the SLAC accelerating structure are shown in Fig. 20. We will note for later reference that a bunch with an arbitrary current density as a function of length can be constructed from a superposition of point charges.

6.4. *The longitudinal wake potential*

6.4.1. *Wake potential for a point charge*

The integrated effect of the wake fields produced by a driving point charge on a trailing test particle that follows a distance s behind it is usually of greater interest than the details of the wake fields themselves. Called the wake potential (or wake function), it is given in units of volts per coulomb. An analytic evaluation of the wake functions is possible in only a few simple geometries, such as a closed pillbox cavity. In most practical cases wake functions are evaluated using computer codes (see Ref. 20, App. 11.A). The

wake potential (volts/coulomb) is given in terms of the wake fields by

$$W_z(\mathbf{r}, \mathbf{r}_q, s) = -\frac{1}{q} \int_{-\infty}^{\infty} dz\, E_z \left[\mathbf{r}, \mathbf{r}_q, z, t = \frac{z+s}{c} \right]$$

Here $\mathbf{r}_q$ is the radius of the driving point charge q and $\mathbf{r}$ is the radius of the following test particle. For our purpose here we will assume $\mathbf{r}_q = 0$. The question of the upper and lower limits on the integral was discussed previously. The longitudinal momentum kick experienced by a test particle of charge e is $\Delta P_z = -(eq/c)W_z(s)$. Note the convention that a positive longitudinal wake is retarding if e and q have the same sign.

6.4.2. *Wake potential and loss factor for an arbitrary charge distribution*

Once the response to a point charge has been calculated, the wake potential can be used as a Green function to compute the potential behind an arbitrary charge distribution. If the line density of the charge distribution is $\lambda(s)$ per unit length, the longitudinal potential is

$$V_z(s) = \int_0^{\infty} ds'\, \lambda(s - s') W_z(s')$$
$$= \int_{-\infty}^{s} ds'\, \lambda(s') W_z(s - s').$$

Once the bunch potential is known, the total energy loss is

$$\Delta U = \int_{-\infty}^{\infty} ds\, \lambda(s) V_z(s).$$

The loss factor as previously defined is $K_L = \Delta U/q^2$.

6.4.3. *Wake potential and loss factor for a normal mode*

If we consider a point charge interacting with the nth cavity mode, then along a path behind a point driving charge the longitudinal wake potential has the form

$$W_{zn}(s) = 2K_{\delta n} \cos(\omega_n s/c).$$

To avoid confusion, we have taken the longitudinal loss factor for a point bunch to be $K_{\delta n}$. If the cavity is excited by a Gaussian bunch with an rms length σ, the induced potential can be obtained by

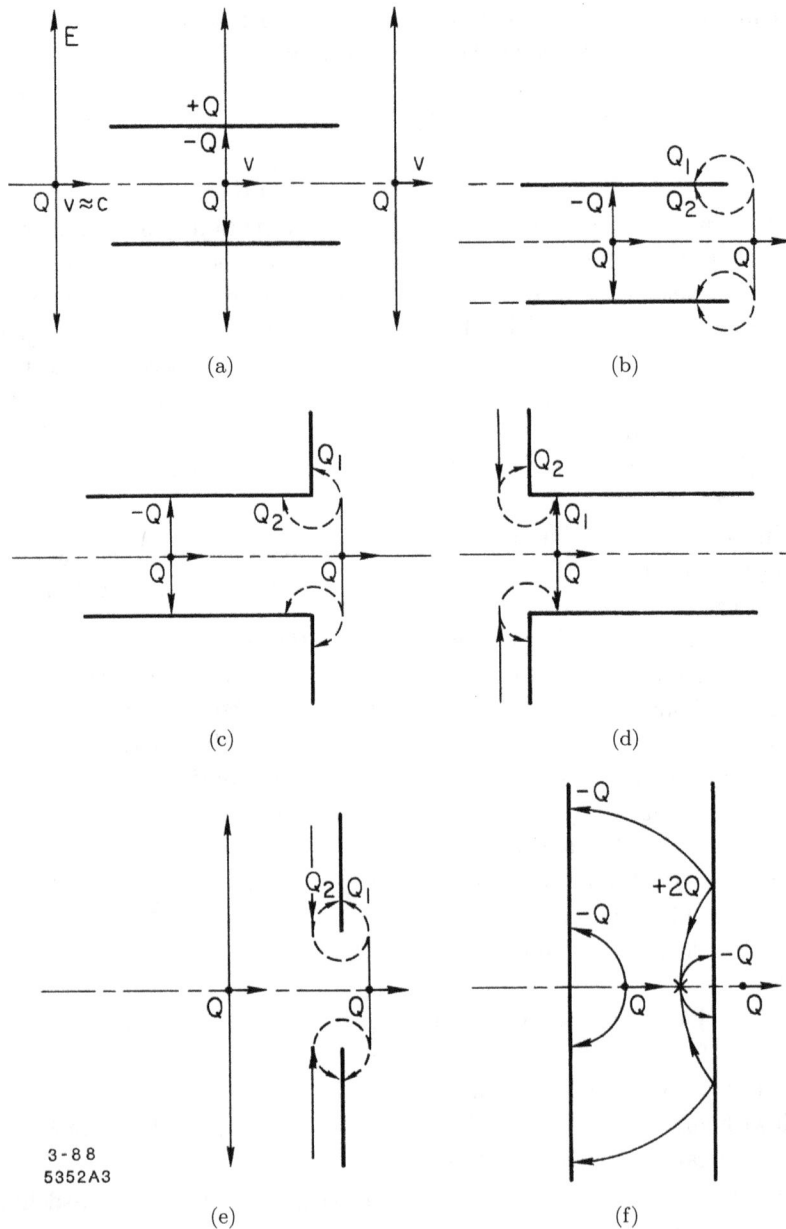

Fig. 19. Wake fields excited by a point charge moving through several cylindrically symmetric discontinuities.

substituting the above wake potential, giving

$$V_{zn}(s) = \frac{-2qK_{\delta n}}{\sqrt{2\pi}\,\sigma} \int_{-\infty}^{s} ds' \cos\left[\frac{\omega_n(s - s')}{c}\right]$$
$$\times \exp\left(-\frac{s'^2}{2\sigma^2}\right).$$

If position s is at least several bunch lengths behind the center of the bunch, the upper limit on the integral can be taken as infinity to give

$$V_{zn}(s) = -2K_{\delta n} \cos\left(\frac{\omega_n s}{c}\right) \exp\left(\frac{-\omega_n^2 \sigma^2}{2c^2}\right).$$

The loss factor for a Gaussian bunch interacting with the nth normal mode is

$$K_{Ln} = \exp\left(-\frac{\omega_n^2 \sigma^2}{c^2}\right).$$

6.5. *Transverse deflection modes*

6.5.1. *TM$_{110}$ mode pillbox cavity*

Just as the cylindrically symmetric TM$_{010}$ mode pillbox cavity in Subsec. 2.2 served as the basic model for the geometry of an accelerating structure,

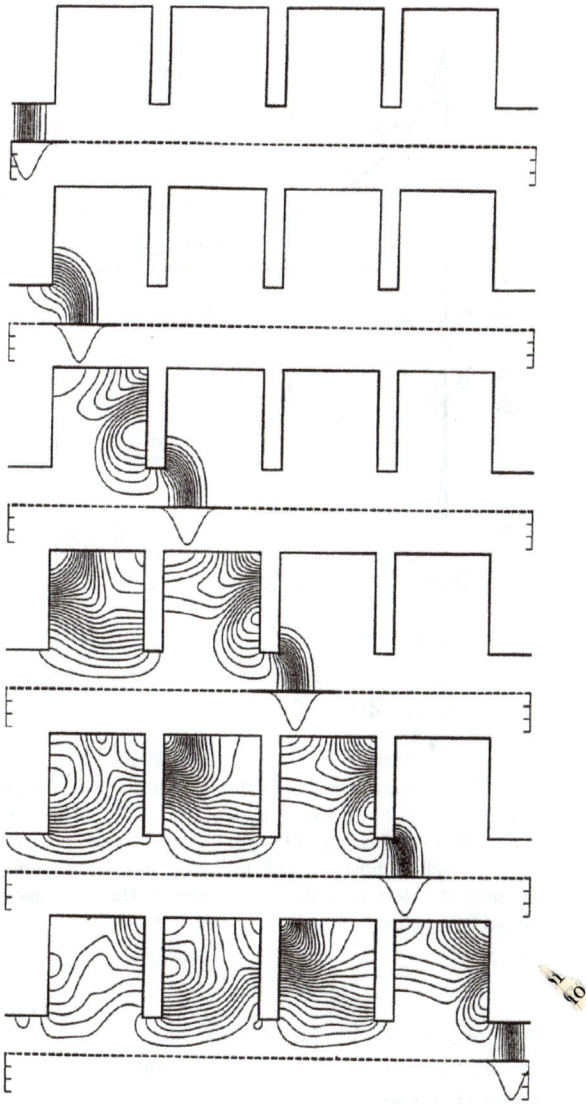

Fig. 20. Wake fields excited by a Gaussian bunch moving through the SLAC structure (*courtesy of T. Weiland*).

a TM$_{110}$ pillbox cavity serves as a model for a deflecting cavity. The fields vary as

$$E_z = E_0 J_1(kr) \sin(\theta) \cos(\omega t),$$

$$H_\theta = \left(\frac{E_0}{Z_0}\right) J_1'(kr) \sin(\theta) \sin(\omega t),$$

$$H_r = \left(\frac{E_0}{Z_0}\right) \left[\frac{J_1(kr)}{kr}\right] \cos(\theta) [-\sin(\omega t)].$$

A sketch of the cavity fields is shown in Fig. 21. But note that E_z and H_θ are not actually maximum at the same time in order to maintain a constant stored energy in the cavity.

The cavity will produce a transverse deflecting force on a charge moving on the axis. If E_z is proportional to $E_0 J_1(kr)$, the deflection voltage is

$$\Delta V_\perp = \left(\frac{c}{e}\right) \Delta p_\perp.$$

A transverse wake potential can be defined as the integral over the transverse EM forces along a straight path at a distance s behind a point charge q traveling at velocity $v \approx c$, normalized by dividing by q:

$$\mathbf{W}_{\perp z} = \frac{1}{q} \int_{-\infty}^{\infty} dz (\mathbf{E} + \mathbf{v} \times \mathbf{B})_\perp.$$

The integration path may be offset from, but parallel to, the path followed by charge q. Following Ref. 20, Subsec. 3.3, the above definition can be used for structures without any symmetry, but in general the transverse wake function depends strongly on the path of integration. It vanishes at a plane or axis of symmetry and starts at zero at the location of a point bunch. Common structures used in linear accelerators, such as accelerating cavities or vacuum chambers, usually have one or more symmetry planes or even an axis of symmetry. In this case the amplitude of the transverse wake function is zero if the exciting charge is on the axis and increases linearly with the displacement of q from the axis. The field induced behind q is uniform throughout the entire beam aperture and is in the direction of the displacement of q.

6.5.2. *The Panofsky–Wenzel theorem*

For the pillbox cavity in Fig. 21, one can work out that the deflecting voltage per unit length at a distance s behind a point unit charge is

$$W_\perp(s) = f(L, b)(kr_q) \sin(ks).$$

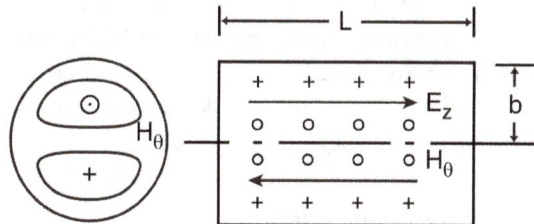

5-2008
8775A4

Fig. 21. Sketch showing the fields in a TM$_{110}$ mode pillbox cavity.

The longitudinal wake potential can be expressed as

$$W_z(s) = f(L,b)(k^2 r r_q) \cos(ks).$$

Note that

$$\frac{\partial W_\perp}{\partial s} = \frac{\partial W_z}{\partial r}.$$

This is a special case of the *Panofsky–Wenzel theorem*. In a more general form it can be stated as [22]

$$\boldsymbol{\nabla}_\perp \int_{-L/2}^{L/2} ds\, F_\parallel = \frac{\partial}{\partial z} \int_{-L/2}^{L/2} ds\, \mathbf{F}_\perp.$$

Here L is the length of the region or component over which a significant beam-induced field exists. See the above reference for details.

6.6. *Longitudinal and transverse wake potentials for the SLAC structure*

Periodic accelerating structures are reasonably well modeled by the three-parameter periodic structure shown in Fig. 6, with the addition of a fourth parameter for the disk thickness. The old computer programs KN7C [23] and TRANSVERS [24] solved for the synchronous longitudinal and dipole modes for the SLAC structure to obtain the figures shown below. Now, of course, there are more modern programs that will do this job (see Ref. 20, App. 11.A). The short range longitudinal wake function for the SLAC linac structure is shown in Fig. 22.

The wake in the top figure is obtained by adding an analytic extension to the sum of 450 normal modes. This extension is discussed in Ref. 25 (p. 592) and is based on the so-called optical resonator model.

The long range longitudinal wake for the SLAC structure is shown in Fig. 23. It was calculated using a very short bunch ($\sigma_z = 0.025\,\lambda$), by two different methods (see Ref. 26 for details). In this figure the wake is shown as negative immediately after the bunch. Note how the wake settles down into an oscillation at the fundamental mode frequency.

The short range dipole wake for the SLAC structure is shown in Fig. 24. Again, the dashed curve is the sum of 450 modes and the dotted–dashed curve is the lowest frequency dipole mode.

6.7. *The frequency domain*

6.7.1. *The impedance function*

We have already met impedances in the parallel resonant circuit model for a resonant mode in a cavity.

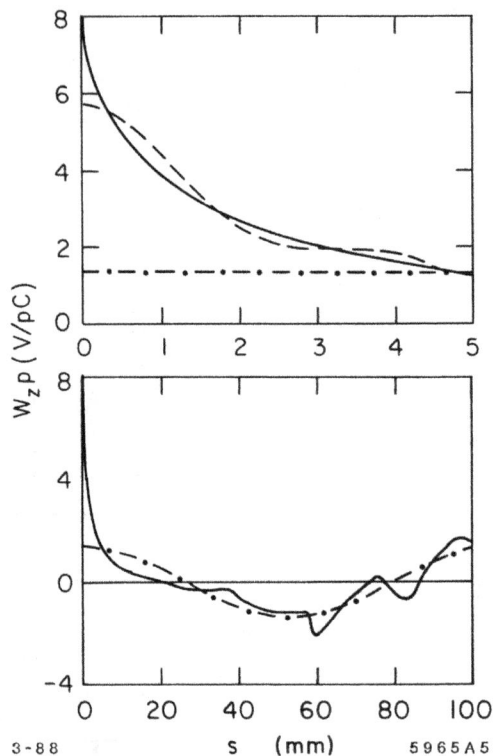

Fig. 22. Short range longitudinal wake for the SLAC structure ($a = 1.17\,\text{cm}$, $b = 4.13\,\text{cm}$, $d = 3.50\,\text{cm}$ and $g = 2.92\,\text{cm}$). Here d is the period and g is the period minus the disk thickness. The dashed curve in the top plot shows the sum of 450 modes; the dotted–dashed curve shows the accelerating mode only (*courtesy of K. Bane, SLAC*).

Now imagine a line current on the axis of a vacuum chamber component or an accelerating cavity. Assume that the current has a charge density modulation proportional to $\sin(\omega t)$. The fields induced in the component by this modulated line current, $I(\omega)$, will produce a voltage, $V(\omega)$, on the axis that carries information about the component. The relationship $V(\omega) = Z(\omega)I(\omega)$ then defines an impedance. The time and frequency domains are related by the fact that the impedance is the Fourier transform of the wake function for a point charge. For a more general discussion on the concept of impedance, see Ref. 19, Subsec. 2.3, and Ref. 20, Chap. 4. There is no room here to go into detail, but we will note a few things about impedances. First, causality is explicitly preserved in the time domain, but only implicitly in the frequency domain. More visually, as a point charge moves across a cavity, at every instant the real component of the induced field extracts energy from the charge in order to preserve conservation of

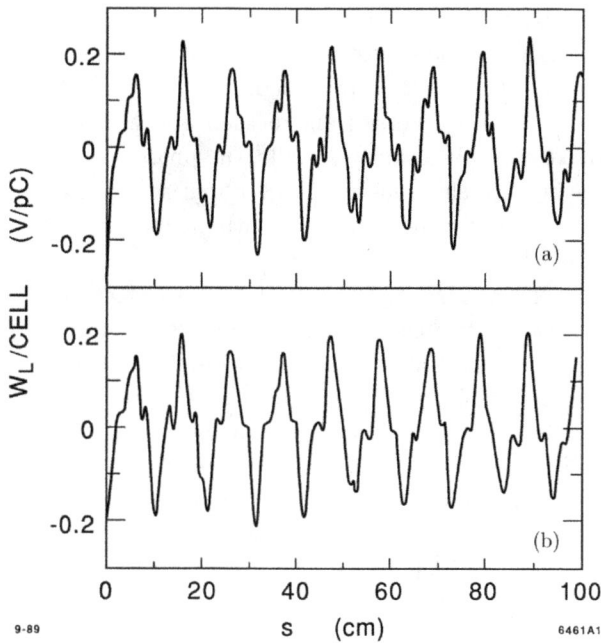

Fig. 23. Long range longitudinal wake for the SLAC structure produced by a short bunch.

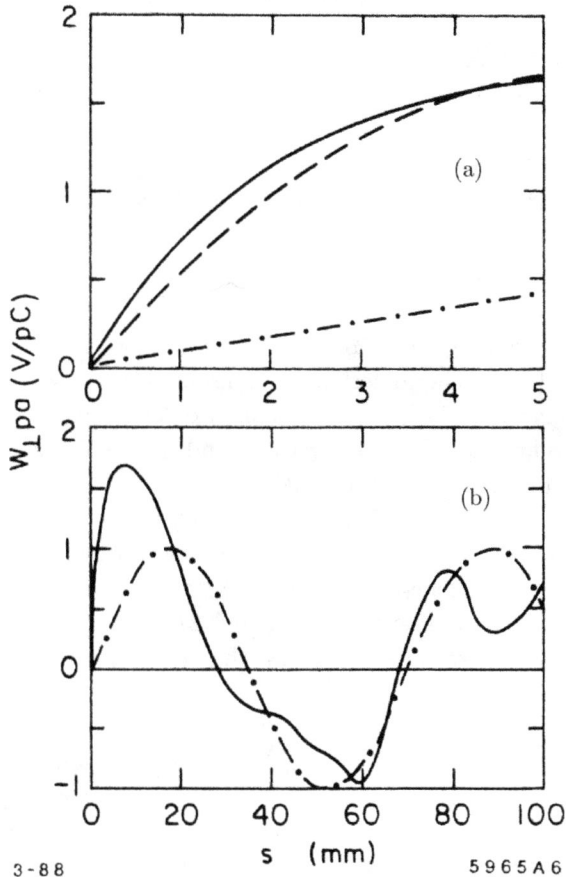

Fig. 24. Short range dipole wake for the SLAC structure (*courtesy of K. Bane, SLAC*).

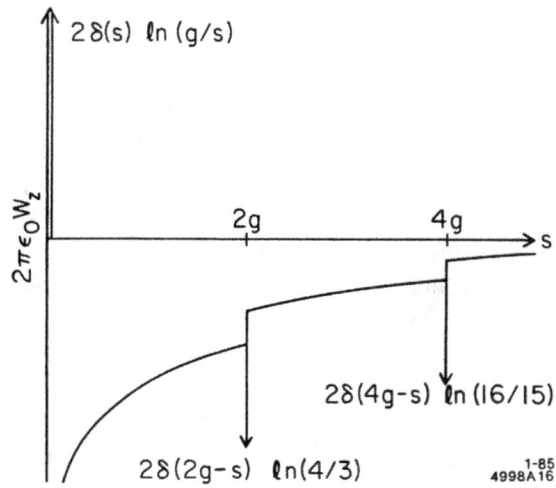

Fig. 25. Longitudinal wake potential for a point charge moving perpendicular to two parallel metallic plates.

energy, as was shown in Subsec. 5.1. However, two equal and opposite imaginary components are also induced. The function of these components is to preserve causality. More formally, the real and imaginary parts of the impedance are related by *Hilbert transforms* (see Ref. 20, Subsec. 4.A.2).

6.7.2. *The Condon method*

Wake functions are often more readily calculated by applying the *Condon method* to the *eigenmodes* of a structure in the frequency domain. An excellent discussion on the method is given in Ref. 22, Subsec. 2.1. To apply the Condon method, the vector and scalar potentials are expanded in terms of the eigenmodes of the empty cavity. The coefficients of the eigenmodes are functions of the position of the charge along the axis of the cavity. Once these coefficients are obtained, the fields can be calculated. In particular, the fields shown in the parallel plate geometry of Fig. 7(f) can be obtained by first calculating the fields and wake potential for a pillbox cavity following the formalism in the reference given above. For $s < (4R^2 + g^2)^{1/2} - g$, where R and g are respectively the cavity radius and length, reflections from the outer cavity wall cannot reach the axis before the charge has left the cavity. The cavity is then reduced to a parallel plate geometry. The wake potential shown below can then be calculated. Details are given in Ref. 22.

7. Transverse Beam Instabilities

7.1. *Introductory remarks*

We will not cover here the extensive subject of transverse beam dynamics in a high energy electron linac. An excellent introductory treatment of this subject is given by Helm and Miller in Ref. 27, Subsec. B.1.2, although this is an out-of-print book that may be difficult to find. A more modern treatment is given in Ref. 28, Chapter 2. However, it is useful to make a few general comments concerning the beam in an electron linac.

Following Subsec. B.1.2 in Ref. 27, in an electron machine of even modest energy, the particles are highly relativistic over most of the length. Transverse forces mostly disappear in the structure itself because of the cancellation of electric and magnetic forces at relativistic velocities. Longitudinal bunching motion becomes negligible because of the increasing "longitudinal mass," proportional to γ^3. The problems of acceleration and beam transport are essentially decoupled. There is no first order coupling between longitudinal and transverse motions. Bunching action virtually ceases, and the energy spread depends primarily on the second order effects of bunch width and on transient rf and beam loading effects. Finally, transverse rf forces essentially vanish so that the transverse momentum (in the absence of external focusing) may be considered a constant of the motion; hence divergence angles decrease as γ^{-1} and transverse displacements increase only logarithmically.

This slow divergence of the transverse orbits is sometimes interpreted in terms of an effective or "contracted" length, $L_{\mathrm{eff}} = (\gamma_0/G_n) \ln(\gamma/\gamma_0)$. Here G_n is the normalized gradient, $G_n = eG/mc^2$. The effective length is defined such that $x - x_0 = x_0' L_{\mathrm{eff}}$, where is x_0 is an initial displacement at an initial angle x_0'. Thus the final transverse displacement is the same as it would be if the electron mass remained unchanged. As an example, the original SLAC linac accelerated electrons to $20\,\mathrm{GeV}$ over a physical length of $3000\,\mathrm{m}$; however, the contracted length is only $33\,\mathrm{m}$.

7.2. *Beam steering and focusing*

There are many treatments of beam optics in a linear accelerator (for example, see Ref. 5, Chap. 7; Ref. 28, Ch. 2; and Ref. 29. Here we will present only a few basic concepts. There are many disruptive forces along the length of a linac that can affect the beam. The include stray magnetic fields, misalignments, rf asymmetries and possible beam breakup (next two subsections). Finally, the phase volume of the injected electrons would fill the entire radial aperture in only a short distance without focusing. It is clear that both magnetic field steering dipoles and focusing quadrupoles are needed. A steering dipole consists of two coils on opposite sides of the beam pipe with currents in the same direction and a reasonably uniform magnetic field in the beam pipe. A quadrupole consists of four magnetic pole pieces wrapped with coils having currents that, viewed from the end of a pole piece, flow alternately in the clockwise and counterclockwise directions. A quick sketch of the magnetic field lines will show that a quadrupole is focusing in one plane, for example the x plane, but defocusing in the y plane. The focal length is $F = p/Q$, where p is the beam momentum and Q is the quadrupole strength, defined as the radial magnetic field gradient times the axial length of the quadrupole. As in regular optics, focusing and defocusing quadrupoles placed fairly close together have a net focusing effect in both planes. More complicated quadrupole arrangements of so-called doublets and triplets are actually used as the actual focusing elements in linear accelerators.

7.3. *Regenerative beam breakup*

Regenerative beam breakup (regenerative BBU), once called pulse shortening, was first described in 1961 [30]. The underlying physics of this transverse oscillation of a beam in a structure deflection mode was first calculated analytically, and the starting current for the onset of the oscillation given, in 1963 [31], and later in Ref. 32. Using a more elegant approach, in 1962 Gluckstern also calculated the threshold current for regenerative BBU [33]. It was later analyzed by Helm and Loew in Ref. 27, Subsec. B.1.4. Here, however, we will follow the analysis in Ref. 4, Subsec. 8.1.

Regenerative BBU is due to an oscillation in a single structure resulting from an interchange of energy between the beam and a dipole (deflecting) mode. In these modes the E_z field component linearly with transverse distance y close to the axis, and as cosine θ in the azimuthal direction (see Subsec. 6.5). In making the calculation, the structure is assumed to be a periodic chain of coupled TM_{110} pillbox cells. We assume that the beam starts as a continuous line current on the cavity axis, ignoring the rf structure. In a typical constant gradient structure the higher frequency lowest dipole mode is trapped in a relatively short region near the front end of the structure. This trapping results from the relatively rapid change in the iris radius with length for a constant gradient structure. Assume next that the structure dipole mode fields are excited to a very low amplitude, perhaps by the noise component at this frequency. Assume that the electrons enter the resonant region when H_x is maximum. Here they receive a small transverse momentum kick in the y direction and begin to move away from the axis as they travel downstream. The important thing is that the electron beam must slip ahead of the cavity modulation by about half a wavelength so that the electrons will see a retarding electric field toward the end of the resonant region. Here energy is extracted from the beam, producing a slight increase in the amplitude of cavity fields. If the power extracted from the beam is equal to the power dissipated in the structure walls, then an oscillation will occur. The starting current for the oscillation as time approaches infinity is

$$I_S(SW) = 0.03 \left(\frac{V_0}{Q} \right) \left(\frac{\lambda}{L} \right)^2 A.$$

In this expression V_0 is the beam energy in volts, λ is the wavelength of the oscillation and L is the effective length over which the oscillation occurs. Note that the starting current decreases as L^2 at a given frequency. This expression assumes that the beam energy is constant over the resonnt length. If there is substantial acceleration over this length, and if $V_0 = V_f$ is the beam energy at L, then if V_f is large compared to the initial beam energy the starting current must be reduced by a factor of 3.0. A table for this factor for an intermediate energy gain is given in Ref. 31.

Actual rf pulses are usually on the order of a microsecond. The starting current I at the end of a pulse of length T_P depends on the number of e-folding times (defined as $(A = \exp[F_e])$ needed for the oscillation to increase in amplitude by a factor of A from the amplitude assumed for the initial noise level.

$$F_e = \left(\frac{T_P}{\tau} \right) \left[\frac{I}{I_S} - 1 \right].$$

Here τ is the decrement time due to wall losses. Roughly, F_e should be kept below about 15.

7.4. *Cumulative beam breakup*

Cumulative beam breakup is a more difficult physical phenomenon both to visualize and to calculate analytically. Helm and Loew, in Ref. 27, Subsec. B.1.4, give a detailed analytic calculation of cumulative BBU. A more visually oriented calculation is given in Ref. 34 for cumulative BBU for the steady-state case. An alternative approach to cumulative BBU is given in Ref. 35; the references in this paper lead to other studies of cumulative BBU. Cumulative BBU can again be modeled using a periodic chain of coupled TM_{110} mode pillbox cavities. Assume that a given cavity has a certain transverse magnetic field strength in the x direction. The field strength increases linearly with y (see Fig. 21). The increase in the amplitude of the fields in the cavity will be proportional to y offset of the oncoming beam as it passes through it, since the electrons will put energy into the cavity through E_z. The change in field amplitude is assumed to be small compared to the existing amplitude. We assume a phase shift θ between the beam and the cavity, and also some cavity detuning off-resonance by a phase angle ψ so that there is an H_x and an E_z in the cavity at the same time. The y kick given to the electrons in the beam will be proportional to the existing H_x amplitude. Assuming that the breakup mode exists only over a fraction of the total structure length, then the change in y the amplitude in the next downstream cavity will be proportional to the H_x field times the drift length between cavities (roughly the structure length). Putting all this together [33], the optimum phase shift $\theta = -\pi/6$ and optimum tuning angle ($\psi = \pi/6$) can be calculated for the maximum gain in the field amplitude. The gain between each cavity is small — on the order of 1.02. But there are about 1000 structures in the SLAC linac, and $(1.02)^{1000}$ is an F_e of 20. The exact equations for F_e are given in the two references above. However, there are two problems with this calculation. First, rf pulse lengths for room temperature accelerators are in the microsecond region — much shorter than the time to reach a steady state condition. Second, we have assumed that there in no focusing. The problem of calculating F_e for the transient case was first solved by Panofsky and Bander (see Ref. 27, Subsec. B.1.4).

The result is

$$F_e = 2.6(FI_0\tau z/G)^{1/3}.$$

Here I_0 is the beam current, τ is the pulse length, z is the distance along the accelerator, $G(V/m)$ is the accelerating gradient and

$$F = \frac{\omega_1^2(r_\perp/Q)(l/L)}{(4V_e c)}.$$

In this expression ω_1 is the breakup mode frequency, l is the length over which the deflecting mode exists, L is the length between structures and $V_e = 0.51\,\text{MeV}$ is the rest mass of the electron. $(r_\perp/Q) \approx 100/\lambda_1$ is a reasonable approximation (accelerator definition of $r_\perp$). A loss term given by $\omega_1\tau/2Q$ must be subtracted from F_e. Using SLAC parameters (see Ref. 27, Subsec. B.1.4), the beam broke up at 600 m for a beam current of 25 mA.

Focusing can be taken into account approximately by replacing distance z with $\Delta z = 2.5\,\lambda_\beta^2/f$, where λ_β is the wavelength *for betatron oscillations* [29] and f is the focal length of a quadrupole doublet.

In summary, fixes to increase the starting current of cumulative BBU are: first, increase the focusing strength as much as possible by adding additional quadrupoles; and second, tune the deflecting mode frequency differently from one accelerating structure to the next. At SLAC, dimples were made in a few cavities in the deflection mode region at the front end of the structures to change the frequency of the transverse mode while maintaining the frequency of the accelerating mode. Three different tunings were made in successive structures along the linac.

8. RF Sources

8.1. *Klystrons*

All linear electron accelerators for particle physics are powered by klystrons. High power klystrons were initially developed at Stanford University's High-Energy Physics laboratory to drive the 1 GeV Mark III linear accelerator. These tubes, working at 2856 MHz, had an output power of about 20 MW. This was about a thousand times the power delivered by any previous klystron — certainly a quantum jump in klystron technology. Since the tubes were designed to run at beam voltages up to 400 keV, relativistic effects had to be taken into account for the first time. A description of the design and performance of the klystron is given in Ref. 36. When the

20 GeV, 3000 m SLAC linac was completed in 1966, it was powered by rf from an improved version of the earlier Mark III klystrons. The nominal rf power output was 24 MW at a beam voltage of 250 keV and a pulse width of 2.5 μs. Later, around 1985, the klystron was redesigned to power the SLC (SLAC linear collider) that was being planned at SLAC (see Sec. 9). This so-called 5045 klystron ran at a peak power of about 65 MW with a pulse width of 3.6 μs [37]. The longer pulse width was required to accommodate the SLED rf pulse compression scheme used to increase the peak power delivered by the klystron (see next subsection). A cut-away view of the 5045 klystron is shown in Fig. 26.

In Fig. 26, number 1 shows the cathode and dc gun producing the flow of electrons, and number 2 shows the bunching cavities. The first cavity modulates the dc beam from the gun at the rf frequency (input coupling not shown), while the intermediate cavities amplify the input rf signal. Number 3 shows the output cavity, which is excited by the strongly bunched beam; the high power microwaves extracted from the output cavity are fed into a waveguide number 4, which transports them to the accelerator. The

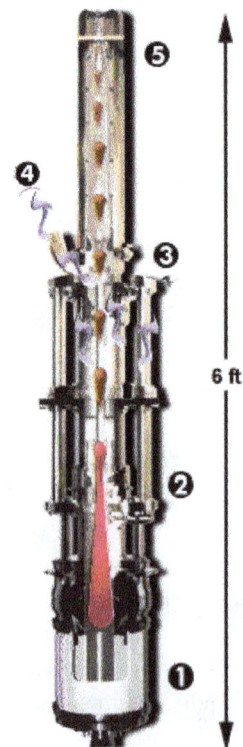

Fig. 26.   Cut-away view of the SLAC 5045 klystron [38] (*courtesy of SLAC*).

spent beam beyond the output cavity is absorbed in a water-cooled collector number 5.

8.2. *RF pulse compression*

Rf pulse compression is a method for increasing the peak power of an rf source. The compression device always trades a longer, lower power pulse from the source for a shorter, higher power output pulse. Unfortunately, the relative increase in peak power is always less than the relative decrease in pulse length by an efficiency factor. There are two types of compression systems. The compression device used at SLAC, called SLED, emits an exponentially decaying pulse; this compression system will be described shortly. Other pulse compression systems are capable of emitting a flat-top output pulse. They consist of various arrangements of multimoded delay lines. Some of them are described in Refs. 39 and 40. All pulse compression systems require an energy storage device, either high Q cavities or delay lines.

The theory of the SLED pulse compression system is given in Ref. 41. How SLED works is best understood by referring to Fig. 27.

The power from the klystron is directed to a 3 db coupler that splits the power flow between two TE_{015}

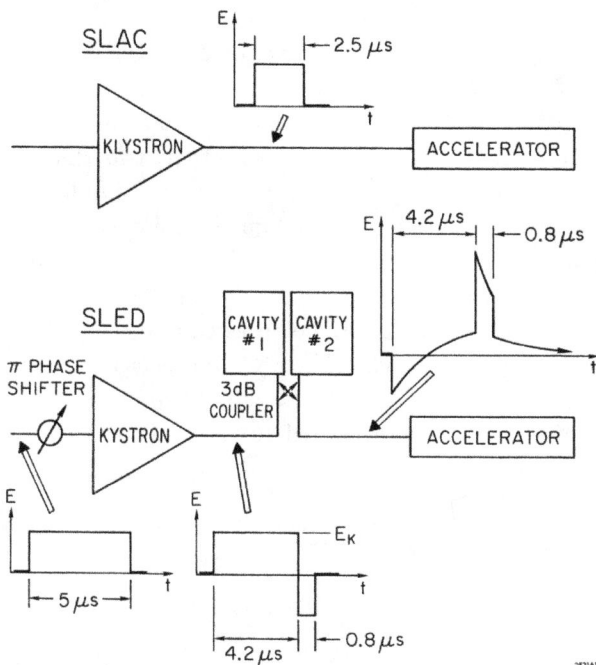

Fig. 27. Comparison between the original SLAC rf system and the system after SLED was installed.

mode overcoupled cavities with a Q_0 of about 10^5. The directional coupler insures that all the energy reflected or emitted from the cavities is directed toward the accelerator. The key to understanding SLED is once again the principle of superposition. For a high Q cavity, all the klystron power (and an equivalent field parameter) is reflected as a unit wave from the cavities during the first part of the klystron pulse. But a wave, proportional to the fields stored in the cavity, is also emitted from the cavity-coupling apertures. This emitted wave has a sign that is opposite to the unit wave reflected from the klystron. If the cavity is critically coupled, the cancellation is exact (no reflected power). If the cavity is heavily overcoupled, the emitted wave can reach twice the amplitude of the reflected klystron wave. If the klystron is suddenly turned off at this point, the power initially emitted from the cavity is then four times the klystron power. However, the basic principle of SLED is not just to turn off the klystron power, but to leave it on and to *reverse the phase of the klystron power*. The emitted wave and the klystron wave are now added together to give a net field amplitude of 3 for a potential power gain that is nine times the klystron output power.

The installation of SLED was completed by 1978. The name SLED (SLac Energy Doubler) was based on the assumption that the power from SLED, integrated over a structure filling time, would increase the klystron power by a factor of 4. This would have taken a $5\,\mu s$ dc pulse from the modulator. However, it was found that the modulators ran better at a shorter pulse width of about $3.5\,\mu s$. This pulse width gave an increase in power by only a factor of 3. However, the 5045 klystrons also put out about three times more power than earlier klystrons. The bottom line is that the linac energy was increased from $20\,GeV$ to about $60\,GeV$. This higher energy made the SLC linear collider possible (see next subsection). A klystron with the added SLED cavities is shown in Fig. 28.

8.3. *Other rf sources*

All past high energy electron linear accelerators were driven by klystrons. A linac useful for high energy would have many such microwave tubes and would consequently require a lot of electric power. For example, the yearly cost of providing electric power

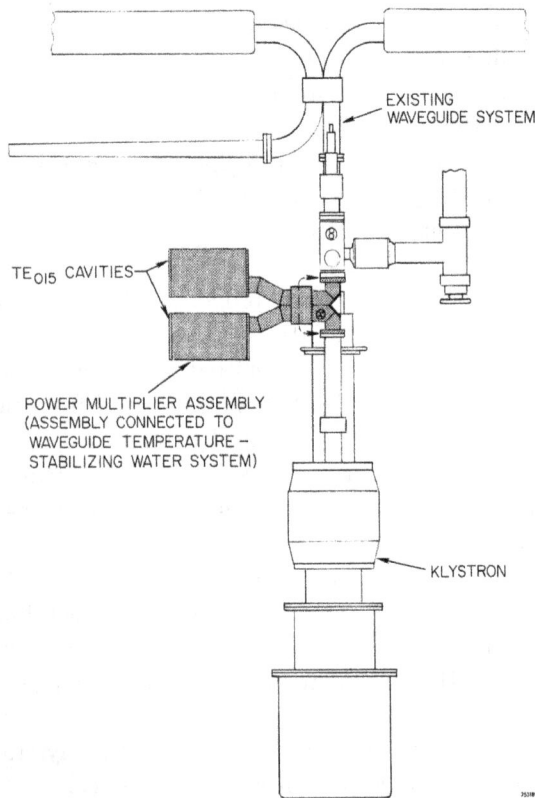

Fig. 28. SLED cavities attached to a klystron.

for the operation of the SLAC linac at 120 pps was on the order of US$10 million. Because future high energy linear colliders will run at a higher gradient than SLAC's 20 MV/m and are considerably longer, they will need an enormous amount of electric power, in the range of 300 MW. The maximum allowable power will be determined by social and political constraints. Since klystrons are expensive and only about 60% efficient, other rf sources need to be considered for powering these colliders. The use of magnetrons or a drive accelerator (two-beam scheme) for an rf source are possibilities (see discussion in the next section).

9. Linear Colliders

9.1. *Overview*

The future of high energy electron particle physics lies in the linear collider. The original SLAC linac accelerated electrons, directed them in a beam switch-yard at the end of the linac into one of several beam lines, and then smashed the electrons

into a fixed target or a bubble chamber, both surrounded by a large detector. The detector was capable of measuring the emission angle, the energy and the momentum of particles emitted from the target. Because of conservation of momentum, most of the initial electrons went through the target and dissipated their energy in a beam dump. Only a small fraction of the initial electron energy was transferred into the rest mass and kinetic energy of particles created in the target. However, by firing electrons from one linac to make a head-on collision with positrons from a second linac, all the initial particle energy is available for creating particles and giving them momentum; furthermore, the electrons and positrons annihilate. All that remains after the collision is an extremely dense region of energy that can give birth to a variety of new particles.

9.2. *Emittance preservation in a linear collider*

The main linacs (both electron and positron) in a linear collider must transmit the beams from the two injector complexes into the smallest possible spot sizes at the interaction point. The overlapping portion of the two spots produces e^+e^- collisions at a rate measured by the *luminosity* per square centimeter per second. Each injector complex delivers a beam having a certain *emittance*. The concept of emittance is explained, for example, in Ref. 5, Subsecs. 9.1 and 9.4. In general, if the rms displacements, x and y, and the rms divergence angles, x' and y', are plotted against each other, two ellipses are obtained. If the accelerator is composed of linear elements, such as dipoles and quadrupoles, the emittance is an invariant. In general these *phase space* ellipses do not have an axis that lies along x or y. However, in principle the ellipses can be rotated so that they are upright. Then, at any position along the linac, the horizontal and vertical spot sizes and angular spreads are related to the emittances and *beta functions* by

$$\sigma_x = \left(\frac{\varepsilon_x \beta_x}{\gamma}\right)^{1/2},$$

$$\sigma'_x = \left(\frac{\varepsilon_x}{\gamma \beta_x}\right)^{1/2},$$

and similar expressions for y. As explained in Ref. 28, the beta functions are a measure of focusing strength.

They are also the wave numbers for local betatron oscillations.

Dipole deflecting modes (see Subsec. 6.5) are the worst offender in increasing the transverse emittances along the linac. If the bunch is displaced from the axis, the head of the bunch will then kick the tail of the bunch in the same direction as the displacement. There are many clever ways to minimize the effect of the dipole mode wake field. See, for example, Refs. 42 and 43 for room temperature colliders. The damping problems are quite different for superconducting linacs [44].

9.3. *Beam–beam effects*

When electron and positron beams pass through each other at the interaction point, several physical effects can degrade the luminosity. First of all, since the horizontal and vertical betatron functions both reach minimums at the interaction point, the longitudinal bunch length must be comparable to or shorter than the waists of these minimums in order to avoid the "hour glass" effect. Second, the collective fields from the particles in the oncoming bunch act like a lens to focus the particles in the opposing bunch toward the axis. This *disruption* of the beams may or may not be helpful. Third, as the electrons in one bunch are deflected by the collective field of the oncoming bunch, they emit synchrotron radiation called *beamstrahlung*. As the energy of the colliding electrons

increases, at some critical energy one radiated photon would have to carry more than the entire energy of the radiating electron, which is clearly impossible. Quantum considerations therefore impose an absolute cutoff on the radiation spectrum at $h\nu = \gamma mc^2$. The total fraction of the energy radiated by the bunch is called the *beamstrahlung parameter*, which must be kept small. Another important beam–beam parameter is the number of beamstrahlung photons produced per incident electron. This number must also be kept small. A good review of all these beam–beam effects in given in Ref. 45.

9.4. *The SLC*

B. Richter came up with the idea of converting most of the SLAC linac into a linear collider; more importantly, he directed the hard work needed to put this idea into practice. The key to making head-on collisions between electrons and positrons using the linac available at SLAC was to accelerate both types of particles in the linac to about 55 GeV, separate them at the end of the linac and then inject them into two arcs. The arcs curved around by 270° and brought the two types of particles into collision. The layout of the Stanford Linear Collider (SLC) is shown in Fig. 29.

The contruction of the SLC was completed in 1987, and commissioning was completed in 1998. The energy was chosen to create Z^0 mesons with a rest mass of about 90 GeV. Because so many new and

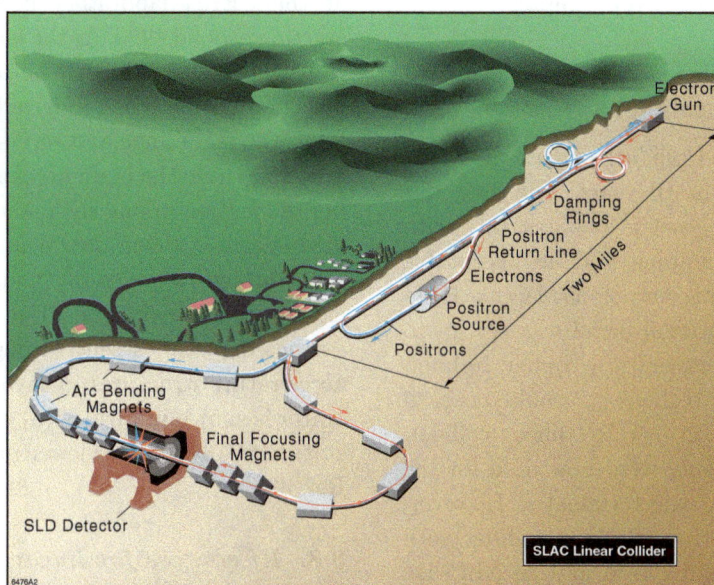

Fig. 29.   Layout of the SLC [46].

unexpected accelerator physics problems had to be solved, it took about five years of hard work before the SLD detector began to deliver an acceptable Z^0 production rate. The SLC was the world's first electron linear collider; however, its operation was terminated in 1998. The success of the SLC, and the accelerator physics knowledge created in making it work, have led to designs for a new generation of linear collides in the 0.5–2.0 TeV center-of-mass energy range. A more detailed description of the SLC is found Ref. 46.

9.5. *The ILC linear collider*

The ILC linear collider was designed for an initial c.m. energy of 0.5 TeV, with a possible upgrade to higher energies by increasing its length. The ILC would use superconducting niobium rf cavities; the accelerating gradient is therefore limited by the critical magnetic field of niobium. Although this critical magnetic field limits the gradient to about 45 MV/m, in practice an average gradient of about 30 MV/m has been achieved for the nine-cell accelerating cavities. At this gradient, the total active length (ignoring the space between cavities) of both linacs is 17 km. The ILC design is an upgraded version of the TESLA design for a linear collider based on superconducting rf cavities. The design for TESLA was developed at the DESY laboratory, Hamburg, Germany. Details of the design are given in Ref. 47. See Ref. 48 for an overview of the ILC collider.

9.6. *The NLC linear collider*

The NLC (Next Linear Collider) project at SLAC was active for more than 10 years. The goal was to build an operational accelerating structure with transverse wake field damping and an acceptable breakdown rate at a gradient of 65 MV/m. However, in 2005 an international panel was set up to decide whether room temperature copper technology or superconducting technology would be used to build a next generation 500 GeV linear collider. Superconducting technology won (now used for the ILC) and the NLC project was terminated. However, there will eventually be interest in a collider with a c.m. energy greater than 1 TeV, and an NLC-like technology would be needed. A layout of the NLC collider is shown in Fig. 30. The structure would

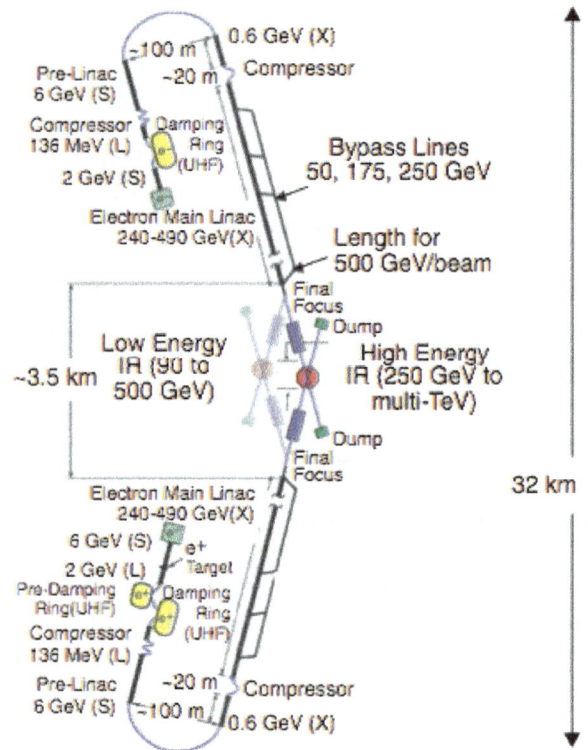

Fig. 30.   Layout of the proposed NLC linear collider [47].

operate at a gradient greater than 100 MV/m, perhaps using new materials for the iris tips.

9.7. *The CLIC linear collider*

The two-beam accelerator concept was developed at the CERN laboratory in Geneva, Switzerland, starting around 1980. As the idea took shape it began to be called CLIC (CERN Linear Collider). In CLIC, a second beam, called a drive beam, runs antiparallel to the main linac and serves as a source for the rf power to drive the main accelerator linac. At a number of locations along the main linac a portion of the drive beam is taken out, bent around by 180°, and sent through a series of "transfer structures" that extract power from the drive beam electrons. The extracted power is sent through waveguides to the accelerating structures. A layout of the CLIC accelerator is shown in Figs. 31(a) and 31(b). The current status of CLIC can be viewed on the CLIC home page Ref. 49.

9.8. *Rf sources for linear colliders*

A linear collider operating at 11.4 GHz at a gradient of 100 MV/m would require about 150 MW/m.

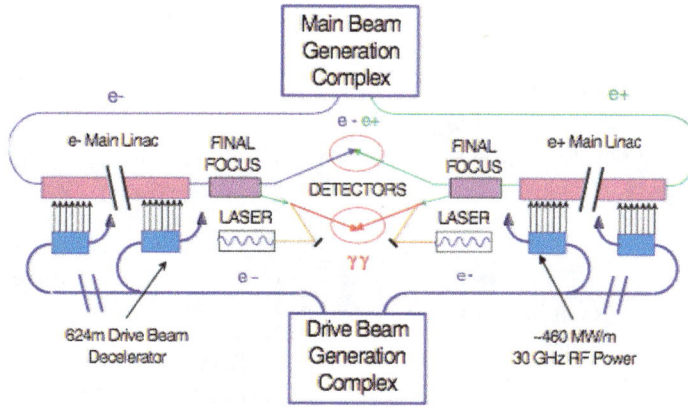

Fig. 31(a).   Layout of the CLIC accelerator [47].

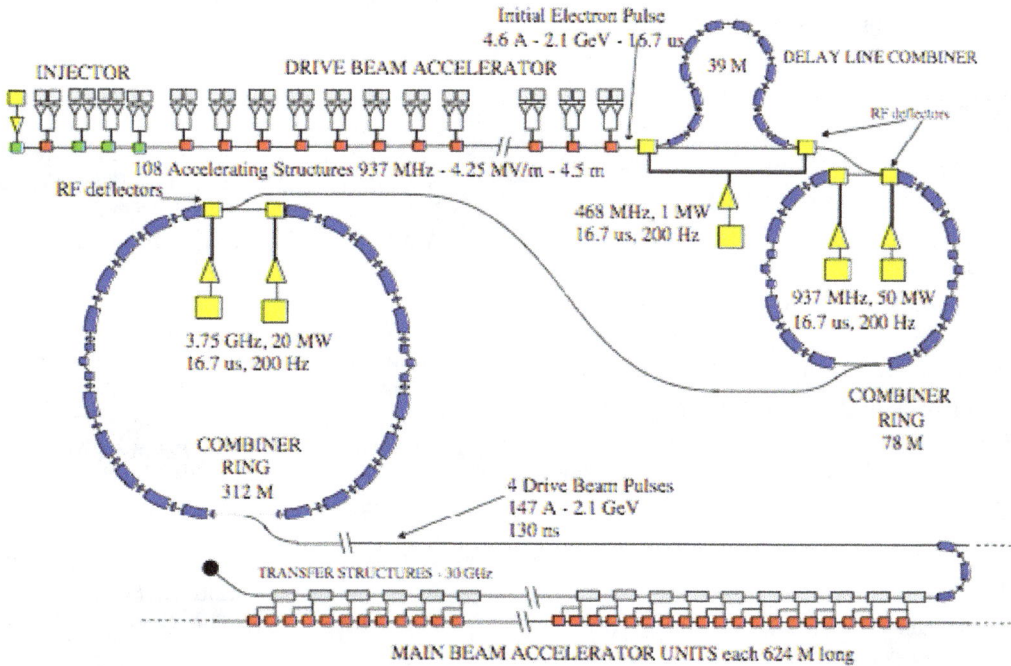

Fig. 31(b).   CLIC drive beam generation complex [47].

From the preceding layout diagrams, it is seen that there are two ways to provide the rf power for such a collider. The first possibility is a modular system, like the SLAC linac. Spare modules would power a separate, short length of linac at the end of the accelerator. These modules would be running, but their output would be out of time by a few microseconds. If an accelerator klystron fails, for example because of a burst in vacuum pressure, that module shuts off and one of the special module pulses is switched between rf pulses to the correct timing. One disadvantage of

the drive beam scheme is that it lacks this ability — the all-the eggs-in-one-basket syndrome. Proponents of the idea claim that a modular scheme cannot provide the required peak power per meter for a high gradient structure. Proponents of a modular system will counter that no real research in this area has been done, but is badly needed before the idea of a modular system is discarded.

Phase-locked magnetrons are one possible rf source for a high gradient collider [50]. They will be considerably cheaper than klystrons and are very efficient (about 80%). They would run at a voltage of around 40 kV, at least a factor of 5 lower than the voltage needed by a klystron. This lower voltage also makes the modulator considerably cheaper. However, much more R&D is needed to develop a practical rf system based on magnetrons.

9.9. *Muon colliders*

The examples of linear colliders given in this section have all been electron–positron colliders. However, electrons and positrons are not the only particles that could be made to collide. As explained in Ref. 3, Subsec. X.7, muon colliders are also a possibility. Quoting from the preceding reference, "Muons are leptons, like electrons, and share with electrons the advantage of an unambiguous point-like structure." Because the muon is 200 times heavier than the electron, it will not radiate significant amounts of synchrotron light, even at an energy of 10 TeV. The muons can therefore be accelerated in a storage ring, like the LHC at CERN. In a storage ring the rf cavity voltage is used many times over; the diameter of the storage ring will be much less than the length of a linear accelerator with the same final energy. However, it is difficult to produce an intense muon beam; furthermore, the muons decay in flight. These problems can be overcome, but much R&D needs to be carried out before a muon collider design is ready to move forward into the construction phase.

References

[1] G. A. Loew and R. Talman, Elementary principles of linear accelerators, SLAC-PUB-3221, Sec. 2 (Sep. 1983).

[2] P. J. Bryant, A brief history and review of accelerators. http://omnis.if.ufrj.br/~mms/lab4/p1.pdf (undated).

[3] A. Sessler and E. Wilson, *Engines of Discovery* (World Scientific, Singapore, 2007).

[4] P. B. Wilson, High energy electron linacs, SLAC-PUB-2884-Rev. (Nov. 1991).

[5] T. P. Wangler, *Principles of RF Linear Accelerators* (John Wiley & Sons, New York, 1998).

[6] R. E. Collin, *Foundations for Microwave Engineering*, 2nd edn. (IEEE Press, New York, 2001).

[7] R. P. Feynman, R. B. Leighton and M. Sands, *The Feynman Lectures on Physics*, Vol. 1 (Addison-Wesley, Reading, Massachusetts, 1963).

[8] A. E. Vlieks *et al.*, Development of an X-band photoinjector at SLAC, SLAC-PUB-9365 (Aug. 2002).

[9] P. B. Wilson, CERN Report CERN/MPS/MU-EP (69-4).

[10] C. Nantista, Radio-frequency pulse compression for linear accelerators, SLAC-R-95-455, App. (Jan. 1995).

[11] R. B. Neal, ed., *The Stanford Two-Mile Accelerator* (W. A. Benjamin, 1968).

[12] J.-W. Wang, SLAC (private communication).

[13] C. G. Montgomery, R. H. Dicke and E. M. Purcell, *Principles of Microwave Circuits*, Radiation Laboratory Series, Vol. 8, Ch. 7 (McGraw-Hill, New York, 1948).

[14] R. Boxman, D. Sanders and P. Martin, eds., *Handbook of Vacuum Arc Science and Technology* (Noyes, Park Ridge, New Jersey, 1995).

[15] P. B. Wilson, A theory for the rf surface field for various metals at the destructive breakdown limit, SLAC-PUB-12354 (2007).

[16] N. S. Xu and R. V. Latham, *J. Phys. D* **27**, 2547 (1994).

[17] Reprinted with permission from Gabovich and Poritskii, *JETP Lett.* **33**(6), 306 (1981). Copyright 1981, American Institute of Physics.

[18] Reprinted with permission from G. Taylor, *Proc. R. Soc. London, Ser. A, Math. Phys. Sci.* **280**(1382) (1964), facing p. 394.

[19] A. W. Chao, *Physics of Collective Beam Instabilities in High Energy Accelerators* (John Wiley & Sons, New York, 1993).

[20] B. W. Zotter and S. A. Kheifets, *Impedances and Wakes in High-Energy Particle Accelerators* (World Scientific, Singapore, 1997).

[21] P. B. Wilson, Introduction to wake fields and wake potentials, SLAC-PUB-4547 (Jan. 1989).

[22] K. L. F. Bane, P. B. Wilson and T. Weiland, Wake fields and wake field acceleration, SLAC-PUB-3528 (1984).

[23] E. Keil, *Nucl. Instrum. Methods* **100**, 419 (1972).

[24] K. Bane and B. Zotter, *Proc. 11th Int. Conf. High Energy Accelerators* (Birkhäusen Verlag, Basel, 1980), p. 581.

[25] K. Bane and P. Wilson, *Ibid.*, p. 592.

[26] D. U. L. Y. Yu and P. B. Wilson, Long range wake potentials in disk-loaded accelerating structures, SLAC-PUB-3528 (1989).

[27] P. M. Lapostolle and A. L. Septier, *Linear Accelerators* (North-Holland, Amsterdam, 1970).

[28] S. Y. Lee, *Accelerator Physics* (World Scientific, Singapore, 1999).

[29] K. L. Brown and R. V. Servranckx, First and second-order charged particle optics. SLAC-PUB-3381 (Jul. 1984).

[30] M. C. Crowley-Milling *et al.*, *Nature* **191**, 483 (1961).

[31] P. B. Wilson, A study of beam blow-up in electron linacs. Internal Memorandum HEPL-297, High-Energy Physics Laboratory, Stanford University (Jun. 1963).

[32] P. B. Wilson, Type-I beam break-up in superconducting linacs. Report HEPL-TN-67-2, High-Energy Physics Laboratory, Stanford University (Mar. 1967).

[33] R. L. Gluckstern, Transverse beam blow-up in standing-wave linacs. BNL Internal Report AADD-38 (Jul. 1964); also published in *Proc. MURA Conference on Linear Accelerators* (Stoughton, Wisconsin, 1964), p. 186.

[34] P. B. Wilson, A simple analysis of cumulative beam breakup for the steady state case. HEPL-TN-67-8, High Energy Physics Laboratory, Stanford University (Sep. 1967).

[35] R. L. Gluckstern, F. Neri and R. K. Cooper, *Particle Accelerators* **23**, 53 (1988).

[36] M. Chodorow, E. L. Ginzton, I. R. Neilson and S. Sonkin, Microwave Laboratory Report No. 212, W. W. Hansen Laboratories of Physics, Stanford University (1953).

[37] M. A. Allen *et al.*, *1987 IEEE Particle Accelerator Conference* (IEEE, New York, 10017), p. 1713.

[38] Virtual Visitor Center at SLAC, http://www2.slac.stanford.edu/vvc/accelerators/klystron.html.

[39] S. G. Tantawi *et al.*, *Phys. Rev. Special Topics — Accelerators and Beams* **8**, 042002 (2005).

[40] S. G. Tantawi *et al.*, *IEEE Trans. Microwave Theory Tech.* **47**(12) (1999).

[41] Z. D. Farkas. H. A Hogg, G. A. Loew and P. B. Wilson, SLED: A method of doubling SLAC's energy, SLAC-PUB-1453 (Jun. 1974).

[42] R. Jones *et al.*, *Phys. Rev. ST Accel. Beams* **9**, 102001 (2006).

[43] M. Minty, Emittance preservation in linear accelerators. DESY report DESY M 04-02 (Mar. 2005).

[44] International Linear Collider Reference Design Report, http://www.slac.stanford.edu/pubs/slacreports/slac-r-857.html.

[45] P. B. Wilson, Applications of high power microwaves to TeV linear colliders. In: A. V. Gaponov-Grekhov and V. L. Granatstein (eds.), *Applicatons of High-Power Microwaves* (Artech House, Norwood, MA, 1994), Ch. 7, Subsec. 7.2.5.

[46] The SLAC (Stanford) Linear Collider: (1) open the SLAC home page, www.slac.stanford.edu; (2) search the SLAC Web for SLC; (3) open the third listing, which will direct you to SLAC Linear Collider (SLC).

[47] *The International Linear Collider Technical Review Committee Second Report*, SLAC report SLAC-R-606 (2003).

[48] International Linear Collider Reference Design Report, http://www.slac.stanford.edu/pubs/slacreports/slac-r-857.html.

[49] CLIC homepage, http://clic-study.web.cern.ch/CLIC-Study/Welcome_frame.html.

[50] S. Tantawi, private communication.

Perry B. Wilson is an Emeritus Professor (Applied Research) at the Stanford Linear Accelerator Center at Stanford University. In the 1960's, while at the High Energy Physics Laboratory, he made important contributions to the early technology of superconducting accelerator cavities. At SLAC, he made early contributions to the concept of wakefields and impedances. He is currently working on the theory of rf breakdown in accelerator structures. He is a Fellow of the American Physical Society. In 1991 he received the IEEE Particle Accelerator and Technology Award as the co-inventor of the SLED principle.

Reviews of Accelerator Science and Technology
Vol. 1 (2008) 43–64
© World Scientific Publishing Company

World Scientific
www.worldscientific.com

The Development of High Power Hadron Accelerators

G. H. Rees

ASTeC Division, Rutherford Appleton Laboratory, STFC,
Chilton, Didcot, Oxon OX11 0QX, UK
ghrees@ukonline.co.uk

A brief review is given of the development of linacs, cyclotrons, synchrotrons (also accumulators) and FFAG accelerators for a variety of high power hadron beam applications.

Keywords: ADSR; APT; ATW; FFAG; spallation; neutrino factory; muon collider.

1. Introduction

The history of particle accelerators extends over the last 75 years. For the first half of this period, the main developments were directed towards achieving higher and higher particle energies, but in the latter half, a parallel development has been the study and design of hadron accelerators over a beam power range from 0.1 to > 100 MW and a beam kinetic energy range from 0.5 to ≈ 30 GeV/amu.

The first high power accelerators were developed for the meson factories. These included the proton linear accelerator at Los Alamos National Laboratory (LANL) in the USA [1], the sector-focused H⁻ cyclotron at TRIUMF in the University of British Columbia, Canada [2], and the sector-focused proton cyclotron at The Paul Scherer Institute (PSI), Villigen, Switzerland [3].

The LANL 0.8 GeV proton linac was designed for a beam power of 1 MW at a 120 Hz pulse repetition rate. The TRIUMF 0.5 GeV cyclotron chose to accelerate H⁻ ions in a large, low field magnet to limit the Lorentz force stripping, prior to foil-stripping the ions for extraction as protons. At PSI, two cyclotrons act in series, with kinetic energies of 0.072 and 0.590 GeV, and both have enhanced outer orbit separation to allow high efficiency proton extraction. It is planned to increase the beam power from 1.2 to 1.8 MW.

It became apparent, after studies of high intensity beam behaviour and various technical advances, that the power limits of hadron accelerators could be raised above the levels reached at the meson factories. This was not only the case for linacs

and cyclotrons but also for accelerators such as synchrotrons and fixed field, alternating gradient (FFAG) rings. The choice between these four different types of accelerator depends very much on the parameters of the proposed application.

All four types need a front end with a stable, high current, high brightness ion source. A test stand is needed for research and development on topics such as reliability, beam current stability, lifetime, charge states, and space charge and neutralization effects in the output line. The ion source is on a platform at a voltage of up to 750 kV dc for some applications, where it is fed from a generator, such as a Cockroft–Walton or a Haefely rectifier set. Most high power accelerators, however, use a lower voltage platform, at ≈ 100 kV, as the stage ahead of a radio frequency quadrupole linac (RFQ).

The RFQ [4] is a resonant structure with electric quadrupole type fields for alternating gradient, transverse ion focusing, and longitudinal electric fields at vane or rod modulations for gradual bunching and acceleration of the ions. It is suitable for most low charge state ion beams and for the front end of most types of accelerator.

High power proposals for hadron beams include:

- A continuous or a pulsed proton beam for the transmutation of radioactive waste (ATW)
- A continuous or a pulsed proton beam for driving a sub-critical reactor (ADSR)
- A continuous or a pulsed proton beam for the target production of tritium (APT)
- A pulsed H⁻ linac and proton ring for development of a short pulse (2 μs) spallation neutron source

- A continuous or a pulsed proton beam for a cw or a long pulse (2 ms) spallation source
- A pulsed H^- linac and H^+ rings for a superbeam or to make intense muon beams for a neutrino factory
- A pulsed H^- linac and proton rings for development of intense μ^+ and μ^- beams for a muon collider
- A pulsed H^- linac and proton rings for development of an intense source of kaons for a kaon factory
- A pulsed ion or radioactive ion beam for studies of nuclear matter and rare isotopes (RIB)
- A pulsed proton or deuteron linac beam for fusion material irradiation (IFMIF)

Some of these are funded; most are at a conceptual stage, but all have a chance of approval at some stage. Many have contributed to the underlying ideas, and a few major contributors will be identified in the next section. Later sections consider, in turn, the four types of accelerator most suited to high power applications. A fifth type, the induction linac, is not included as it has been developed mainly for high power electron beams.

2. Accelerator Milestones to High Power

2.1. *Ion sources*

Several types of ion source have been developed for high power applications. The earliest study for H^- ions was by G. I. Dimov and V. G. Dudnikov, on the Penning type of source, at Novosibirsk [5]. Elsewhere, multi-cusp volume and magnetron H^- sources have been studied in addition to the Penning sources, both with and without addition of caesium. The evaporation of caesium into the hydrogen discharge region causes a partial, surface monolayer to form, giving a lower work function and an enhanced H^- production. An rf-driven, multi-cusp, H^- ion, volume source has been designed by K. N. Leung and R. Keller at Lawrence Berkeley National Laboratory (LBNL), and subsequently developed by M. Stockli [6] at Oak Ridge National Laboratory (ORNL) for use in the spallation neutron source (SNS).

Electron cyclotron resonance (ECR) sources were pioneered by R. Geller [7], C. Lyneis and H. Postma in the 1970s, and are now the predominant choice for light and heavy ions. In ECR sources, microwave power is injected into a volume of gas or vapour at a frequency defined by the applied magnetic field and the cyclotron resonance involved so that free electrons are heated to ionize the gas atoms and molecules. An ECR source at 2.45 GHz may produce more than 100 mA of H^+ and D^+ ions in a dc mode. For other charge states, a higher frequency is used and ions are confined in the low pressure gas long enough for multiple collisions and ionizations to occur without significant recombination.

2.2. *Radio frequency quadrupole linac*

The RFQ was invented by I. M. Kapchinskii and V. A. Teplyakov [4] at Novosibirsk in 1970, and the invention was soon introduced at LANL by J. Brady, after which the device spread rapidly to accelerator laboratories worldwide. The RFQ structures may take the form of four vanes or four rods or, for high mass ions, a split coaxial shape, at low rf frequency. The early history was described during EPAC '06, in an invited talk for V. A. Teplyakov [8], presented by R. A. Jameson.

2.3. *Cyclotrons*

Though an ion source and RFQ may act as the front end of most accelerators and is the preferred choice for high power linacs, it was not chosen for the upgrading of the PSI cyclotrons. An ion source on a high voltage platform was chosen instead, as it allows a lower longitudinal emittance, which has advantages for the later cyclotron beam dynamics. Further schemes are under way at PSI [9] to raise the output beam power from 1.2 to 1.8 MW and the experience gained suggests that a future cyclotron could reach a cw beam power of up to 10 MW.

2.4. *Proton and H^- linacs*

Linacs allow the most direct route to high power but need to feed an accumulator-compressor ring or a ring accelerator chain for some applications. The design of a high power proton or H^- linac includes a front end, a medium energy beam transport (MEBT) section and various standing wave accelerating structures [10].

These may include the Alvarez type of drift tube linac (DTL, invented in 1947); the cavity-coupled drift tube (CCDTL) and cavity-side-coupled (CCL) linacs, of the LANL type; the disk-and-washer structure, employed at a 0.6 GeV linac at TROISK, near Moscow; the annular-ring-coupled structure (ACS) for the 0.4 GeV upgrade of the J-PARC linac in Japan; the inter-digital (IH) design of the University of Frankfurt; and the superconducting (SC) cavities used successfully in the high energy part of the linac of the SNS project. Future hadron linacs may choose to use SC cavities for most stages after the RFQ.

2.5. *Heavy ion linacs*

A history of heavy ion linacs spans the building [11] of HILAC at LBNL in 1955–57; the construction of the UNILAC [12] at GSI Laboratory, Darmstadt, Germany, in 1966–69; the upgrading of HILAC to SuperHILAC in 1971–72 [13]; the building of a low frequency, SC linac at Argonne National Laboratory (ANL) [14] in 1977–78; a linac construction [15] at the Institute of Physical and Chemical Research (RIKEN), Waco, Japan, in 1975–77; and completion by 2011 of a linac producing exotic isotopes for the GANIL, SPIRAL 2 project [16] in Caen, France. There has also been a detailed conceptual design study [17] at ANL and Michigan State University (MSU) for a rare ion facility (RIF), using two superconducting linacs — one a 400 kW heavy ion driver for ions from protons to uranium, and the other for rapid acceleration of the radioactive ions produced.

2.6. *High energy heavy ion rings*

The SuperHILAC and the UNILAC came to serve as injectors for high energy rings, the former for a revamped Bevatron and the latter for the SIS synchrotron and its associated ESR storage ring [18]. The UNILAC and the SIS and ESR rings are next to serve as the injectors for the newly approved GSI, FAIR project (Facility for Antiproton and Ion Research). FAIR [19] has a double synchrotron ring of superconducting magnets of 1200 m circumference and, with electron and stochastic cooling rings, is to be used for studies of antiprotons, nuclear matter, rare isotopes, and atomic and plasma physics.

2.7. *Radioactive ion beam factory*

In the period 1997–2007, RIKEN has developed a radioactive ion beam factory (RIBF), by construction of three new ring cyclotrons [14] to boost the output beam energies from its existing K540 MeV cyclotron and linac. The new cyclotrons have K values of 570 MeV, 980 MeV and 2600 MeV, and the

third is the first cyclotron to be built with SC sector magnets. Light ions may be accelerated up to 440 MeV per nucleon and very heavy ions to 350 MeV per nucleon. The heavy ions are used to form intense beams of radioactive ions, either by projectile fragmentation or by in flight fission of uranium isotopes.

2.8. *High power synchrotron and ring design issues*

The main issues involved in the design of synchrotrons, accumulators and FFAG rings for high power are the beam dynamics, the reduction of injection and extraction beam losses, the protection of the ring against accidental losses, the collimation of the halo and lost beam at localized regions of the ring to allow hands-on and "active" maintenance, and the ring reliability and availability. (The use of synchrotrons and accumulators for high intensity protons was previously reviewed by J. Wei in 2003 [20].) The key issue of H^- ion injection in rings is outlined next.

2.9. *Charge exchange injection*

Charge exchange injection has had a major influence on the development of ring accelerators to high power as it circumvents limitations on beam phase space densities that arise from Liouville's theorem. Proposals date from the 1950s, when ideas arose for stripping a H_2^+ ion beam to protons (P. B. Moon, Birmingham University, UK [21]), and development of H^- ion sources for H^- ion acceleration, followed by foil-stripping injection to protons (G. Budker and G. I. Dimov, at Novosibirsk [22]). Both these schemes have been developed, but the latter is now the recognized route to high power.

2.10. *H^- injection system development*

The Novosibirsk scheme was first adopted in the USA by R. L. Martin [23], who converted the 12 GeV ZGS ring accelerator at ANL for H^- injection. Later development, in many countries, of long life stripping foils and improved H^- sources has allowed conceptual designs of circular accelerators to be made for pulsed proton beam powers up to 5 MW, and possibly higher.

Associated with these advances have been studies for the "painting" of the injected beam distributions, to reduce transverse space charge forces and to limit foil temperature rises caused by proton foil traversals during injection. An extreme case is the KAON Factory Study at TRIUMF (1991), where extraction of H⁻ ions from the 450 MeV ring cyclotron was envisaged prior to the 16,000-turn injection into an accumulator ring, at 50 Hz.

2.11. H⁻ beam "painting" distributions

For acceptable proton foil traversals in KAON, there was need for a foil with two unsupported edges and use of correlated, longitudinal and transverse injection beam amplitudes. Small amplitude longitudinal motion was correlated with transverse motions large in the horizontal (h) and small in the vertical (v) plane at the start of injection, transposing to opposite size amplitudes by the end [24]. The cyclotron and accumulator had rf frequency ratio 1:2, and simultaneous "painting" in the longitudinal and h planes resulted from use of finite dispersion at the stripping foil together with a ramped, injected beam momentum, and in the v plane by using a programmed localized vertical orbit bump.

Later, in US "painting" distribution studies of the h and v planes only, the term "anti-correlated" came to replace "correlated" and the use of "correlated" came to mean the opposite of its original definition. Some details of operating systems are given in later sections.

2.12. Influence of superconducting technology

Already noted are the use of SC rf cavities in the high power SNS H⁻ linac at ORNL and the use of SC sector magnets in the K 2600 MeV ring cyclotron at RIKEN. SC technology is also planned for the FAIR project at GSI and is being studied at BNL for the "harmonic number jump" type of acceleration in high power, proton and heavy ion, FFAG rings.

2.13. FFAG accelerators

FFAG rings, of a particular, nonlinear scaling form, originated in the 1950s and interest in them returned in the 1990s [25]. By "scaling", it is meant that the beam orbits for different momenta are scaled replicas of one another. Non-scaling rings have been studied since the late 1990s and may be designed to have smaller orbit separations. Both scaling and non-scaling rings are suitable for high power accelerators

as they may operate at a high repetition rate or, possibly, in cw mode. They require an upstream injector ring for very high power, however, as their short straight sections are not suited for optimized, low loss H⁻ injection.

3. High Power Linacs

3.1. High power applications

A high power proton beam of ≈ 1 GeV is required for the transmutation of radioactive waste (ATW), as a driver for a sub-critical reactor (ADSR) and for the production of tritium (APT). All three applications require a similar pulsed or continuous proton beam but the beam powers may range from 10 to 150 MW or higher. A linac is the most direct method of providing such high beam power and many countries (the USA, the Soviet Union, Japan, France, the UK, China and Korea) have studied, but not built, such a linac, and some of the studies have remained classified. The first of the designs was for an APT facility and was developed in the years between 1995 and 1997 at LANL [26].

3.2. Studies for an APT proton linac

The LANL linac design study was for a cw proton beam power of 170 MW at 1.7 GeV. In the final design, a normal conducting (NC) linac using copper cavities accelerated a 100 mA proton beam to 217 MeV, after which a SC niobium cavity linac continued the acceleration to 1.7 GeV. The NC linac consisted of an ion source, a 0.075–6.7 MeV RFQ linac at 350 MHz, a 6.7–100 MeV, 700 MHz CCDTL structure and a 100–217 MeV, 700 MHz side-coupled CCL stage. The output beam of the RFQ was matched into the CCDTL, which employed a coupled sequence of two- and three-gap, short DTLs. Both CCDTL and CCL used an 8 $\beta\lambda$ FODO focusing cell with F, D quadrupoles external to the cavities and O their separations.

Five-cell, 700 MHz niobium cavities were used for the high energy, SC linac. Two types of cryomodules and cavity shapes were proposed — one for the medium beta range from 217 to 469 MeV and one for the high beta range from 469 to 1700 MeV. The shapes of the cavities were optimized at a beta of 0.64 for the former and 0.82 for the latter. In the cryomodules were either three or four of the five-cell cavities. Each cavity was fitted at its downstream end with two diametrically opposite, high power, antenna-type coaxial couplers, and dual warm coaxial ceramic windows were proposed for the input feeder lines. Conventional quadrupole doublets were planned for the focusing and were external to the cryostats to simplify alignment and diagnostic designs.

3.3. *Features of superconducting cavities*

Use of wide bore, SC cavities allows lower length of linac, beam halo loss and radio frequency (rf) structure power. An SC linac requires more rf generators, however, than an NC linac as the input couplers [27] have a lower rf power capability and for adequate rf control only one or two cavities may be used per driver. This situation will change in the future as higher power couplers and phase shifters are developed.

Though the rf structure power is small in an SC linac, significant power is used in switching on the cavities for pulsed operation. The use of a high cavity frequency is advantageous in this respect.

3.4. *Studies for 25 MW proton linacs*

Some classified studies have considered a lower proton beam power of 25 MW at 1 GeV, in a pulsed operating mode at 50 Hz. Designs similar to the LANL APT linac were assumed but the 12.5% duty cycle for available klystrons required 200 mA pulsed beam currents. This influenced the low energy design, and two parallel low energy stages have been envisaged, each consisting of a 2.5 MeV RFQ and a 20 MeV DTL, the output beams of which are merged in a complex funnel section.

3.5. *The superconducting H$^-$ linac for the SNS*

The only SC H$^-$ linac built has been that for the short pulse spallation neutron source at ORNL [28]. It was designed at LANL soon after the APT proton linac study had been completed. Initial specifications were a 7% on duty cycle and a beam power of up to 1.4 MW at 60 Hz and 1 GeV. A revised upgrade specification for an energy of 1.3 GeV, at beam powers of $\geq$ 3 MW, has subsequently been approved.

The upgrade involves the development of a higher current H$^-$ source and addition of a 300 MeV, SC linac section.

The linac design differed from that for APT in several ways. The rf frequencies for the low and high energy stages were increased from 350 to 402.5 MHz, and from 700 to 805.0 MHz, respectively, while the RFQ output energy was lowered from 6.7 to 2.5 MeV. The MEBT energy then became 2.5 MeV and the following CCDTL was replaced by a 2.5–87.0 MeV DTL. In addition, a H$^-$ ion source was developed in collaboration with LBNL, as noted earlier, and two beam choppers were introduced — one for the LEBT between the ion source and RFQ, and a faster rise time unit for the MEBT.

The LEBT uses electrostatic elements for focusing and chopper beam deflections. Most of the unwanted H$^-$ ions are removed by fields applied to four deflecting plates near the LEBT output. The plates are powered with pulses of 40 ns rise and fall times at an associated ring frequency of $\approx$ 1 MHz. This reduces the power to be removed in the MEBT where a second chopper [29] trims the pulses for 10 ns rise and fall times.

The MEBT section has to provide the following: a smooth focusing match for longitudinal and transverse beam motion from RFQ to DTL, a full set of diagnostics for characterizing the RFQ output beam, space for the fast beam chopper and its associated loss collector, and collimators for removing the beam halo ahead of the DTL. The transverse and longitudinal focusing is provided by quadrupoles and buncher cavities, respectively.

Seven 2.5 MW pulsed klystrons are used to power the 402.5 MHz stages — one for the four-vane RFQ, and one for each of the six DTL tanks. Four 5 MW pulsed klystrons power the four 805 MHz CCL stages. A much larger number of klystrons is needed for the SC part of the linac, as each SC cavity is fed by a single klystron. For the initial 1 GeV stage of the SNS linac, 81, 0.6 MW pulsed klystrons are used to power the 805 MHz cavities. The early commissioning is reported in Ref. 30.

A digital field control module is employed for each system to stabilize the rf cavity field amplitude to better than $\pm$0.75% and the rf phase to better than $\pm$0.75°. Adaptive feed-forward is used to aid the compensation for cavity beam loading. Piezo-electric tuners on the SC cavities are programmed to assist in compensation for the Lorentz force detuning [27], caused after switching on the rf field in each cycle. There has been some evidence of limiting cavity field gradient variations over time.

There are 11 cryomodules, each housing 3 medium beta SC cavities and 12 cryomodules, each housing 4 of the high beta SC cavities. All units have been supplied by Jefferson Laboratory (JLAB). Warm regions between the cryomodules contain ion pumps, optical viewing ports, beam position monitors and main focusing quadrupoles.

The linac has a comprehensive set of diagnostics, which contributes to the rapid setting-up of the beam. An important development has been that for a beam profile monitor, based on laser stripping [31] of the H^- beam.

3.6. *Proposals for other high energy H^- linacs*

3.6.1. *The linac proposed in the ESS study*

Prior to the SNS, studies had started on a high power H^- linac for a short pulse European spallation source (ESS). An initial design considered a pulse duration ≤ 0.75 ms, an energy of 1.334 GeV and a beam power of 5 MW at 50 Hz. A room temperature and an SC linac design and a pair of 50 Hz accumulator rings were evaluated. The beam power was later raised from 5 to 10 MW, with the extra 5 MW for the feeding of a long pulse (2.5 ms) target station, in an interleaved mode of operation at 16.667 Hz. In an effort to improve approval prospects for the ESS, the 5 MW, 50 Hz short pulse facility may be postponed until a future phase of the project.

3.6.2. *The 8 GeV H^- linac under study at FNAL*

An 8 GeV H^- linac has been proposed at FNAL (Fermi National Accelerator Laboratory), based on the SC technology of the ILC ($e^\pm$ International Linear Collider). Standard ILC, 1.3 GHz, SC structures are proposed from 1 up to 8 GeV, and $8\pi/9$ mode, 1.3 GHz, SC structures from 0.603 to 1 GeV. The lower energy linac section plans to use 325 MHz (one-quarter of 1300 MHz) for the 2.5 MeV RFQ, the 2.5–10 MeV, cross-bar, spoke-type NC copper structure and the 10–603 MeV, spoke-type SC structures, one design of which is shown in Fig. 1.

Basic parameters are a 1 ms pulse extent, a 9 mA current and a 360 kW beam power at 5 Hz. Other

Fig. 1. Superconducting 0.325 GHz spoke cavity (*courtesy R. Webber, FNAL*).

design features are the SC solenoids for transverse focusing from 2.5 to 120 MeV and the use of single high power klystrons to drive several accelerating cavities with high power vector modulators for separate control of each rf amplitude and phase [32]. A major linac function is use with the Fermilab main injector to develop a superbeam.

3.6.3. *The 3.5 GeV H^- linac under study at CERN*

Parameters of 40 mA average pulse current, 0.57–0.71 ms pulse duration and 4 or 5 MW beam power at 50 Hz are compatible with superbeam or neutrino factory designs. Included are three 352.2 MHz NC low energy stages [a 3.0 MeV RFQ, a 3–40 MeV DTL and a 40–90 MeV CCDTL (or equivalent stage)], and three 704.4 MHz, higher energy structures [an NC, 90–180 MeV CCL, an SC ($\beta = 0.65$), 180–645 MeV sequence of cavities and an SC ($\beta = 1.0$), 645–3500 MeV further series of cavities].

For the high energy stages, it is planned to use 44 specially built 5 MW pulsed klystrons, with fast, phase and amplitude shifters for the CCL sections and the five-cell SC cavities. The total length of the linac is 430 m [33].

4. High Power Cyclotrons

4.1. *Developments of the PSI cyclotrons*

Higher current is needed for the 590 MeV spallation neutron target. The increase in current from

2 to 3 mA is to be achieved by a continued installation of new, high gradient resonators in the main and injector cyclotrons. This gives increased proton turn separations and results in lower beam losses. Supporting measures are installation of harmonic bunchers for both rings. In the injector, two new 50 MHz, 400 keV resonators are to replace the 150 MHz cavities, despite the limited space.

The multi-cusp proton source is to be replaced in 2008 by a compact, permanent magnet ECR ion source. The 2.5 GHz microwave injection system needs a power of 400 W to create a stable beam current of 12 mA.

4.2. *Developments for the TRIUMF cyclotron*

The TRIUMF 500 MeV H$^-$ cyclotron delivers proton beam currents of ≤ 0.3 mA to multiple users, including 0.1 mA for a new target, developed for the radioactive ion beam facility ISAC. A new beam position monitor system is to be installed from the ring to the ISAC target, and a forced electron beam induced arc discharge source (FEBIAD) and a 2.8 GHz ECR ion source are being developed [34] to complement existing surface and laser ion sources. Source designs for on-line applications have to allow for a large range of gas pressures and for high radiation fields due to the target proximity.

4.3. *Development of the RIKEN K2600 cyclotron*

The large K2600 ring for the RIBF at RIKEN is the first cyclotron to be built with SC sector magnets, the weight of which is approximately 8000 tons. The first beam [15] was extracted at the end of 2006 and the first experiment was conducted before the end of May 2007. A new neutron-rich isotope, Pd-125, was discovered in the in-flight fission of U-238 isotopes, accelerated up to 345 MeV per nucleon.

4.4. *Advancement of SPIRAL 2 at GANIL*

SPIRAL 2, now under construction [16], consists of a multi-beam driver (5 mA, 40 MeV deuterons; 5 mA, 33 MeV protons; and 1 mA of 13.5 MeV/u heavy ions), an ISOL scheme for radioactive ion beam production, and ion post-acceleration, in an existing cyclotron, CIME.

5. High Power Synchrotrons and Accumulators

5.1. *Early development of spallation sources*

The initial development of the rapid cycling synchrotrons (RCSs) to high power was to provide the beams of protons required for short pulse spallation neutron sources. First proposed in 1972 by J. Carpenter and then developed at ANL, short pulse sources have become increasingly important for neutron scattering research. The success of the 450 MeV, 6.4 kW, 30 Hz RCS of the IPNS source [35] at ANL continued for over 26 years and has been followed by major advances elsewhere.

The KENS source at KEK, Tsukuba, Japan, was soon to follow, using protons from a 20 Hz, 500 MeV, 3.5 kW synchrotron to feed a new spallation target. Though ANL had plans for a higher intensity source, the next advances were the building of a 50 Hz, 800 MeV, 160 kW proton synchrotron at RAL (1979–84) for the ISIS source and the addition in 1981–86 of a 12 Hz, 797 MeV, 80 kW compressor ring (PSR) to the LAMPF linac at LANL for the LANSCE source.

5.2. *The 800 MeV proton synchrotron at ISIS*

The factor-of-25 increase in beam power for the ISIS synchrotron [36] presented new challenges. Amongst these were developments for: a reliable 50 Hz H$^-$ ion source, long-life H$^-$ stripping foils, rf shields for ceramic vacuum chambers, a glass bonding technique for ceramic sections, easily moved rf cavities, quick release systems for flanges and water cooling fittings, a loss protection system, low loss fast extraction, and a momentum and betatron beam loss collimation system (for active and hands-on maintenance).

The ISIS 70 MeV linac and 800 MeV synchrotron (RCS) had to be developed by making use of a lot of existing equipment. A H$^-$ upgrade, from 1 up to 50 Hz, of an old 70 MeV proton linac involved revamped rf systems, new modulators and new sets of diagnostics and loss monitors for the linac and 70 MeV beam line.

The ring design was constrained by having to fit in an existing hall, with injection and extraction directions defined, and by the need to be compatible with a choke and capacitor banks inherited from an earlier, 50 Hz, resonant magnet power supply. The parameters of the choke set both the acceptances of the ring magnets and the lattice periodicity of 10. As a result, the lattice cell chosen had a quadrupole doublet (DF), a long straight section (O), a single quadrupole (d), a 36° gradient-bending magnet (Bf) and a shorter straight section (o). The stronger-focusing DF doublet served to reduce the peak stored energy in the Bf gradient magnet.

All H$^-$ charge exchange injection units were within quadrupoles of a long (O) section, allowing separate beam line and ring magnet tuning. Included were four symmetrical orbit bump magnets, a central injection stripping foil with a foil change mechanism, a dc injection septum magnet, a loss collector for H° and H$^-$ beams, a stripped electron collector and diagnostic units.

The bump magnets were made identical, with a one-turn, high current septum so that the upstream unit and the H$^-$ injection magnet could have adjacent septa. A section of the vacuum chamber wall and a magnetic shield buttressed these septa and diverted the stray field of the dc multi-turn septum magnet from the ring beam. A vertical sweeper magnet, upstream of the dc septum, completed the injection system.

The injection elements enabled 2D, anti-correlated betatron distributions to be painted, with large vertical and small horizontal oscillation amplitudes at the start of injection and the opposite size amplitudes at the end. Horizontal painting was made by injecting at a point of finite dispersion on an inner closed orbit as the ring's magnetic guide field fell towards its minimum value. The 24 mA, 70 MeV H$^-$ linac beam was injected by charge exchange stripping for $\approx$ 150 turns. An Al$_2$O$_3$ foil was developed to span the vertical aperture of the ring. It had an area of 120 mm × 40 mm and one unsupported edge. The 0.25 micron foils were made by anodizing an unmasked area of aluminium and dissolving the backing from the oxide film, leaving masked limbs as a support.

All main magnets were designed so that they could be powered in series by the single resonant power supply. A common lamination design was obtained for D and F quadrupoles by adjusting their lengths. The d unit was given a figure-of-eight design to reduce requirements for vertical beam extraction. Each cell had trim quadrupoles adjacent to the DF

doublet for rapid changes of betatron tunes. These proved important for reaching the design intensity of 2.5×10^{13} protons per pulse at 50 Hz.

Limited space at the linac input led to the decision to dispense with a beam chopper. The beam injected in the ring had then to be trapped with low initial rf fields in the cavities at a time near the minimum of the biased sine waveguide field. A rapid rf field rise had to follow for the first ms of acceleration, placing demands on the feed-forward control for the six harmonic number 2, ferrite-tuned cavity systems.

A key feature was localization of beam loss in one of the ten ring superperiods. Momentum loss collectors were used for the first time in a synchrotron and proved very effective in reducing the extent of the activated areas and in protecting components. Untrapped beam was intercepted with high efficiency since the internal rf shields were provided with rectangular apertures and the inward beam movement was more than 1 mm per turn.

5.3. *The 797 MeV PSR accumulator ring at LANL*

The PSR accumulator was conceived for weapons-related research but was converted to joint use as a spallation source. Extensive changes were made to the LAMPF linac and beam lines to provide the H$^-$ beam for the PSR.

A ring lattice was chosen with a circumference of 90.2 m and nominal betatron tunes, $Q_h = 3.23$, $Q_v = 2.22$. It consisted of ten identical FODB cells. The B dipoles had a 2.55 m length, a 280 mm by 105 mm aperture and a 1.2 T field. The inscribed radius for the quadrupoles was 181 mm and the F unit gradients were 3.95 T/m and those for the D, 2.35 T/m. A single 2.8 MHz ferrite-loaded cavity ($h = 1$) provided up to a 14 kV peak per turn.

The ring initially employed a two-stage scheme for injection, with external Lorentz stripping of H$^-$ ions to H° atoms, followed by foil stripping of the H° to protons. Beam losses were high with this scheme, so a direct H$^-$ injection system was later introduced [37]. A beam power of $\approx$ 80 kW was then realized, limited by injection and ejection losses and by fast coherent beam motion.

During H$^-$ transit of the foil, partial stripping left a variety of H° excited states whose lifetimes depended on the magnetic fields encountered, and this was found to contribute to the 0.3% injection beam losses. Other ring designs have benefited from the PSR studies of injection and the observed fast coherent beam motion, which was later identified, using electron cloud diagnostics [38], as an electron–proton $(e^-–p)$ instability.

5.4. *The 1 (1.3) GeV SNS accumulator at ORNL*

The SNS proposed a factor-of-9 increase in beam power compared with ISIS and a factor-of-18 increase compared with the PSR. A further increase by more than a factor of 2 was later planned for a 1–1.3 GeV energy upgrade. The need to achieve low loss, multi-turn H^- injection in the SNS accumulator ring [39] was thus a major issue.

A hybrid cell structure was selected for the four-superperiod lattice of the ring. It was formed with four FBDB cells in each arc and two back-to-back doublet quadrupole cells in each long straight. This allowed integer tunes for the arcs which created zero dispersion straights and simplified the designs for both the orbit and the sextupole chromaticity corrections.

The doublets enabled flexible matching to the arcs and provided long dispersion-free regions for the H^- injection, collimation, cavity and ejection system designs. The partially separated injection system included two septum magnets, four dc orbit bump magnets and two stripping foils, all in one 12.5 m central region, and two horizontal and two vertical adjustable low field magnets in the 6.85 m drift regions of adjacent cells. These units allowed correlated and anti-correlated painting schemes.

The fringe fields of the central bump magnets were designed for the stripped electrons to spiral down to a collector. After the first foil, the unstripped H^- ions and partially stripped $H^°$ atoms separated from the proton beam and were stripped at a second foil, for ejection by a septum magnet. The beams to be removed were affected by nonlinear fringe fields of the bump units and some beam was lost before reaching an external dump [40, 41].

An important feature of the SNS ring was the multi-stage betatron collimation system, designed to localize most of the beam loss in shielded ring locations. Design values for fractional loss levels were set at 10^{-3} at the collimators and 10^{-4} elsewhere. Other key features were the use of titanium nitride wall

coatings and a beam-in-gap kicker to stop development of the $(e^-–p)$ instability.

5.5. *The 3 and 40 (50) GeV J-PARC synchrotrons*

The Japan Proton Accelerator Research Complex provides high power proton beams for a variety of groups [42]. A 3 GeV beam is used for a spallation neutron source, for low energy muon production, and for feeding a 50 GeV ring, used for hypernuclei, neutrino oscillation, and rare decay experiments. A linac will be developed to 0.6 GeV for studies of nuclear waste transmutation.

The initial phase includes a 25 Hz, 0.18 GeV H^- linac, a 25 Hz, 3 GeV RCS (now being commissioned) and a 0.3 Hz, 40 GeV ring (now nearing completion). In this phase, beam power is 0.5 MW for the RCS and 0.4 MW for the main ring. In a future phase, the H^- injector linac energy is to be raised to 0.4 GeV and the main ring energy to 50 GeV, to double the beam power levels. For the 0.6 GeV linac, a 0.2 GeV, SC stage is to be added.

The RCS has a 348.3 m circumference and three superperiods, each with three straights of DOFO cells and two sets of three DBFB arc cells. The centre cell of a set omits dipoles to create high gamma-t and high dispersion, for momentum collimators and chromaticity-adjusting sextupoles. The injection system occupies three half cells in a dispersion free straight. The central half cell has a symmetrical, four-bump fixed magnet chicane, with a central H^- stripping foil. The other half cells each have a septum magnet and two horizontal beam painting magnets. Two external magnets create vertical painting.

The main ring has three superperiods, a 1567.5 m circumference and uses rf harmonic 9. It is 4.5 times as large as the RCS, allowing box-car stacking of eight RCS ($h = 2$) bunches over four cycles. An MA (magnetic alloy) loaded cavity is the basis of the rf systems in both rings. Each arc has eight groups of three cells, with missing bend centre cells to create an imaginary transition lattice. There are fast and resonant (slow) beam extractions.

5.6. *The synchrotron of the Chinese spallation source*

The CSNS [43] is to build a 25 Hz, 120 kW, 1.6 GeV proton synchrotron with an 81 MeV H^- injector linac. Later linac energies of 134 and 230 MeV allow output beam powers of 240 and 500 kW, respectively. The H^- linac, of length 41.5 m, has a 324 MHz RFQ and DTL.

The RCS ring uses four superperiods, with a hybrid lattice of doublet straights and FBDB arc cells, but with one missing B dipole, to locate a momentum collimator. In the doublet straights are 6, 9 and 6 m drift lengths to house injection, ejection, collimation and rf components. Injection is similar to that used for the SNS accumulator.

5.7. *Other high power ring and synchrotron studies*

5.7.1. *The ESS 5 MW accumulator ring study*

A possibility for the continuing (since 1991) ESS design [44] is to build a 5 MW, 2.5-ms-long pulse source and, in parallel, upgrade the ISIS short pulse source. In the ESS study, a 50 Hz, 5 MW, 1.334 GeV H^- linac fed two compressors and the single proton bunch accumulated in each was sent to a 10 or a 50 Hz target.

Low loss injection dominated the ring designs. As a separated injection system had worked well in ISIS, it was decided to adopt one for the ESS rings, though one of very different form. Instead of merging the H^--injected beam and the protons at the centre point of an orbit bump system, a two-free-edge stripping foil was introduced in a low field lattice dipole, at the centre of one of the arcs. It was located at the aperture corner, as shown in Fig. 2.

The accumulator lattice used three superperiods of triplet cells, with three cells for an arc and three for a straight section. The centre cell of the arc had the low field dipole suitable for injection, and the adjacent cells, the main high field dipoles. Arrangement of the arc was similar to that for an ISIS upgrade ring (as in Fig. 3), except for the different bend angles. The cells proposed for the straights were triplets for the ESS and doublets for the upgrade. A common quadrupole field gradient and lamination profile was obtained by varying the lengths.

In contrast to the SNS scheme, a finite dispersion region was proposed for the injection system design. The dispersion increased from zero in the straight section to a maximum near the centre of the arc, where the foil was positioned. Though this led to

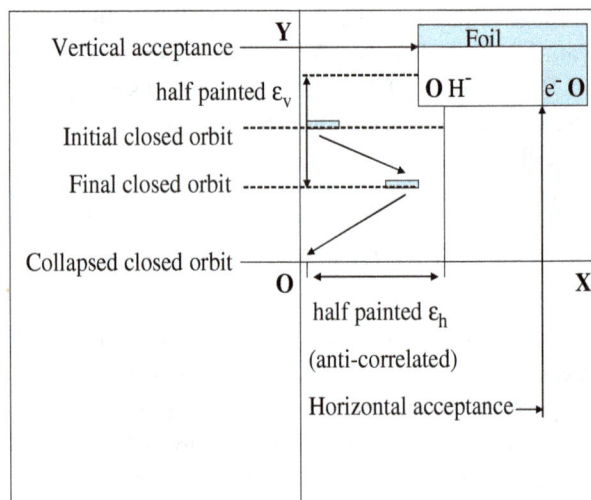

Fig. 2. Aperture cross section at foil position, showing orbits for painting, positions of the e^- collector and foil, and the rectangular, vertical and horizontal acceptances.

some loss of flexibility for longitudinal beam painting, other benefits ensued.

There was no need for a fixed four-bump chicane or an injection septum unit or horizontal painting magnets. The total length was ≈ 10 m, allowing a fully separated design, while SNS used ≈ 25 m for a partially separated scheme. Stripped electrons were deflected in the injection dipole to a loss collector placed outside the ring acceptances, as in Fig. 2. The injection dipole field was chosen to give an optimal gap in lifetimes of excited H° Stark states 4 and 5 for the partially stripped beam between the main and secondary stripper foils [44, 45].

Anti-correlated or correlated painting was arranged using a programmed, localized vertical orbit bump and a ramped, injected beam momentum. Figure 2 shows the preferred, anti-correlated scheme. The correlated painting used a reversed direction for the momentum ramp.

Beam loss collimation used three adjacent straights. Ring and collimator acceptances corresponded to 4σ and 2.6 to 3.0σ, respectively, with σ the rms beam amplitude.

5.7.2. *The AUSTRON 1.6 GeV synchrotron study*

Detailed feasibility studies [46] were made at Vienna and CERN in 1993–94 as part of a Central European

Initiative for a spallation source and medical facility, AUSTRON. A 0.13 GeV H$^-$ injector linac and a 1.6 GeV, 213 m circumference proton RCS were to deliver a beam output power of 205 (410) kW for the neutron targets, when operating at 25 (50) Hz.

A three-superperiod, symmetric triplet lattice was used for the ring, similar to that of the ESS accumulator. The betatron tunes were set at $Q_h = 4.20$ and $Q_v = 4.356$. Acceleration in the ring was at harmonic number 2, with a frequency range from 1.35 to 2.62 MHz, and with maximum voltages of 202.5 (270) kV peak per turn.

Despite the extent of the study, the project was not approved, though the medical design survived and plans have been made for building the facility south of Vienna.

5.7.3. *A new, 3.2 GeV synchrotron possibility at ISIS*

A preliminary study has been made for a 2 MW upgrade of the ISIS spallation source. The energies chosen have been 0.8 GeV for a H$^-$ injector linac, and 3.2 GeV for a new synchrotron ring, but direct injection of 0.8 GeV protons from the ISIS ring may also be considered.

The lattice for the new ring has been designed to minimize the bending magnet power, and to provide sufficient straight sections for economical rf cavity design. A 67% lower stored energy is found for a choice of five instead of four superperiods, due to lower dipole fields and bend angles. Thus, five are proposed, which also gives more scope for orienting the ring at the RAL site.

Each superperiod has two sets of triplet, and four sets of doublet, quadrupoles, with two of the doublet sets in a back-to-back arrangement at the long straight centre. Use of five superperiods allows lower arc dispersion and better lattice optimization for dispersion-free straights.

A beam power of 2 (or 5) MW may be obtained with 1.3 (or 2) 10^{14} protons per pulse at 30 (or 50) Hz and the lower figures are chosen initially, for cost reasons. A harmonic number 4, 2.73–3.16 MHz rf system occupies 57 m of the 370 m circumference. Assuming a cosine guide field wave form with no second harmonic component, a peak voltage per turn of 422 kV is needed for the acceleration of four, 1.8 eV sec bunches.

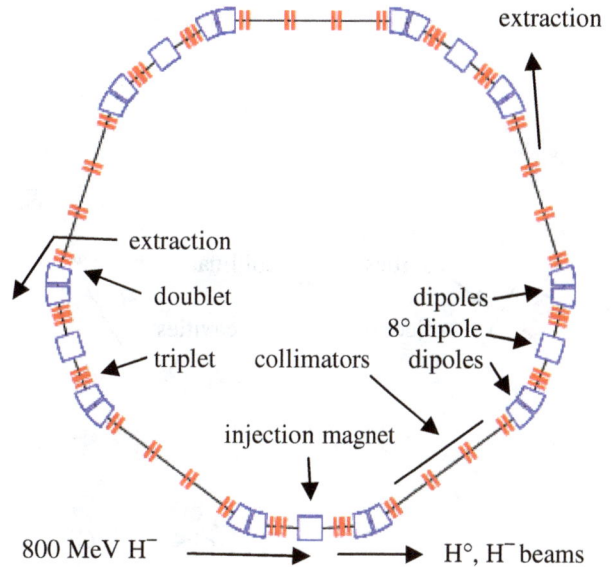

Fig. 3. Lattice option for a 3.2 GeV ISIS upgrade.

Components are shown in Fig. 3. The fields in the 8° dipoles cycle between 0.125 and 0.35 T, and those in the main, 16° dipoles, from 0.36 to 0.99 T, while the normalized gradients of the 40 doublet and 30 triplet quadrupoles track from 2.17 to 5.97 T m^{-1}. The H$^-$ injection scheme is similar to that for the ESS rings.

5.7.4. *A 50 Hz, 4 MW H$^+$ driver for a neutrino factory*

Proton driver requirements for a neutrino factory (NF) are more demanding than those for a pulsed spallation source. An International Scoping Study (ISS) of 2006 recommended a 50 Hz pulse repetition rate, a 4 MW beam power, a 10 ± 5 GeV energy, and three bunches per cycle, each of time extent ≤ 3 ns rms [47]. Several driver options were considered, and three of them are:

- A 0.2 GeV linac, a 3 GeV RCS and a 10 GeV NFFAG
- A 0.2 GeV linac, a 3 GeV RCS and a 10 GeV RCS
- A 5 or 8 GeV linac, with bunch compression rings

The first option is deferred to Subsec. 6.3, while the third option is omitted apart from noting that it needs a high linac pulse current. The second is outlined here and in Fig. 4. The 10 GeV RCS has three

G. H. Rees

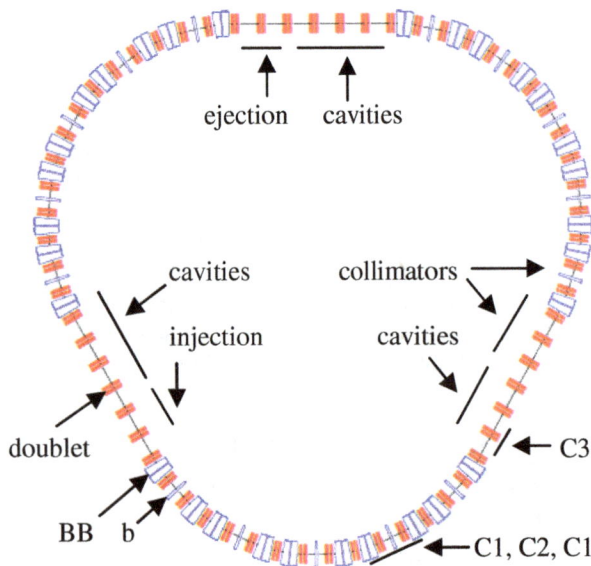

Fig. 4.   Layout of a 50 Hz, 4 MW, 3–10 GeV RCS.

superperiods, and a ring circumference of 801.45 m. Each straight has six C3 doublets and each arc has five groups of (C1, C2, C1) doublet cells. All cells have (Q_h, Q_v) tunes of $\approx (0.8, 0.5)/3$, with arc tunes of (4.0, 2.5) and ring tunes of (16.79, 10.39). The bending (b) perturbation of the C2 cells and the choice for a Q_h of 4.0 in the arcs cause an increase in the value of gamma-t to 19.9.

The RCS booster accelerates and compresses three 1.1 eV sec bunches, using harmonic number 3. The driver may use a higher harmonic ($\times 8$) and so has $h = 24$ (8.72–8.94 MHz). The peak voltage per turn for an adiabatic bunch compression to 3 ns rms at 10 GeV is 1.26 MV but acceleration, in the dual harmonic magnet field, requires 2.82 MV. An NFFAG ring needs a much lower voltage, and so is preferred to the RCS option.

5.7.5. A 50/3 Hz, 4 MW H^+ driver for a $\mu^\pm$ collider

A proton driver for a muon collider (MC) is the most complex driver mentioned. This is because it needs a single proton bunch per cycle (to create single, initial and final μ^+ and μ^- bunches). The beam power and final bunch duration have to be the same as in the NF driver but both the repetition rate and the bunch number are a factor-of-3 less. The beam energy may be raised from 10 up to 30 GeV to recover one factor

of 3 and this allows the NF driver to be developed into a driver for an MC.

It is assumed that NF operation is over by the time the MC is needed. This enables a modified NF driver to act as the injector chain for a new 30 GeV ring. As only a single proton bunch may be employed, the rf harmonic numbers must be reduced from 3 to 1 in the 3 GeV RCS booster, and from 24 to 8 in the 10 GeV ring. The bunch phase space area must also be increased to 3.3 eV sec, to preserve the levels of space charge in the 3 GeV ring. At the repetition rate of 50/3 Hz, the NF rings require only one third of their previous rf acceleration voltages. The rf ferrite-loaded cavities are thus adequate for the lower frequency ranges but need to be modified by increasing the ferrite permeability and the capacitor loading.

There is the choice between an RCS and an NFFAG for the 10–30 GeV ring, and the former is preferred as an NFFAG of similar size needs SC magnets. A 50/3 Hz, 4 MW MC driver may thus be a linac with a 3 GeV RCS, a 10 GeV NFFAG and a 30 GeV RCS. The final ring requires a high value of gamma-t (~ 40), to avoid crossing transition during acceleration, and this may be achieved with five groups of three cells for the arcs, as used for the 10 GeV RCS lattice in Fig. 4.

A concentric arrangement for the NFFAG and larger 30 GeV ring is assumed, with the NFFAG layout as in Fig. 6. The 30 GeV ring lattice is similar to that in Fig. 4 but has four superperiods. Its circumference is 25% larger than that of the NFFAG. The C1 arc cells use combined function magnets to keep the ring size small, while only quadrupoles are used for the C2 cells. Doublet cells, similar to C2 cells, are used for the C3, C4, C4, C3 cell arrangement of the straights. The (Q_h, Q_v) cell tunes are $\approx (0.8, 0.6)/3$, giving arc tunes of (4.0, 3.0) and ring tunes of (20.39, 15.30), an arrangement which is good for the adjustment of the ring chromaticity.

All quadrupoles have the same gradients. Positions are adjusted for a gamma-t of 38.1 and for maximum matched values of $\beta_h = 19$ m, $\beta_v = 20$ m and $D_h = 2.76$ m in the arcs. The long drifts in C2, C3 and C4 cells are 7.21, 5.76 and 6.56 m, respectively. Peak rf voltages are 3.52 MV for acceleration, and 1.38 MV for compression of the single bunch to 2.5 ns rms, at 11.96 MHz ($h = 40$).

6. High Power FFAG Accelerators

6.1. *Scaling FFAG rings at KEK and KURRI*

FFAG rings of a particular, nonlinear scaling form were invented by T. Ohkawa [48] in 1953 and subsequently prototyped at the Midwestern Universities Research Association (MURA), Wisconsin. The rings used combined function gradient magnets, which either had spiral edges or formed F(+) O D(−) O type lattice cells, where + and − indicate bend directions. In the 1990s, interest in them returned, mainly in Japan, where a number have now been built. These FFAGs have either a spiral or D(−) o F(+) o D(−) O triplet cell design (with short o and longer O drift lengths).

A chain of three proton FFAGs is being built at the Kyoto University Research Reactor Institute (KURRI) [49], to allow basic ADSR studies to be made at the Institute's research reactor. The proton chain consists of a 0.1–2.5 MeV spiral ring, a 2.5–20 MeV triplet ring and a final, 20–150 MeV triplet FFAG. The output energy of the first ring is adjustable and fields may be lowered in the other rings to provide scaled momentum ranges. The first ring uses induction acceleration, while the second and third rings use broadband magnetic alloy (MA) cavities for the acceleration. These have been developed for use in FFAGs and in the J-PARC ring accelerators (C. Ohmori, Y. Mori). The KURRI design aim is to operate at 120 Hz, with a 1 μA beam current at energies up to 150 MeV, and a longer term aim of the Institute is to construct a fourth FFAG ring for an energy of 500 MeV, and to develop the beam current up to 1 mA for a 0.5 MW source. Higher powers may be considered for a scaling FFAG by increasing the repetition rate and the output beam energy.

6.2. *Non-scaling FFAG rings*

A scaling ring is one in which the beam closed orbits at different momenta are scaled replicas of one another. Non-scaling rings may be designed to have smaller orbit separations by using different magnet arrangements. The fields may fall off outwardly so that the bend radii and associated beam dynamics alter for each orbit. It proves feasible to design a non-scaling ring which is non-isochronous or nearly, or fully, isochronous. Two types of non-scaling FFAG ring have been considered, together with scaling rings, as the drivers for high power applications. Designs are reviewed in Refs. 50 and 51.

6.3. *Non-scaling, 10 GeV neutrino factory driver*

One type of non-scaling FFAG ring has a lattice cell of a d(−) o F($\pm$) o D(+) o F($\pm$) o d(−) O form, and uses three types of nonlinear magnet for the five needed. The improved control of focusing with orbit radius allows both a non-isochronous ring, with constant cell tunes, and an isochronous ring, of fixed vertical tune, to be realized. The name "pumplet" is used for this symmetrical group of five magnets. It is possible to match two pumplet cells which have different lengths for their long straight section and so form bending-type insertions. The non-isochronous and the isochronous pumplet rings are identified as NFFAG and IFFAG, respectively, and when insertions are included, as NFFAGI and IFFAGI.

A nonlinear, non-scaling, non-isochronous NFFAG was proposed as one of the options for the 50 Hz, 4 MW, 3–10 GeV proton driver for an ISS NF study in 2006 [47]. The ring orbit circumference needed to be 801.45 m, at 10 GeV, to be compatible with the associated, 20 GeV, $\mu^{\pm}$ decay rings. Insertions were not required for a ring of this size, so the lattice used only identical, "pumplet" cells of room temperature magnets. The number of cells (see Fig. 5) was chosen to be 66, for a cell orbit length at 10 GeV of 12.14 m.

The nonlinear d and D units of the dFDFdO cell are vertically focusing, parallel-edged, combined function magnets, but the d have − bends and the D have + bends. There are zero entry and exit, edge angles, respectively, for the input and the output d magnets. The nonlinear, horizontally focusing, (+) bend, combined function F units have edges parallel to those of d and D magnets, as indicated in Fig. 5.

The injector chain is a 50 Hz, 0.2 GeV H⁻ linac, followed by a 0.2–3 GeV booster synchrotron, which

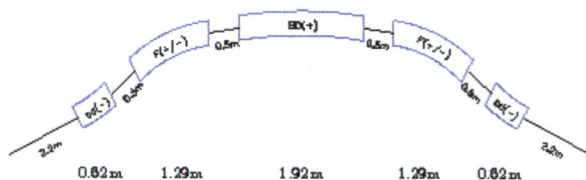

Fig. 5.   Single cell for a 4 MW, 10 GeV proton driver.

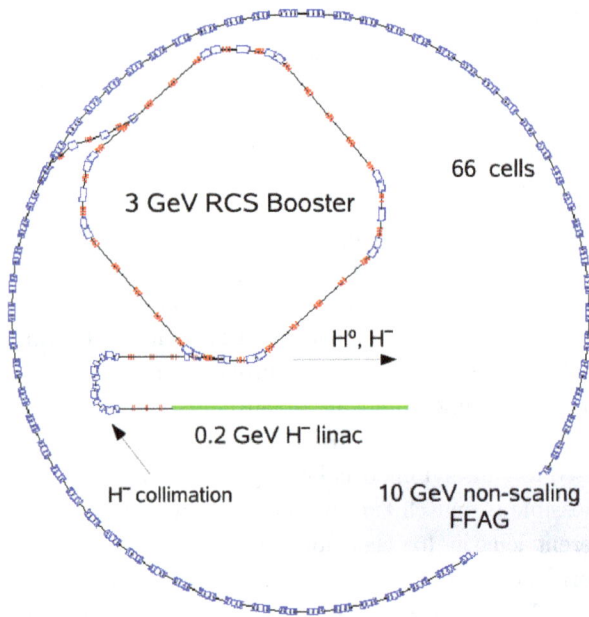

Fig. 6. Layout of the 4 MW NFFAG driver.

is preferred to an NFFAG booster, as it allows a superior H$^-$ injection system. The layout of the complete 4 MW driver is shown in Fig. 6.

For the short proton bunches required, adiabatic bunch compression begins in the booster, with a harmonic number of $h = 3$, and continues in the driver with $h = 24$ (8.72–8.94 MHz). A voltage of 1.3 MV peak per turn is needed to provide bunches at 10 GeV of duration 3 ns rms. The use of ferrite-tuned rf systems is proposed, distributed in the 4.4 m ring straight sections.

A single-pulse injection and a three-pulse extraction kicker and septum system are needed in the NFFAG, with extensive, primary and secondary, collimation systems, to protect against losses of the high power beam. Though progress has been made in understanding the nonlinear ring behaviour, the

construction and study with a small, 3–5.447 MeV electron model is considered necessary for advancing the design and proving its viability.

6.4. Non-scaling, linear focusing FFAG drivers

Another type of non-scaling FFAG accelerator [52] was proposed in 2004 by A. G. Ruggiero (BNL), for use in proton or heavy ion, high power driver applications. It employed magnets which had linear focusing gradients, as in the FFAG ring designs that had been introduced by C. J. Johnstone, W. Wan and A. Garren [53] in 1997, for the rapid acceleration of high energy, $\mu^\pm$ beams.

The beam dynamics is different from that of the muons, however, as the normalized beam emittances are smaller, space charge issues are involved, acceleration is always in the stable region, and operation is far away from isochronism. The two designs are similar, though, in that the negative chromaticities give large variations of tune with momentum, so that integer and half integer betatron resonances are crossed in the 10–100 turns of acceleration and yet give only a small growth of the emittances.

One example, which is indicated schematically in Fig. 7, is a design for the acceleration of uranium 238 ions, to produce rare isotopes. The values proposed for the average beam current and average beam power are 4.2 particle μA and 0.4 MW, respectively. The charge states planned are 30$^+$ for the output of the ion source, 70$^+$ after stripping at 15 MeV/u and 90$^+$ after the final stripping at 80 MeV/u. A second example is that of a proton driver, used to feed a sub-critical reactor (ADSR) for energy production. The driver employs two FFAG rings in series, to accelerate a 10 mA proton beam from 50 to 1000 MeV, for 10 MW of beam power.

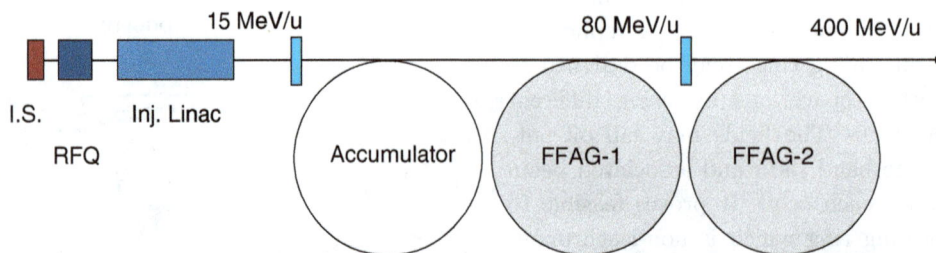

Fig. 7. Schematic drawing of a U-238 accelerator for radioactive ion production.

The two FFAG accelerators of each design employ a high periodicity lattice, which consists of linear focusing, F(−)D(+)F(−), triplet cells. The compact cells give a ring circumference of 204 m, and the F and D magnets may have small apertures. A repetition rate of 1–10 kHz, or a cw operating mode, is needed for both designs.

Three, non-isochronous modes of acceleration have been considered for the linear, non-scaling FFAG rings:

- Use of broadband rf cavities for acceleration at a pulse repetition rate of 1 kHz
- Use of "harmonic number jump" acceleration at a pulse repetition rate of 10 kHz
- Use of "harmonic number jump" acceleration for a continuous wave (cw) mode

Magnetic alloy (MA) loading may be used to obtain low frequency, broadband cavities, as at the J-PARC and the KURRI accelerators. These cavities do not exhibit the dynamic loss effects found in ferrite-tuned units, and are suitable for use at repetition rates up to ≈ 1 kHz.

A higher rate allows a lower beam current, less space charge forces and a faster crossing of betatron resonances in acceleration (when the tunes fall). A technique known as harmonic number jump (HNJ) acceleration is being studied for rates up to 10 kHz and for cw operation.

The HNJ technique [54] requires fixed frequency cavities, with longitudinal fields which increase almost linearly with radial position. A frequency ≥ 500 MHz is proposed, using high field, SC rf cavities. The radial field profile may be obtained by the use of adjacent TM_{11}, TM_{01} and TM_{11} mode cavities, provided that the deflecting mode of the TM_{11} unit is not a problem.

The rf system may not be switched on and off in the 100 μs cycle time at 10 kHz, so cw klystrons are used, without pulse modulators. In a system using circulators, which is optimized for beam loading, the klystrons work at full power while beam circulates, and at ≈ half power while beam does not circulate (to prevent field fall-off). For acceleration then to be efficient, most of the cycle (≥ 150 turns for protons) has to be used, and the resonance crossing rates have still to be compatible with low FFAG beam loss.

As the ion velocity, βc, rises, the harmonic number must fall, almost in proportion to $1/\beta$, with a jump for each turn of one or more integers. For the two ring schemes, the number of bunches injected into the first ring has to be ≤ the harmonic number in the second, at final energy (<, if the beam has a time gap for an extraction kicker). There may thus be only a partial ring filling at injection, since the chosen rings have the same circumference.

For the semi-continuous, cw mode, the ions are pre-chopped prior to the partial injection filling of the first ring, and then accelerated rapidly by the HNJ method. The orbits of both rings are occupied by beam, and the separations of the injection and extraction orbits from adjacent orbits must be sufficient for low loss operation.

A small electron model is required for the linear, non-scaling rings in order to advance design ideas and to study beam dynamics issues, as in the case of the nonlinear, non-scaling NFFAG rings.

7. Beam Dynamics Issues

7.1. *Linac issues*

7.1.1. *Beam loss levels and reliability*

A major design issue is the reduction of beam losses. The high energy losses must be under 1 nA/m for the hands-on maintenance of the linac stages and for short notice repair. This corresponds to a fractional beam loss of 10^{-6} over a 100 m length of a 100 mA (average), high energy linac stage.

A linac may run with some SC cavities compensating for others that are off, but the extent to which it may do so depends on how well particular sections are matched. Some loss of the beam halo may result, which affects the long term linac maintenance and reliability. A better option for high power linacs is the use of modest, SC field gradients. Protons are chosen for high power, except when H$^-$ ion ring injection is needed. The use of H$^-$ ions places constraints on linac vacuum system design and on the maximum quadrupole fields at high energies.

7.1.2. *MEBT chopper, collimator and matching section*

The MEBT has to restrict the growth of the halo in the beam received from the strongly focusing RFQ. The ratio between the nonlinear and linear space charge forces is a function of the beam distribution and varies continuously in the MEBT focusing channel. Fractional, space charge phase shifts per cell, $\Delta\mu/\mu$, are typically about -0.45 (e.g. μ values from $80°$ down to $44°$).

Linac stages and inter-stage transitions need smooth changes of focusing. Cell lengths are given in units of $\beta\lambda$, where β is ion velocity relative to that of light and λ is the free space wavelength at f, the top linac frequency [e.g. $f = 560$ MHz for an ESS linac; 280 MHz stages have a $2\beta\lambda$ RFQ (FODO), an $80\beta\lambda$ MEBT, a $4\beta\lambda$ DTL (FODO) and a $9\beta\lambda$ funnel, while the 560 MHz stages have a $10\beta\lambda$ CCDTL and $11,12\beta\lambda$ CCLs]. In the $80\beta\lambda$ MEBT, there are four doublet cells for the choppers, beam dumps, and input and output matching quadrupoles and bunchers.

To restrict beam halo growth in the MEBT, changes in the ratio of nonlinear to linear space charge forces must be minimized. Hence, horizontal to vertical beam aspect ratios are kept constant on average, and maximum and minimum beam size changes are avoided if possible. A more complex chopper is needed when there is no electrostatic LEBT chopper helping to lower the MEBT beam dump heat loads.

An R&D dual chopper system at RAL includes a fast and a slow beam deflector [55], with the former removing three adjacent beam bunches to create a 14 ns time gap for the rise time of the latter, which has a longer pulse duration. There is chopping for the start and the end of associated ring revolution periods. A slow-wave electrode structure is used for a fast chopper, and water-cooled, lumped electrode deflectors for a slower main system, which also serves as a beam dump. The 25 Hz repetition rate is half that of the linac and an identical dual chopper unit is used downstream to allow 50 Hz operation with reduced beam dump heat loads.

A chopper is needed for progressively raising the beam power even when a linac does not feed a ring. One possibility for easing the chopper design is to double the MEBT drift lengths (for the same lattice γ parameters) by the addition of a solenoid. This creates space for both permanent diagnostic equipment and the scrapers. The scheme is being considered at RAL, but requires next a detailed study of the solenoid. The scrapers are needed to remove any beam halo that does arise in the MEBT.

7.1.3. Momentum correction and ramping cavity

Adaptive feed-forward helps to reduce errors in the linac output energy, due to amplitude and phase errors in the beam-loaded linac fields. External beam line cavities may also be used, with a first cavity system to raise the beam momentum spread (and hasten the debunching) and a second system to aid the energy correction and give a reduced momentum spread and (or) a ramping of the momentum of the beam.

A second cavity, at peak voltage V, correctly phased, at distance ℓ downstream, removes the energy error existing after the first cavity if $\ell V = (\beta\gamma)^3 (E_o/e) (c/2\pi f)$, where β and γ are relativistic beam factors, e and E_o are the charge and rest mass of the ions concerned, c is the speed of light and f is the correction cavity excitation frequency.

The ℓV setting gives a momentum spread reduction and allows momentum ramping via a cavity phase ramp. The latter may also correct an energy error at the second cavity caused by slowly varying phase errors at the first. Low-Q cavities are needed for ramping, and the use of wide aperture, externally loaded SC cavities is preferred. For a very high energy linac, the ℓV value is not practical for energy correction, due to the $(\beta\gamma)^3$ dependence. At the output of the linac, the beam debunches rapidly in the absence of rf fields and the output focusing match must take account of the reduced space charge forces.

7.1.4. Coherent beam envelope motion

A combination of three coherent longitudinal-transverse coupled envelope modes may arise due to quadrupole or rf focusing field errors. There is a pure quadrupole mode which has out-of-phase, coherent h and v beam betatron motions and a high and a low, longitudinal-transverse mode, each with in-phase coherent h and v motions.

Approximate formulae are available for the coherent mode cell tunes, $Q_c = \mu_c/2\pi$. Resonance effects may arise if n or $Q_c = (kQ_h + \ell Q_v + mQ_z)$, where k, ℓ, m, n are integers and Q_h, Q_v, Q_z are incoherent cell tunes under space charge. Low order resonances, for the high or low coupled modes or with n, the cell focusing periodicity (e.g. $0 = Q_h - 2Q_z$), may be avoided if tunes are set with approximate energy equipartitioning [56, 57] in the three oscillation planes. The pure quadrupole mode, given by $Q_c = Q_h + Q_v$, cannot be avoided, however, so matching of a space-charge-dominated linac is an important issue.

7.1.5. *Output beam transport lines*

A linac output line to a ring is very different from a linac output line to a high power target. In the former case, the low emittance, 1–5 MW beam is kept small for the H^- injection, whereas the cross-sectional area of a 25 MW proton beam has to be expanded by a factor of ≈ 700 at a target. Large temperature variations over the target must be avoided. Thus, the dense core of the linac beam with its surrounding halo needs to be transformed in a beam expansion line [58] for a uniform target irradiation.

To achieve this, nonlinear magnets may be added to a quadrupole focusing channel and a long drift length used after a final quadrupole triplet lens. Included ahead of the triplet are four quadrupole doublets and two octupole–dodecapole units, whose fields have an odd multipolarity, allowing edge particle focusing towards the beam centre. As h–v motion is coupled, one nonlinear unit is set where beam is small in the h plane and relatively large in the v plane, and the other where the aspect ratio is reversed. Two-D Gaussian distributions of an elliptical cross-section beam may be transformed to uniform distributions of an expanded, rectangular beam, or to 2D elliptical or 2D parabolic distributions of an enlarged elliptical cross-section beam. Displacements of beam at the nonlinear units must be corrected.

For a H^- injection line, an achromatic section is needed for collimation and beam monitoring. Upstream of the achromat are cavities for momentum spread reduction, ramping and correction, and also horizontal halo collimators. The SNS uses conventional collimators but the ESS planned, for momentum and vertical halo collimation, the use of stripping foils in the low field, combined function achromat magnets. These were to be open on the outer radius of their C-shaped cores for easy exit of stripped beam to external, shielded beam dumps.

Vertical halo collimation was planned for both beam edges at four achromat positions and momentum edge-collimation at two further locations. An initial momentum halo scraping preceded the main momentum collimation, which was arranged to be at the very high normalized dispersion, centre point of the arc. Bunchers at quarter and three-quarter arc positions were to restore the upright orientation of the momentum phase space ellipses at both the achromat centre and the end.

7.1.6. *Funnel section*

A funnel doubles the beam current by an interleaving of two low energy linac beams prior to further acceleration. The first funnel design is from LANL [59] and a later one is for the ESS linac [60]. In the latter, beams from two 57 mA, 280 MHz, 20 MeV, H^- DTLs are merged into a 114 mA, 560 MHz CCDTL. Each leg of the funnel has six $9\beta\lambda$ FODO cells which include bunchers and septum magnets. The layout of the scheme proposed for the ESS is shown in the schematic drawing of Fig. 8.

There is an 840 MHz buncher cavity in the upstream part of the first five cells and a two-channel, asymmetric, common 560 MHz buncher for the sixth cell. Also, there are septum magnets in the first, second and fifth cells of each funnel leg, and a $\pm$ 2°, transverse deflection cavity in the final common section of the sixth cell. In one leg, septum magnets bend the beam successively through the angles of 10°, 2° and −10°, while in the other leg the equivalent bending is in the opposite direction. The two input DTL beams are in parallel, and the common 280 MHz deflection cavity at the funnel exit bends the beam from each leg alternately in opposite directions to bring the merged, interleaved beams on axis for the following CCDTL.

Buncher fields are adjusted so that the beam bunches are upright in longitudinal phase space at the septum magnets and final deflector. The $9\beta\lambda$ cell lengths allow a natural matched bunch length for the beams that emerge from the two DTL linacs. The average transverse phase advance per cell under space charge is approximately 45°, which gives achromatic focusing for the bunching and the bending that is provided. The funnel is optimized for a minimum beam halo by adjusting the buncher voltages and focusing gradients for periodic beam sizes in all the cells, compensating for increased dispersive effects with a reduction in beam betatron amplitudes.

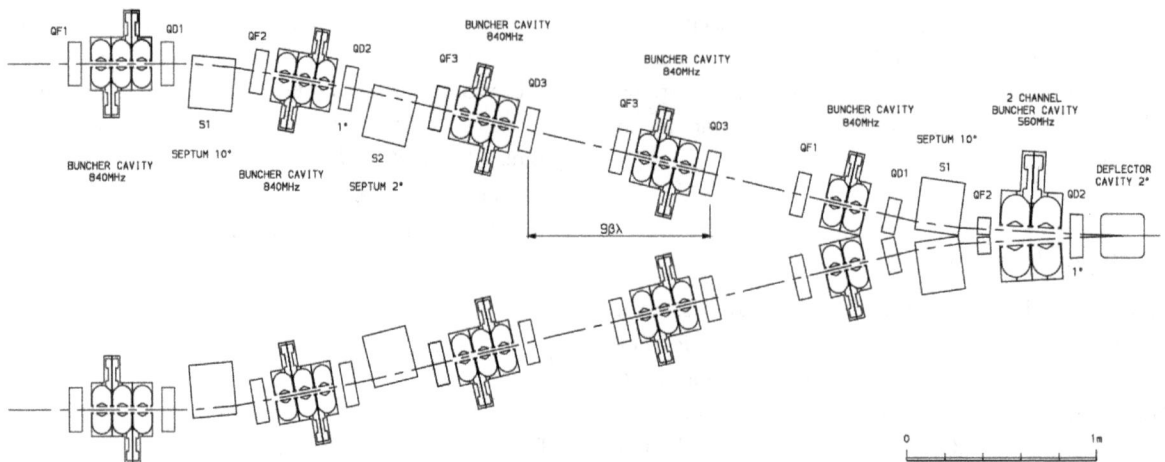

Fig. 8. Schematic drawing of the 280, 560 and 840 MHz, 20 MeV funnel proposed for an ESS linac.

7.2. Ring design issues

7.2.1. Ring beam loss collimation systems

Some beam loss at injection is inevitable but most other ring losses are localized by beam collimation systems. Fractional loss design levels are set at $\approx 10^{-3}$ at the collimators and 10^{-4} elsewhere in the ring, with the latter equivalent to a beam loss of ≈ 1 watt per metre [61]. Momentum and betatron collimators are used, and the former is important in an RCS for ring protection.

An RCS needs rectangular vessels to avoid vertical loss of any untrapped beam particles which spiral to a momentum collector (set at high, normalized dispersion). Rectangular vessels also aid transverse halo collection. Betatron collectors are in three adjacent straight sections, with a primary unit and three long secondary units in the h and v planes, at $17°$, $90°$ and $163°$ betatron phase shifts after the primary unit. The collection efficiency is $> 90\%$. A sub-tunnel may be used within the main ring tunnel to reduce the air activation near the collectors.

7.2.2. Resonance crossing

Limitations arise in a slow-cycling high intensity ring at incoherent space charge tune depressions of ≈ 0.6, due to the effects of the integer and the half integer betatron resonances. The associated injection beam losses are too large for a high power ring where, to limit the beam halo, typical design values for space charge tune depressions are < 0.2 (SNS) and < 0.1 (ESS).

In the case of a linear, non-scaling FFAG ring, an even lower tune depression may be needed, depending on the speed of crossing of many integer and half integer betatron resonances during the rapid beam acceleration.

7.2.3. Electron–proton (e^-–p) instability

The e^-–p instability is a serious effect for a high power proton ring. The e^- are created by residual gas ionization, by H^- stripping and by e^- emission when ions or e^- hit the collimators or the vacuum chamber walls. Above a H^+ threshold, a multipactor regime develops, as e^- that leave the end of a proton bunch pass quickly to the walls and create copious prompt secondary e^- (and tertiaries).

Cures include active damping systems [62], coating of vacuum units with TiN or NEG material (to reduce e^- yields), collection of the foil-stripped e^-, and the use of clearing electrodes and also a beam-in-gap kicker.

7.2.4. H^- charge exchange injection

Anti-correlated painting is the preferred injection scheme as correlated painting produces some particles with both large h and v amplitudes and these may be lost by $h - v$ coupling. The H^- beam (10σ) is scraped in the input line at 5σ and the beam centre at the foil is then set at a distance 5σ (≈ 5 mm) from the nearby foil corner edges. A steering error (≈ 0.5 mm) causes some beam to miss the primary foil and to exit the ring via a septum magnet,

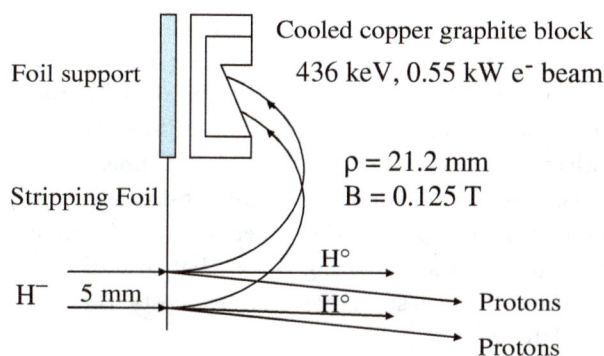

Fig. 9. Stripped electron collection for 800 MeV H^- injection into a potential 2 MW upgrade ring for ISIS.

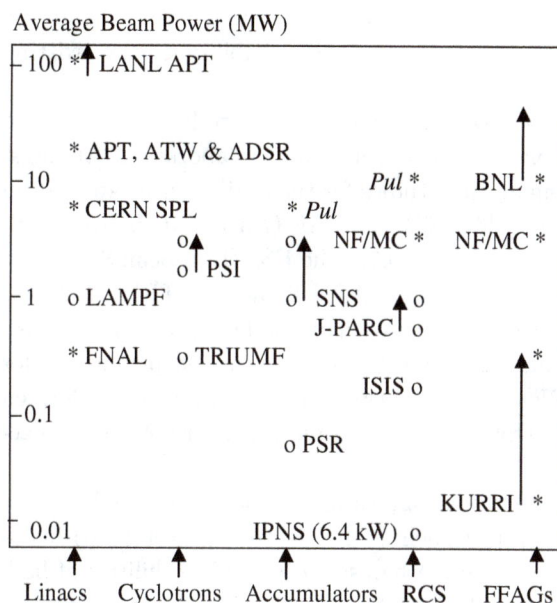

Fig. 10. Beam power plot for H^-/proton accelerators, with o for operating machines, * for conceptual designs, *Pul* for probable upper limits and NF/MC referring to a proton driver for a neutrino factory or a muon collider.

after being stripped at the downstream secondary stripper foil.

The scheme proposed for collecting the 436 keV, 0.55 kW stripped e^- in a 2 MW ISIS upgrade ring design is shown in Fig. 9. The electron bending radius of 21.24 mm is large enough for the collector to be positioned outside the horizontal acceptance of the ring. The arrangement is different from that used at the SNS ring, where the e^- spiral downwards in the fringe field of a central bump magnet to a collector located at the bottom of the aperture.

Unstripped H^- ions and partially stripped $H°$ atoms (in a range of excited states) may lead to ring activation, so the positioning of the secondary stripper unit and its associated extraction septum magnet is an important design issue.

A feature of the injection scheme that employs finite dispersion (and a ramped beam momentum spread) is some broadening of the injected beam microbunches. This helps to reduce the transverse space charge forces in the initial stages of the injection process.

8. Beam Power

An exact definition of high beam power in the context of a hadron ring is hard to obtain. In this article, the view has been taken that the term refers to those rings where the practical operating limit is set by the beam power and not by a space-charge-limited beam intensity per pulse. Most slow-cycling high intensity rings then fall outside the definition.

A chart for high power H^- or proton accelerators, with this definition, is given in Fig. 10, both for operating machines and for various conceptual designs.

The high power proton linacs for APT, ATW and ADSR present an entirely new range of challenges. The very stringent requirements on the beam reliability and availability are well beyond present capabilities and will require large component redundancy, sophisticated rf systems and control systems that are able to respond safely on msec time scales. These extreme requirements are set by the need to avoid stress-induced damage of the core components of the targets, due to interruption of beam current. The developments and the progress in designs are documented in a series of OECD/NEA workshop proceedings.

9. International Collaborations

Many types of international collaborations have assisted the development of the high power hadron accelerators. A series of meetings organized by ICANS (International Collaboration on Advanced Neutron Sources) has been held at irregular intervals since 1977, and the most recent one, the 18th [63], was in April 2007, in the city of Dongguan, near to the site proposed for the new Chinese spallation source. At ICANS, recent advances in neutron scattering science are reported together with new accelerator and target design issues and the status, performance and reliability of the existing sources.

Several high power design studies have involved an element of international collaboration. Amongst them are the studies for the EHF (European Hadron Facility [64, 65]), the AHF (Advanced Hadron Facility at LANL [66, 67]), the ESS (European Spallation Source [44]), the KAON Project at TRIUMF, Vancouver [68], the proton driver designs at Fermilab [69, 70 and 32], the ISS (International Scoping Study for a Neutrino Factory [47]), and the recently formed IDS (International Design Study for a Neutrino Factory [71]).

International workshops have been held on relevant topics such as the electron–proton (e–p) instability drive mechanism [72], the instabilities of high intensity hadron beams in rings [73] and the collimation of the beam halo in designated areas of high power linacs or rings [74]. The International Committee on Future Accelerators (ICFA) has set up a working group to establish parameter databases and to help organize a series of ICFA workshops (HB02, HB04, HB06 and HB08).

FFAGs have been studied at workshops held in Japan (KEK, KURRI), the USA (FNAL, BNL), Canada (TRIUMF), France (Grenoble) and the UK (DL). Those in Japan since 2004 [75] have formal proceedings. *ICFA Newsletter* No. 43 [50] reviews the various FFAG designs.

10. Future Developments

Most high power applications make use of a linac, either directly or for H^- injection into accumulator–compressor rings or the first ring of an accelerator chain. The next major advance for hadron linacs is expected to be for an ADSR, APT or ATW project, using a 10–25 MW, 1 GeV proton beam. The design is likely to rely heavily on the experience gained in commissioning the SC stages of the SNS H^- injector linac to 1.4 and 3.0 MW. A future design issue is the possible extension of the SC cavity stages down to 10 MeV, as envisaged at FNAL.

Commercial criteria hold for ADSR, APT and ATW projects. Forty-year operating costs, including those for staff, power, accelerator maintenance, recurrent spares and consumables, are a predominant factor. An estimate for a UK 1.0 GeV, 25 MW conceptual proton linac and its beam line, assuming 200 staff, 5000-hour operation yearly with 1000 hours for beam set-up, is £ 1500, million at 1998 prices. The cost for the linac, beam line, buildings, tunnels, shielding, electrical substations, component handling and the other services is about one third of the operating costs. Higher staff costs in the US suggest higher running costs than for the UK.

A choice between a linac, a cyclotron and an FFAG for a proton beam power >10 MW at 1 GeV favours the linac due to its less complexity, shorter setting-up times and simpler collimation, shielding and maintenance.

Ring advances will depend on approval of specific projects. Short pulse proton drivers with a beam power of 4–5, or possibly 10 MW, are needed for spallation neutron sources, neutrino factories and muon colliders (listed here in order of increasing complexity). The first two drivers need a 50 Hz repetition rate, while current muon collider designs consider a single proton bunch of time extent $\leq$ 3 ns rms at a rate of $\approx$ 50/3 Hz, requiring a driver kinetic energy of $\approx$ 30 GeV, or more.

A choice between compressor rings, synchrotrons and FFAG accelerators for short pulse, high power proton drivers is not obvious, as indicated in the review. For spallation sources, powers up to $\approx$ 1 MW are best provided by a linac and an RCS, powers of $\approx$ 5 MW by a linac and two accumulator–compressor rings, and powers of $\approx$ 10 MW by a linac, an accumulator and either two NFFAG or two RCS accelerators. NFFAG rings would be favoured if a low energy electron model was first developed and proved to be satisfactory.

The proton driver for a neutrino factory may be a 50 Hz, 4 MW, 10 GeV accelerator chain, consisting of a 0.2 GeV linac, a 3.0 GeV RCS booster and a 10 GeV NFFAG or RCS ring. Again, the NFFAG would be the preferred choice but with the same proviso as above. The 10 GeV driver may be developed for a muon collider (as discussed in Subsec. 5.7.5) by reducing its pulse rate from 50 to 50/3 Hz and adding a 50/3 Hz, 4 MW, 30 GeV RCS ring. An alternative driver based on a high energy H^- linac, an accumulator, a compressor and a 30 GeV RCS appears to be a difficult option, as it requires a high pulse current for the high energy linac.

References

[1] E. A. Knapp, Status report on LAMPF, in *Proc. US National PAC71* (1971), pp. 508–511.

[2] J. R. Richardson *et al.*, Prod. of var. ener. beams from TRIUMF cyclotron, in *Proc. PAC73* (1973), p. 1403.

[3] J. P. Blaser and H. A. Willax, First tests with cyclotron at SIN, in *Proc. 9th I .H. E. Conf. Stanford* (1974), p. 643.

[4] I. M. Kapchinski and V. A. Teplyakov, *Prib. Tekh. Eksp.* **19**(2), 19 (1970).

[5] Y. I. Bel'chenko, G. I. Dimov and V. G. Dudnikov, *IZV. Acad. Sci. USSR* **37**, 2573 (1973).

[6] R. F. Welton *et al.*, *Rev. Sci. Instrum.* **5**, 1793 (2004).

[7] R. Geller, *Proc. First Int. Conf. Ion Sources* (Saclay, France, 1969), p. 287.

[8] V. A. Teplyakov, The first cw accelerator in USSR and a birth of AFF, in *Proc. EPAC '06*, THPPA03 (2006).

[9] M. Seidel, Upgrade of the PSI cyclotron facility to 1.8 MW, in *Proc. Cyclotrons 2007* (Sicily, 2007).

[10] J. M. Potter *et al.*, *Handbook of Accelerator Physics and Engineering*, eds. A. Chao and M. Tigner, pp. 517–537.

[11] A. Ghiorso *et al.*, *IEEE Trans. Nucl. Sci.* 155 (1973).

[12] N. Angert, Status report on the upgraded UNILAC, in *Proc. PAC83* (1983), p. 2980.

[13] A. Ghioran, Progress with the SUPERHILAC, in *Proc. PAC73* (1073), p. 152.

[14] K. W. Shepard, *IEEE Trans. Nucl. Sci.* **NS-26**, 3659 (1979).

[15] Y. Yano, Commissioning of the RIBF at RIKEN, in *Proc. Cyclotrons 2007* (Sicily, 2007).

[16] P. Bertrand, The advancement of Spiral 2 Project, in *Proc. Cyclotrons 2007* (Sicily, 2007).

[17] J. A. Nolen, The U.S. Rare Isotope Accelerator Project, in *Proc. LINAC 2002* (Korea) (2002), ROPA002.

[18] K. Blasche *et al.*, Status report on SIS-ESR, in *Proc. EPAC '94* (London, 1994), p. 133.

[19] P. J. Spiller *et al.*, Approaches to high intensities for FAIR, in *Proc. EPAC '06* (2006), MOZAPA01.

[20] J. Wie, *Rev. Mod. Phys.* **75**, 1383 (2003).

[21] P. B. Moon, Prop. for inj. of H^+ by diss. of mol. H_2^+, in *CERN Symp. High Energy Physics* (1956), pp. 231–233.

[22] G. I. Budker *et al.*, *J. Nucl. Energy, Part C* **8**, 692 (1966).

[23] R. L, Martin, *IEEE Trans. Nucl. Sci.* 953 (1971).

[24] C. W. Planner, G. H. Rees and G. H. MacKenzie, A separated H^- injection system. In TRI-DN-89-K98 (1989).

[25] M. K. Craddock, *CERN Courier* 44 (2004).

[26] G. P. Lawrence, High power proton linac for APT, in *Proc. LINAC98* (1998), MO2002, pp. 26–30.

[27] D. Proch, *Handbook of Accelerator Physics and Engineeing*, eds. A. Chao and M. Tigner, pp. 534–535.

[28] I. Campisi *et al.*, Status and performance of the SNS SC linac, in *Proc. PAC07* (2007), WEPM5072, pp. 2502–2504.

[29] A. Aleksandrov *et al.*, New design of the MEBT chopper, in *Proc. PAC07* (2007), TUPAS073.

[30] S. Henderson, Status of the SNS; machine and science, in *Proc. PAC07* (2007), MOXK103.

[31] S. Assadi *et al.*, The SNS laser profile monitor and implementation, in *Proc. PAC07* (2007), WPPG054.

[32] R. Webber, S. Nagaitsev and R. McGinnis, in *Accelerator Physics and Technology Workshop for Project X* (FNAL, 2007).

[33] F. Gerigk *et al.*, Conceptual design of the SPL II, a high power SC H^- linac at CERN-2006-006 (2006).

[34] P. G. Bricault, ISOL RIB ion sources and targets for high power, in *Proc. Cyclotrons 2007* (Sicily, 2007).

[35] G. E. McMichael *et al.*, Accelerator research on the RCS at IPNS, in *Proc. EPAC '06* (2006), MOPCH126.

[36] G. H. Rees, Status of the SNS (now ISIS), in *Proc. PAC83, IEEE Trans. Nucl. Sci.* **NS-30**(4), 3044 (1983).

[37] D. H. Fitzgerald *et al.*, PSR injection upgrade, in *Proc. PAC99*, Vol. 1 (1999), pp. 518–520.

[38] R. J. Macek *et al.*, Electron cloud diagnostics at the PSR, in *Proc. PAC03*, Vol. 1 (2003), pp. 508–510.

[39] J. Wei *et al.*, Injection choice for Spallation Neutron Source ring, in *Proc. PAC01* (2001).

[40] M. A. Plum *et al.*, SNS ring commissioning results, in *Proc. EPAC '06* (2006), MOPCH131.

[41] J. A. Holmes *et al.*, Orbit injec. dump simulations of $H°$, H^- beams, in *Proc. PAC07* (2007), THPA5076.

[42] JAERI-KEK Project Team, Accelerator Technical Design Report for J-PARC, J-PARC 03-01 (2003).

[43] J. Wei, CSNS design, research and development, in *Proc. EPAC 2006* (2006), MOPCH136, pp. 366–368.

[44] The ESS (European Spallation Source) Project, Vol. III, Technical Report, pp. 2–4 - 2-56 (2002).

[45] M. S. Gulley *et al.*, *Phys. Rev. A* **53**(5), 3201 (1996).

[46] P. Bryant, M Regler and M. Schuster, eds., *The AUSTRON Feasibility Study* (Vienna, 1994).

[47] G. H. Rees, A proton driver report for the Neutrino Factory ISS. In RAL-TR-2007-023 (Dec. 2007).

[48] T. Ohkawa, *Proc. Annual Meeting of the JPS* (Japanese Physical Society, 1953).

[49] Y. Mori, Development of FFAG Accelerators at KURRI, in *Proc. Int. Workshop on FFAG Accelerators* (KURRI, 2006).

[50] C. R. Prior, issue ed., *ICFA Beam Dynamics Newslett.* No. 43, 12 (2007).

[51] G. H. Rees, Non-isochronous and isochronous, non-scaling FFAGs, in *Proc. Cyclotrons 2007* (Italy, 2007).

[52] A. G. Ruggiero, FFAG-based high intensity proton drivers, in *Proc. ICFA-HB2004* (2004), p. 324.

[53] C. J. Johnstone *et al.*, Fixed field circular accelerator designs, in *Proc. PAC99* (1999), pp. 3068–3070.

[54] A. G. Ruggiero, *ICFA Beam Dynamics Newslett.* No. 43, 84 (2007).

[55] M. A. Clarke-Gayther, A fast beam chopper for next generation proton drivers, in *Proc. EPAC '04* (2004).

[56] Ll. M. Young, Equipartitioning in a high current proton linac, in *Proc. PAC97* (1997), pp. 1920–1922.

[57] I. Hofmann and R. W. Hasse, Equipartitioning and halo due to anisotropy, in *Proc. PAC97*.

[58] B. Blind, *Nucl. Instrum. Methods* **B56/57**, 1099 (1991).

[59] S. Nath, Funnelling in high intensity linac designs, in *AIP Proc. 346, ADTT Conf. Nevada* (1994), pp. 397–396.

[60] C. R. Prior, Funnel studies for the ESS (European Spallation Source). In Report ESS-99-96-A (1999).

[61] J. Wei, Cleaning in high power proton accelerators, in *ICFA, AIP Conference Proc.*, Vol. 69 (2003), pp. 38–46.

[62] R. McCrady *et al.*, Active damping of the e–p instab. at the LANL PSR, in *Proc. PAC '07* (2007).

[63] S. Henderson, *ICFA Beam Dynamics Newslett.* No. 43, 136 (2007).

[64] F. Bradamante *et al.*, Proposal for a European hadron facility, in *EHF 87-1* (May 1987).

[65] F. Bradamante *et al.*, *XIVth workshop for the EHF, Eindhoven University of Technology, EHF 88/49* (1989).

[66] A proposal to extend the intensity frontier of nuclear & particle physics to 45 GeV, in *LA-UR-84-3982* (1984).

[67] *Proc. the 1988 Advanced Hadron Facility Workshop, LA-11432C* (Feb. 1988).

[68] TRIUMF-KAON Project: KAON Factory Study; accelerator design report. TRIUMF report (1991).

[69] The Proton Driver Design Study (16 GeV, 15 Hz RCS). *Fermilab-TM-2136* (Dec. 2000).

[70] Proton Driver Study II, Part 1 (8 GeV, 15 Hz RCS). *Fermilab-TM-2169* (May 2002).

[71] K. Long, ed., International Design Study (IDS), in *Neutrino Factory Workshop, RAL* (Jan. 2008).

[72] R. J. Macek, ed., *Workshop on two-stream instabilities* (Santa Fe, New Mexico, 1999).

[73] T. Roser and S. V. Zhang, eds., *Workshop on Instabilities of High Intensity Hadron Ring Beams, AIP 496* (1999).

[74] N. V. Mokhov and W. Chou, eds., *Workshop on Beam Halo and Scraping* (Fermilab, 2003).

[75] Y. Mori *et al.*, *The International Workshop on FFAG Accelerators, KEK* (2004), *KURRI* (2005, 2006, 2007).

Grahame Rees has worked on a range of accelerator topics. After early work on a neutron source for UK weapons' research, he helped build a proton synchrotron at Princeton University. He then joined Rutherford Laboratory, where he contributed to studies of superconducting synchrotrons, heavy ion fusion rings, an e–p collider EPIC, medical rings, a radioactive ion facility, an APT project, a FFAG design, and a Neutrino Factory. At Rutherford Laboratory, he initially worked on the Nimrod synchrotron, then on the ISIS spallation neutron source, and recently on ISIS upgrades. He has spent periods at several laboratories in the world, and has served as a member of many machine advisory committees.

Reviews of Accelerator Science and Technology
Vol. 1 (2008) 65–97
© World Scientific Publishing Company

Cyclotrons and Fixed-Field Alternating-Gradient Accelerators

M. K. Craddock

*Department of Physics and Astronomy,
University of British Columbia, and
TRIUMF, 4004 Wesbrook Mall,
Vancouver, BC V6T 2A3, Canada
craddock@triumf.ca*

K. R. Symon

*Physics Department,
University of Wisconsin–Madison,
Madison, WI 53706-1390, USA
krsymon@wisc.edu*

This article describes particle accelerators using magnets whose field strengths are fixed in time to steer and focus ion beams in a spiral orbit so that they pass between (and can be accelerated by) the same electrodes many times. The first example of such a device, Lawrence's *cyclotron*, revolutionized nuclear physics in the 1930s, but was limited in energy by relativistic effects. To overcome these limits two approaches were taken, enabling energies of many hundreds of MeV/u to be reached: either frequency-modulating the rf accelerating field (the *synchrocyclotron*) or introducing an azimuthal variation in the magnetic field (the *isochronous* or *sector-focused cyclotron*). Both techniques are applied in *fixed-field alternating-gradient accelerators (FFAGs)*, which were intensively studied in the 1950s and '60s with electron models. Technological advances have made possible the recent construction of several proton FFAGs, and a wide variety of designs is being studied for diverse applications with electrons, muons, protons and heavier ions. All fixed-field accelerators offer high beam intensity: classical and isochronous cyclotrons operate in cw mode and in some cases deliver beams of 2 mA; synchrocyclotrons and most FFAGs operate in pulsed mode, but are capable of much higher pulse repetition rates ($\leq$ kHz) than synchrotrons.

Keywords: Cyclotron; FFAG; synchrocyclotron; isochronous cyclotron; Thomas focusing; bucket; phase displacement; beam stacking; colliding beams.

1. Introduction

It was nearly 80 years ago that the 27-year-old Ernest Lawrence first conceived of the cyclotron [1] and, with his graduate student Stanley Livingston, provided the first experimental demonstration [2] of the magnetic resonance principle. Flegler [3] and several others [4] — Gabor, Szilard, Steenbeck — had similar ideas around the same time, but only Thibaud [5] carried out experiments and these proved unsuccessful (in spite of a superior design with the ion source outside the vacuum chamber). Like them, Lawrence, with personal experience of the difficulties of producing high voltages, had concluded that the simplest way of accelerating ions to high enough energies to induce nuclear reactions would be to raise their energy in many small steps using low voltages.

As he explained in his Nobel Prize lecture [6], the idea had come to him while seeking inspiration from recent journals in the University of California library, where he had come across Widerøe's report [7] of his successful acceleration of alkali ions to 20 keV in a two-gap linear accelerator using a 10 kV alternating voltage. Although Lawrence knew little German, the diagrams and photos were sufficient for him to grasp the resonance principle involved and to speculate on the possibility of using a magnetic field to bend the ions' paths into circular arcs so that they could be repeatedly accelerated by the same electrodes, excited by an alternating voltage to ensure an energy gain at each gap crossing.

When he worked out the dynamics of the orbits he found an unexpectedly favorable result. For an

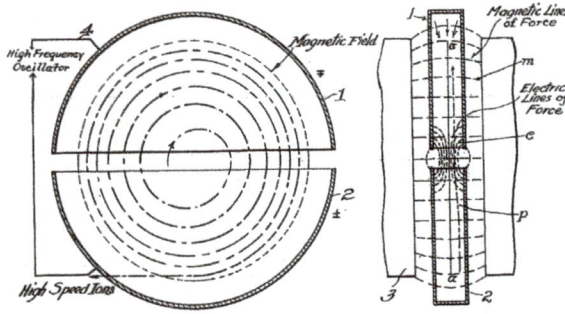

Fig. 1.   The cyclotron concept, from Lawrence's 1934 patent.

ion of mass m and charge q moving with velocity $\mathbf{v}$ normal to magnetic induction $\mathbf{B}$, the Lorentz force qvB perpendicular to both $\mathbf{v}$ and $\mathbf{B}$ produces circular motion with a radius of curvature R determined by the momentum:

$$qBR = mR\omega, \tag{1.1}$$

where $\omega = v/R$ is the angular velocity.

Canceling out R then gives

$$\omega = \frac{qB}{m}, \tag{1.2}$$

$$R = \frac{mv}{qB}. \tag{1.3}$$

For constant q and m, and uniform B, the angular frequency ω (and orbital period $2\pi/\omega$) remain constant, independent of ion speed — meaning, crucially, that the frequency of the accelerating voltage need not be adjusted as the ion gains energy. The orbit radius, on the other hand, does increase in proportion to the speed v, so that the orbit takes the form of a spiral.

Lawrence's graduate student J. J. Brady [8] later recalled his young supervisor's excitement following his eureka moment in early 1929:

"He came bursting into the lab..., his eyes glowing with enthusiasm, and pulled me over to the blackboard. He drew the equations of motion in a magnetic field. 'Notice that R appears on both sides,' he said. 'Cancels out. R cancels R. Do you see what that means? The resonance condition is not dependent on the radius... *Any* acceleration!'... 'R cancels R,' he said again. 'Do you see?'... He left in a rush, I suppose to tell other people that R canceled R."

Three-quarters of a century later, Lawrence's excitement seems fully justified. Magnetic resonance is the basis for all "circular" (or, more correctly, recirculating) accelerators except the betatron, and his original constant-field fixed-frequency concept

has proved remarkably adaptable in the face of the inertial mass m *not* remaining constant, as assumed, but increasing in proportion to the ion's total energy E, in accordance with the relativistic equation $E = mc^2$. The various adaptations have led to four distinct accelerator types:

(1) *Synchrocyclotrons*, where the rf frequency is modulated to match the decreasing angular frequency ω at higher energies (at the expense of pulsed operation and low beam intensity);

(2) *Synchrotrons*, where the magnetic field is ramped to keep the accelerating beam at a fixed radius (also pulsed, but outside the scope of this article);

(3) *Isochronous cyclotrons* (sometimes referred to as azimuthally varying-field (AVF) or sector-focused cyclotrons), where the magnet is split into radial or spiral sectors to create extra focusing forces, and the field strength is varied radially to regain isochronism, cw operation and high beam intensity;

(4) *Fixed-field alternating-gradient (FFAG) accelerators*, where the rf frequency is usually modulated and the magnet is broken into sectors.

The three varieties (1, 3, and 4) with fixed magnetic field and spiral orbits can all be regarded as members of the cyclotron family, as indicated in Table 1.

However, from the perspective of Fig. 2, where the range of rf modulation is plotted against the

Table 1.   The cyclotron family.

Magnetic field–azimuthal variation	Fixed frequency (cw beam)	Frequency-modulated (pulsed beam)
Uniform	Classical cyclotrons	Synchrocyclotrons
Periodic	Isochronous cyclotrons & cw FFAGs	FFAGs

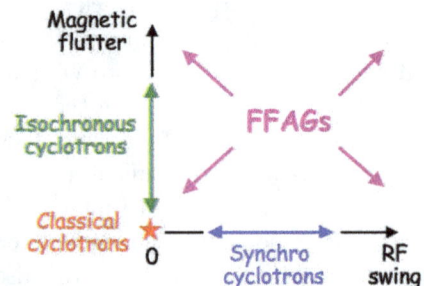

Fig. 2.   Cyclotrons as special FFAGs.

magnetic flutter (a measure of the azimuthal field variation — see below), cyclotrons can be viewed as just special cases of the FFAG.

Classical cyclotrons (i.e. those following Lawrence's basic design) could provide beams of 10–20 MeV/u, and in their heyday in the 1930s and '40s played a major role in the development of nuclear physics. Synchrocyclotrons overran this energy frontier in the late '40s, eventually delivering proton beams of up to 1 GeV (albeit pulsed and a thousand times weaker) and allowing artificial production and study of pions and muons. These machines in turn were supplanted in the '50s and '60s by isochronous cyclotrons, which cover virtually the same energy range, and can offer very high intensity cw beams — for protons up to 2 mA. Over the last 50 years, several hundred isochronous cyclotrons have been built for low- and medium-energy nuclear and particle physics, and for a variety of other applications. Chief among these are the production of proton-rich isotopes and particle-beam cancer therapy — the latter being a field pioneered by John Lawrence in 1939, using one of his brother Ernest's cyclotrons, and continued with various synchrocyclotrons.

FFAGs were intensively studied in the 1950s and '60s, and several electron prototypes were successfully built and tested. For energies up to 10 GeV or so, they offer much higher acceptances and repetition rates than synchrotrons — and therefore higher beam intensities — at the cost of more complicated magnet and rf cavity designs. Perhaps because of the difficulty and expense anticipated, early studies never progressed to the construction of proton FFAGs, but since 2000, with improvements in magnet and rf design technology, five have been built and brought into operation. In addition, more than 20 designs are under study for the acceleration of protons, heavy ions, electrons and muons, with applications as diverse as treating cancer, irradiating materials, driving subcritical reactors, boosting high-energy proton intensity, and producing neutrinos. Moreover, it has become apparent that FFAG designs need not be restricted to the "scaling" approach explored in the '50s. Dropping this restriction has revealed a range of interesting new design possibilities.

In the following sections the technicalities of the different fixed-field accelerator designs are covered in more detail. For further reading, Livingston and Blewett's book *Particle Accelerators* [9] provides a very full treatment of classical cyclotrons and synchrocyclotrons. The best introductory treatment of isochronous cyclotrons and FFAGs probably remains Livingood's *Cyclic Particle Accelerators* [10]. A more detailed account of isochronous cyclotrons is given in a review article by Richardson [11]. Less complete but more recent accounts are given by Stammbach and others [12] and by Blosser [13]. For FFAGs an interesting anecdotal account of the work at MURA is provided by Cole [14]. Recent work is summarized in a series of articles in the *ICFA Beam Dynamics Newsletter* [15]. Details of these studies are available from the presentations and proceedings of the 16 FFAG workshops (see e.g. Refs. 16–21) that have been held since 1999. (Web links to all the early workshops may be found at FFAG04 [18], and to the later ones at FFAG2007 [20].) Recent developments on both cyclotrons and FFAGs may be followed in the proceedings of the Cyclotron, APAC, EPAC, and PAC conferences on the JACoW website [22].

2. Classical Cyclotrons

2.1. *Berkeley beginnings*

Along with his first statement [1] of the cyclotron principle in late 1930, Lawrence described the experimental arrangement: "Semicircular hollow plates in a vacuum... are placed in a uniform magnetic field which is normal to the plane of the plates. The diametral edges of the plates are crossed by a grid of wires so that inside each pair of plates there is an electric field free region." He then went on to describe how the plates form part of a resonant circuit tuned to oscillate in synchronism with the circulating ions, and to give an example of the method's power: "... oscillations of 10,000 V and 20 m wavelength impressed on plates of 10 cm radius in a magnetic field of 15,000 G will yield protons of about one million electron volts of kinetic energy."

No specific apparatus was mentioned in this brief paper. For the first tests, earlier that year, he had enlisted the help of a graduate student who had just completed his thesis, Nels Edlefsen, who put together a small model based on a vacuum chamber formed from a flattened glass flask silvered to form the electrodes, a laboratory magnet, and a low-power rf generator. Edlefsen observed currents on an electrode inserted at the outer edge of the chamber, but was

Fig. 3. The 4-inch cyclotron vacuum chamber, showing the single dee and electrostatic deflector.

unable to demonstrate resonance convincingly. With a nice combination of optimism and caution, their paper concluded: "Preliminary experiments indicate that there are probably no serious difficulties in the way of obtaining protons of high enough speeds to be useful for studies of atomic nuclei."

In the summer the project was taken over by a new graduate student, M. Stanley Livingston, whose initial experiments convinced him that an improved apparatus was necessary. "This thin flat glass chamber defied our technical skills," so instead he built a cylindrical brass chamber, using red sealing wax as a vacuum seal (Fig. 3). The rf electrode was a single hollow D-shaped half-pillbox facing a dummy dee in the form of a slotted bar; a 10 kW tube yielded voltages up to 2 kV on the dee at a tunable frequency in the MHz region. Electrons from a tungsten wire cathode near the center of the chamber were used to ionize hydrogen gas and provide H_2^+ ions for acceleration. A shielded Faraday cup behind slits was used to detect ions reaching the edge of the chamber.

The new apparatus quickly proved successful [2]. By November Livingston was able to observe sharp resonance peaks at the values of magnetic field predicted by Eq. (1.2), and with the expected linear dependence on rf frequency. At first the 4-inch-diameter, 5.2 kG magnet limited the ion energies to 13 keV, but early in 1931 they were able to borrow a 13 kG magnet and reach 80 keV. Moreover, Livingston, concerned that the grid wires might be intercepting the beam, had taken advantage of his

supervisor's absence for a few days to try the effect of removing them, and had found a more than tenfold increase in beam current to the 1 nA level. Quickly writing up his thesis, he achieved the distinction of being awarded his Ph.D. in less than a year from starting his research. Those were the days!

With these successes Lawrence was able to raise US$1200 from the university and the National Research Council to build a machine large and powerful enough to reach the 1 MeV goal thought necessary to induce nuclear reactions. A magnet with an 11-inch diameter pole was acquired and Livingston began constructing a scaled-up version of the 4-inch cyclotron, but now with a 20 kW oscillator capable of producing 4 kV across the dee gap (Fig. 4). Slits and an electrostatic deflector were installed for direct energy measurement and by January 1932 they were able to announce [23] production of 1.22 MeV protons — a world record. Livingston noted that his supervisor "danced with glee" as the signal appeared. Unhappily, though, no equipment was available to put the ions to use, and they were chagrined to discover in the Spring that Cockcroft and Walton had beaten them in the race for artificial disintegration, using a 600 kV cascade generator. Indeed, protons of a mere 125 keV had been found energetic enough to break up lithium nuclei. Berkeley quickly acquired detectors, and later that year they were able to confirm the Cambridge results and extend them to higher energies.

Lawrence moved quickly to increase the Berkeley lead. With US$12,000 obtained from the Research

Fig. 4. The 11-inch cyclotron installed in LeConte Hall, UCB.

Fig. 5. Livingston (*left*) and Lawrence with the 27-inch (later 37-inch) cyclotron.

Corporation and the Chemical Foundation, Livingston and he were able to convert a redundant 80-ton "Poulsen arc" magnet from the Federal Telegraph Company into a cyclotron with 27-inch diameter poles (Fig. 5). By March 1933 they were producing nanoamps of 5 MeV H_2^+ and 1.5 MeV He^+ ions, and a year later 5 MeV deuterons [24] with beam currents as large as $0.3\,\mu A$. (Proton energies were limited by the higher frequencies required of the oscillators.)

In 1936, Lawrence and Cooksey [25] made various improvements (including a vacuum chamber sealed by gaskets rather than wax) that enabled the deuteron energy to be raised to 6.3 MeV and the current to $20\,\mu A$. Finally, in 1937, the pole diameter was increased to create the "37-inch" cyclotron, which, within a year, had been tuned up to produce reliable $50\,\mu A$ beams of 8 MeV deuterons [26], and a year later $100\,\mu A$ beams at 8.5 MeV.

2.2. *Isochronism and focusing*

Whether the cyclotron concept could be developed into a practical tool clearly depended on what fraction of the initial beam survived hundreds of turns and accelerations to reach top energy. As Lawrence and Livingston pointed out in their first paper [2], there were two critical aspects to this — isochronism and focusing.

Regarding the first, if the beam was to remain in synchronism with the rf over several hundred cycles, "the magnetic field should be constant to

about 0.1% from the center outward." The magnet pole faces were therefore accurately machined, but it was nevertheless usually necessary to add some thin iron shims, especially near the outer edges of the poles, to compensate for the drop in field strength there caused by the field's leakage beyond the poles (Fig. 6). Much of the Berkeley group's success in getting their beams to full energy and achieving the high currents mentioned above was due to careful shimming of the magnet — one of the essential arts of "cyclotroneering."

In a perfectly isochronous cyclotron there is no phase wander, or any phase focusing, so that after n turns an ion arriving at midgap at rf phase φ will have kinetic energy

$$T_n = T_0 + 4nqV_0\cos\varphi, \qquad (2.1)$$

where $\pm V_0$ is the peak dee voltage. In principle, all phases $-\pi/2 < \varphi < \pi/2$ are accelerable, but larger values of $|\varphi|$ require larger n, leading to lower beam quality and greater losses. In practice, the phase spread in a beam $\Delta\varphi$ should be $\leq 40°$ to ensure good beam quality and turn separation, and low losses. The spread is defined in the central region, either naturally by obstacles or electric focusing effects (see below), or artificially by slits.

As to the second aspect, focusing, they pointed out that both the magnetic and electric fields provided natural focusing effects that help to prevent the beam's being spread out transversely over its long spiral path, thereby losing intensity. Axial deviations were of particular concern because of the smaller aperture available.

In the magnetic case, Fig. 6 shows how the bowing of the lines of force results in ions above or

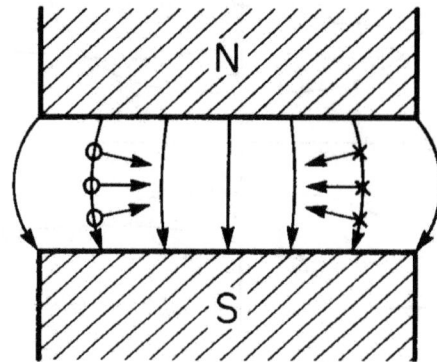

Fig. 6. Field pattern and axial focusing between parallel pole faces.

below the midplane experiencing a restoring force back toward that plane. In an axially symmetric field **B**, and using cylindrical polar coordinates (r, θ, z), the axial force component is given by

$$F_z = qvB_r = qv\frac{\partial B_r}{\partial z}z. \qquad (2.2)$$

But for stable betatron oscillations,

$$F_z = -m\omega^2\nu_z^2 z, \qquad (2.3)$$

where ν_z represents the number of oscillations per turn, i.e. the "betatron tune." Since div **B** $= 0$ in the absence of magnetic monopoles, and writing the logarithmic field gradient

$$k(r) \equiv \frac{r}{B_z}\frac{\partial B_z}{\partial r} \qquad (2.4)$$

(so that locally $B_z \sim r^k$), we find that

$$\nu_z = \sqrt{-k}, \qquad (2.5)$$

a result first obtained by Rose [27] and Wilson [28] in 1937. Similarly, for horizontal motion,

$$\nu_r = \sqrt{1+k}. \qquad (2.6)$$

The complementary dependence of axial and radial tunes on k is a direct consequence of **B**'s being divergenceless. For the field to be focusing in both planes, B_z must decrease with the radius, but only weakly:

$$-1 < k < 0. \qquad (2.7)$$

The electric field **E** at the dee gap also exerts focusing forces (Fig. 7), ions off the midplane being deflected toward it on their approach, but away as they leave. In general the two effects do not exactly

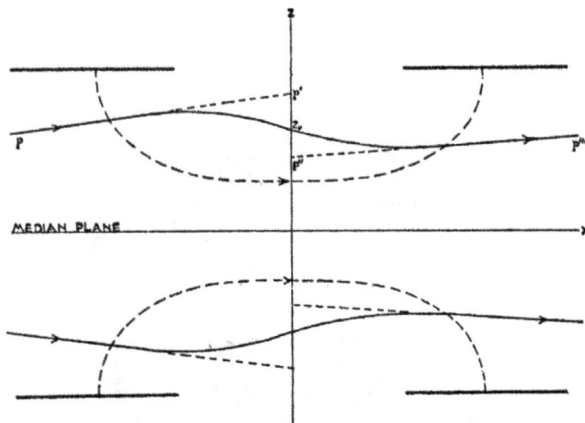

Fig. 7. Ion trajectories through the electric field at a dee gap. (Reprinted with permission from M. E. Rose, *Phys. Rev.* **53**, 392 (1938). ©1938 by the American Physical Society.)

balance: even if the field were static, acceleration would result in ions spending a shorter time leaving than entering, so that there would be an overall axial focusing effect. With an alternating field there is an additional and strongly phase-dependent effect caused by the field change during the ions' transit. This is focusing for ions that are late ($\varphi > 0$, when the field is falling) and defocusing for those that are early ($\varphi < 0$, when it is rising). Rose [27] showed that a dee gap whose geometry is independent of the radius acts as a convergent lens of focal length f_z, where

$$\frac{1}{f_z} = \frac{h}{R}\left[\frac{qV_0}{T}\sin\phi + b\left(\frac{qV_0}{T}\right)^{3/2}\cos^2\phi\right]; \qquad (2.8)$$

h is the ratio of the rf to the ion frequency, and b is a geometry-dependent constant of order unity. Repeated passage through these gap lenses provides additional axial focusing, $\Delta\nu_z^2 = R/\pi f_z$. In (2.8) the first (and stronger) term describes the transit-time effect, and the second the acceleration effect. Because of the (V_0/T) terms, electric focusing is of significance only in the central region, where T is low — but there it can be dominant in the absence of magnetic focusing, restricting the acceptance to positive phases. Rose's work was later extended by Cohen [29] and, for radially varying gaps, by Dutto and Craddock [30], who showed that

$$\Delta\nu_r^2 + \Delta\nu_z^2 = -\frac{h}{2\pi T}\frac{\partial(\Delta T)}{\partial\phi}, \qquad (2.9)$$

demonstrating the complementary relationship between transverse and longitudinal focusing (due to the electric field's being divergenceless if space charge can be neglected), a topic further developed by Gordon and Marti [31].

2.3. Discoveries and applications

With a source of proton, deuteron, and alpha beams of steadily increasing intensity, Lawrence's rapidly expanding group began exploring the new fields of nuclear reactions, induced radioactivity, and neutron production and experiment. By 1938 it could be said [26], "More new isotopes have been made artificially than there are stable ones in nature" — of which notable examples discovered at Berkeley were ^{14}C, ^{24}Na, ^{32}P, ^{59}Fe, and ^{131}I. Indeed, the first artificial element, technetium, was produced on the 37-inch cyclotron.

The value of radioisotopes as tracers was quickly realized, and they were soon put to use in studies of chemical reactions and biological processes, both in plants and in animals. Medical research was also undertaken, led by Lawrence's brother, John, who joined the group in 1935. Ingestion of ^{32}P proved beneficial for leukemia and polycythemia, and studies of the neutron irradiation of mice showed that neutrons caused significantly more damage to malignant cells than x-rays, for the same damage to healthy cells. This and some promising initial results on human cancer patients persuaded a local philanthropist to help fund a more powerful facility, the 200ton, 60-inch Crocker cyclotron and medical laboratory (Figs. 8 and 9). This was commissioned [32] in 1939, providing 100 μA beams of 16 MeV deuterons

Fig. 8. The 60-inch Crocker cyclotron with (*left to right*) Cooksey, Lawrence, Thornton, Backus, Salisbury and (*above*) Alvarez, McMillan.

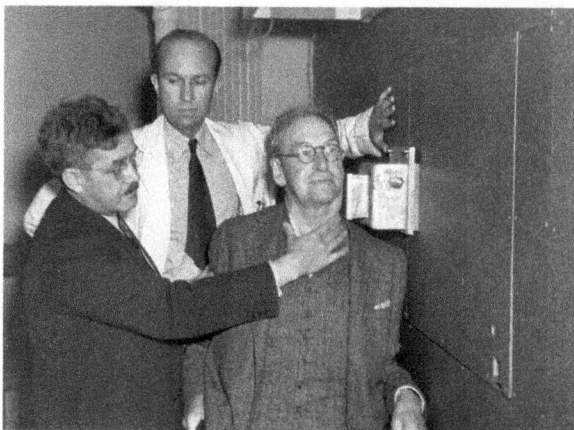

Fig. 9. Robert Stone and John Lawrence setting up a patient for neutron therapy at the 60-inch cyclotron.

and, later on, 200 μA at 20 MeV, and 40 MeV alpha beams.

The 60-inch was also available for nuclear physics and was quickly put to use to study the newly discovered fission process, and to test the viability of magnetic separation of ^{235}U from ^{238}U. It was also the tool for the discovery of new elements, no less than three being found in 1940: astatine, neptunium, and plutonium. From 1944 to 1955 it served as the source of four more transuranic elements (numbers 96, 97, 98, and 101).

2.4. *Beyond Berkeley*

The potential of cyclotrons for producing high-energy ion beams without the problems associated with high voltages was quickly recognized outside Berkeley — but so were the special skills apparently needed to commission the temperamental early cyclotrons and keep them running reliably. But as Lawrence's PhD students graduated and visitors to Berkeley returned to their home institutes, those skills gradually spread across the US and around the world.

The first missionary was Livingston, who in 1934 left for Cornell, where he built a 16-inch cyclotron providing 2 MeV protons (and later a 42-inch at MIT for 11.5 MeV deuterons). He was soon followed by others, and by 1940 there were no less than 24 cyclotrons at academic institutions across the US, the majority being based on the 37-inch, with poles ranging from that size to 42 inches. The first overseas cyclotron after Thibaud's was started in Leningrad in 1932, and by 1940 there were three in Japan, six in Europe and two in the USSR.

Several important developments arose at the US missionary outposts, and then spread rapidly around the diaspora via the Berkeley hub. One example was Livingston's use of a capillary ion source to increase the beam current — a technique he had adopted from the Carnegie Van de Graaff — in place of the simple filament used in the early Berkeley cyclotrons to ionize the residual gas in the vacuum chamber. Another was the replacement of inductive coupling of power to the dees, supported by unreliable insulators, by the use of quarter wave coaxial transmission lines that themselves provided the dee support. This scheme, pioneered at Illinois and Columbia, greatly improved reliability and also increased the dee voltages attainable.

More cyclotrons were built during and after World War II, including several with 60-inch poles, some produced by commercial firms. The most ambitious were the 86-inch at Oak Ridge [33] and the 225-cm (89-inch) at Stockholm [34], both of which delivered 22 MeV protons and 24 MeV deuterons. Oak Ridge also developed the production and acceleration of heavy ions such as C^{3+} and N^{4+}, a technique pioneered on the Berkeley 37-inch in 1940. Another innovation in some cyclotrons was a variable-frequency rf system, allowing the final beam energy to be varied; this required somewhat lower than usual magnetic fields, so as to avoid saturation effects that would have called for reshimming at each energy.

Many of these early machines were subsequently converted to synchrocyclotrons or isochronous cyclotrons, or used as bending magnets at higher-energy facilities. A few, like the 37-inch, have become science exhibits. Others were simply dismantled, but the fate of none is so poignant as that of the four Japanese cyclotrons that were summarily dumped in the waters of Tokyo and Osaka Bays at the end of the war [35].

2.5. *Limitations*

Careful readers will have noticed a conflict between the requirements for isochronism and axial magnetic focusing described in Subsec. 2.2 above, even for ions of constant mass. Isochronism requires the magnetic field to remain constant as an ion's energy and radius increase, while focusing calls for it to decrease. The relativistic increase of mass with energy, following $E = mc^2$, only makes matters worse, requiring the field to increase with the radius. In practice, a compromise was necessary. The magnetic field was shimmed to give the minimum gradient providing acceptable focusing, while the ion's phase was allowed to drift, first from near zero to near $-\pi/2$, and then back again as its angular frequency fell. How far this compromise could be pushed depended on how many turns the phase drift could accommodate. So the maximum energy achievable depended on the dee voltage available. Bethe and Rose's initial estimate [36] for the deuteron energy limit on the 37 inch was 8 MeV for 50 kV on the dees, with a $\sqrt{V}$ voltage dependence. The 25 MeV deuterons obtained at Oak Ridge and Stockholm were achieved with dee voltages over 200 kV.

Fig. 10.   The dee stem of a RIKEN cyclotron being committed to Tokyo Bay in 1945.

In the 1930s, though, as Livingston noted [9], "this theoretical threat to the future of the cyclotron was not taken seriously by experimentalists, who were convinced that technical improvements would raise the limits almost indefinitely." This was certainly Lawrence's attitude in 1939 as he planned to breach the 100 MeV barrier with his next monster cyclotron and create mesons artificially: "...there are no practical difficulties in the way of applying 1 or 2 million volts to the dees...." [37] The following year, as a fresh Nobel laureate, he was awarded an unprecedented US$1.4 million to build it, and construction began of a 4300-ton magnet with 184-inch diameter poles. The war (Fig. 10), however, interrupted its completion as a classical cyclotron. Instead the magnet served as a testbed for the calutron electromagnetic separators used to produce enriched ^{235}U for the Manhattan Project.

In fact, as was mentioned in the Introduction, more convenient ways were found of reaching higher energies than applying enormous voltages to the dees, and the first of these will be described in the next section.

3. Synchrocyclotrons

The most obvious way to avoid losing resonance between the oscillating electric field and the orbiting ion as the latter gains energy and mass is simply to reduce the radio frequency at the same rate as the orbital frequency:

$$\omega_{rf}(t) = \frac{qB}{m(t)}, \qquad (3.1)$$

where B has been assumed to be uniform. This is the basic principle of the synchrocyclotron [also known as the frequency-modulated (fm) cyclotron, or the phasotron in Russia].

But the condition (3.1) can only be satisfied for an ion at one particular "synchronous phase" ϕ_s, whose energy gain per gap is exactly matched to the drop in frequency:

$$2qV_0 \cos\phi_s = -\frac{\pi E}{\omega^2}\frac{d\omega}{dt}. \qquad (3.2)$$

The practicality of basing an accelerator on this principle depends on whether the motion of neighboring nonsynchronous particles is stable or unstable. This question of "phase stability" was examined at the end of World War II by Veksler [38] in the USSR and McMillan [39] at Berkeley, and is described more fully in an accompanying article by Courant [40]. It turns out that the phase motion is stable only if the synchronous phase is on the falling side of the rf voltage wave; if it is on the rising side, the motion is unstable. Moreover, the area of stable motion in longitudinal phase space (the "bucket") is finite, being largest for $\phi_s = \pi/2$ and falling to zero as $\phi_s \to 0$ at peak voltage. A compromise choice of $\phi_s \approx \pi/3$ provides both acceleration and acceptance.

With the 184-inch cyclotron still awaiting completion at the end of the war, phase stability must have been a welcome discovery at Berkeley. But it was at first thought that a moderately high injection energy would be required, so that it could not be readily applied to cyclotrons. Reg Richardson, however, from his calutron experience, was able to show that 1–5% of the ions emerging from an ion source

could be captured in the stable bucket. In 1946 he led a small *ad hoc* group in converting the 37-inch cyclotron to fm operation as the first synchrocyclotron [41]. This involved shimming the magnetic field down toward the outside (to simulate the anticipated 13% fall-off in deuteron orbital frequency in the 184-inch), and installing a frequency-modulated rf system. This not only provided the first demonstration of the phase-stability principle (Goward and Barnes' first operation of a synchrotron followed a few months later), but also confirmed the feasibility of operating the 184 inch as a synchrocyclotron. Furthermore, much lower dee voltages were needed than in classical cyclotrons, and the beam remained in resonance for many thousands of turns. Richardson records that when Lawrence learnt of their success, he "became very excited and rushed out to drive up the hill... [As he] passed a truck carrying one of the huge dee stems necessary for 1 MV... he told the driver to turn around and take the tank back to storage — or maybe the dump!" [42].

The rf system for the 184-inch was immediately redesigned, and by the end of 1946 it had been brought into operation [43] as a synchrocyclotron, delivering 190 MeV deuterons, 380 MeV alphas, and (later) 350 MeV protons — revolutionary steps in ion energy. (In 1957 the magnet was reconfigured, raising the field from 1.5 T to 2.3 T and the proton energy to 720 MeV.)

Berkeley's success, together with the ease of operation compared to classical cyclotrons, encouraged other laboratories to follow suit. Over the next

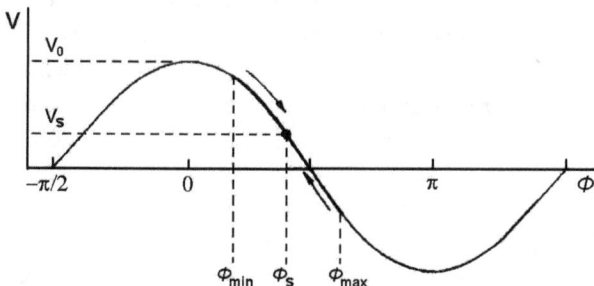

Fig. 11. The principle of phase stability in the synchrocyclotron.

Fig. 12. The 184-inch cyclotron with Lawrence (seated) and the UCRL staff.

Table 2.   Large synchrocyclotrons.

	Pole diameter (m)	Magnet wt. (t)	Proton energy (MeV)	Date first operated
UCRL, Berkeley	4.70	4300	350	1946
			740	1957
U. Rochester	3.30	1000	240	1948
Harvard U.	2.41	715	160	1949
AERE, Harwell	2.80	660	160	1949
Columbia U.*	4.32	2487	380/560*	1950
McGill U.	2.29	216	100	1950
U. Chicago	4.32	2200	450	1951
GWI, Uppsala*	2.30	650	187	1951
Carnegie I. T.	3.61	1500	450	1952
U. Liverpool	3.96	1640	400	1954
DLNP, Dubna*	6.00	7200	680	1954†
CERN, Geneva	5.00	2560	600	1958
NASA SREL	5.00	2765	590	1965
PNPI, Gatchina	6.85	7874	1000	1967†
IPN, Orsay	3.20	927	200	1977†

*Later modified with spiral sectors.
†Still in operation.

12 years, 14 large synchrocyclotrons, capable of delivering $\geq$ 100 MeV protons, were built around the world (see Table 2). The largest is that at Gatchina, near St. Petersburg, with a magnet weighing 7800 t (Fig. 13), which provides 1 μA beams of 1 GeV protons.

In adopting frequency modulation to reach higher energies, however, an important beam property has to be sacrificed, namely intensity. In a classical cyclotron, particles can be accepted for acceleration from the ion source on every rf cycle, so that the output is a cw stream of bunches at the radio frequency. The synchrocyclotron, by contrast, is a batch device, producing a pulsed beam.

Fig. 13.   The Gatchina 1-GeV synchrocyclotron.

A group of ions leave the source, and the frequency of the accelerating system is then steadily lowered to stay matched with the group while it is being accelerated. During this relatively long period, ions emerging from the source cannot be accelerated, since their orbital frequency is no longer in synchronism with the rf. Ions can be captured and accelerated again only after the first group has reached full energy and the radio frequency has returned to its initial value. As a result, the ion current delivered by a synchrocyclotron is typically $\leq$ 1% of that from a classical cyclotron under comparable conditions.

In practice, pulse repetition rates are in the range 60–2000 Hz. To modulate the rf at such rates, it is tuned with a large multibladed capacitor, usually rotating, though in some cases a vibrating reed is used. Over such long pulse periods the ions make around 10^5 revolutions, and so the dee voltages required are very modest, just a few kilovolts.

Synchrocyclotrons had their heyday in the postwar years. The lower-energy machines were mostly used to study nuclear reactions and scattering, while the higher-energy ones opened up the field of particle physics by elucidating the properties, decay modes and interactions of pions and muons. Synchrocyclotrons have also been used to produce protons for cancer therapy, a technique pioneered at Berkeley and Uppsala and pursued most intensively on the Harvard machine, where over 9000 patients were treated between 1961 and 2002.

Largely because of their low beam intensity, only three large synchrocyclotrons remain in use; for physics applications all but those at Gatchina and Dubna have been supplanted by their higher-intensity cousins, isochronous cyclotrons — and even that at Dubna has been modified with spiral-sector pole pieces to increase beam intensity.

4. Isochronous Cyclotrons

The idea of maintaining both isochronism and axial focusing to high energies by introducing an azimuthal variation in the magnetic field was proposed in 1938 by Llewellyn Thomas [44] (perhaps better known in physics for Thomas precession and the Thomas–Fermi statistical model of the atom). The suggestion has been extremely fruitful and has led to hundreds of isochronous cyclotrons being built in a broad spectrum of sizes and types for a wide variety of applications. [Note that these machines are also sometimes referred to as azimuthally-varying-field (AVF) or sector-focused cyclotrons.] Over the years, dramatic improvements in performance have been achieved through various technical innovations, including the use of:

- Spiral sectors;
- Powerful ion sources, mounted externally;
- Beam extraction by resonance or stripping;
- Accurate computation of fields and orbits;
- Superconducting magnets;
- Multistage designs using ring cyclotrons.

4.1. *Radial-sector cyclotrons*

Although Thomas's original proposal was to add a single harmonic component to the field, the mechanism is easier to understand in terms of a square-wave perturbation, such as might be produced by adding N sector-shaped pole pieces symmetrically around the magnet.

Figure 14 shows how the alternation of strong field B_h in the "hills" and weak field B_v in the "valleys" leads to a scalloped rather than circular orbit, which crosses the sector edges at a small "Thomas angle" κ, given, in this hard-edge approximation, by

$$\kappa = \frac{\pi}{N} \frac{(B_h - B_m)(B_m - B_v)}{B_m(B_h - B_v)}, \qquad (4.1)$$

where B_m is the mean field value along the orbit. As a result there is a positive edge-focusing effect when

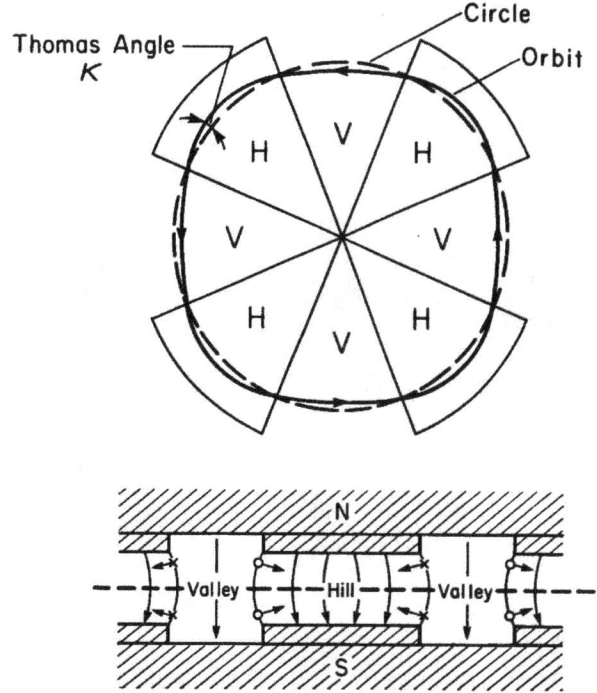

Fig. 14. A radial-sector cyclotron, showing how the Thomas contribution to axial focusing arises.

both entering and leaving a hill, each edge acting as a thin lens with a focal length given by

$$\frac{1}{f_z} = \frac{q}{mv}(B_h - B_v)\tan\kappa. \qquad (4.2)$$

The net result is a contribution to the axial focusing,

$$\Delta\nu_z^2 = \left(\frac{B_h}{B_m} - 1\right)\left(1 - \frac{B_v}{B_m}\right) = F^2, \qquad (4.3)$$

where the magnetic flutter F^2 is defined as the fractional mean square deviation of the field:

$$F^2 \equiv \left\langle \left(\frac{B(\theta)}{B_m} - 1\right)^2 \right\rangle. \qquad (4.5)$$

From Eq. (2.5) we see that the total axial tune will be given by

$$\nu_z^2 = -k + F^2. \qquad (4.6)$$

where we now understand

$$k \equiv \frac{r}{B_m}\frac{\partial B_m}{\partial r}. \qquad (4.7)$$

If the average field is increased in proportion to the ion's total energy $E = \gamma m_0 c^2$, so as to maintain isochronism,

$$B_m = \gamma B_c, \qquad (4.8)$$

then Eq. (2.4) yields

$$k = \beta^2 \gamma^2 \qquad (4.9)$$

and the expression for the total axial focusing becomes

$$\nu_z^2 = -\beta^2 \gamma^2 + F^2. \qquad (4.10)$$

Thus axial focusing can be maintained provided that the flutter exceeds $\beta^2 \gamma^2$. For the radial focusing, Eq. (2.6) gives

$$\nu_r = \gamma. \qquad (4.11)$$

More general treatments [45–49], covering periodic fields of any shape, show that Eqs. (4.10) and (4.11) give just the leading terms in a series, but for most cyclotron applications they are adequate approximations.

The designer must not only ensure that the focusing is real, but that the working point (ν_r, ν_z) does not cross any dangerous betatron resonances during acceleration. With sufficient energy gain per turn, most resonances can be crossed quickly and safely, but the half-integer intrinsic resonance $\nu_r = N/2$, where the phase advance per sector equals π, creates an uncrossable stopband. The radial tune's low-energy value, $\nu_r = 1$, then implies that

$$N \geq 3, \qquad (4.12)$$

as Thomas had pointed out in his original proposal.

Although his proposal was noted in many prewar cyclotron papers, no attempt was made to put it into practice at that time. This was probably mainly due to concern about the complexity of the scheme's magnet design, but may also have been influenced by its rather abstract mathematical formulation. It was not until 1950, when a need arose for a high flux of 300 MeV deuterons for neutron production, that a test was made at Berkeley at the instigation of Richardson, who argued that a Thomas cyclotron would be superior to the 18 m (*sic*) diameter Materials Testing Accelerator linac being built at Livermore. He then led the construction of two three-sector electron models (Fig. 15) and although, as in classical cyclotrons, the magnetic field required painstaking trimming (using 54 circular trim coils), they operated as predicted and electrons were successfully accelerated to $\beta \approx 0.5$, as required for the deuterons. Extraction efficiencies over 90% were shown to be possible. Less happily, because the work

Fig. 15. One of the magnet poles for the second electron model of an isochronous cyclotron, showing the harmonically contoured hills.

was classified, it had to wait five years for publication [50]. An electron model was also successfully tested at Oak Ridge in 1957.

The first isochronous proton cyclotron was completed by Heyn and Khoe [51] at Delft in 1958. It had three sectors, 86-cm-diameter poles and a top energy of 12.7 MeV. This was followed in the early 1960s by radial-sector machines at Birmingham (12 MeV d), Moscow (32 MeV d), Karlsruhe (48 MeV d), Orsay (9 MeV/u N^{5+}), Dubna (8 MeV/u Ne^{4+}), and Milan (45 MeV p). The reason for these relatively low energies (only the last had reached an energy beyond the capability of classical cyclotrons) is that it is hard to achieve a flutter F^2 much greater than 0.1 in a magnet with a single yoke. Equation (4.10) shows that in this case the axial focusing will vanish for $\gamma \geq 1.05$, i.e. for proton energies over 50 MeV. (Note, however, that this limit is currently being challenged by a group from CIAE, Beijing, which is building a 100 MeV H^- cyclotron with radial sectors [52].)

In spite of this limitation, radial sectors are convenient and sufficient if only low energies are required. Over the last 40 years, around 200 "compact" radial-sector cyclotrons in the 10–40 MeV range have been built by commercial firms for production of proton-rich isotopes (mainly for medical diagnostics and treatment) and for industrial applications (Fig. 16). Top-of-the-line instruments will deliver proton currents of 1 mA or more.

4.2. *Spiral-sector cyclotrons*

To obtain more axial focusing than radial sectors alone could provide, cyclotron designers turned to a

Fig. 16. The four radial sectors and two wedge-shaped rf cavities of a 13 MeV H$^-$ cyclotron used to produce isotopes for a PET scanner.

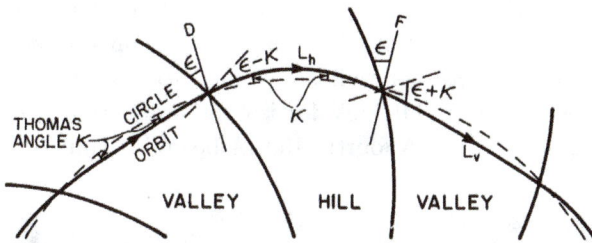

Fig. 17. The geometry of scalloped orbits in a spiral-sector cyclotron.

suggestion made by Donald Kerst [53, 54] for FFAG accelerators in 1955 — spiral sectors. If the sector edge crosses the radius vector at an angle ε (Fig. 17), the ion orbit will no longer cross both edges at the Thomas angle κ, but will cross one edge at $\kappa + \varepsilon$ and the other at $\kappa - \varepsilon$. The Thomas focusing will thus be augmented by a strong focusing effect: if $\varepsilon > \kappa$, one edge will act as an F lens, and the other as a (weaker)

D lens. The overall effect is to make the axial tune strongly dependent on the spiral angle:

$$\nu_z^2 \approx -\beta^2\gamma^2 + F^2(1 + 2\tan^2\varepsilon). \qquad (4.13)$$

For a 35° spiral the effect of the flutter is doubled, and for 45° trebled. The spiral has no effect on the radial tune, for which Eq. (4.11) remains a good approximation.

From the beginning, isochronous cyclotron designers took advantage of the additional focusing provided by spiral sectors. It was first employed at JINR, Dubna (13 MeV d) and Urbana (15 MeV p; a classical cyclotron conversion), both machines being commissioned in 1959. Looking to higher energies, all but one of the first ten isochronous cyclotrons to operate with $\gamma \geq 1.03$ (i.e. outside the classical range) employed spiral sectors (see Table 3).

Of these, Richardson's UCLA machine [55] is notable for its compact design and the adoption of Colorado's innovation [56] of accelerating H$^-$ ions rather than protons. This made it possible to extract protons (by stripping the H$^-$ ions in a thin foil) with around 99% efficiency, rather than the 80–90% characteristic of more conventional techniques. Moreover, the extracted beam energy could be varied by simply moving the foil to a different radius. (On the debit side, however, H$^-$ ions are rather fragile, with a binding energy of only 0.75 eV; so it is necessary to maintain a vacuum of 10^{-7} Torr or better; also, to prevent the ion from tearing itself apart, the magnetic field must not exceed a critical limit — though as the limit $\propto 1/\beta\gamma$, it is only of concern above 50 MeV.)

In cyclotrons not accelerating strippable ions, variable energy beams are obtained by altering the radio frequency. This is achieved by using either a movable shorting plane or movable panels within the resonant lines supporting the accelerating electrodes.

Table 3. The earliest isochronous cyclotrons operating beyond the classical range.

	Pole diameter (m)	Sectors	Maximum spiral	Energy (MeV)	Date first operated
UCLA	1.25	4	47°	50 H$^-$	1960
UCRL, Berkeley	2.24	3	56°	60 p	1961
U. Colorado	1.32	4	45°	30 H$^-$	1962
Oak Ridge N. L.	1.93	3	30°	75 p	1962
U. Michigan	2.11	3	43°	37 p	1963
U. Manitoba	1.17	4	48°	50 H$^-$	1964
U. Milan	1.66	3	0°	45 H$^-$	1965
Michigan S. U.	1.63	3	10°	56 p	1965
V. U. Amsterdam	1.40	3	37°	33 p	1965
AERE, Harwell	1.78	3	45°	53 p	1965

Fig. 18. Spiral pole pieces of the Oak Ridge Isochronous Cyclotron.

It is also possible to make large jumps in energy by changing to a different harmonic of the ion frequency.

In the late 1960s and the 1970s another 30 or so "compact" spiral-sector cyclotrons were built for nuclear research, 15 of them with top energies ≥ 50 MeV for protons. ("Compact" is used in the rather stretched sense of the magnet's being built as a single entity.) Three sectors were usually chosen for lower-energy machines and four sectors for the higher-energy ones. The largest spiral-sector cyclotrons are listed in Table 4. The most relativistic "compact" machine to date is the IBA 235 MeV proton therapy cyclotron [57].

Heavy ion cyclotrons are generally characterized by their K value, defined (nonrelativistically) by the maximum kinetic energy per nucleon achievable:

$$\frac{T_{\max}}{A} = K \left(\frac{Q}{A}\right)^2, \tag{4.14}$$

where Qe is the ionic charge and A the atomic number. Thus for a fully stripped ion $T_{\max}/A \approx$

$K/4$ (MeV/u). In terms of the cyclotron's maximum bending power $B\rho$,

$$K = \frac{(eB\rho)^2}{2m_u}, \tag{4.15}$$

where m_u is the atomic mass unit. (Note that, while such machines have enough bending power for protons of energy K, they may not have sufficient axial focusing.)

The physically largest cyclotron, that at TRI-UMF [58], is also of spiral-sector design, though certainly not compact. Its great size (Fig. 19) is a result of its being designed to accelerate H^- ions to 520 MeV ($\beta = 0.77$): to restrict the total electromagnetic stripping to 6% the maximum field is limited to 0.6 T. This low field requirement means that no iron is needed in the valleys, but they are not field-free as the whole magnet is excited by a single circular coil. Extraction by foil stripping provides three simultaneous beams: ≤ 150 μA for pion and muon production, ≤ 100 μA for radioactive ion production, and ≤ 50 μA at 70–110 MeV for isotope production and cancer therapy. A fourth, 100 μA beam is planned.

Fig. 19. A partially assembled magnet for the 520 MeV TRI-UMF cyclotron, showing the transition from radial to spiral sectors as the ion energy and orbit radius increase.

Table 4. The largest spiral-sector cyclotrons.

	Pole diameters (m)	Sectors	Maximum spiral	Proton/ion energy (MeV)	First beam
U. Maryland	2.67	4	52°	100/K180	1970
CGR-MeV 930	2.16	4	53°	95/K115	1972
RCNP, Osaka	2.30	3	52°	85/K140	1974
TRIUMF, Vancouver	17.17	6	70°	70–520/—	1974
INP, Kiev	2.40	3	45°	80/K140	1976
Dubna U400M	4.00	4	40°	—/K540	1991
IBA/SHI C235	2.24	4	60°	232/—	1998
PNPI, Gatchina	2.05	4	60°	45–80/—	

The most relativistic spiral-sector cyclotron ever to operate was the Oak Ridge Analogue II [59], an electron model for the proposed Mc^2 Cyclotron meson factory, in which protons were to be accelerated to 810 MeV in an eight-sector magnet with 12-m-diameter poles. In the model the electrons reached the corresponding energy of 465 keV ($\beta = 0.86$), and were extracted with 80% efficiency with the help of the $\nu_r = 2$ resonance.

4.3. *Separated-sector ring cyclotrons*

In a separated-sector cyclotron the magnet consists of hill sectors only, with no iron in the valleys, and with a separate exciting coil wound around each sector magnet. Such an arrangement is not compatible with low-energy injection, so a preaccelerator (sometimes a smaller cyclotron) is used, making the initial orbit radius quite large, and leading to the term "ring cyclotron." Separating the sectors offers two advantages:

(1) Separation of functions — the rf accelerating cavities, diagnostics, and injection and extraction equipment can be mounted in the virtually field-free space between the sectors, allowing much greater freedom in their design;

(2) The pole gap can be smaller, reducing the magnet power requirement and increasing the flutter. The magnetic flutter is no longer restricted to ≈ 0.1. If the hills occupy a fraction H of the orbit, in the hard-edge approximation the flutter

$$F^2 = \frac{1}{H} - 1, \qquad (4.16)$$

allowing radial sectors to provide adequate axial focusing up to $\gamma = 1/\sqrt{H}$ (e.g. $\gamma = 1.41$ for $H = 0.5$), at the expense of a radius proportional to $\sqrt{H}$.

The idea was first introduced by Willax [60] in his design for the 590 MeV SIN (now PSI) meson factory, an eight-sector proton cyclotron with spiral angle $\leq 35°$ and flutter ≈ 1.4, fed by a 72 MeV cyclotron. Four symmetrically placed rf cavities (Fig. 20) provide an energy gain of 2.9 MeV per turn. With the help of the $\nu_r = 3/2$ resonance an extraordinary extraction efficiency of 99.97% is achieved, enabling a 2 mA proton beam to be delivered to a spallation neutron source.

This machine holds the record for proton energy and current in an isochronous cyclotron, and for beam power (1.2 MW) in any type of proton accelerator.

Fig. 20. The 590 MeV separated-sector ring cyclotron at PSI, showing the eight spiral-sector magnets and four rf cavities.

Around a dozen more separated-sector cyclotrons have been built (see Table 5), mostly for heavy ions, as the emphasis in nuclear research moved in that direction in the 1970s and '80s. For these slower particles, and for lower-energy protons, radial sectors were able to provide sufficient focusing, and so only one (at Osaka, designed for 400 MeV protons) required spiral sectors.

The latest and most powerful of these heavy-ion machines is the K2600 SRC [61] at RIKEN, Tokyo (Fig. 21). This has six sectors and is the first separated-sector cyclotron to have magnets excited by superconducting coils. It is only slightly smaller than the TRIUMF cyclotron and has more steel (8300 t) than the Gatchina synchrocyclotron (7800 t) — partly because of the magnetic shielding required. It was commissioned in 2006 with a 345 MeV/u Al^{6+} beam, and has subsequently accelerated and delivered U^{86+} beams of the same energy.

A number of designs have been published for ring cyclotrons delivering very-high-intensity beams at GeV energies. Joho's 16-sector ASTOR proposal [62] would have boosted the 2 mA beam from the PSI cyclotron to 2 GeV. An interesting feature was the option to run it in a pulsed mode at 1500 Hz for neutrino experiments, stacking the beam by a "phase expansion" effect. A more recent PSI proposal [63] is for a 10 MW separated-sector cyclotron (1 GeV, 10 mA) for an accelerator-driven subcritical reactor (ADSR). It would be a scaled-up version of the 590 MeV cyclotron, but with 12 sectors and 8 rf cavities (instead of 8 and 4). Others have made similar proposals [64].

Table 5.   Large separated-sector ring cyclotrons.

	Pole diameter (m)	Magnet wt. (t)	Sectors	Spiral	Proton/ion energy (MeV)	First beam
PSI, Villigen	9.30	1990	8	35°	590	1974
Indiana UCF	6.92	2000	4	—	200/K210	1975
HMI, Berlin	3.80	360	4	—	72/K130	1977
ISN, Grenoble	4.50	400	4	—	K160	1981
GANIL, Caen	6.90	1700	4	—	K380	1982
NAC, Stellenb.	9.09	1400	4	—	220/K220	1985
RIKEN RRC	7.9	2100	4	—	210/K540	1986
IMP, Lanzhou	7.17	2000	4	—	K450	1988
RCNP, Osaka	8.90	2200	6	30°	400/K400	1991
GANIL CIME	3.50	550	4	—	K265	1998
RIKEN fRC	7.4	1320	4	—	K570	2006
RIKEN IRC	9.3	2720	4	—	K980	2006
RIKEN SRC	12.4	8300	6	—	K2600	2006

Fig. 21.   The SRC superconducting ring cyclotron at RIKEN.

The CANUCK proposal [65] at TRIUMF called for 3 GeV and 15 GeV cyclotrons using separate spiral-sector magnets excited by superconducting coils to accelerate a $100\,\mu$A proton beam for a kaon factory. The tolerances required to cross the many integer and half-integer radial resonances were determined [66], and it was shown that the final integer resonance could be used to develop a large-enough coherent amplitude to allow an extraction septum to be inserted. The major challenge would be to ensure rigidity of the superconducting coils around the spiral-sector magnets.

Rees has recently proposed isochronous versions of his "pumplet" FFAG lattices to accelerate muons from 3.2 to 8 GeV and from 8 to 20 GeV. These will be discussed below, in Subsec. 5.7.

4.4. *Superconducting compact cyclotrons*

Superconducting coils were first used for cyclotrons in compact spiral-sector machines, where they provided a powerful extension of capabilities. The superconducting coils allow the magnetic field strength to be increased to around 5 T, or three times higher than was previously typical of cyclotrons. This results in the linear dimensions of the cyclotron being reduced by about the same factor, areas by a factor of 9, and so on, compared to a room-temperature cyclotron of the same energy. The result is a large reduction in the cost of many cyclotron components — and in the power bill.

The pioneers responsible for this development were Bigham *et al.* [67] at AECL, Chalk River, and Blosser *et al.* [68] at MSU, East Lansing. The main coil is typically housed in an annular cryostat (Fig. 22), while the conventional room-temperature components, including pole tips, rf accelerating system, vacuum system, and ion source, are inserted in the warm bore of the cryostat from top and bottom. The largest superconducting compact cyclotron, at NSCL, MSU, is only 2.9 m in overall diameter, but can accelerate heavy ions to energies as high as 200 MeV/u.

The various superconducting compact cyclotrons are listed in Table 6. All but the two low-energy machines use spiral sectors. Besides those intended for nuclear research, several have been designed specifically for cancer therapy. The first was a 50 MeV MSU-designed deuteron cyclotron for neutron therapy, so small (1.65 m in overall diameter)

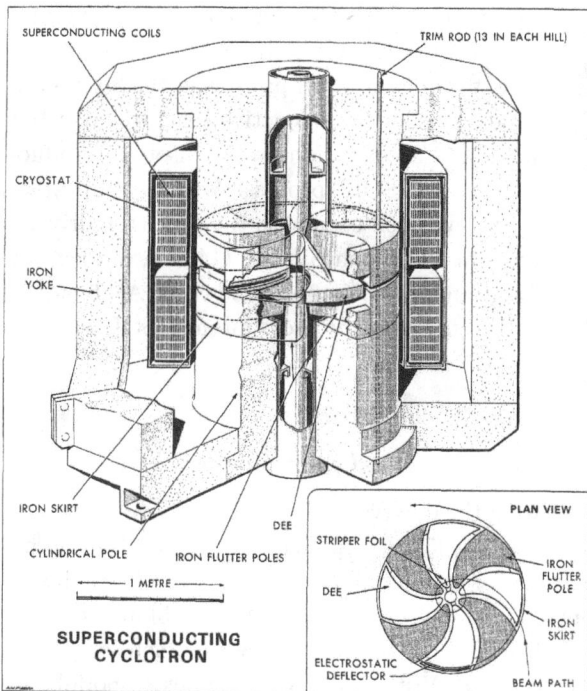

Fig. 22. The first superconducting cyclotron design — the K520 at AECL Chalk River (©Atomic Energy of Canada Limited, 1996).

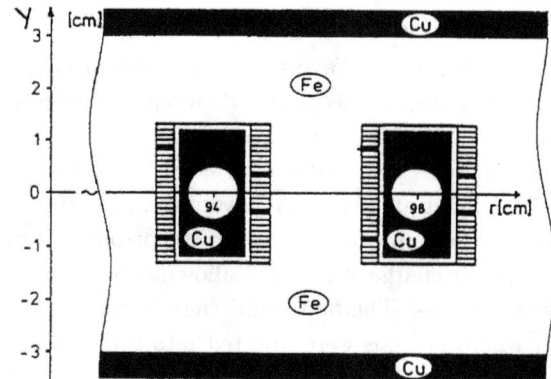

Fig. 23. A vertical cross-section through one of the TRITRON sector magnets, showing two adjacent beam channels and their superconducting coils and copper shielding.

that it is mounted on a gantry and rotated around the patient [69]. More recently, two ACCEL 250 MeV superconducting cyclotrons have been brought into operation [70] for proton therapy, and two separate 300 MeV designs are being developed.

4.5. *Separated-orbit cyclotrons*

The separated-orbit cyclotron (SOC), in which the bending and focusing fields of the sector magnets are specially tailored for each orbit, was proposed by Russell [71] in the early 1960s for the acceleration of high-current (milliampere) proton beams to GeV energies. Its complexity and cost, using normal magnets, deterred potential builders. However, a prototype superconducting SOC, the K85 TRITRON, was built and demonstrated in Munich [72]. This was the ultimate in superconducting cyclotron design, since the six rf accelerating cavities as well as the magnet were superconducting. The entire cyclotron was enclosed in a 3.5-m diameter vacuum chamber, with the vacuum being maintained by cryogenic pumping. The magnets were very modest in size, consisting of 12 sectors, each 6 cm high, about 90 cm in radial extent, and containing 20 2 cm × 2 cm channels containing the coils, copper shielding, and a 1-cm-diameter beam aperture (Fig. 23). A 40 MeV S^{14+} ion beam from an MP tandem was accelerated through six turns to 72 MeV.

Table 6. Superconductiong compact cyclotrons.

	Pole diameter (m)	Magnet wt. (t)	Sectors	Proton/ion energy (MeV)	First beam
NSCL MSU	1.42	90	3	K520	1982
AECL, Chalk R.	1.39	170	4	K520	1985
NSCL MSU	2.20	265	3	K1200	1988
Harper Hosp.	0.64	22	3	50 (d)	1988
Texas A&M U.	1.42	90	3	K520	1988
Oxford Instr.	0.50	1.5	3	12 (H$^-$)	1990
LNS, Catania	1.80	176	3	K800	1994
KVI, Groningen	1.88	320	3	200/K600	1994
ACCEL (PSI)	1.69	90	4	250/—	2006
Kolkata	1.42	90	3	K520	(2008)
IBA/JINR	3.74	660	4	260/K1600	
LNS, Catania	≈ 2.7	350	4	260/K1200	

4.6. *Multistage cyclotron systems*

Several laboratories have multistage cyclotron systems. These became necessary when separated-sector cyclotrons were introduced in an effort to reach higher energies, especially with heavy ions. In this case they offer an additional advantage, in that electrons can be stripped off between cyclotrons to produce higher-charge states and allow acceleration to higher energies. The first multistage system was at PSI, where protons were injected into the 590 MeV ring cyclotron from a 72 MeV injector cyclotron. The latest example is the K2600 SRC at RIKEN (Fig. 24) — fed by a chain of no less than three lower-energy ring cyclotrons from either a linac or a compact spiral-sector cyclotron! Other cyclotron laboratories also employ linacs as injectors, or in some cases electrostatic accelerators, again using electron stripping to increase the charge state.

Cyclotrons have also been used to inject ions into synchrotron storage rings, which can further accelerate, store, and cool the ions for precision experiments. Examples are the 500 MeV cooler ring at IUCF, the 1300 MeV CELSIUS ring in Uppsala, the 220 MeV/u TARN-II ring in Tokyo, the 2.9 GeV COSY ring in Jülich, and the K4400 CSRm and K3000 CSRe rings at HIRFL, Lanzhou.

An important development in nuclear science research is the acceleration of radioactive ion beams to enable the interactions of unstable isotopes to be studied. Cyclotrons are actively used in this area in several ways. One is in the production of radioactive species by bombardment of a target with a cyclotron beam. If the beam consists of heavy ions, a thin target is used, allowing high-energy radioactive ions to emerge for capture and transport directly to the experimental apparatus. This "nuclear fragmentation" process is used for the cyclotrons at GANIL;

KVI, Groningen; FLNR, Dubna; LNS, Catania; MSU; RCNP, Osaka; and RIKEN.

If the beam consists of protons, a thick target is used, and low-energy radioactive ions are collected for subsequent acceleration — the "isotope production on line" or "ISOL" process. A variety of postaccelerators have been used with the cyclotron drivers: cyclotrons at UCL, Louvain and GANIL, tandems at Oak Ridge and LNS, Catania, and the ISAC linacs at TRIUMF (currently the most intense ISOL source).

5. FFAG Accelerators

5.1. *Invention and principles*

Following the discovery of alternating-gradient (AG) focusing in 1952, its application to fixed-field accelerators was proposed independently by Haworth and Snyder [9, 73] and by Symon [45, 74, 75] in the USA, by Ohkawa [76] in Japan, and by Kolomensky [77, 78] in the USSR. With fixed magnetic fields, modulated radio frequency, and pulsed beams, FFAGs operate in the same mode as synchrocyclotrons, and like them allow higher pulse rates and larger acceptances (and therefore higher intensities) than synchrotrons. What was new was to break the magnet into sectors with strong radial field gradients, thus providing edge and strong focusing, much as in sector-focused cyclotrons, but free of the requirement for isochronism. (The original suggestions were for radial sectors, but as mentioned in the section on isochronous cyclotrons above, Kerst [53, 54] then proposed spiral sectors as a more compact alternative.) A further distinction from synchrocyclotrons was to remove the central region, creating a ring- rather than disk-shaped machine (Fig. 25). This requires injection of a preaccelerated beam, but drastically reduces the width and cost of the vacuum chamber and magnets. These can be further restricted by choosing a design with high-momentum compaction, creating an orbit with a very tight spiral. These features made it possible to design machines with high pulse repetition rates for much higher energies than was practical for synchrocyclotrons.

The advantage of a fixed-field machine is that it separates the guide field from the acceleration process. This allows a great variety of acceleration schemes and simplifies accelerator experiments. High beam currents can be achieved in FFAG accelerators, and very large circulating beams can be stacked,

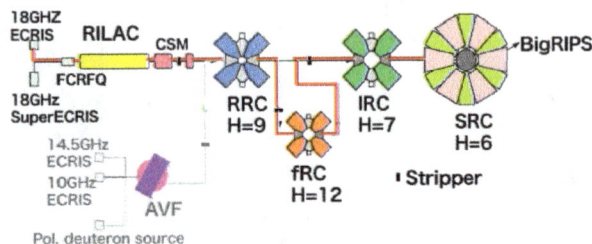

Fig. 24.   The accelerator chain at RIKEN RIBF, as used to produce 345 MeV/u ions over the mass range $A \approx 50$–92.

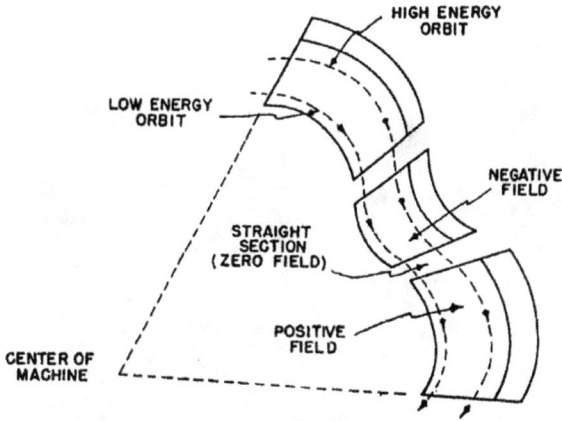

Fig. 25. Radial-sector FFAG magnets and orbits.

with currents high enough to make colliding beams practical.

The most intensive studies were carried out by Symon, Kerst, Ohkawa, and others at MURA (the Midwestern Universities Research Association) in Wisconsin, who gave FFAG its name. Their work culminated in the construction and successful testing of electron models of radial- and spiral-sector designs, as described in detail below. Electron models were also built in the USSR [79], while at CERN studies proceeded as far as testing model magnets [80].

5.1.1. *Scaling*

Resonance crossing was a major concern in the early days of AG focusing, because of the low energy gain/turn envisaged. The *scaling* principle was therefore adopted, whereby the field pattern is kept the same at all radii, making the orbit shape, optics, and tunes independent of energy. To first order the tunes are given by the same equations as those for imperfectly isochronous cyclotrons [cf. Eqs. (2.6), (4.11), and (4.13)]:

$$\nu_r^2 \approx 1 + k, \tag{5.1}$$

$$\nu_z^2 \approx -k + F^2(1 + 2\tan^2\varepsilon). \tag{5.2}$$

Clearly, achieving *constant* ν_r requires that

$$k = \text{constant}, \tag{5.3}$$

implying magnetic field and momentum profiles of the forms

$$B_m = B_0 \left(\frac{r}{r_0}\right)^k, \tag{5.4}$$

$$p = p_0 \left(\frac{r}{r_0}\right)^{k+1}. \tag{5.5}$$

Constant k also means that to achieve *constant* ν_z we must keep

$$F^2(1 + 2\tan^2\varepsilon) = \text{constant}. \tag{5.6}$$

This quantity must also be given a high value, since usually $k \gg 0$ in order to minimize the radial aperture. MURA's recipe was to keep the flutter

$$F^2(r) = \text{constant}, \tag{5.7}$$

by using a constant field profile $B(\theta)/B_m$ and:

- for spiral sectors: choosing *constant* ε, so the sector axis is a logarithmic spiral,

$$R = R_0 e^{\theta\cot\varepsilon}; \tag{5.8}$$

- for radial sectors: boosting the flutter F^2 by alternating positive-bending magnets [usually with positive k and so radially focusing (F)] with shorter reverse-bending defocusing (D) magnets, initially with

$$B_D = -B_F. \tag{5.9}$$

Of course, reverse fields raise the average radius: its ratio to the local radius of curvature, the "circumference factor," is ≥ 4.45 in the absence of straights [52], but smaller in their presence. The radial-sector design is shown schematically in Fig. 25.

5.1.2. *Recent developments*

Progressing from electron models to full-scale FFAGs took over 30 years. MURA's proposals for proton FFAGs were not funded in the 1960s, nor were those put forward by the Argonne [81] and Jülich [82] laboratories for 1.5 GeV machines as spallation neutron sources in the 1980s. While the major challenge was to obtain funding, there was also a technical challenge that had been absent for the models, where the electrons could be injected at quite low energies while already moving near the speed of light. Protons, by contrast, would have to be speeded up in the FFAG itself, requiring the frequency of the rf accelerating field to be varied over a wide range.

These challenges were finally overcome by Mori and his group at KEK in 2000, when they commissioned the 1 MeV PoP (proof of principle) proton FFAG [83]. Mori's group has gone on to build three more radial-sector and one spiral-sector machine, ranging in energy from 2.5 to 150 MeV. Their work stimulated renewed interest in FFAGs for applications requiring large acceptances and very high repetition rates (> 50 Hz).

Indeed, it was pointed out by Mills [84] and Johnstone [85] in 1997 that FFAG designs need not be restricted to the "scaling" approach if resonance crossing is not an issue. Dropping this restriction has revealed a range of interesting *non-scaling* design possibilities, and a 10–20 MeV electron model, EMMA, is now being built at Daresbury to confirm the feasibility of this approach.

Altogether, about a dozen scaling and a dozen non-scaling designs are currently under study for the acceleration of protons, heavy ions, electrons, and muons, with top energies ranging from 1 MeV to 20 GeV, and diameters from 5 cm to 400 m. This diversity simply reflects the variety of their applications, which range from treating cancer to irradiating materials, driving subcritical reactors, boosting high-energy proton intensity, and producing neutrinos.

We now describe the various FFAG designs in more detail, beginning with the early work at MURA.

5.2. MURA contributions[a]

MURA was incorporated in 1954 with 15 universities as members. Its purpose was to promote a large accelerator in the Midwest. In 1956 the MURA working group was located in Madison, Wisconsin, the chosen site for a MURA accelerator. During the next 13 years some 74 MURA employees, graduate students, and staff from MURA universities participated in the working group, making many important contributions to accelerator science, although MURA never built a high-energy accelerator. An extensive informal account of the work at MURA is given by Cole [14].

5.2.1. *The radial-sector model*

The radial-sector FFAG model accelerator [86] shown in Fig. 26 was built by the MURA group. It began operation in 1956. The injection energy at the inner orbit was 20 keV; the energy at the outer

[a]Portions reprinted with permission from K. R. Symon, *Proc. PAC' 03* (IEEE Press, 2003), pp. 452–456, ©2003 IEEE.

Fig. 26.　The radial-sector model.

orbit radius of 54 cm was 400 keV. There are eight sectors, each consisting of a large and a small magnet. The magnetic field increases with radius as r^k, with $k = 3.36$. The field in the smaller magnet is reversed, providing AG. This of course makes the orbit circumference about five times larger than for a uniform magnetic field.

The betatron core visible in the figure provided a very easy way to accelerate electrons. In a typical experiment, a beam could be injected and betatron accelerated to an intermediate energy. The experiment, such as an rf acceleration process, could then be carried out, and the result observed by betatron accelerating the resulting beam onto a detector.

5.2.2. *The spiral-sector model*

Figure 27 shows a spiral-sector model [87–89] which began operation in 1957. Each sector has just one magnet whose edges spiral out in the radius. The magnets are separated by field-free straight sections which also spiral with the radius.

5.2.3. *Nonlinear orbits*

In most accelerators, magnetic fields are made to vary as linearly as possible, so that nonlinear effects are small perturbations. In scaling FFAGs, nonlinear effects are inherent, particularly because of the increase of the magnetic field with the radius. Nonlinear effects determine the maximum allowed oscillation amplitudes.

Analysis of the equations of motion leads to the prediction of resonances where the betatron

Fig. 27. The spiral-sector model.

amplitudes can grow to large values if the tunes ν_x, ν_z satisfy the resonance equation:

$$n_x\nu_x \pm n_z\nu_z = m, \qquad (5.10)$$

where n_x and n_z are zero or positive integers and m is any integer. A resonance of order n (defined by $n \equiv n_x + n_z$) is driven by terms of order $n - 1$ in the equations of motion. Higher-order resonances appear only when appropriate nonlinear terms are included. Second-order terms in the equations of motion can drive third-order resonances even at infinitesimal amplitudes of oscillation. This is surprising, since we ordinarily think that for small-enough amplitudes we get a sufficiently good approximation if we include only linear terms in the equations of motion. Resonances of order higher than 4 appear in the solutions only at finite amplitudes. Difference resonances do not cause amplitude growth, but they result in coupling between the x and z motions.

Experiments were done on both models to check theoretical predictions regarding orbit stability as a function of betatron oscillation frequencies. Figure 28 is a contour plot [86] showing beam intensity in the radial-sector model as a function of the tune ν_x plotted horizontally and the tune ν_z plotted vertically. Theoretically predicted linear and nonlinear resonances lie along the straight lines shown. One can see the wide stopband along the linear resonances $\nu_x = 3$ and $\nu_z = 2$, as well as reductions in intensity along other linear and nonlinear resonances (though, as

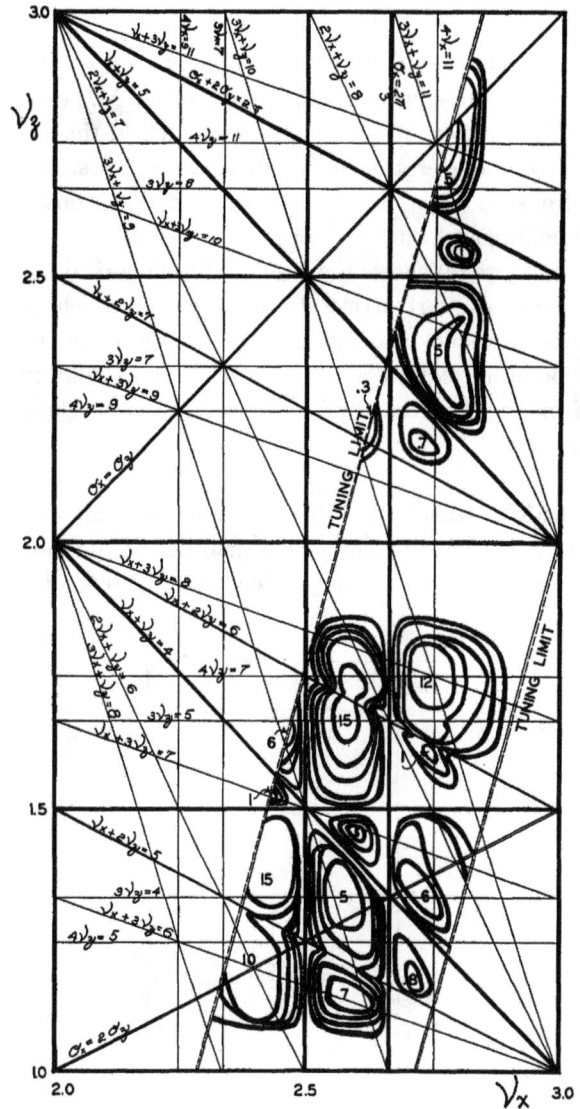

Fig. 28. A resonance survey, for the radial-sector model.

expected, not along the difference resonances). Similar measurements [88, 89] made with the spiral sector model also confirm the predictions of orbit theory.

Numerical calculations of FFAG orbits often showed apparently random behavior which was called "stochastic." Such behavior would now be called "chaotic." At first the MURA group were not sure whether these effects were real or artifacts of the numerical calculation. They devised exactly canonical numerical algorithms to eliminate the possibility of nonphysical features of the algorithm. They also made extensive checks to guard against round-off errors. They thus convinced themselves that these stochastic effects are real.

5.2.4. *RF acceleration*

Fixed-field accelerators allow a great variety of rf acceleration schemes. K. R. Symon and A. M. Sessler wrote a paper [90] in which they analyzed the process of rf acceleration in fixed-field accelerators. They were able to write the acceleration equations in Hamiltonian form.

Figure 29 shows the results of a numerical simulation of an rf acceleration process in which the radio frequency and voltage are fixed. Once per turn a point is plotted at the particle energy and the rf phase when the particle arrives at the accelerating gap.

There is a fixed point at phase π, energy 500 MeV, where the radio frequency is nine times the revolution frequency, and another at 814 MeV, where the radio frequency is ten times the revolution frequency. Both points are surrounded by trapping regions where the points lie on closed curves surrounding the fixed points. If we were to change the radio frequency slowly, the trapped phase points would be carried up or down in energy. This suggested to Symon using a high harmonic number so that there are a number of trapping regions between the injection and the output energy. By modulating the frequency, these regions could be moved upward past the injector so as to carry injected particles to the output energy. He called this scheme a "bucket lift" in analogy with the devices used by farmers to load hay or grain into their barns. The trapping regions were then called 'buckets," a name which is still in use, although no bucket lift accelerator was ever constructed.

Because of Liouville's theorem, the phase points in any rf acceleration process move like an incompressible (two-dimensional) fluid. This makes the name "bucket" even more appropriate. An interesting consequence is that if the buckets are moved upward, the surrounding untrapped phase space must on average move downward. This is called "phase displacement."

Among the topics studied theoretically and experimentally by the MURA group were acceleration of buckets, phase displacement, capture of a beam in an expanding bucket, beam stacking, and acceleration across the transition energy.

With high rf voltages, stochastic phenomena near the boundaries of a bucket were observed, as shown in Fig. 30. On the hypothesis that stochastic phenomena occur when bucket boundaries overlap, a case was run with two nearby rf frequencies with voltages such that the predicted buckets would overlap. The results in Fig. 31 show totally chaotic orbits. The solid curves are the predicted bucket boundaries.

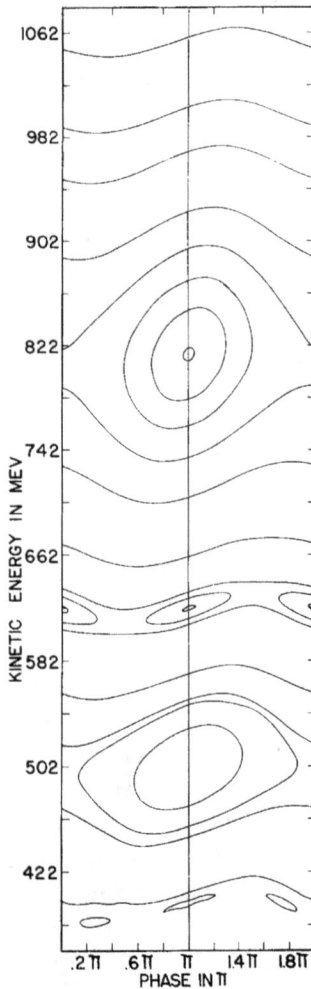

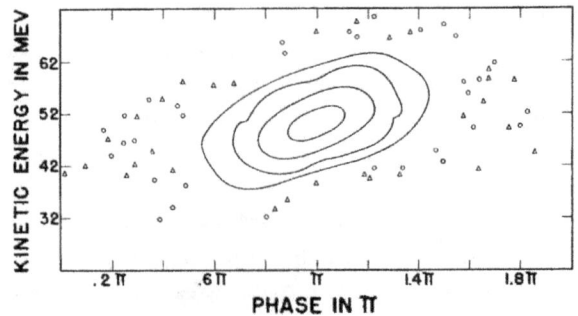

Fig. 29. Numerical simulation of rf acceleration.

Fig. 30. Stochastic behavior near the bucket boundary.

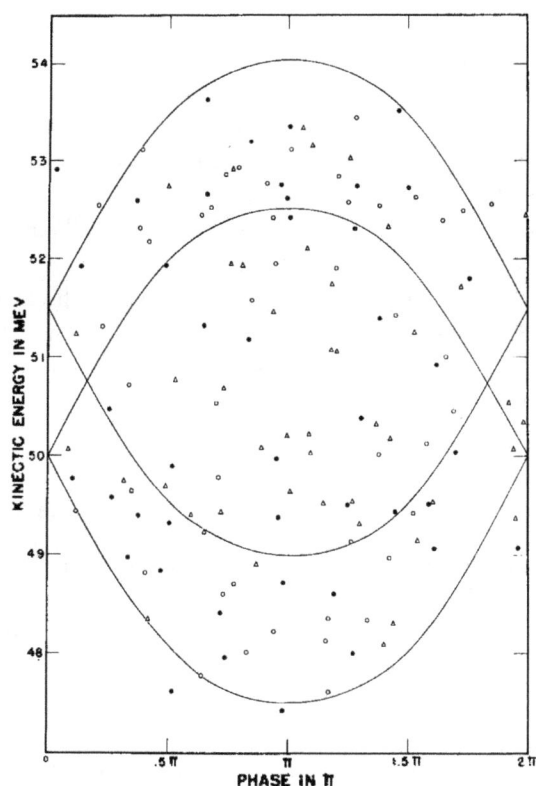

Fig. 31. Scattered orbits in two overlapping buckets.

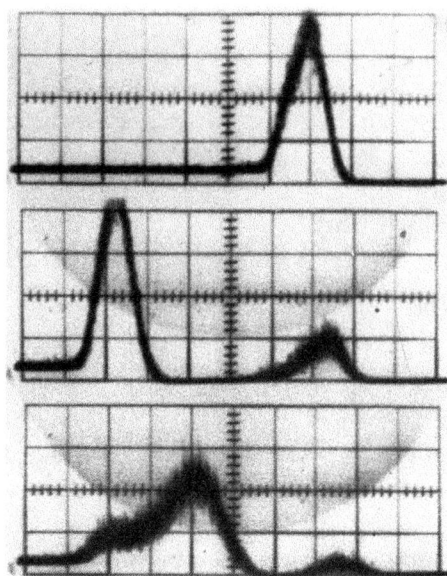

Fig. 32. A beam stacking experiment.

5.2.5. *Beam stacking*

One possible procedure in an FFAG machine is to accelerate buckets of particles successively to an intermediate energy, a process called "beam stacking." Since the space charge limit is higher at higher energies, it is possible to build up a large circulating beam at an intermediate energy. The stacked beam could then be accelerated to a high energy, thus achieving a large accelerated beam. Alternatively, it turns out that Liouville's theorem allows stacked beams of sufficient intensity to make it practical to consider colliding equal and opposite beams, as will be discussed below.

Beam stacking experiments [91] were carried out on the FFAG models, using the method described above (see Subsec. 5.2.1). Figure 32 shows the results for the radial-sector model. These are oscillographs of beam intensity versus time of flight of the accelerated particles (so that energy increases toward the left). The first trace shows an injected beam at an initial energy. The beam is captured in a bucket and accelerated up to a higher energy. The result is shown in the second trace, where we also see a little untrapped

beam remaining at the initial energy. In contrast to actual beam stacking, in this experiment successive beams were not injected, but successive rf cycles were carried out with empty buckets. The result after four cycles is shown in the third trace. We see that the result of four cycles is to accelerate most of the remaining beam and to displace the first beam down in energy. From these results, we would be able to predict what would be the result if we carried out a real stacking with full buckets in each rf cycle.

5.2.6. *Colliding beams*

The center-of-mass energy for a collision of a particle of relativistic energy E with a stationary particle is proportional to $E^{1/2}$. This suggests the energy advantage in letting two equal-energy particles collide, in which case the center-of-mass energy is the sum of the energies of the colliding particles. For example, two 15 GeV protons colliding head-on produce a center-of-mass energy of 30 GeV. A single proton hitting a stationary proton would need to have an energy of 450 GeV to produce the same center-of-mass energy.

Unfortunately the cross sections are such that the event rate for accelerator beams achievable before 1950 would have been impractically low. It was D. W. Kerst who observed that with the intensity achievable with stacked beams, colliding beam experiments become practical [92].

Keith Symon remembers being invited to give a colloquium on this subject at the University of Illinois. When he mentioned colliding beams, the audience burst out laughing. He was somewhat taken aback, until he learned later that the week before professors Kerst and Kruger had used peashooters to shoot at each other from opposite sides of the stage.

5.2.7. *The 50 MeV model*

A 50 MeV electron model [93] was constructed, and it first operated in 1961. It was a radial-sector machine (see Fig. 33) which had two identical magnets in each sector, with oppositely directed magnetic fields. Ohkawa [94, 95] and Kolomensky [96] first pointed out that particles in such a machine can circulate in either direction, and that the orbits are closed because they are at larger radii in the positive magnets. This configuration would allow colliding beams in a single machine. However, the ratio of the circumference to that for a uniform field is about 8.

The machine was successfully operated in two-way mode. However, most of the experiments were performed in one-way mode, with one of each pair of magnets excited to a higher field than the other. In case of trouble crossing the transition energy at 1.13 MeV, betatron cores were installed; they can be seen in the figure. However, it was possible to accelerate over the transition energy with the rf cavity, so most experiments were carried out using rf buckets from the injection energy (100 keV). The cavity on the right powers the rf accelerating gap. It was necessary to make up for radiation loss of a few volts per turn in the stacked beam. Due to its energy spread, a high voltage on an accelerating gap would have been required in order to trap the stacked beam in a bucket. The MURA group therefore chose to make up for the radiation loss by phase displacement, using a low voltage supplied by the cavity on the left. Its frequency was modulated so as to move a small empty bucket down through the stacked beam from above.

After compensating for positive ions with clearing electrodes, and for instabilities with feedback, as well as for the effect of the stacked-beam current on the magnets, they succeeded in stacking a beam of over 10 A.

5.2.8. *The end of MURA*

During its 13-year life, MURA submitted some half-dozen proposals to the AEC for FFAG accelerators in the 10–20 GeV range, none of which were approved. Some emphasized colliding beams and high-intensity single beams, and some proposed only high-intensity single beams.

In 1967 MURA disbanded and sold its site and laboratory to the University of Wisconsin–Madison. The lab became the Physical Sciences Laboratory. Ednor Rowe built a small storage ring, Tantalus, initially for orbit studies, but later converted into a synchrotron radiation source — the first dedicated synchrotron radiation source. Tantalus started the Synchrotron Radiation Center. An 800 MeV storage ring, Aladdin was later added; it is still serving many users from around the country and the world.

5.3. *FFAGs in the 1980s*

The 1980s saw a revival of interest in FFAGs, with major proposals in the USA and Germany for spallation neutron sources. The Argonne team's design aim was to provide a time-averaged proton beam of 3.8 mA at 1.5 GeV, consisting of intense bursts, 325 ns long, at a repetition rate of 45–50 Hz. For this Kustom *et al.* [81] proposed a 20-sector FFAG "ASPUN" with a 61° spiral and gradient $k = 14$, the minimum and maximum orbit radii being 25.9 m and 28.1 m (Fig. 34). The injector was a 200 MeV H⁻ linac. Acceleration was to be provided by 10 cavities delivering 40 kV peak voltages and operating at 1.545–1.566 MHz.

The KFA Jülich proposal [82] was for a 5 mA proton beam at 1.5 GeV and called for an FFAG

Fig. 33. The 50 MeV model.

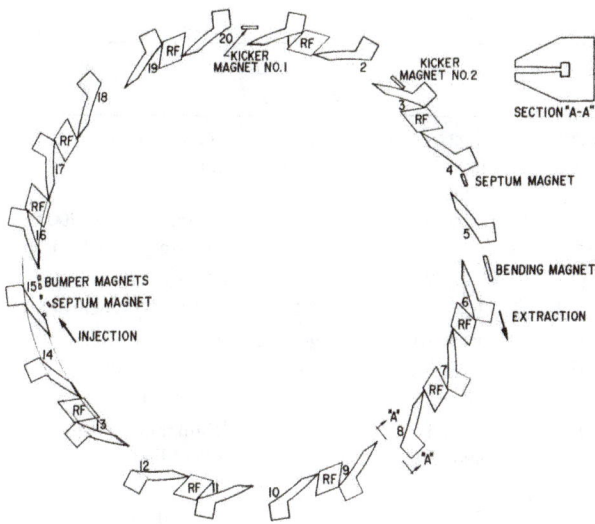

Fig. 34. ASPUN 1.5 GeV spallation-neutron-source FFAG.

having 20 radial sectors with gradient $k = 13.4$, operating at 100 Hz. To compensate for the dilationary effects of the reverse bends, superconducting magnets were to be used, restricting the injection and extraction radii to 26.4 m and 28.1 m. The injector was a 350 MeV linac, and the rf system consisted of 10 cavities providing 200 kV/turn at 1.244–1.566 MHz.

Unhappily, neither of these projects was funded.

5.4. *Recent scaling FFAGs*

Recent years have seen the construction and successful operation of the first-ever FFAGs for protons (with energies of 1 MeV [83] and 150 MeV [97, 98]) by Mori's group at KEK, and the initiation of several more (Table 7). All follow scaling principles and all but one (with spiral sectors) employ radial-sector triplet magnets.

The KEK machines introduced important innovations in both magnet and rf design. The DFD triplets are built and powered as single units, without a steel return yoke, forcing the return flux through the Ds and automatically providing a reverse field. The open structure also facilitates injection and extraction. The rf innovation (avoiding the cumbersome rotary capacitors on synchrocyclotrons) was to load the cavities with Finemet metallic alloy, providing:

- low $Q(\approx 1)$, allowing broadband operation without any need for active tuning over the 1.5–4.6 MHz range of orbital frequency;
- high permeability, and so short cavities with high effective fields;
- high repetition rates (250 Hz or more).

A 150/200 MeV FFAG of similar design is being installed at the Kyoto University reactor, together with injector and booster FFAGs [99, 100] (Fig. 35). This will eventually provide 100 μA beams to test Accelerator-Driven Subcritical Reactor (ADSR) operation. The Kyoto group is also building a proton FFAG ionization-cooling ring, ERIT (Emittance–Energy Recovery Internal Target[101]), as a neutron source for boron neutron-capture therapy. This uses FDF rather than DFD triplets.

FFAGs are of interest for muons too. PRISM (Phase-Rotated Intense Slow Muon source), based on a 10-cell DFD radial-sector FFAG of 6.5 m radius, is under construction at RCNP, Osaka for eventual installation at J-PARC [102]. It will collect muon bunches at 68 MeV/c and rotate them in phase space, reducing the momentum spread from $\pm 30\%$ to $\pm 3\%$. With a repetition rate of 100–1000 Hz, the intensity

Table 7. Scaling FFAGs operating or under construction.

	Ion	Energy (MeV)	Cells	Spiral angle	Orbit radius range (m)	First beam
KEK-PoP	p	1	8	0°	0.8–1.1	2000
KEK*	p	150	12	0°	4.5–5.2	2003
KURRI-ADSR	p	2.5	8	40°	0.6–1.0	2006
"	p	20	8	0°	1.4–1.7	2006
"	p	150	12	0°	4.5–5.1	2008
NEDO-ERIT	p	11	8	0°	2.35	2008
PRISM study	α	0.8	6	0°	3.3	2008
NHV	e	0.5	6	30°	0.19–0.44	2008
Radiatron	e	5	12	0°	0.3–0.7	(2008)

*To be moved to Kyushu University in 2008.

Fig. 35. The three proton FFAG rings for ADSR studies at the Kyoto University Research Reactor Institute.

Table 8.   Scaling FFAGs — design studieds.

Accelerator	Ion	Energy (MeV/u)	Cells	Spiral angle	Radius (m)	Repetition rate (Hz)	Comments
MElCo laptop [104, 105]	e	1	5	35°	0.023–0.028	1000	Hybrid; magnet built
eFFAG [106]	e	10	8	47°	0.26–1.0	5000	20–100 mA
Ibaraki med. accel. [107]	p	230	8	50°	2.2–4.1	20	0.1 μA
LPSC RACCAM [108]	p	180	10	54°	3.2–3.9	>20	Magnet sector 2008
MElCo p therapy [104, 105]	p	230	3	0°	0–0.7	2000	SC, quasi-isochronous
MElCo C therapy [104, 105]	C^{6+}	400	16	64°	7.0–7.5	0.5	Hybrid
	C^{4+}	7	8	0°	1.35–1.8	0.5	FFAG/synchrotrons
NIRS Chiba ion therapy	C^{6+}	400	12	0°	10.1–10.8	200	Compact
accelerators [109]	"	100	12	0°	5.9–6.7	"	radial-sector
	C^{4+}	7	10	0°	2.1–2.9	"	designs
PRISM [102]	μ	20	10	0°	6.5		Emittance rotator
Mu cooling ring [110, 111]	μ	160	12	0°	0.95 ± 0.08		Gas-filled
J-PARC neutrino	μ	20,000	120	0°	200		$\Delta r = 0.5$ m
factory accelerators [112, 113]	"	10,000	64	0°	90		≈ 10 turns
	"	3000	32	0°	30		SC magnets
	"	1000	16	0°	10		Broadband rf

will be high enough to allow ultrasensitive studies of rare muon decays. The first six magnets will initially form an α-particle test ring to demonstrate the principle.

Finally, RadiaBeam Technologies is building "Radiatron," a compact 5 MeV high-power electron FFAG for medical and industrial applications [103]. Like ERIT, this uses FDF triplets, but employs a betatron core to accelerate by induction rather than by rf.

5.4.1. Scaling FFAG studies

In addition, more than a dozen different scaling FFAG designs have been published (Table 8), mostly in Japan, but also in France (RACCAM [108]) and the USA (a muon cooling ring [110]). They range from a fist-sized 1 MeV prototype for electron irradiation [104, 105], to medium-sized sources for proton and ion therapy (for which the high pulse repetition rates are clinically advantageous), to the 400-m-diameter 20 GeV muon ring proposed for a neutrino factory. Both spiral- and radial-sector designs are employed, the latter all using DFD triplet cells. The proposals include some "hybrid FFAG/synchrotron" designs by MElCo (Mitsubishi Electric Co.) [104, 105], where a limited field rise is permitted.

The KEK/Kyoto group's most ambitious plan is to build a neutrino factory [112, 113] at J-PARC based on a chain of four FFAGs accelerating muons from 0.3 to 20 GeV. The largest would have a radius

of 200 m (with a total orbit spread of 50 cm) and consist of 120 cells, each containing a superconducting DFD triplet. Most cells would also contain rf cavities to provide an overall energy gain of around 1 GeV per turn, restricting the losses through muon decay to 50% overall. The use of low-frequency rf (24 MHz) keeps the buckets wide enough to contain the phase drift occurring as the orbit expands. A major advantage of FFAGs over linacs — either single or recirculating — is that their large radial and momentum acceptances obviate the need for muon cooling or phase rotation. There are also significant cost savings on the accelerators themselves.

5.5. Linear non-scaling (LNS) FFAGs for muons

In a study of FFAG arcs for recirculating muon linacs in 1997, Mills [84] and Johnstone [85] noted that the rapid acceleration (< 20 turns) essential for muons allows betatron resonances no time to damage beam quality, and so scaling can be abandoned, the tunes allowed to vary, and a wider variety of lattices explored. Moreover, using constant-gradient *linear* magnets greatly increases the dynamic aperture and simplifies construction, while employing the strongest possible gradients minimizes the real aperture. Johnstone *et al.* [114] applied this *non-scaling* approach to a complete FFAG ring, showing that it would be advantageous to use superconducting magnets with positively bending Ds and negatively

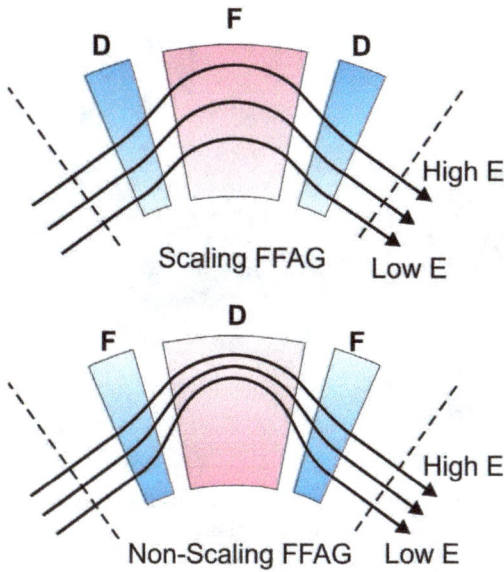

Fig. 36. Scaling and non-scaling FFAG magnetic fields and orbit patterns. Positive-bending fields are shown in pink, negative in blue, with the color density indicating the field strength. (Note that FDF scaling and DFD non-scaling lattices are also possible.)

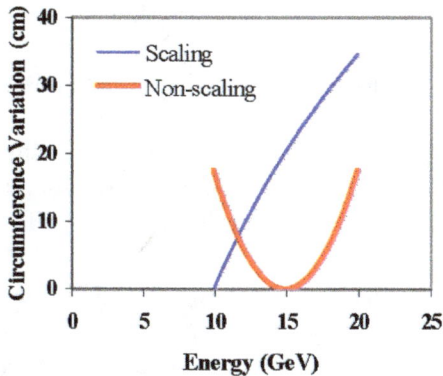

Fig. 37. Circumference variation with energy compared for scaling and linear non-scaling FFAGs.

bending Fs, i.e. both B_D and $|B_F|$ decrease outward (Fig. 36). The radial orbit spread would be reduced (allowing the use of smaller vacuum chambers and magnets), and the orbit circumference $C(p)$ shortened and made to pass through a minimum instead of rising monotonically as $p^{1/(k+1)}$ (Fig. 37). The variation in the orbit period is thereby reduced, allowing the use of high-Q fixed-frequency rf.

$C(p)$'s parabolic variation and its parametric dependence can be derived using a simple model [115, 116], treating the F and D magnets as thin lenses of equal strength S (gradient × length). For symmetric F0D0 or triplet cells, and assuming that $S_F = S_D \equiv S$,

$$C(p) = C(p_m) + \frac{12\pi^2}{e^2 S^2 N L_{FD}}(p - p_m)^2, \quad (5.11)$$

where N is the number of cells, and L_{FD} is the (shorter) F–D spacing. The minimum is at $p_m = (4p_c + eSL_{FD})/6$, where the p_c closed orbit is such that $B_F = 0$. The orbit radii $r(p)$ show similar dependence, with distinct p_{min}.

Lattices along these lines were developed [16–18] by Johnstone at Fermilab, by Berg, Courant, Trbojevic and Palmer at Brookhaven, by Keil at CERN and Sessler at LBNL, and by Koscielniak at TRIUMF. Results from a cost-optimization study of muon acceleration from 2.5 to 20 GeV by Berg *et al.* [18] favored a chain of three rings using doublet cells with superconducting magnets and high-field 200 MHz superconducting cavities. Their top energies would be 5, 10, and 20 GeV, with circumferences of 246, 322, and 426 m, and 64, 77, and 91 cells, respectively. The smallest would be similar in price to a linac, but those above 10 GeV less costly. Recent tracking studies, however, suggest that time-of-flight variations for particles with large transverse amplitudes may make it difficult to match multiple FFAG stages. The International Scoping Study for a Neutrino Factory (ISS) has therefore recommended FFAGs only for the higher-energy stages of muon acceleration, 12.6–25 GeV and 25–50 GeV [117].

The betatron tunes in such lattices fall off very strongly with momentum (Fig. 38). With constant-gradient magnets and high momentum compaction, the behavior of k is dominated by that of B_m,

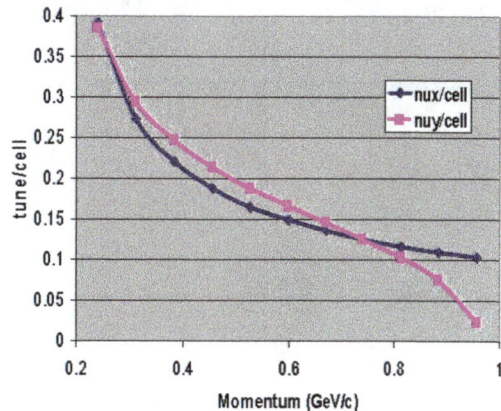

Fig. 38. Dependence of the cell tune on momentum in an LNS FFAG designed for rapid acceleration of muons.

and since B_m rises with momentum, k and ν_r fall [Eq. (5.1)]. In the case of ν_z [Eq. (4.6)], the high k at low momentum is offset by the strong scalloping of the orbit and consequent high flutter, but the latter falls off faster with momentum than k, and so ν_z falls too. Note that neither tune exceeds $N/2$, and so the π stopband is avoided and the only intrinsic resonances that have to be crossed are higher-order ones.

With the orbit length varying by only 20 cm, first falling and then rising, Berg [118, 119] and Koscielniak [120, 121] have shown that, provided a critical rf voltage is exceeded, an acceleration path can be created (Fig. 39) that stays close to the voltage peak (crossing it three times), snaking between neighboring buckets (rather than circulating inside them) just as in an imperfectly isochronous cyclotron [122]. By using high-field superconducting 200 MHz cavities it should be possible to accelerate from 10 to 20 GeV in 17 turns, with a decay loss of 8%.

In order to demonstrate the novel features of such a design — particularly "serpentine" or "gutter" acceleration outside buckets, and the crossing of many integer and half-integer imperfection resonances — a 10–20 MeV electron model (EMMA, with $C = 16.6$ m and 42 doublet cells) [123] is being built at Daresbury, where the 8–35 MeV ALICE energy-recovery linac prototype (previously known as ERLP) will act as injector. Figure 40 shows four cells of EMMA, with the F and D magnets formed by offset quadrupoles. A 1.3 GHz accelerating cavity is installed in every other cell, except in the injection and extraction regions.

In an effort to reduce the number of FFAG stages for muons from three to two, Trbojevic [124] has proposed an oval "racetrack" lattice, with

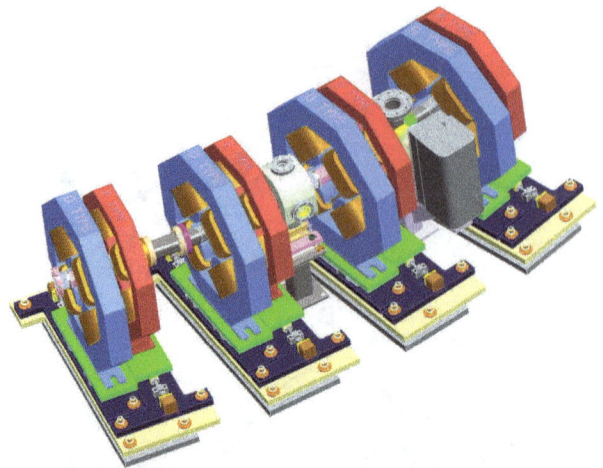

Fig. 40. Four EMMA doublet cells. The drift spaces contain (*left to right*) a resistive wall monitor, an rf cavity, and an ion pump.

two small-radius 90° arcs, where the magnets are close-packed, and two large-radius ones, with straights to accommodate the rf and other equipment. The betatron and dispersion functions are matched at the central energy. The momentum range is increased from $\pm 33\%$ to $\pm 40\%$ per stage, allowing two stages to span the range 3.7–20 GeV — assuming that the accelerating gradient can be raised from 10 to 17 MeV/m. On similar principles, he and colleagues [125] propose an electron FFAG ring composed of six small-radius and six large-radius arcs for e-RHIC, to fit within the existing tunnel.

5.6. *LNS FFAGs for protons and heavy ions*

Using non-scaling FFAGs to accelerate protons or ions that are not fully relativistic introduces two complications: first, the orbit time may vary over a wide-enough range that fixed-frequency operation is not possible; and second, cost considerations may favor a lower energy gain per turn and a longer acceleration time, so that resonance crossing is more of a concern.

Keil, Trbojevic, and Sessler [126] have proposed a system of three concentric LNS FFAGs for cancer therapy. Each is composed of 48 doublet cells, the largest ring having $C = 52$ m. The smaller pair would accelerate protons to 250 MeV and the larger pair C^{6+} ions to 400 MeV/u. FM operation is envisaged, with pulse rates up to ≈ 500 Hz. As in the muon machines, the tune per cell drops from ≈ 0.4 to

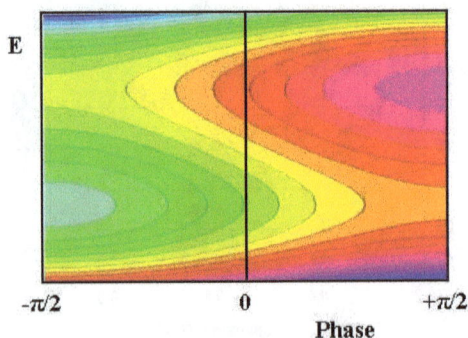

Fig. 39. The acceleration path (yellow) in longitudinal phase space in a linear non-scaling FFAG.

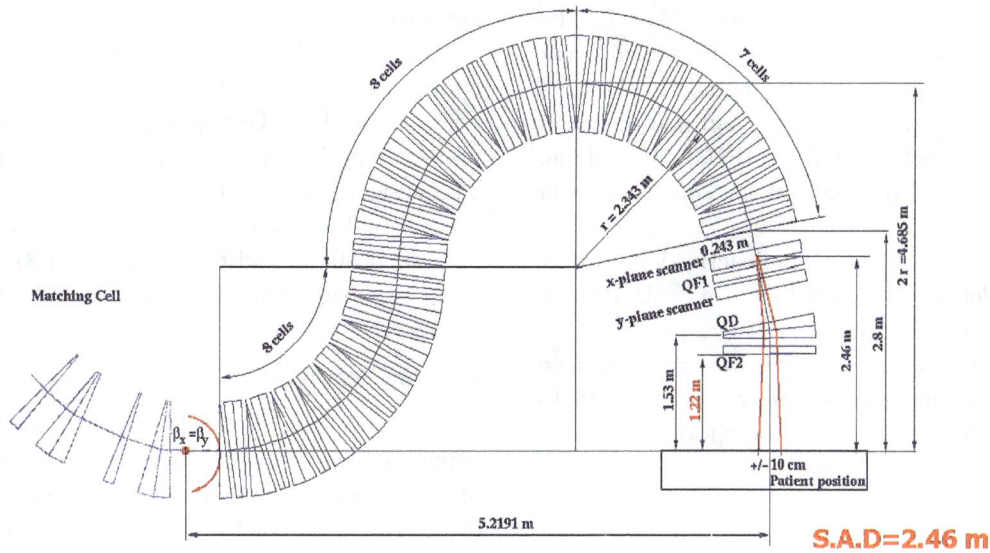

Fig. 41. A lightweight superconducting LNS FFAG gantry (S.A.D. = effective Source to Axis Distance). Note that the scale transverse to the beam is greatly exaggerated.

≈ 0.15 during acceleration, but quite modest rf voltages ($\leq 220\,\mathrm{kV}$ at $\approx 20\,\mathrm{MHz}$) are sufficient to retain good beam quality while crossing more than a dozen integer and half-integer imperfection resonances (the design avoids all intrinsic resonances below the third order). They also propose a lightweight LNS FFAG gantry [127, 128] (Fig. 41), composed of 23 superconducting triplets, capable of accepting the whole extracted momentum range at fixed field. These magnets would weigh 1.5 t, rather than the 130 t of comparable conventional gantry magnets. Permanent magnet designs are also being considered.

Also for cancer therapy, Johnstone and Koscielniak [129] propose a single-stage LNSFFAG (with 14 F0D0 cells and $C = 40\,\mathrm{m}$) accelerating carbon ions from $18\,\mathrm{MeV/u}$ to $400\,\mathrm{MeV/u}$ (and protons too). Their design introduces a powerful new feature: the magnets are wedge-shaped, with opening angles chosen so that the edge focusing compensates for that from other sources, minimizing the tune variation (Fig. 42). The dynamic aperture at injection is 10–$20\,\pi\mu\mathrm{m}$ (95%, geometric).

A further study of an LNS FFAG for hadron therapy (PAMELA) has been funded in the UK [130], in association with the EMMA project mentioned above. Its aims are to find an optimal scheme for a $450\,\mathrm{MeV/u}$ carbon machine, together with designs for the magnets and rf system and a preliminary cost estimate, then to scale it down to $70\,\mathrm{MeV}$ and $230\,\mathrm{MeV}$ proton machines as possible prototypes.

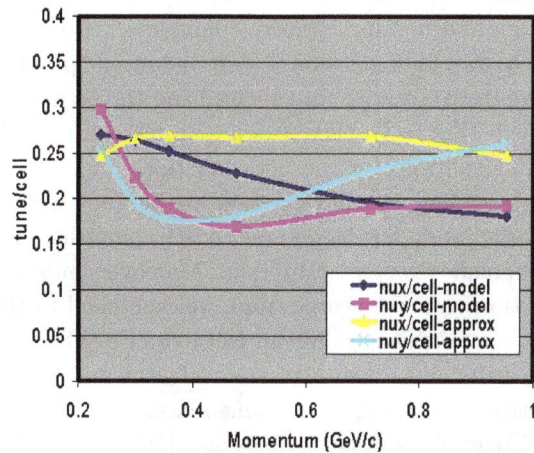

Fig. 42. Dependence of the cell tune on momentum in an LNS FFAG tune-stabilized for medical therapy [127, 128]. In the legend, "approx" refers to an analytic solution, and "model" to one obtained using MAD.

LNS FFAGs have also been considered by Ruggiero *et al.* [131, 132] for higher-energy proton and heavy-ion drivers. For a $10\,\mathrm{MW}$ proton source they propose a 50–$250\,\mathrm{MeV}$ ring followed by a 250–$1000\,\mathrm{MeV}$ one, both with $C = 204\,\mathrm{m}$ and 80 FDF cells. The frequency swing during acceleration could be accommodated by using either broadband (magnetic alloy) cavities at a few MHz ($1\,\mathrm{kHz}$ pulse rate) or the harmonic number jump (HNJ) technique [133] with $804\,\mathrm{MHz}$ cavities (either cw or $10\,\mathrm{kHz}$). To ensure regular integer jumps per turn in the HNJ case, the accelerating field must change with energy (i.e. vary across the aperture). This would be achieved by grouping cavities operating in different TM modes together. An FFAG upgrade to the AGS Booster is also being considered, and one octant of a 0.2–0.8 MeV electron model (MINHA, $C = 18$ m, $N = 48$) for that is scheduled for construction in 2007–8.

The heavy-ion driver for radioactive-ion production is to deliver 4.2 particle-μA of ^{238}U ions at 400 MeV/u. It would also use two rings with $C = 204$ m and 80 FDF cells (though with different field strengths), the first stage accelerating from 15 to 80 MeV/u. The same acceleration options and pulse rates are suggested as for protons.

5.7. *Nonlinear non-scaling FFAGs*

Rees [134–138] has designed several NS FFAGs using nonlinear field profiles and slightly more complicated dFDFd cells (termed "pumplets," from the Welsh word for five, "pump" — pronounced "pimp"), where the d/Ds are parallel-edged and the Fs wedge-shaped (Fig. 43). The extra parameters provide greater control over the beta functions and dispersion, enabling the vertical tune to be kept almost constant and, for highly relativistic particles, the orbits made exactly isochronous. Moreover, pumplet insertions can be incorporated, well-matched to the arcs — a notoriously difficult feat in an FFAG. Their long drifts allow the use of more efficient multicell rf cavities, shortening the circumference.

To accelerate muons, Rees uses just two FFAGs, the first from 3.2 to 8 GeV, the second from 8 to 20 GeV ($C = 920$ m with 4 superperiods, each composed of 20 normal and 10 insertion cells), both rings being isochronous — muon cyclotrons! Although isochronous cyclotron designs in this energy range have been reported before [65], they relied on spiral-edge focusing to counteract the high k values. What is remarkable here is that the spiral is not needed. Méot *et al.* have carried out tracking studies [134, 135] to confirm the optics using realistic magnetic fields. Rees has also designed a nonisochronous 10 GeV pumplet FFAG as a 4 MW proton driver for a neutrino factory or muon collider. This design, which has been recommended by the ISS [117], would have 66 cells, a circumference of 801 m, operate at 50 Hz, and be injected by a 3 GeV rapid-cycling synchrotron.

6. Conclusions

The history of fixed-field ion accelerators now extends over a period of almost 80 years. Cyclotrons, in their modern isochronous form, supply beams of unrivaled intensity in the energy range 10–600 MeV. The larger ones serve as important sources of protons, neutrons, pions, muons, and heavy ions (both stable and unstable) for particle, nuclear, and condensed-matter physics. There seems to be an unquenchable market for small cyclotrons for radioisotope production, and proton cyclotrons are becoming increasingly popular for cancer therapy.

FFAGs have been the subject of renewed interest in recent years for a wide variety of particles and applications. Technological advances have made possible the construction and successful operation of proton FFAGs for the first time, all following the traditional scaling principle. In addition, a variety of non-scaling designs have been proposed and two prototypes are under construction. The exploration of non-scaling lattices is in its early days, but the initial efforts have already yielded some remarkable results. Who knows, maybe there are yet more varieties of FFAG waiting to be discovered?

Acknowledgments

We would like to thank the numerous fellow workers who have kindly provided data or figures for this article. We also wish to express our gratitude to IEEE for permission to include some material previously published [139] in the *Proceedings of the 2003 Particle Accelerator Conference*, and to the editors of the *ICFA Beam Dynamics Newsletter* for material [140] from Newsletter 43. The figures are reproduced courtesy of: American Physical Society (7); Argonne National Laboratory (34), managed and operated by UChicago Argonne, LLC, for the

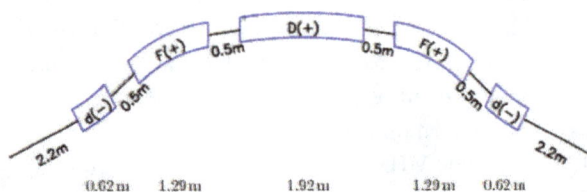

Fig. 43. A five-magnet pumplet cell.

U.S. Department of Energy under contract No. DE-AC02-06CH11357; Atomic Energy of Canada Limited (22); Brookhaven National Laboratory (41); Ernest Orlando Lawrence Berkeley National Laboratory (3–5, 8–9, 12, 15); Fermi National Accelerator Laboratory (38, 42); Kyoto University Research Reactor Institute (35); Oak Ridge National Laboratory, managed for the US Department of Energy by UT-Battelle, LLC (18); Paul Scherrer Institut (20); Petersburg Nuclear Physics Institute (13); RIKEN (10, 21, 24); STFC Daresbury Laboratory (40); STFC Rutherford Appleton Laboratory (43); Technische Universität München and CERN (23, ©1996 CERN); TRIUMF (16, 19); and the University of Wisconsin–Madison Physical Sciences Laboratory (25–33).

References

[1] E. O. Lawrence and N. E. Edlefsen, *Science* **72**, 376 (1930).

[2] E. O. Lawrence and M. S. Livingston, *Phys. Rev.* **37**, 1707 (1931).

[3] R. Wideröe, *The Infancy of Particle Accelerators* (Vieweg, Braunschweig, 1994), p. 41.

[4] J. L. Heilbron and R. Seidel, *Lawrence and His Laboratory: A History of Lawrence Berkeley Laboratory*, Vol. 1 (University of California Press, Berkeley, 2000), pp. 80–82. http://ark.cdlib.org/ark:/13030/ft5s200764

[5] J. Thibaud, *Structure et propriétés des noyaux atomiques: Rapports et discussions du septième conseil de physique Solvay, Bruxelles, 1933* (Gauthier-Villars, Paris, 1934), pp. 72–75.

[6] E. O. Lawrence, *Les prix Nobel de 1951* (Norstedt, Stockholm, 1952), pp. 127–140.

[7] R. Wideröe, *Arch. f. Elektrotechnik* **21**, 387 (1928).

[8] N. P. Davis, *Lawrence and Oppenheimer* (Simon and Schuster, New York, 1968), pp. 19–20.

[9] M. S. Livingston and J. P. Blewett, *Particle Accelerators* (McGraw-Hill, New York, 1962).

[10] J. J. Livingood, *Cyclic Particle Accelerators* (Van Nostrand, New York, 1961).

[11] J. R. Richardson, *Prog. Nucl. Tech. Instrum.* **1**, 1 (1965).

[12] *Proc. CERN Accelerator School: Cyclotrons, Linacs and Their Applications* (La Hulpe, Belgium, 1994), CERN 96-02 (1996).

[13] H. G. Blosser, *Handbook of Accelerator Physics and Engineering*, eds. A. Chao and M. Tigner (World Scientific, 2006), pp. 13–16.

[14] F. T. Cole, *O Camelot! A Memoir of the MURA Years* (1994), in *Proc. Cyclotrons 2001*, Suppl.: http://accelconf.web.cern.ch/accelconf/c01/cyc2001/extra/Cole.pdf

[15] C. H. Prior (ed.), *ICFA Beam Dynamics Newslett.* **43**, 19 (2007); http://www-bd.fnal.gov/icfabd/Newsletter43.pdf

[16] *FFAG 2003 Workshop* (BNL), http://www.cap.bnl.gov/mumu/conf/ffag-031013/

[17] *FFAG 2004 Workshop* (TRIUMF), http://www.triumf.ca/ffag2004/programme.html

[18] *FFAG04 Workshop* (KEK, 2004), http://hadron.kek.jp/FFAG/FFAG04_HP

[19] *FFAG05 Workshop* (KEK, 2005); http://hadron.kek.jp/FFAG/FFAG05_HP

[20] *FFAG 2007 Workshop* (LPSC, Grenoble), http://lpsc.in2p3.fr/congres/FFAG07

[21] *FFAG07 Workshop* (KURRI, Osaka), http://hadron.kek.jp/FFAG/FFAG07_HP

[22] JACoW, http://jacow.org

[23] E. O. Lawrence and M. S. Livingston, *Phys. Rev.* **40**, 19 (1932).

[24] M. S. Livingston and E. O. Lawrence, *Phys. Rev.* **45**, 608 (1934).

[25] E. O. Lawrence and D. Cooksey, *Phys. Rev.* **50**, 1131 (1936).

[26] F. N. D. Kurie, *J. Appl. Phys.* **9**, 691 (1938).

[27] M. E. Rose, *Phys. Rev.* **53**, 392 (1938).

[28] R. R. Wilson, *Phys. Rev.* **53**, 408 (1938).

[29] B. L. Cohen, *Rev. Sci. Instrum.* **24**, 589 (1953).

[30] G. Dutto and M. K. Craddock, *Proc. 7th Int. Conf. Cyclotrons* (Birkhauser, Zürich, 1975), pp. 271–274.

[31] M. M. Gordon and F. Marti, *Part. Accel.* **11**, 161 (1981).

[32] E. O. Lawrence *et al.*, *Phys. Rev.* **56**, 124 (1939).

[33] R. S. Livingston, *Nature* **170**, 221 (1952).

[34] H. A. Atterling and G. Lindstrom, *Nature* **169**, 432 (1952).

[35] Y. Hirao, *Proc. 11th Int. Conf. Cyclotrons* (Tokyo, 1986), pp. 761–767.

[36] H. A Bethe and M. Rose, *Phys. Rev.* **52**, 1254 (1937).

[37] E. O. Lawrence, Foreword to *The Cyclotron*, by W. B. Mann (Methuen, London, 1940).

[38] V. I. Veksler, *J. Phys. USSR* **9**, 153 (1945).

[39] E. M. McMillan, *Phys. Rev.* **68**, 143 (1945).

[40] E. D. Courant, in this issue.

[41] J. R. Richardson, K. R. MacKenzie, E. J. Lofgren and B. T. Wright, *Phys. Rev.* **69**, 669 (1946).

[42] J. R. Richardson, *Proc. 10th Int. Conf. Cyclotrons* (IEEE Press, 1984), pp. 617–622 .

[43] W. M. Brobeck, *et al.*, *Phys. Rev.* **71**, 449 (1947).

[44] L. H. Thomas, *Phys. Rev.* **54**, 580 (1938).

[45] K. R. Symon, D. W. Kerst, L. W. Jones, L. J. Laslett and K. M. Terwilliger, *Phys. Rev.* **103**, 1837 (1956).

[46] L. C. Teng, *Rev. Sci. Instrum.* **27**, 1051 (1956).

[47] W. Walkinshaw and N. M. King, *Linear Dynamics in Spiral-Ridge Cyclotron Design*, AERE GP/R 2050 (1956).

[48] L. Smith and A. A. Garren, *Orbit Dynamics In The Spiral-Ridged Cyclotron*. UCRL-8598 (1959).

[49] H. L. Hagedoorn and N. F. Verster, *Nucl. Instrum. & Methods.* **18–19**, 201 (1962).

[50] E. L. Kelly, R. V. Pyle, R. L. Thornton, J. R. Richardson and B. T. Wright, *Rev. Sci. Instrum.* **27**, 492 (1956).

[51] F. A. Heyn and K. T. Khoe, *Rev. Sci. Instrum.* **29**, 662 (1958).

[52] T. Zhang, Z. Li, C. Chu, *Proc. 18th Int. Conf. Cyclotrons* (Giardini Naxos, 2007), pp. 33–38.

[53] D. W. Kerst, K. M. Terwilliger, K. R. Symon and L. W. Jones, *Phys. Rev.* **98**, 1153 (A) (1955).

[54] D. W. Kerst, *Proc. CERN Symposium on High Energy Accelerators and Pion Physics* (CERN, Geneva, 1956), Vol. 1, pp. 366–375.

[55] D. J. Clark, J. R. Richardson and B. T. Wright, *Nucl. Instrum. Methods* **18–19**, 1 (1962).

[56] D. A. Lind, J. J. Kraushaar, R. Smythe and M. E. Rickey, *Nucl. Instrum. Methods* **18–19**, 62 (1962).

[57] Y. Jongen *et al.*, *Proc. PAC' 97* (IEEE, 1998), pp. 3816–3818.

[58] J. R. Richardson *et al.*, *Proc. PAC' 75, IEEE Trans.* **NS-22**, 1402 (1975).

[59] J. A. Martin and J. E. Mann, *Nucl. Instrum. Methods* **18–19**, 461 (1962).

[60] H. Willax, *Proc. Int. Conf. Sector-Focused Cyclotrons & Meson Factories* (1963), CERN 63-19, pp. 386–397.

[61] Y. Yano, *Proc. PAC' 07* (2007), pp. 700–702.

[62] W. Joho, *Proc. PAC' 83, IEEE Trans.* **NS-30**, 2083 (1983).

[63] T. Stammbach, *Proc. 15th Int. Conf. Cyclotrons* (IoP, London, 1998), pp. 369–376.

[64] T. Stammbach *et al.*, *Proc. EPAC' 02* (2002), pp. 159–163.

[65] J. I. M. Botman *et al.*, *Proc. PAC' 83, IEEE Trans.* **NS-30**, 2007 (1983).

[66] R. Baartman *et al.*, *ibid.*, 2010.

[67] J. H. Ormrod *et al.*, *Proc. PAC' 77, IEEE Trans.* **NS-24**, 1093 (1977).

[68] H. G. Blosser and F. Resmini, *Proc. PAC' 79, IEEE Trans.* **NS-26**, 3653 (1979).

[69] H. Blosser *et al.*, *Proc. PAC'85, IEEE Trans.* **NS-32**, 3287 (1985).

[70] A. E. Geisler *et al.*, *Proc. 18th Int. Conf. Cyclotrons* (Giardini Naxos, 2007), pp. 9–14.

[71] F. M. Russell, *Nucl. Instrum. Methods.* **23**, 229 (1963).

[72] A. Cazan, P. Schütz and U. Trinks, *Proc. EPAC' 98* (1998), pp. 556–558.

[73] H. S. Snyder, private communication.

[74] K. R. Symon, *Phys. Rev.* **98**, 1152(A) (1955).

[75] D. W. Kerst, K. R. Symon, L. J. Laslett, L. W. Jones and K. M. Terwilliger, *Proc. CERN Symposium on High Energy Accelerators and Pion Physics* (CERN, Geneva, 1956), Vol. 1, pp. 32–35.

[76] T. Ohkawa, *Bull. Phys. Soc. Jpn.* (1953).

[77] A. A. Kolomensky, V. A. Petukhov and M. Rabinovich, *Lebedev Phys. Inst. Report*, RF-54 (Moscow, 1953).

[78] A. A. Kolomensky, *Proc. CERN Symposium on High Energy Accelerators and Pion Physics* (CERN, Geneva, 1956), Vol. 1, pp. 66–67.

[79] V. V. Kannunikov *et al.*, *Proc. Int. Conf. High Energy Accelerators & Instrumentation* (CERN, Geneva, 1959), pp. 89–98.

[80] M. Barbier *et al.*, *ibid.*, pp. 100–114.

[81] R. L. Kustom, T. K. Khoe and E. Crosbie, *Proc. PAC' 85, IEEE Trans.* **NS-32**, 2672 (1985).

[82] P. F. Meads and G. Wüstefeld, *PAC'85, IEEE Trans.* **NS-32**, 2697 (1985).

[83] M. Aiba *et al.*, *Proc. EPAC' 00* (2000), pp. 581–583.

[84] F. Mills, *Proc. 4th Int. Conf. Physics Potential and Development of $\mu^+\mu^-$ Colliders* (San Francisco, 1997) (1998), pp. 693–696.

[85] C. Johnstone, *ibid.*, pp. 696–698.

[86] F. T. Cole *et al.*, *Rev. Sci. Instrum.* **28**, 403 (1957).

[87] D. W. Kerst *et al.*, *Rev. Sci. Instrum.* **28**, 970 (1957).

[88] R. O. Haxby *et al.*, *Proc. Int. Conf. High Energy Accelerators & Instrumentation* (CERN, Geneva, 1959), pp. 75–81.

[89] D. W. Kerst *et al.*, *Rev. Sci. Instrum.* **31**, 1076 (1960).

[90] K. R. Symon and A. M. Sessler, *CERN Symposium on High Energy Accelerators and Pion Physics* (CERN, Geneva, 1956), pp. 44–58.

[91] K. M. Terwilliger *et al.*, *Rev. Sci. Instrum.* **28**, 987 (1957).

[92] D. W. Kerst, *CERN Symposium on High Energy Accelerators and Pion Physics* (CERN, Geneva, 1956), pp. 36–39.

[93] F. T. Cole *et al.*, *Rev. Sci. Instrum.* **35**, 1393 (1961).

[94] T. Ohkawa, *Scaled Radial Sector FFAG for Intersecting Beams.* MURA-124 (1956, unpublished).

[95] T. Ohkawa, *Rev. Sci. Instrum.* **29**, 108 (1958).

[96] A. A. Kolomensky, *Sov. Phys.*, *JETP* **6**, 231, (1958), translated from *Zh. Eksp. Teor. Fiz. SSSR* **33**, 298 (1957).

[97] Y. Yonemura *et al.*, *Proc. PAC' 03* (2003), pp. 3452–3454.

[98] M. Aiba *et al.*, *Proc. EPAC' 06* (2006), pp. 1672–1674.

[99] M. Tanigaki *et al.*, *ibid.*, pp. 2367–2369.

[100] T. Uesugi, in *FFAG07 Workshop* (KURRI, Osaka), http://hadrom.kek.jp/FFAG/FFAG07_HP

[101] K. Okabe, M. Muto and Y. Mori, *Proc. EPAC' 06* (2006), pp. 1675–1677.

[102] A. Sato, Y. Kuno *et al.*, *ibid.*, pp. 2508–2510.

[103] S. Boucher *et al.*, *Proc. PAC' 07*, (2007), pp. 3035–3038.

[104] H. Tanaka, in *FFAG04 Workshop* (KEK, 2004), http://hadron.kek.jp/FFAG/FFAG04_HP

[105] H. Tanaka and T. Nakanishi, *Proc. 17th Int. Conf. Cyclotrons* (Tokyo, 2004), pp. 238–240.

[106] Y. Yuasa, *WG3 Summary*, in *FFAG04 Workshop* (KEK, 2004), http://hadron.kek.jp/FFAG/FFAG04_HP

[107] T. Yokoi, *WG3 Summary*, in *FFAG04 Workshop* (KEK, 2004), http://hadrom.kek.jp/FFAG/FFAG04_HP

[108] J. Pasternak *et al.*, *Proc. PAC' 07* (2007), pp. 1404–1406.

[109] T. Misu *et al.*, *Phys. Rev. ST AB* **7**, 094701 (2004).

[110] A. Garren, H. Kirk and S. Kahn, in FFAG 2004 Workshop, TRIUMF, http://www.triumf.ca/ffag2004/programme.html

[111] H. Kirk *et al.*, *Proc. PAC' 03* (2003), pp. 2008–2010.

[112] Y. Mori, *Proc. EPAC' 02* (2002), pp. 278–280.

[113] S. Machida, *Nucl. Instrum. Methods* **A503**, 41 (2003).

[114] C. Johnstone, W. Wan and A. Garren, *Proc. PAC'99*, (1999), pp. 3068–3070.

[115] M. K. Craddock, in FFAG 2003 Workshop, BNL, http://www.cap.bnl.gov/mumu/conf/ffag-031013/.

[116] S. Koscielniak and M. Craddock, *Proc. EPAC' 04* (2004), pp. 1138–1140.

[117] C. R. Prior, *Proc. PAC' 07* (2007), pp. 681–685.

[118] J. S. Berg, *Proc. Snowmass 2001* (2001), p. T503.

[119] J. S. Berg, *Proc. PAC' 05* (2005), pp. 1532–1534.

[120] S. Koscielniak and C. Johnstone, *Proc. Snowmass 2001* (2001), p. T508.

[121] S. Koscielniak and C. Johnstone, *Nucl. Instrum. Methods* **A523**, 25 (2004).

[122] M. K. Craddock, C. J. Kost and J. R. Richardson, *Proc. PAC' 79, IEEE Trans.* **NS-26**, 2065 (1979).

[123] R. Edgecock, *Proc. PAC' 07* (2007), pp. 2624–2626.

[124] D. Trbojevic, *ibid.*, pp. 3202–3204.

[125] D. Trbojevic *et al.*, *ibid.*, pp. 3205–3207.

[126] E. Keil, A. M. Sessler and D. Trbojevic, *Phys. Rev. ST AB* **10**, 054701 (2007).

[127] D. Trbojevic, B. Parker, E. Keil and A. M. Sessler, *ibid.*, **10**, 053503 (2007).

[128] D. Trbojevic *et al.*, *Proc. 18th Int. Conf. Cyclotrons* (Giardini Naxos, 2007), pp. 207–209.

[129] C. Johnstone and S. R. Koscielniak, *Proc. PAC' 07* (2007), pp. 2951–2953.

[130] K. Peach *et al.*, *ibid.*, 2886–2888.

[131] A. G. Ruggiero, in ICFA Beam Dynamics Newsletter **43**, 84–94 (2007).

[132] A. G. Ruggiero *et al.*, *Proc. PAC' 07*, 1880–1882 (2007).

[133] A. G. Ruggiero, *Phys. Rev. ST AB* **9**, 100101 (2006).

[134] G. H. Rees, ICFA Beam Dynamics Newsletter **43**, 74 (2007).

[135] G. H. Rees, ICFA Beam Dynamics Newsletter **43**, 102 (2007).

[136] G. H. Rees, Proc. FFAG04 Workshop, KEK (2004); http//hadron.kek.jp/FFAG/FFAG04-HP/.

[137] G. H. Rees, FFAG05 Workshop, KEK (2005); http//hadron.kek.jp/FFAG/FFAG05-HP/.

[138] F. Lemuet, F. Méot and G. Rees, *Proc. PAC' 05* (2005), pp. 2693–2695.

[139] K. R. Symon, *Proc. PAC' 03* (2003), pp. 452–456.

[140] M. K. Craddock, ICFA Beam Dynamics Newsletter **43**, 19 (2007), pp. 19–26.

Michael Craddock received his training at Oxford University and the Rutherford High Energy Laboratory, joining the faculty at the University of British Columbia in time to contribute to the TRIUMF meson factory proposal. He served as group leader for beam dynamics during the early years of the 520-MeV cyclotron, and then led the design of the accelerators for the 30-GeV 0.1-mA KAON Factory project — a 17-year unfulfilled dream! More recently he has participated in the Canadian contribution to the LHC and to studies of non-scaling FFAGs.

Keith R. Symon is Professor of Physics Emeritus at the University of Wisconsin-Madison. He is the author of an intermediate textbook Mechanics. He specializes in the design of particle accelerators, and in plasma theory. He is credited with the co-invention of FFAG accelerators, the development of the smooth approximation in orbit theory, and contributions to the theory of rf acceleration and to the study of instabilities in accelerators and plasmas. He received the 2003 IEEE Particle Accelerator and Technology Award and the 2005 Wilson Prize of the American Physical Society.

Reviews of Accelerator Science and Technology
Vol. 1 (2008) 99–120
© World Scientific Publishing Company

Particle Colliders for High Energy Physics

D. A. Edwards* and H. T. Edwards†

Deutsches Elektronen Synchrotron (DESY),
85 Notkestr., 22607 Hamburg, Germany
*edwards@desy.de
†hedwards@fnal.gov

The purpose of this article is to outline the development of particle colliders from their inception just over a half-century ago, expand on today's achievements, and remark on the potential of coming years. There are three main sections, entitled "Past," "Present," and "Future." "Past" starts with the electron and electron–positron colliders of the 1950s, continues through the proton rings at CERN, and concludes with LEP. Technology development enters the section Present, "which includes not only the major colliders in both the lepton and baryon worlds, but also recognition of the near-immediate entry of the Large Hadron Collider. "Future" looks at the next potential steps, the most prominent of which is an electron–positron partner to the LHC, but there are other very interesting propositions undergoing exploration that include muon storage and even conceivably departure from reliance on radio frequency acceleration.

Keywords: Accelerator history; particle colliders.

1. Introduction

The discoveries of the electron and the proton were achieved through the use of gas discharge tubes and radioactive sources, in experiments conducted by such illustrious names as J. J. Thompson and Lord Rutherford. But, by the second decade of the 20th century, it was clear that higher beam energies were needed to carry particle physics forward. One direction was continuation of work on electrostatic fields, and that avenue led to the Cockroft–Walton and van de Graaff developments. The single-pass character of these devices utilizing static fields precludes particle energy above the 10 MeV or so range. The other path, put forward at roughly the same time, was the use of acceleration through the use of radio frequency fields, and was the direction that resulted in the outstanding doctoral dissertation of Wideroe. The few-MHz frequency limitation of that period limited him to rather heavy ions such as those of sodium and potassium; the speeds of electrons put them out of the question. The development of the cyclotron by Lawrence and his colleagues followed almost immediately, leading to proton acceleration, and about a decade thereafter the radar developments of the Second World War permitted the construction of electron synchrotrons with energies in the 300 MeV neighborhood.

The period under the heading "Past" begins with the mid-1950s recognition that higher center-of-momentum system (cms) energies would be necessary to carry HEP research forward, and that the electron synchrotrons constructed in preceding years would provide a natural direction for either electron–electron collisions or, potentially more interestingly, electron–positron collisions. This opportunity was seized upon immediately in the US and Europe, leading to construction of a number of lepton colliders during the 1960s. The following decade saw the achievement of proton–proton collisions in the Intersecting Storage Rings (ISR) at CERN, and the discovery of J/ψ at SPEAR. Of course, there is some arbitrariness in defining where the Past ends and the Present begins, but for our purposes we will choose the accomplishment of proton–antiproton collisions, followed by the achievement of electron–positron collisions above 200 GeV cms in LEP as the events defining the termination of this period.

The term *luminosity* had entered the language of the field as the ratio of interaction rate to cross-section at the time of the earliest collider suggestions. The Present may be characterized as a period of thirst for ever-higher luminosity from colliders of both principal varieties. The B factories at KEK and SLAC are particularly noteworthy in this regard, as performance advances combine with energies that do not challenge the synchrotron radiation barrier. For hadrons, we will stretch the Present a bit by starting with the Tevatron, including HERA despite the recent termination of its operation, and ending with the LHC, which is undergoing commissioning at the time of this writing.

The section entitled "Future" attempts to outline an engaging uncertainty. There is widespread support for an electron–positron linear collider as a logical partner to the LHC, but it is likely that approval of such a project must await initial results from this new proton collider. Presently a collider design based on superconducting rf technology is receiving intense effort; however, supporters of a version employing "conventional" rf systems continue their activities. In this report, we should comment on both. The reason for the modifier "linear" is the barrier to electron synchrotrons of higher-than-LEP energies presented by synchrotron radiation. As a consequence, research into a muon collider continues with undiminished interest. To close this section, we will comment on the research aimed at ending the hegemony of rf as the principle acceleration mechanism through the use of plasma fields.

A final, brief section comments on the technology which supports accelerator science and which has been the main interest of these authors in recent years. The advances of Wideröe and Lawrence used the technology base of their time. Since then the interests of accelerator physics have influenced technology development, most importantly in this context as an aspect of materials science.

2. Past

2.1. *The situation in the mid-1950s*

The beam physics basis was in place for the next step toward colliding beams. The invention of the synchrotron by McMillan [1] and Veksler [2] with its near-constant orbit radius had started the process of separation of design energy from site demand to a very large extent. The invention of alternating gradient focusing by Courant, Livingston, and Snyder [3] provided the next critical step of dissociating beam aperture requirements from accelerator scale, leading first to the higher energy synchrotrons and, of significance in this context, to the use of modules

in the design process as concerns about symmetry diminished.

That the first colliders employed electrons and/or positrons is understandable. Such authors as Schwinger [4] had addressed synchrotron radiation, and Robinson [5] had articulated the relationship between stability and orbit parameters. The comparative ease of electron, and even positron, manipulation played a role in the choice of the early collider designs.

Linear accelerators were in rapid development. For example, using radar components available after the conclusion of the Second World War, Luis Alvarez and associates constructed a 30 MeV proton linear accelerator [6]. The high frequency limitation — only 2 MHz or so — of Wideröe's time was gone and electron linacs developed swiftly and would support the concentration on electron colliders that followed, principally due to the ease of electron and positron production, and assistance in their manipulation due to synchrotron radiation.

However, reliance on existing technology was a necessity. For example, if the goal was to construct a rapid cycling magnet, the materials specification to insert on the purchase request would be "Armco A6 29 gauge punch type" or an equivalent transformer steel of that time. Today's interactions concerning the production of niobium sheets were scarcely thinkable at that time. The primary power system familiar to these authors in our learning process at Cornell was a 30 Hz motor-generator set obtained from the city of Buffalo after the light-flickers at that frequency were found objectionable.

2.2. *The first steps*

A Letter Physical Review by G. K. O'Neill [7] characterizes the situation of the late 1950s very well. While citing the suggestions of Kerst, Brobeck, and others toward beam–beam facilities, O'Neill offered a particular proposal: build a proton–proton collider using the 3 GeV Princeton–Pennsylvania Accelerator as the source. Since protons are intrinsically more difficult to deal with than electrons (if suitable rf frequency for electrons is available), this suggestion led toward the electron–electron collider at Stanford. Designated CBX, this two-ring system provided collisions at a cms energy of 1 GeV.

Developments in Italy, France, and Siberia were also underway. At Frascati, AdA (Anello di Accumulazione) was under construction, as an electron–positron collider.[a] After initial studies at Frascati, AdA was shipped to Orsay because of the more intense electron beam from the linear accelerator available at that institution, and a weekly cross-border commute began to develop performance of this device. At Frascati, design and construction of the next step, Adone, was already in progress. In Novosibirsk, VEP-1 was underway as the first of the long series of electron colliders at what was soon to be recognized as the Budker Institute.

By today's standards, luminosities were scarcely impressive, in the 10^{25}–10^{27} cm^{-2} s^{-1} range. Further comment will appear in the next subsection. More to the immediate point, the insights gained into collider progress were very important. The necessity of preparation of accelerator components for the vacuum environment needed for long-term storage was recognized; diffusion pumps with liquid nitrogen traps were no longer sufficient. This subject will return throughout this paper. Here, we just note that at the 10^{-6} Torr pressures typical of diffusion pumps, the mean free path for ionization would be about 300 km, acceptable in the early days of rapid cycling synchrotrons but an immediate concern for storage. The role of intrabeam scattering in the lifetime was identified and these considerations remain of importance today. Then there is the matter of dust. Installation of AdA at Orsay, as shown in Fig. 1, was accompanied by an ingenious injection mechanism that inverted the ring between electron and positron entry. Dust in the vacuum chamber fell from up to down and disposed of the beam.

A surprise among the intrabeam scattering processes was the mechanism named the Touschek effect, in honor of its interpreter [8]. A single particle–particle scattering can drive the pair outside of the stable aperture. In the laboratory frame, deviations of both transverse and longitudinal momenta from that of a reference particle are comparable in magnitude. In a frame moving with the center of the particle distribution, the transverse momentum remains unchanged under the Lorentz transformation, but the longitudinal deviation is reduced. If, in this bunch rest frame, two particles undergoing purely

Fig. 1. AdA as installed at Orsay. This photograph may be found in the article "AdA: The First Electron–Positron Collider," by Carlo Bernardini in *Phys. Perspect.* **6** (2004).

transverse motion undergo forward–backward deflection, then the longitudinal momentum in that frame is increased in magnitude for both particles. The factor of γ attendant upon conversion of the longitudinal momentum back to the laboratory frame put both particles outside of the phase space acceptance of the rf system for longitudinal motion.

At this point, we recommend that the reader look at Table 1 on p. 12 of the *Handbook of Accelerator Physics and Engineering* [9] (frequently referred to as *HAPE* in the following pages), which lists colliders in order of construction.

2.3. *Luminosity I*

The principal challenge of the collider approach was (and remains) the production of a sufficient interaction rate to take advantage of the cms energy of the two beams. Luminosity is comparatively easy to achieve in the fixed target approach. For example, if a beam with a flux of 1.5×10^{13} protons per second were incident on a liquid hydrogen target 1 m in length, the luminosity would be about 10^{38} cm^{-2} s^{-1}.

---

[a]At that time, we were associated with Cornell, and because of a Cornell–Frascati collaboration, we heard more about AdA than the activities elsewhere. No bias is intended; this is only the normal influence of recollection.

This was the proton flux originally specified for the 200 GeV proton synchrotron to be constructed at the newly (as of 1967) approved National Accelerator Laboratory. Such is the attraction of higher cms energy that R. R. Wilson, the first director of the facility that became Fermilab, immediately raised the energy goal to 500 GeV, with inevitable luminosity reduction. One frequently hears the term "discovery reach" characterizing the conflict between cms energy on the one hand and luminosity on the other while realizing that production cross sections for the processes of interest typically vary as the inverse square of energy.

An elementary expression for the relationship between luminosity and cross section for colliding beams may be obtained as follows. Suppose that the bunches contain n particles in a transverse area A, and collide with frequency f. The fraction of the area obscured by a cross section of interest, σ_{int}, is σ_{int}/A, so the event rate is $R = f(n^2/A)\sigma_{\text{int}}$. For many colliders, the transverse distribution may be approximated by a Gaussian. So starting points for luminosity can be written as

$$\mathcal{L} = f\frac{n^2}{A} \Rightarrow f\frac{n_1 n_2}{4\pi\sigma_x\sigma_y}, \qquad (1)$$

where in the second form σ_x and σ_y characterize the Gaussian transverse beam profiles in the horizontal (bend) and vertical directions. For many facilities, the Gaussian usage is quite good, given that whatever the distribution at the source, by the time the beam reaches high energy, the normal form is a reasonable approximation thanks to the central limit theorem of probability and the diminished importance of space charge effects. When synchrotron radiation is a dominant influence in establishing the phase space distribution, the progress can be demonstrated explicitly [5]. A variety of expressions for luminosity may be found in the article by Furman and Zisman in *HAPE* [10].

The beam size can be expressed in terms of two quantities — one termed the *transverse emittance*, ε, and the other, the *amplitude function*, β. The transverse emittance is a beam quality concept reflecting the process of bunch preparation, extending all the way back to the source for hadrons and, in the case of electrons, mostly dependent on synchrotron radiation. The term came into usage during the 1960s as

an alternative to the phase space area or volume language familiar from statistical mechanics. The meaning is the same, and both will be used here. The amplitude function is a beam optics quantity and is determined by the accelerator magnet configuration. It is discussed extensively in textbooks [11] and courses [12] on accelerator physics, and a capsule summary appears in Subsec. 2.1.1 of *HAPE*. For our immediate purposes here, we note only that β plays the role of a position-dependent $\lambda/2\pi$ for the transverse betatron oscillations, and so has a typical magnitude on the scale of the alternating gradient lattice. When expressed in terms of σ and β, a commonly used definition[b] of the transverse emittance is

$$\varepsilon_x \equiv [\langle x^2\rangle\langle x'^2\rangle - \langle xx'\rangle^2]^{1/2} \Rightarrow \frac{\sigma_x^2}{\beta_x}, \qquad (2)$$

where $x' \equiv p_x/p_s$. In suppressing the other coordinates, y and s, in the integral of Eq. (2) there is the presumption that the motion is uncoupled.

Of particular significance is the value of the amplitude function at the interaction point, β^*. Clearly, one wants β^* to be as small as possible; how small depends on the capability of the hardware to make a near-focus at the interaction point. Equation (1) can now be recast in terms of emittances and amplitude functions as

$$\mathcal{L} = f\frac{n_1 n_2}{4\pi\sqrt{\varepsilon_x\beta_x^*\,\varepsilon_y\beta_y^*}} . \qquad (3)$$

2.4. *The next generation*

Even as the first electron colliders of the preceding paragraphs were underway, CERN had embarked on the first proton–proton system. Designated the Intersecting Storage Rings (ISR), this facility consisted of two interlaced beam paths horizontally offset from each other and crossing at eight symmetrically located points. The proton source was the 28 GeV CERN proton synchrotron (PS), which shared with the Brookhaven AGS the role of the major step in proton acceleration following the invention of the alternating gradient structure. The ISR was the reason for our hesitancy in going immediately to a Gaussian form in Eq. (1), because the ISR used stacking of many PS pulses to create the

---

[b]This definition represents 15% of the one-degree-of-freedom phase space area. Other definitions often seen contain a multiplicative factor to increase the area of inclusion.

unusual DC beams of this collider; the beam profile resembled a rectangle. The luminosity of 1.3×10^{32} cm^{-2} s^{-1} for hadron collisions was not surpassed until over 30 years later by the Tevatron. Well worth reading is an account of the ISR written by its leader, Kjell Johnsen [13].

Long term storage of intense beams is always accompanied by new interesting and unpredicted phenomena. For instance, there was the "pressure bump" in the ISR. The need for a vacuum at the 10^{-10} Torr level was well understood and addressed at the outset. Clearing electrodes for electron removal had been installed. But then ionization of residual gas followed by repulsion by the intense beam of those ions into the wall of the vacuum system resulted in a feedback cycle with consequent pressure escalation. Fortunately, *in situ* baking of the vacuum chamber was possible, and an increase in the baking temperature aided in the suppression of this difficulty, abetted by the addition of more vacuum pumps. This facility remains a unique achievement; it is unlikely that this impressive storage of two high intensity DC beams will recur in the future.

The ISR developed luminosity by concentrating on the n^2 term in Eq. (1). To our knowledge, the initial try at low β^* was undertaken at the Cambridge Electron Accelerator (CEA). The CEA was a high energy electron counterpart to the Brookhaven AGS and the CERN PS. This 7 GeV Harvard–MIT collaborative product was the forefront electron synchrotron of 1961, and was closely followed by the 7 GeV electron synchrotron at DESY. Like its proton synchrotron cousins, the CEA had a highly symmetric focusing structure motivated by both resonance avoidance and cost considerations. At that time, there was considerable discussion of the need for symmetry preservation in focusing optics, but all that went away with the conversion of the CEA to an electron–positron collider. In a remarkable structure modification, a bypass was created into which the beams could be switched for collision. A β^* as low as 5 cm was achieved, in contrast to the 10 m level of amplitude functions typical of a synchrotron of that scale. We recommend that the reader look at a summary written at that time by Livingston [14].

Note the differing topology of these systems. Two rings were unavoidable if like-charge particles were to be collided, while electron–positron collisions were possible within a single synchrotron. The

potential disadvantages of the single ring case were yet to be recognized, for additional bunch–bunch collisions could occur elsewhere than at the desired interaction points. The adverse effect of these extra collisions is discussed below when we mention the beam–beam tune shift, as is the method of their avoidance.

The late 1960s, early 1970s descendants of VEP-1, AdA, and CBX were VEPP-2, Adone, and SPEAR, at cms energies of 1.4 GeV, 3.0 GeV, and 5.0 GeV respectively. All electron–positron systems, they were designed for particle physics research. At its luminosity of up to 2×10^{28} cm^{-2} s^{-1}, VEPP-2 carried out studies of ρ, ω, and ϕ mesons. Later, in the mid-70s, VEPP-2 became the injector for VEPP-2M, which reached a luminosity of 5×10^{30} cm^{-2} s^{-1}. Adone is especially memorable to accelerator physicists for the identification of the single-bunch instability known as the head–tail effect, and the consequent extension of required magnet types beyond dipoles and quadrupoles to include sextupoles. That circumstance invites some remarks on "chromaticity."

Chromatic aberration takes place equally well in focusing magnets as it does in ordinary lenses. The number of betatron oscillations per turn in a synchrotron, the *tune*, would change depending on the particle momentum if the focusing strength of the magnetic lenses were independent of momentum. The derivative of the tune with respect to momentum was called the chromaticity, and to restrain the limits of tune excursion versus momentum, a variation of the focusing gradient as a function of the transverse position was introduced in, for example, the 10 GeV synchrotron at Cornell. The usual focusing moment in the magnetic field is called the quadrupole term with its linear dependence on x, so this next step of differing symmetry with an x^2 variation represented the intentional introduction of a sextupole variation in the field. Higher order magnetic field dependencies entered as primary design aspects.

The HEP program at SPEAR shared honors with researchers at the Brookhaven AGS for the discovery of J/ψ. The recognition that colliders had become the essential vehicle for HEP research spread quickly, with the construction of DORIS at DESY, DCI at Orsay, and planning for CESR at Cornell. The word "factory" entered the lexicon to characterize a collider that not only produced new particles but produced them with sufficient frequency to permit study of their properties and decay modes.

2.5. *Luminosity II*

The luminosity gains described in the previous subsection were a result of increased beam intensity and use of low β^*. Inevitably, a true limitation reflecting the influence of particles in one beam on particles of the other was encountered. A particle bunch presents a nonlinear lens to particles in the other beam and so leads to perturbation of the linear motion. A parameter characterizing the strength of this effect is obtained by calculating the linear tune shift associated with periodic passage through the field gradient at the bunch center. When initially introduced, it was often called the "beam–beam tune shift," but now terms such as "space-charge parameter" or "beam-beam parameter" are used as they are more appropriate to the complexity of the process. For head-on collision of Gaussian bunches, the quantity is

$$\xi_{x,y} = \frac{n r_e}{2\pi\gamma(\sigma_x^* + \sigma_y^*)} \left(\frac{\beta^*}{\sigma^*}\right)_{x,y}, \qquad (4)$$

where n and r_c are the number of particles in a bunch and the classical radius of the particle species respectively of the "other" beam and this form assumes only a single collision point.

Space charge influences on the betatron oscillation tune were already familiar from single-beam synchrotrons, and had been found to be surprisingly forgiving. Tune shifts of 1/4 or more did not lead to beam loss, and this was attributed to the strong dependence of the tune on amplitude. There was optimism that the bunch–bunch circumstance would exhibit similar behavior, but such proved not to be the case. Electrons are quite tolerant of abuse because of the forgetfulness provided by synchrotron radiation; even so the limits on ξ are typically below 0.1. For protons, the region of concern is almost an order of magnitude less.

The topology of the early collider systems was a consequence of the charges of the species involved; if the charges were the same, then two rings were needed as in CBF and the ISR. The opposite charges of particle–antiparticle collisions made a single ring attractive for a variety of reasons, not the least of which was cost. However, as concern about beam–beam effects grew, so did the recognition that the particles had to be kept apart at other than the locations at which collisions were desired. Today, in single-ring colliders for both leptons and hadrons steering systems are installed near the interaction points to keep the beam bunches apart elsewhere.

As colliders became more relevant to HEP, a change in the quotation of luminosity became common. Certainly orders of magnitude like 10^{32} cm^{-2} s^{-1} looked impressive, but another way of stating the same luminosity is 1 inverse microbarn per second or, more concisely, $1\,\mu b^{-1} s^{-1}$. In a similar spirit, luminosity integrated over time suggests the potential yield during an experimental running period. An operational year of length 10^7 s would give an integrated luminosity for the example in the last sentence of 10 pb^{-1}. These other units of luminosity will be used frequently below.

2.6. *Concentration on colliders*

From the mid-1970s on, the physics that could be investigated with colliding beams became more and more of interest. This last phase of the Past requires comment on two aspects: first, the invention that enabled proton–antiproton collisions and its consequences; and second, LEP, perhaps the highest cms electron–positron circular collider to ever be built.

Positrons are rather easily produced; a foil in a few-MeV electron beam will deliver at least some within the typical $1/\gamma$ production angle. Antiprotons are a different matter. The 30-GeV-scale proton beams provided by the CERN proton synchrotron and the Brookhaven AGS certainly could yield significant antiproton flux, but the lesser production cross section and angular divergence characterized by the 0.3 GeV/c Cocconi angle [15] made accumulation a challenge. That challenge was met by the invention of stochastic cooling [17]. The SPS at CERN was converted into the the first proton–antiproton collider, the SP$\bar{\text{P}}$S, at which the intermediate bosons were detected.

Emphasis on colliders was underway elsewhere. The 10 GeV electron synchrotron at Cornell was adapting to its new role as the injector to CESR. Tristan was in progress at KEK. To our minds, a major signpost of the transition from Past to Present was the replacement of LEP by the LHC in the same tunnel.

It is a bit difficult to imagine a higher energy electron synchrotron than LEP. Construction of the

SPS had already crossed the Swiss–French border.[c] LEP required an even larger step — a tunnel under the Jura mountains. Tunnels leak, as we learned during construction of the 10 GeV synchrotron at Cornell and while wading around in the Main Ring enclosure at Fermilab. The inrush of water interfered with installation at LEP. The initial goal of the electron–positron collider was 50 GeV per beam, and after upgrades that figure was more than doubled to just over 100 GeV per beam. The circumference of the tunnel is 27 km, and with a bend radius of 22 km that meant some 500 MeV per turn of synchrotron radiation loss. LEP reached luminosity at the $10^{32}\,\mathrm{cm}^{-2}\mathrm{s}^{-1}$ level. In doing so, use was made of phase space manipulation to manage the beam–beam tune shift limit. Further comment on this subject will appear below. There are many interesting papers concerning LEP on the CERN website. This impressive facility operated only during the years 1989–2000. It incorporated many of the advances in accelerator physics of the preceding half-century.

Particulary noteworthy was the progress in use of superconducting rf. Tristan was constructed with niobium cavities yielding 400 MV. LEP started operation with a conventional 352 MHz system using copper structures that delivered 340 MV; then, for the several upgrades to the W pair level and beyond, superconducting structures were chosen, finally capable of delivering some 3600 MV.

During the same time period as LEP, a signpost to the Future appeared, embodied in the SLAC Linear Collider (SLC) [16]. This remarkable facility collided 50 GeV electron and positron beams from the SLAC linac, and in so doing successfully confronted many of the challenges of this new direction. Further discussion of this approach will be deferred until Sec. 5.

How do we characterize the Past as to technology? Particle colliders used existing technology coupled with the synchrotron invention to advance the field. This statement is an oversimplification, to be sure. New technologies, particularly those associated with superconductivity, were already at play. Increased emphasis on these technologies was to come.

3. Present

As this is written, the high energy end is still held by the aging Tevatron, and comments on this device follow below. Although the unique electron–proton collider, HERA at DESY, has recently been shut down, the overlap of technology makes inclusion here appropriate. The B factories at KEK and SLAC have achieved new highs in luminosity. The LHC is undergoing commissioning, and it is today's major Earth-based HEP endeavor. The Relativistic Heavy Ion Collider (RHIC), in operation at Brookhaven is not discussed here; rather, the reader is referred to the excellent and instructive website [18]. Polarized proton collisions are underway at RHIC, and the production and maintenance of proton polarization is an interesting subject in itself.

3.1. *The Tevatron*

Scarcely had the Main Ring at Fermilab come into operation for HEP in 1972 when Robert Wilson announced that a 1 TeV synchrotron based on 4 T superconducting magnets would be our[d] next goal. The principal challenge, magnet development, is described elsewhere in this publication. In the initial design steps, this new ring was to double the proton energy available for fixed target physics, but immediately the attraction of proton–antiproton collisions entered the picture. The Tevatron was commissioned at 800 GeV in mid-1983, and the first collider attempts began in 1985.

The Tevatron was designed as a fixed target accelerator, and so a reasonably rapid cycle time was important. A 60 s cycle time at 1 TeV was set as the goal, with the intent that the event rate in the neutrino program would be about the same as that familiar from the preceding 400 GeV source with its 10 s cycle. The dipole magnet cross section is shown in Fig. 2. The major compromise as a result of its dual role was the necessity for it to share the same limited space in the tunnel with its earlier brother. The suspension of the cold coil cryostat within a warm iron enclosure and the Tevatron magnets remain unique in this design choice.

---

[c]One of the border guards, recognizing the problem that the "language-impaired" faced in our twice-daily journey between Previssin and the main CERN site, instructed us to just say "*aucune.*"

[d]We joined Fermilab in 1969 in order to participate in the construction and commissioning of the initial program of the laboratory. Our intent was to move on upon completion of that mission. The attraction of 1 TeV was irresistible.

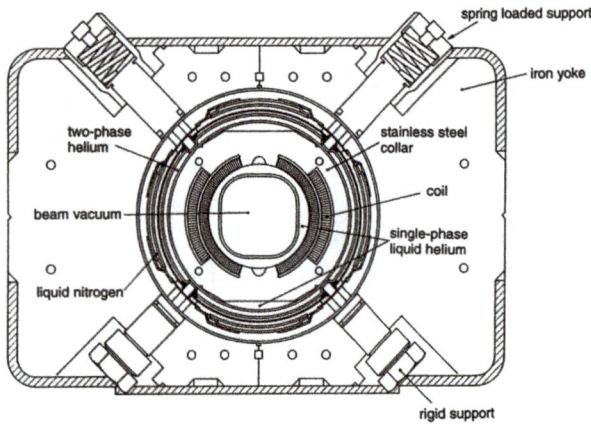

Fig. 2. Cross section of the Tevatron dipole magnet. This drawing is typical of many produced at Fermilab during the Tevatron design process. A similar though not identical version may be found in the article cited in the text [19].

In order that the Tevatron could transform into its role as a collider, two other synchrotrons were constructed at Fermilab to capture and collect antiprotons, and that was achieved during a very valuable collaboration with CERN. Recently,

another ring was added to carry out electron cooling of the antiprotons. Today, two decades later, the Tevatron remains the highest-cms-energy collider in the world, and with an initial luminosity in storage of up to 2.6×10^{32} cm^{-2} s^{-1} the annual integrated luminosity exceeds an inverse femtobarn, over two orders of magnitude beyond the original design specification.

The design and construction of the Tevatron ring has been described by one of us [19] and need not be discussed in any detail in this report. The accelerator occupancy of the Fermilab site is illustrated in Fig. 3. Recalling the concerns of the time, there are three topics that deserve special emphasis, and we limit our comments below to these.

3.1.1. *Quench sensitivity*

How much beam loss could a superconducting magnet sustain before a quench? A quench means conversion to the normal conducting state because of a temperature, rise, and without some corrective

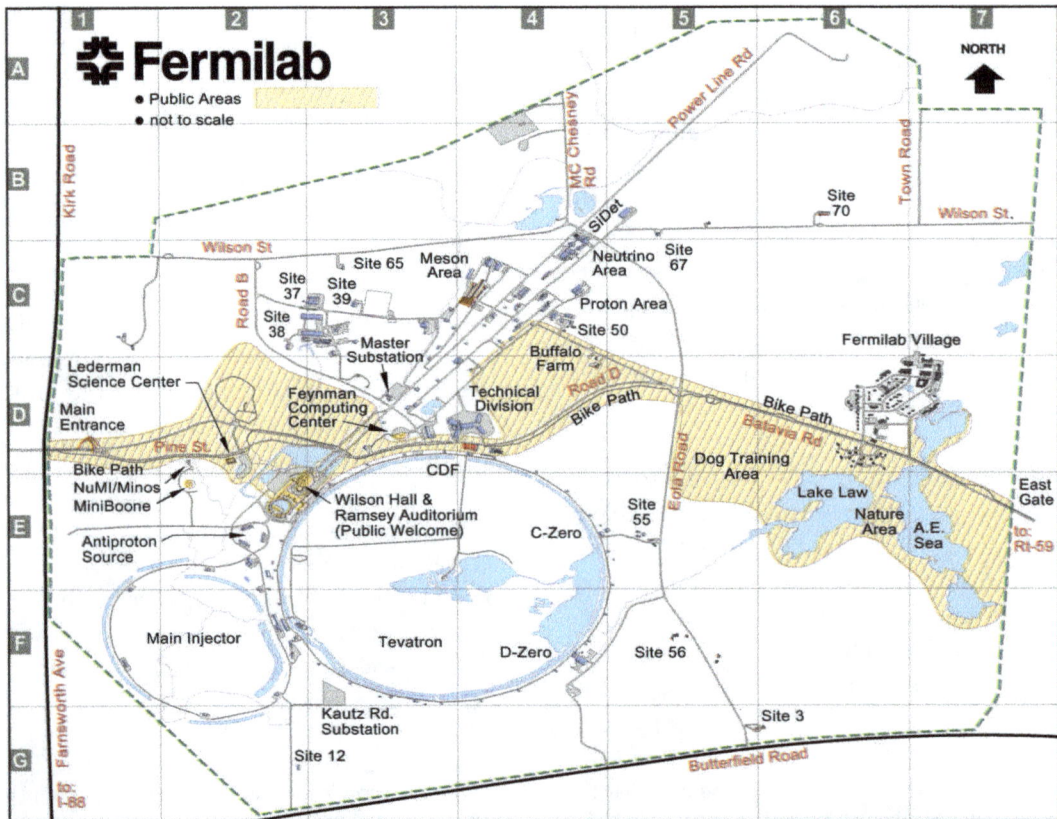

Fig. 3.   Fermilab site. From http://www.fnal.gov/pub/visiting/map/site.html

measure a local quench may propagate throughout the structure, resulting in damage due to resistive heating. The superconducting magnets in the bubble chambers of the 1970s used so-called cryostatic stabilization: sufficient normal conductor such as copper nearby to accept the deflected current without risk. The space constraints of the Tevatron magnets did not permit this approach. A series of experiments [19] established a limit of 4 mJ/g, some five orders of magnitude below the specific ionization loss if a substantial fraction of the 10^{13} particles in the anticipated beam were locally misdirected. This was initially a major worry for fixed target physics, because of the inevitable beam loss due to the extraction process. Another figure to keep in mind was the 6 J/cm^3 of stored energy in the 4 T magnetic field. These concerns led to the rather elaborate and generally successful quench protection system at the Tevatron. The stored beam energy at the LHC will be over two orders of magnitude higher, and this is a major design issue for that facility.

3.1.2. *Field quality and storage*

In the earliest synchrotrons, these subjects did not impact design. At that time, there was continuity from the 300 MeV scale accelerators of the 1950 period toward the next steps, and the estimates associated with the transition from weak to strong focussing had been successful. In retrospect, it was fortunate that the Tevatron was to enter operation as a source for fixed target physics — a very familiar role in which to initiate operation. The new problem was field quality of the superconducting magnets. We were less concerned at the time about subtle dynamical issues of long term storage, but rather the challenge of bringing this rather complicated collection of hardware into operation. An accelerator of this sort does not permit easy modification; it is a long, cold enclosure. We installed an unusually large host of correction and adjustment magnets that have remained useful to this day. An aggravating surprise was the time dependence of persistent currents. That persistent currents existed within superconductors was well known; in simplistic terms, if the wire is no longer required to convey a net current, currents still exist in both directions. The result is a time-dependent sextupole term in the field the details of which were not detected in the strict cycles of fixed

target physics but with the varied injection period of the collider operation turned into a puzzle. A spectrum analyzer with an input from a transverse beam position monitor revealed the onset of the head–tail instability when the chromaticity became negative — a circumstance that was easily corrected once recognized.

3.1.3. *Improvements*

The luminosity stated for the Tevatron project at its outset was $10^{30} \text{ cm}^{-2} \text{ s}^{-1}$, as a reasonable goal based on the SP$\overline{\text{P}}$S performance. Over the two decades of operation, the achieved luminosity has exceeded the original goal by over two orders of magnitude. Of course, a very substantial investment beyond the original complex was required, two additional synchrotrons being the most obvious objects on the landscape. One of these, named the Main Injector, replaced the original major Fermilab synchrotron and provides protons for the collider program and for neutrino physics. The other, called the Recycler, is an interesting device in that it is an 8 GeV synchrotron based on permanent magnets. As the name implies, the original intent was to receive aging antiprotons from the Tevatron, reduce their phase-space volume, and return them to the Tevatron ring. Today, its mission is to serve as a storehouse for antiprotons not yet delivered to the collider and reduction of their emittance by electron cooling [20]. Of equal importance were the improvements in beam management. Luminosity diminishes during storage for a variety of reasons, but abrupt terminations from beam loss are brought under control so that the majority of stores are terminated intentionally.

3.2. *HERA*

There was a basic compromise in the design of the Tevatron: a fortunate compromise in a certain sense; its initial use as a source for fixed target physics made it possible to avoid certain irritating questions concerning magnetic field quality as related to storage. But the support of the cold coil within the warm steel envelope remains to this day a subject of continual concern. The DESY laboratory in Hamburg, long a site for electron accelerators, had an electron–proton collider under design, named HERA. DESY personnel had participated very effectively in the Tevatron magnet design, construction, and commissioning, as

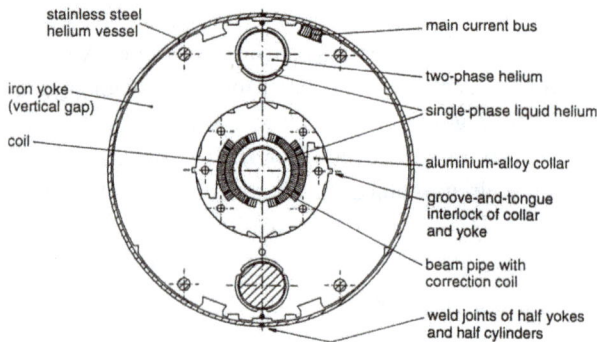

Fig. 4. Cold mass of the HERA proton ring dipole magnet.

Fig. 5. HERA in its surroundings. From http://m.desy.de/

part of a collaboration that has continued to this day. The HERA proton ring contained superconducting magnets which represented the next major step in this technology. The cycle time associated with fixed target physics was no longer a concern, and a "cold iron" cryogenic enclosure replaced that of the Tevatron. The result was a considerably more robust design. A cross section of the HERA dipole "cold mass" as it is placed within its cryostat is shown in Fig. 4.

The magnets of the proton ring at HERA were capable of operation above fields suitable for above 1 TeV, but in normal operation the proton energy was 920 GeV. The electron ring in the same tunnel operated at 30 GeV, providing a cms energy of 235 GeV. Luminosity at the 10^{32} cm^{-2} s^{-1} level was achieved, as was polarization of the colliding particles. In the context of facility construction, there are three additional aspects which we consider noteworthy.[e] The advance in superconducting magnet technology has been noted above. Another of these aspects is not technical — rather, a recognition of association of a laboratory with its surroundings. The HERA tunnel would extend beyond the boundaries of the DESY site, and even though underground, that expansion was naturally a matter of concern in the neighborhood. Members of the DESY Directorate literally went door to door to explain the nature of the facility and allay worries about radiation. Figure 5 illustrates the expansion of the laboratory underneath nearby Hamburg.

The third aspect relates to rf superconductivity. The experience in development of superconducting accelerating structures for the HERA project led almost naturally to the formation of the TESLA collaboration in 1992 [21]. The goal was to raise the 5 MeV per meter gradient provided by the technology at that time to 25 MeV per meter, close to what was regarded as the limit for niobium. At the time, one of the major motivations was to advance rf superconductivity to the point at which it could be considered as an alternative to the more familiar "room temperature" technology of copper for a potential linear collider. Further comment on the linear collider subject will be reserved until the section entitled "Future." HERA operation was terminated in mid-2007 and, with that, DESY's on-site role in HEP concluded. Now the progress achieved in superconducting rf technology is in use at the 1 GeV SASE FEL dubbed FLASH and in the European XFEL project just getting underway, both at DESY.

3.3. Synchrotron radiation

Before turning to factories based on electron–positron collisions, a few words on synchrotron radiation are in order. Suppose that an electron is traveling on a circular path. With respect to an inertial frame moving tangent to the circle at the same speed as the electron, the electron will be briefly at rest at the point of tangency. So, in the inertial frame, we can apply the Lorentz formula for the radiated power:

$$P' = \frac{1}{6\pi\epsilon_0}\frac{e^2 a'^2}{c^3},\tag{5}$$

---

[e]We joined DESY in 1991, and we have a rather murky recall of commissioning shifts, but aside from a different language, such shifts are pretty much the same throughout the world.

where the primes denote quantities measured in the speeding inertial frame. In the laboratory frame, $P = P'$ and $a = \gamma^2 a'$. For a relativistic electron, $a = c^2/\rho$, where ρ is the radius of curvature. The radiated power becomes

$$P = \frac{1}{6\pi\epsilon_0} \frac{e^2 c}{\rho^2} \gamma^4. \qquad (6)$$

In a synchrotron that has a constant radius of curvature, the energy lost from a single particle due to synchrotron radiation is the above expression multiplied by the time spent in bending magnets, $2\pi\rho/c$. In familiar units, we then have

$$W = 8.85 \times 10^{-5} E^4/\rho \; \text{MeV per turn}, \qquad (7)$$

where the energy E is in GeV, the radius ρ is in km, and the expression is evaluated for electrons. The correspondence of the 8.85 here with the 8.85 in ϵ_0 is interesting but accidental; they differ in the next digit. The radiation has a broad energy spectrum which falls off rapidly above the *critical energy*, $E_c = (3c/2\rho)\hbar\gamma^3$. Typically, E_c is in the hard x-ray region.

For the accelerators considered here, this radiation is incoherent. That is, a beam containing n particles experiences an energy loss per turn which is n times that given by Eq. (7), rather than the n^2 dependence characteristic of antenna radiation. The particles are sufficiently far apart in phase space that constructive interference does not take place at the short wavelengths associated with the critical energy. Longer wavelength radiation at which coherence would be anticipated is suppressed by the conducting walls of the beam enclosure. Recently, coherent synchrotron radiation has become a subject of interest in very high phase space density systems such as self-amplified free electron lasers. [22]

The characteristic time for synchrotron radiation processes is the time during which the energy must be replenished by the acceleration system. If f_0 is the orbit frequency, then the characteristic time is given by

$$\tau_0 = \frac{E}{f_0 W}. \qquad (8)$$

Oscillations in each of the three degrees of freedom either damp or antidamp, depending on the design of the accelerator. For a simple separated-function alternating gradient synchrotron, all three modes damp. The damping time constants are related by

Robinson's theorem [5], which, expressed in terms of τ_0, is

$$\frac{1}{\tau_x} + \frac{1}{\tau_y} + \frac{1}{\tau_s} = 2\frac{1}{\tau_0}. \qquad (9)$$

Even though all three modes may damp, the emittances do not tend toward zero. Statistical fluctuations in the radiation rate excite synchrotron oscillations and radial betatron oscillations. Thus, there is an equilibrium emittance at which the damping and excitation are in balance. The vertical emittance is nonzero due to horizontal–vertical coupling.

The time constants in terms of τ_0 are

$$\tau_y = 2\tau_0, \quad \tau_x = \frac{2}{1-\mathcal{F}}\tau_0, \quad \tau_s = \frac{2}{2+\mathcal{F}}\tau_0, \qquad (10)$$

where

$$\mathcal{F} \equiv \frac{\left\langle \frac{D}{\rho^2}\left(\frac{1}{\rho} + 2\frac{B'}{B}\right)\right\rangle}{\left\langle \frac{1}{\rho^2}\right\rangle}. \qquad (11)$$

In this last equation, D is the momentum dispersion function which relates the transverse displacement of a closed orbit to the fractional momentum offset according to $x = D\delta p/p$. The early generations of electron synchrotrons combined bending and focusing in their magnets, so the gradient term B'/B in regions of finite radius of curvature, ρ, resulted in $\mathcal{F} > 1$; thus, radial betatron oscillations were unstable. This was a matter of relatively little concern in these rapid-cycling accelerators designed for external beam physics which spent little time in the neighborhood of low τ_x. However, for a storage ring, it is necessary to separate the roles of focusing and bending. Then $\mathcal{F} \approx D/\rho$. Since one of the virtues of alternating gradient focusing is to decouple local orbit excursions from the scale of the accelerator, for typical designs $\mathcal{F} \ll 1$, and all three modes of oscillation damp.

Polarization can develop from an initially unpolarized beam as a result of synchrotron radiation. A small fraction, $\approx E_c/E$, of the radiated power flips the electron spin. Because the lower energy state is that in which the particle magnetic moment points in the same direction as the magnetic bend field, the transition rate toward this alignment is larger than the rate toward the reverse orientation. An equilibrium polarization of 92% is predicted, and despite a variety of depolarizing processes, polarization above 80% has been observed at a number of facilities.

The radiation rate for protons is of course down by a factor of the fourth power of the mass ratio, and is given by

$$W = 7.8 \times 10^{-3} E^4 / \rho \;\; \text{keV per turn}, \qquad (12)$$

where E is now in TeV and ρ in km. The impact on the LHC design will be discussed below. As an aside, proton synchrotron effects were observed at the Tevatron during the early attempts at storage operation in 1985. A temporary loss of rf power at 900 GeV placed about half of the protons outside of the phase-stable region, with the result that two beams eventually appeared on a transverse profile monitor — one where it should be and the other slowly drifting inward as a result of the 7 eV per turn radiation loss. After an hour or so, the inevitable quench occurred.

3.4. *Factories*

As remarked earlier, a "factory" in the context of this article is an electron–positron collider facility that yields particles of interest in a quantity sufficient for the study of subtle aspects of their characteristics and interactions. The word also conveys a sense of optimization of a limited product line. Such is indeed the case. There are five factories: KEKB and PEP-II, designed around B meson decays; DAΦNE, with its concern clear from the name; and the venerable CESR, after its B meson success in identifying the $\Upsilon(4S)$, has moved in its "c" version to the τ charm regime, where it will soon be joined by BEBC II.

We limit our remarks on the B factories because of their achievement of the highest luminosities, the use of two rings of different energy, and the pursuit of interaction region improvements such as crab cavities.

The advantage of two rings of different energy comes from the competition between the circumstance that just above the threshold for a process the production cross section will reach a maximum, the "resonance", but the production products may decay so quickly that their identities cannot be studied adequately. An energy asymmetry in the collision in the laboratory frame enhances the distinguishability of the products by giving them different velocities within the detection apparatus [23]. However, the energy asymmetry is limited by the detector capability within a rather narrow range, because the same decision that gave the process distinction in the

laboratory frame will, with larger energy asymmetry, put the products outside the detector range.

Both PEP-II and KEKB were designed to make use of the production of B mesons at the $\Upsilon(4S)$ resonance which had been identified at CESR just above the $B\bar{B}$ threshold, with the intent to examine CP violation as the first priority. And both SLAC and KEK already had earlier colliders that could contribute many components to the higher energy members of the two synchrotrons, PEP at SLAC and Tristan at KEK. The energy ratio was substantial: 8 GeV × 3.5 GeV at KEK and 9 GeV × 3.1 GeV at SLAC. SLAC elected to place the lower energy ring above PEP, and KEK decided on a side-by-side arrangement. The luminosity target was an integrated luminosity in the 100 fb^{-1} range.

The initial design reports for these two B factories make interesting reading because of their clear articulation of the challenge of high luminosity in the face of the opposing processes that had been identified over the years [24]. Their achievement in surpassing the 10^{34} cm^{-2} s^{-1} level in instantaneous luminosity is clearly admirable. At KEKB, for example, the integrated luminosity is 820 fb^{-1} at this is written in Spring 2008. These designs saturated the beam–beam tune shift limit. The first of the PEP-II reports cited develops an expression for the luminosity under the assumption that the bunches have the same transverse dimensions and experience the same tune shift during collision:

$$\mathcal{L} = \frac{\xi(1+r)}{2er_e}\left(\frac{I \cdot E}{\beta_y^*}\right)_{+,-}, \qquad (13)$$

where ξ is the beam–beam parameter, designed to be the same for both beams, r is the aspect ratio equal to 0 for flat beams and 1 for round, E is the energy in GeV, I is the average current in amperes, and the subscript on the second parenthetical expression means that the quantities contained can be the parameters associated with either ring. Below the beam–beam limit, luminosity varies as the square of the beam current, and drops to a linear dependence in the ξ-limited regime.

In a single-ring particle–antiparticle collider like the Tevatron, the beam bunches collide head-on. That is not necessarily the case in the two-ring facilities like those discussed here; there is a natural crossing angle and an associated negative impact on luminosity. Therefore a device was proposed to restore the benefits of head-on collision, and this is the role of the so-called crab cavity. The intent of this rf structure is to kick the front and back of bunches destined for collision so that when they interact they are parallel in relative orientation and the lowering of luminosity is avoided. This device, the deflecting mode rf cavity, entered HEP in the context of rf separated kaon beams many years ago. Now, such devices (there are two rings) have been installed at KEKB, with indication of luminosity enhancement.[f]

4. Beam Dynamics and Single-Particle Stability

Before turning to the LHC, some remarks on beam dynamics are in order. A major concern of beam dynamics is stability: conservation of adequate beam properties over a sufficiently long time scale. Several time scales are involved, and the approximations used in writing the equations of motion reflect the time scale under consideration. For example, when we write the equations for transverse motion, no terms associated with phase stability or synchrotron radiation appear; the time scale associated with the last two processes is much longer than that demanded by the need for transverse stability.

The equations of motion of a particle undergoing transverse oscillations with respect to the design trajectory are those of a harmonic oscillator [11]:

$$x'' + K_x(s)\, x = 0, \quad y'' + K_y(s)\, y = 0, \tag{14}$$

where the independent variable s is the path length along the design trajectory.

The functions K_x and K_y reflect the transverse focusing — primarily due to quadrupole fields except for the radius of curvature, ρ, term in K_x for a synchrotron — so each equation of motion resembles that for a harmonic oscillator but with spring constants that are a function of position.

These equations have the form of Hill's equations and so the solution in one plane may be written as

$$x(s) = A\sqrt{\beta(s)}\, \cos[\psi(s) + \delta], \tag{15}$$

where A and δ are constants of integration. In order that Eq. (15) can be the general solution independent of δ, the phase and β satisfy

$$\frac{d\psi}{ds} = \frac{1}{\beta}, \quad 2\beta\beta'' - \beta'^2 + 4K\beta^2 = 4. \tag{16}$$

The first relation confirms the earlier statement that β plays the role of a local $\lambda/2\pi$. The dimension of A is the square root of length, reflecting the fact that the oscillation amplitude is modulated by the square root of the amplitude function.

The wavelength of a betatron oscillation may be some tens of meters, and so typically values on the amplitude function are on the order of meters rather than on the order of the beam size. The beam optics arrangement generally has some periodicity and the amplitude function is chosen to reflect that periodicity. As noted above, a small value of the amplitude function is desired at the interaction point, and so the focusing optics is tailored in its neighborhood to provide a suitable β^*. The second order equation for β simplifies considerably and is differentiated once more to yield

$$\beta''' + 4K\beta' + 2K'\beta = 0. \tag{17}$$

In a drift space, the solution is a parabola resulting in the familiar variation of β in the neighborhood of the interaction point.

The number of betatron oscillations per turn in a synchrotron is called the *tune* and is defined by

$$\nu = \frac{1}{2\pi} \oint \frac{ds}{\beta}. \tag{18}$$

Expressing the integration constant A in the solution above in terms of x, x' yields the *Courant–Snyder invariant*

$$A^2 = \gamma(s)\, x(s)^2 + 2\alpha(s)\, x(s)\, x'(s) + \beta(s)\, x'(s)^2, \tag{19}$$

where

$$\alpha \equiv -\frac{\beta'}{2}, \quad \gamma \equiv \frac{1 + \alpha^2}{\beta}. \tag{20}$$

Because β is a function of position in the focusing structure, this ellipse changes orientation and aspect ratio from location to location but the area πA^2 remains the same.

As noted above, the transverse emittance is a measure of the area in x, x' (or y, y') phase space occupied by an ensemble of particles. The definition used in Eq. (2) is the area that encloses 15% of a Gaussian beam.

---

[f]In the language of cylindrical rf structures, the accelerating mode would be identified as TM010 and the deflecting mode as TM110.

For present-day hadron synchrotrons, synchrotron radiation does not play a role in determining the transverse emittance. Rather, the emittance during storage reflects the source properties and the treatment of the particles throughout acceleration and storage. Nevertheless it is useful to argue as follows: Though x' and x can serve as canonically conjugate variables at constant energy, this definition of the emittance would not be an adiabatic invariant when the energy changes during the acceleration cycle. However, $\gamma(v/c)x'$, where γ is the Lorentz factor, is proportional to the transverse momentum and so qualifies as a variable conjugate to x. Therefore one often sees a normalized emittance defined according to

$$\epsilon_N = \gamma \frac{v}{c} \epsilon, \tag{21}$$

which is an approximate adiabatic invariant, for example during acceleration.

The motion with respect to the design orbit is a collection of straight line segments connected by angles at the quadrupoles, so one might wonder what is gained by introducing all this formalism. One advantage is that we can use familiar techniques associated with harmonic oscillators. Transform the variables in Eq. (15) according to $\zeta = x/\sqrt{\beta}, \phi = \psi(s)/\nu$; then ϕ is an independent variable that increases by 2π in each turn, and the solution looks like an ordinary harmonic oscillator:

$$\zeta(\phi) = A\cos(\nu\phi + \delta). \tag{22}$$

This maneuver was not invented for beam physics, by the way. It is called a Floquet transformation. Next, suppose that there is a difference, perhaps an error, in the magnetic fields used in the equations of motion (14). In the new coordinates, we have

$$\frac{d^2\zeta}{d\phi^2} + \nu^2\zeta = -\nu^2\beta^{3/2}\frac{\Delta B(\zeta,\phi)}{B\rho}, \tag{23}$$

where ΔB is the magnetic field perturbation, usually represented by a multipole expansion:

$$\Delta B_y + i\Delta B_x = B_0 \sum_{n=1}^{\infty}(b_n + ia_n)(x + iy)^{n-1}. \tag{24}$$

Here B_0 is some reference field strength; in a bending magnet, it would be the nominal bend field. The coefficients b_n would represent dipole, quadrupole, sextupole and so on, field contributions for the various values of n, while a_n are the skew components — vertically deflecting dipole, skew quadrupole terms,

etc. The CERN convention for multipole numbering is used.

With insertion of Eq. (24) with a_n set to zero into Eq. (23) to look at one-degree-of-freedom motion, the result is a driven harmonic oscillator with the consequence that resonance with some multipole may result if the tune is any rational number. This does not necessarily mean that oscillations will grow without bound for such tunes — for, after all, the rational numbers are dense — but one ought to be careful. Generalization to two degrees of freedom produces the warning that any sum resonance of the form $M\nu_x + N\nu_y = k$, where M, N, k are positive integers, may be unstable. The lowest order perturbative solution to Eq. (23) would identify $M + N$ as the multipole responsible, but second and higher order solutions break this relationship.

Confidence in this perturbative approach rests on the right hand side of Eq. (23) being "small" and having frequency content close to that of the free oscillation. The constructs that follow from validity of these approximations — clear division of phase space into stable and unstable regions by surfaces known as separatrices, for instance — have been valuable design aids. For sufficiently large values of the transverse coordinates these approximations must become invalid — for, after all, we are dealing with nonlinearities. The notion of a separatrix becomes invalid, and, disagreeably, chaotic motion enters the picture. This circumstance is not a surprise; the difference equations for inclusion of a sextupole magnet in a ring is a variant of the Hénon map found in the literature related to chaotic behavior.

4.1. *The Large Hadron Collider*

The LHC dominates the horizon. This proton–proton collider undergoing commissioning at CERN promises a cms energy of 14 TeV and a luminosity almost two orders of magnitude above that reached at the Tevatron. Like the Tevatron, there was an accelerator enclosure constructed for an earlier project at CERN — the very successful LEP electron–positron storage ring. This 27 km tunnel is just over a factor of 4 greater in circumference than that occupied by the Tevatron, so just with replication of the superconducting magnet technology developed at Fermilab and DESY a significantly higher energy would be possible. A sketch of the CERN site is shown in Fig. 6.

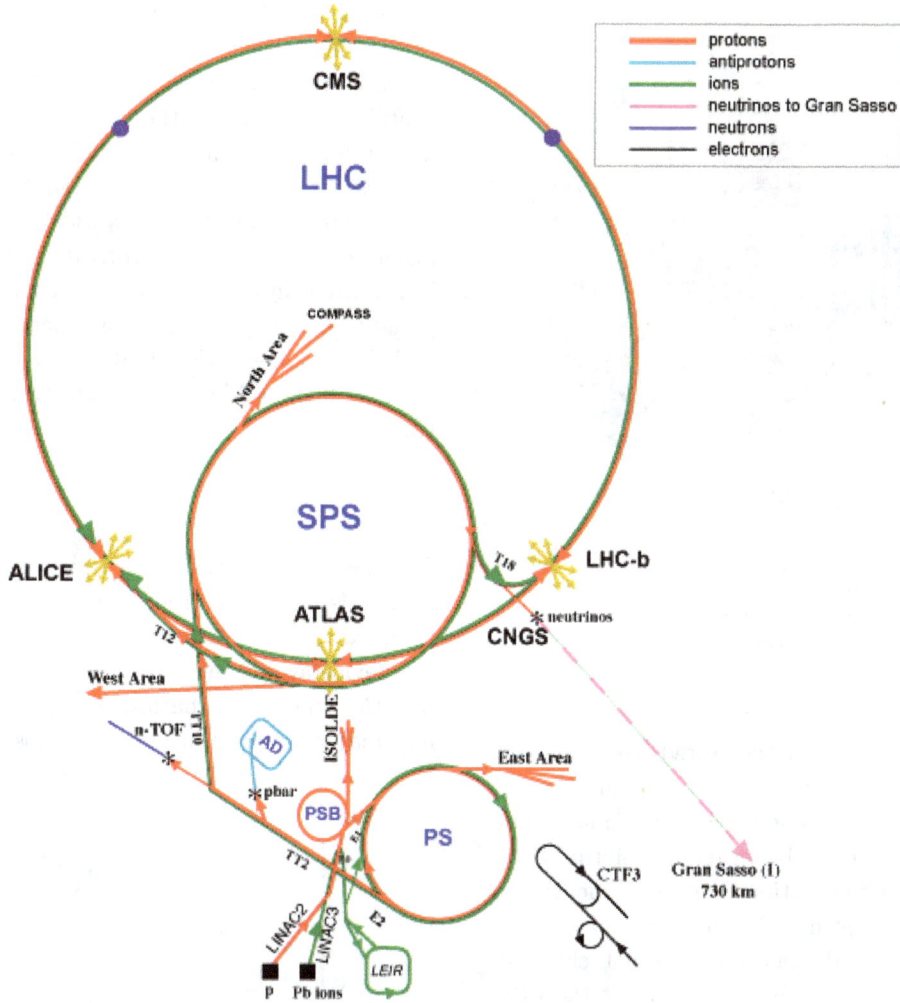

Fig. 6. The CERN accelerator complex. See http://public.web.cern.ch/Public/en/Research/AccelComplex-en.html

But, in the earliest design discussions, a bend field about twice as large — in the 8 T range — was adopted, implying the seven-fold or so increase in collision energy. The space constraints of the LEP tunnel urged the LHC designers to confront the challenge of putting the two beam paths in a single steel enclosure, representing another major step forward in superconducting magnet design. Fortunately, over the years the techniques for analysis of beam dynamics had advanced far beyond the extrapolations that were often necessary in the past.

The three-volume *LHC Design Report* may be downloaded chapter by chapter from the CERN website. [25] It is the most complete and thorough document of this sort that we have seen in the accelerator field. Chapter 2 of Vol. 1 is useful in that

it contains definitions of quantities used throughout the report. Chapter 4 discusses the optics and single-particle dynamics concerns, and Chap. 7 treats the superconducting magnet design.

The challenge of placement of the two beam paths within a single magnetic material enclosure with adequate field quality was successfully met. A cross section of the cold mass of the dipole magnet is shown in Fig. 7, including some field lines. A collection of holes in the steel were inserted to adjust the magnetic field.

We cannot address here the complexity of this facility, but it is worth noting that the massive refrigeration plant is the largest on the Earth. Rather, we limit our further remarks to three subjects of past interest to us concerning beam management.

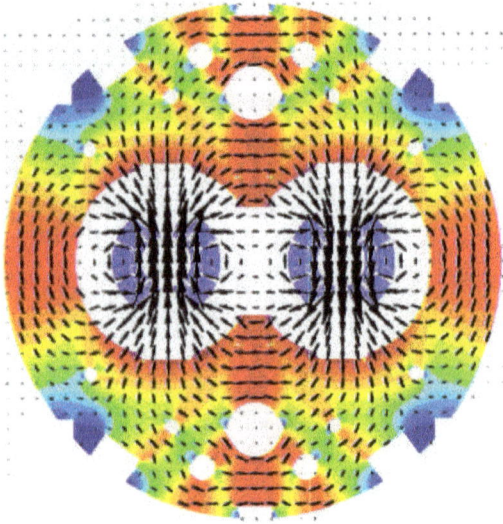

Fig. 7. Cold mass of the two aperture bending magnet for the LHC, including arrows denoting magnetic field direction. Note the use of holes in the steel to adjust magnetic field quality. From Chap. 7 of the *LHC Design Report*.

The kinetic energy of the stored beams will be 362 MJ per beam, roughly the energy equivalent of three gallons of gasoline and two orders of magnitude above the highest total beam energy of the Tevatron in fixed target operation. Earlier we mentioned the 4 mJ per gram quench threshold associated with NbTi technology, and that limit has not changed. Elaborate measures have been installed at the LHC to confront this challenge. It is quite likely that the initial set of absorbers will require modification, and the need for such modification is consistent with the commissioning plan.

In the proton synchrotron world, synchrotron radiation could be ignored until the energy scale of the LHC. For protons, the radiation per turn is given by Eq. (12). For the LHC, synchrotron radiation presents a significant load to the cryogenic system, and impacts magnet design due to gas desorption and secondary electron emission from the wall of a cold beam tube. The 3.7 kW per beam may seem modest, but the photons in this radiation would desorb gases from the wall. This circumstance made the vacuum chamber design more complicated. A 20 K liner was designed to intercept the radiation before it could reach the 2 K tube. The combination of this phenomenon — who could have imagined that synchrotron radiation was a concern in a proton synchrotron? — and the stored energy of the beam are,

for us, among the fascinating aspects of bringing the LHC on line.

Chapter 4 of Vol. 1 of the *LHC Design Report* contains the story of the requirements on magnet design for single-particle stability, and many references to the intense effort mounted not only at CERN but throughout the world to assure that the lessons of the past were absorbed. Computer simulation played an essential role in the variety of ways to address the problem of magnetic nonlinearities and their compensation. Only as one of many examples, we include Fig. 8 as an illustration of particle tracking.

5. Future

The present emphasis for the next major HEP facility is on an electron–positron linear collider to complement the LHC with its hadron basis. A linear device avoids the synchrotron radiation barrier, but the single-pass nature of two colliding linacs has major implications for luminosity. There are

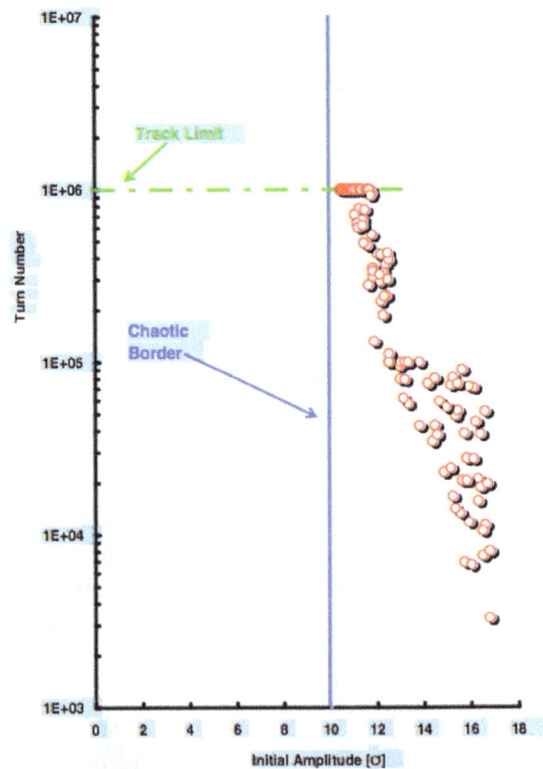

Fig. 8. Boundaries of particle survival during tracking simulation. The reader is invited to read Chap. 4 of the *LHC Design Report* for the story of this exploration of the boundary of stability.

two approaches under examination: the International Linear Collider (ILC), based on superconducting technology; and a development underway at CERN, designated the Compact Linear Collider (CLIC), using room temperature accelerating systems. Both designs rely on the progress achieved at the SLAC Linear Collider (SLC) as a proof of principle. [16] The ILC has been receiving considerable attention, in the hope that this facility will receive near-term approval for construction.

Though the advance in lepton cms energy associated with the linear collider propositions gains much attention, impressive strides in luminosity could be made in a comparatively short time and at less expense in the factory direction. Two possibilities — one a successor to KEKB [26] and the other a descendant of PEP-II [27] — offer luminosity at the 10^{36} cm^{-2}s^{-1} level. That is about 4 fb^{-1} per hour!

For quite some years, the use of muons has been advanced as an alternative to electrons for lepton colliders. This suggestion remains in active R&D, and the advances in beam dynamics associated with it require comment. Then, at the end, has rf acceleration run its course and are we now able to consider laser–plasma combinations?

5.1. *The International Linear Collider*

There had been a general preference in the HEP community for some years that the natural complement to the LHC would be an electron–positron linear collider at a cms energy of 0.5–1 TeV. Two technologies — "warm"' and "cold" — competed for acceptance, each with vocal and persuasive adherents. The copper-based rf structures so successfully used for decades, particularly in the highly successful SLAC program, were a natural contender. The "cold" approach, using superconducting cavities, had developed rapidly in recent years and had emerged as the alternative. In August 2004, an international panel opted for the "cold" approach. This was certainly not an easy decision, and the arguments one way or another make interesting and valuable reading. [28]

With the technology choice in place, significant progress has been made in pulling together past design efforts into a form suitable for worldwide collaboration. The Global Design Effort (GDE) has a website [29] on which developments are posted and

the *Reference Design Report* was presented in October 2007. The luminosity choice at 2×10^{34} cm^{-2} s^{-1} is similar to that of the LHC. The energy is lower; the plan has two 250 GeV linacs colliding initially, with attention given to expansion to twice that energy. It is interesting to scale from the parameters of the LHC to the linear collider circumstance. Let us rewrite Eq. (3) in terms of the normalized emittances,

$$\mathcal{L} = \gamma f \frac{n_1 n_2}{4\pi \sqrt{\epsilon_{x,n} \beta_x^* \, \epsilon_{y,n} \beta_y^*}}, \qquad (25)$$

and below we will assume that the bunch populations are the same: $n_1 = n_2 = n$. Unlike a synchrotron in which the bunches return to the interaction point after every circuit of the ring, in the linac the energy used to accelerate the particles is lost with each passage and must be reinvested anew. The 362 MJ in the LHC orbit becomes 12 MJ at ILC energy, and that gets multiplied by the 10^4 Hz orbit frequency to give a power requirement of 120×10^3 MW, or the output of 120 large power plants. That is obviously out of the question, so the product fn must be reduced by three orders of magnitude.

Beam–beam effects argue for a reduction in the bunch charge by about an order of magnitude. So try a reduction in the collision frequency by a factor of 100, to bring the power in bounds. In the luminosity, this last 100 is largely compensated for by the larger γ of the leptons. But from n^2 we are still down by 100, and this has to be made up for from the emittances and amplitude functions. From Table 1.1 of the 16 March 2006 version of the Basline Configuration Document, $\varepsilon_{y,n} = 0.04$ mm-mrad, $\beta_x = 21$ mm, and $\beta_y = 0.4$ mm for one of the "nominal" configurations. From elsewhere in the document, we infer that $\epsilon_{x,n} = 4$ mm-mrad. These are parameters of the ILC that strike us as adventurous, but fortunately major progress down this road had been made at the SLC, the first of its species. From the collider parameter tables in the *1998 Review of Particle Physics*, we see for the SLC $\varepsilon_{y,n} = 5$ mm-mrad, $\beta_x = 2.5$ mm, and $\beta_y = 1.5$ mm.

The two beam–beam effects that should be mentioned are (the unfortunately bilingual) beamstrahlung and disruption. The first of these refers to the electromagnetic radiation from particles of one bunch due to the fields in the other. The procedure that was followed to obtain Eq. (7) may be used here, at least as a beginning. That is, look in the rest frame of a particle that does the radiating as a bunch charges by in the opposite direction. This approach gets some of the dependencies right, but the averaging over distributions is a bit complicated. The result for the average fractional energy loss is

$$\delta_E \approx 0.86 \frac{r_e^3 n^2 \gamma}{\sigma_z (\sigma_x^* + \sigma_y^*)^2}, \qquad (26)$$

where $r_e = e^2/4\pi\epsilon_0 mc^2$ is the classical radius of the electron. An ambiguity in the collision energy is not what one wants — the lack of such uncertainty is what is hoped for in an e^+e^- collider — so δ_E is limited to a few percent in the design. This measure also avoids bunch distortion during collision. By making $\sigma_x \gg \sigma_y$, one controls δ_E and adjusts the luminosity by σ_y.

Disruption refers to the pinch effect that occurs when two oppositely directed bunches of the opposite sign of charge overlap and the magnetic field acts to compress the bunch density and hence to enhance the luminosity. The disruption parameter, D, is the ratio of σ_z to the focal length of the compression lens, and for the ILC design $D \approx 25$, which brings to mind an interesting picture of particle passage through the overlap region. The associated luminosity enhancement factor, H_D, contributes about a factor of 2.

Finally, there is the so-called hourglass effect, characterizing the circumstance that the beam size grows rapidly with longitudinal distance from the interaction point. The amplitude function in the neighborhood of the IP is a parabola, $\beta(z) = \beta^* + z^2/\beta^*$, and thus there is very limited gain in reducing β_y^* below σ_z. Combination of the above statements about beam power, beam–beam concerns, and the hourglass effect yields a luminosity expression for a linear collider based on that found in the *TESLA Design Report* [30]:

$$\mathcal{L} = \left[\frac{1}{4\pi} \frac{1}{(0.86 r_e^3)^{1/2}} \right] \times \frac{P_b}{E_{cm}} \left(\frac{\delta_E}{\epsilon_{y,n}} \right)^{1/2} \times H_D, \qquad (27)$$

where P_b is the total power of both beams.

The accelerating structures in the current ILC model are the result of the very successful TESLA Collaboration technology development, to which we have referred earlier. A measure of the progress may be found with reference again to the *Reference Design Report*, where the current design gradient is given as 31.5 MV/m based on demonstrated

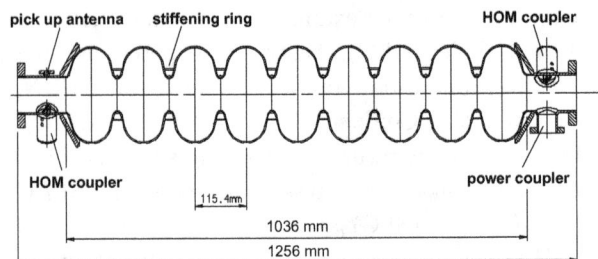

Fig. 9. The 1.3 GHz TESLA cavity. From Part II of the *TESLA Design Report* [30].

performance above this level. Reproducibility of the achieved gradient remains a challenge, and is a major R&D issue.

The TESLA structure is shown in Fig. 9. The shape differs from that of the disk-loaded waveguide in order to increase the ratio of the accelerating field on the axis to the magnetic field on the material surface. Niobium is a Type I superconductor, and hence will revert to the normal state if the magnetic field at the surface exceeds the limit, for Nb, of about 0.2 T. The fabrication techniques are familiar from sheet-metal technology, for Nb is readily worked by standard machining methods. A particular aspect requiring special attention is preparation of the rf surface to eliminate particles or protrusions as sites for field emission. Improvements are still being made as this is written. The ready availability of the "clean-room"' environment developed by the pharmaceutical and microchip industries has been of great value. We commented earlier about the convenience of finding conductors and steel almost in the nearby hardware store; now clean-rooms are catalog items, ultrapure water systems can be rented from your local water source, and a rudimentary 100-atmosphere rinsing pump is available at any supplier of car wash equipment. The development of surface preparation techniques in the pursuit of a higher gradient continues, as does the investigation of other resonator shapes.

5.2. *The Compact Linear Collider*

The CLIC is a two-beam design based on conventional rf systems. In a certain sense, just about all accelerators are two-beam systems if one includes the electron beam in the klystrons. But the CLIC is the real thing; low energy, high current beams drive the high energy, relatively low current linacs. There is no need to reproduce here the fine summary (p. 59) of *HAPE* by Sessler, Westenskow, and Wilson. In a certain sense, the CLIC has combined the instincts of its designers with those of SLAC personnel in their approach to the Next Linear Collider (NLC). The advantage of room temperature rf is clear: a higher accelerating gradient with no Meissner effect barrier associated with superconductivity. CLIC parameters use a gradient of 100 MV/m, almost a-factor-of-3 larger than the ILC figure. On the downside is the limited aperture and relative inflexibility of the repetition rate.

Initially, the CLIC design was based on the K band, 30 GHz, for the high energy linac. Recently the frequency was lowered to 12 GHz, as in the NLC design. Certain beam dynamics considerations are more forgiving at lower frequency, while preservation of the higher frequencies for room temperature structures retains the higher gradient advantage.

5.3. *R&D in particle accelerators*

Three directions come immediately to mind: improvement of existing accelerators; new accelerators such as muon synchrotrons based on present principles; and, to borrow an expression from the Monty Python television show, "now for something completely different." We comment on a few examples; but, to get a broad perspective, take a look at the annual report prepared by the Department of Energy Office of High Energy Physics, which contains summaries of the many activities supported by that agency [32].

5.3.1. *Improvements*

Back when the Tevatron program was in its infancy, about 1972, there were urgings to consider Nb_3Sn for the magnet coils. A premature suggestion then, but now that the technology of this material has advanced considerably, some limited application in the LHC has been made, and an improvement program, the Large Hadron Collider Accelerator Research Program (LARP) [33] is developing high gradient quadrupoles based on Nb_3Sn for a luminosity upgrade. The 18 K transition temperature with its implications for higher magnetic fields makes this material attractive, but it would be nice to find a substance with a transition temperature well

above the boiling point of liquid hydrogen; a possible candidate is MgB_2, which shows potential for technological development as a conductor. To date, broad application in the accelerator field of the the really high T_c, above liquid nitrogen, materials has not been possible other than current input leads for magnets.

The small emittances needed for the ILC are not achieved easily, and the associated damping rings have significant cost and operational impact. A low-emittance source of polarized electrons is an interesting direction of study, and here the photoinjector rf electron gun is a candidate. Whether or not the combination of normalized emittance at the level of the ILC requirements and polarization in the confined vacuum environment of the rf electron gun can be achieved is yet to be determined. That a high transverse emittance ratio can be provided by an rf electron gun has already been demonstrated [31]. Polarization is a different matter; the GaAs cathodes needed for polarization require an excellent vacuum environment at the 10^{-12} Torr level, which is not readily obtained in an rf gun. Incidentally, just as another example of technology obtained from elsewhere, GaAs is the cathode material used in night vision goggles.

5.3.2. *Muon rings*

The synchrotron radiation formula (7) has often been called the "law of inclemency" regarding the design of electron synchrotrons. Muons present themselves as alternatives to electrons; they are still leptons without the burden of substructure, the mass is two orders of magnitude above that of the electron with great benefit to the γ^4 term, and perhaps their limited lifetime in their rest frame of $1.6\,\mu s$ will not prove to be an impediment to their usage provided that high-enough magnetic guide fields are available. Actually, in calculating the number of turns that a muon will survive in a ring, the γ's cancel out with the result $n \approx 300B$, where B is the average value of the guide field in teslas. Of course, getting the muons to maximum energy (and field) influences the preparation and acceleration stages.

There is an extensive literature on muon colliders and storage rings [34]. The subject has attracted bright, inventive people, and their thoughts have been beneficial to other aspects of accelerator physics. There is significant international R&D underway on muon accelerator issues, concentrating, as is to be expected, on emittance reduction following the large phase space occupancy at the muon production source. Given the muon lifetime, methods such as electron and stochastic cooling are not candidates and so ionization cooling has been given prominence [35]. The principle closely resembles that of synchrotron radiation cooling, with absorbers replacing the photon emission process. An absorber reduces momentum in the direction of motion of the muon, while rf cavities provide replacement momentum in the desired direction. The result is net transverse damping. Longitudinal damping can be provided by a wedge-shaped absorber in a region where trajectories of differing momenta are displaced, i.e. in a region of a nonzero dispersion function. Again, there is similarity to the corresponding process in synchrotron radiation.

The finite muon lifetime implies a considerable compression of the energy-loss-and-reacceleration process compared with that of an electron storage ring. With absorbers and accelerating cavities of sufficiently short periodicity, the damping process can be described by differential equations. In the spirit of the synchrotron radiation discussion, a characteristic time is $\tau_0 = E/cE'$, where E is the muon energy and E' is the energy loss due to ionization per unit length, appropriately reduced by the fraction of the path length occupied by the absorber. The muon speed is taken to be c.

The time constants for damping are given by

$$\tau_{\varepsilon_y} = \tau_0, \quad \tau_{\varepsilon_x} = \frac{\tau_0}{1 - \frac{D\delta'}{\delta_0}}, \quad \tau_{\Delta E} = \frac{\tau_0}{\frac{d\ln E'}{d\ln E} + \frac{D\delta'}{\delta_0}}, \tag{28}$$

where the wedge absorbers are located at positions having dispersion function D and they vary in thickness with x according to $\delta = \delta_0 + \delta'x$. The time constants satisfy an invariant resembling Robinson's theorem:

$$\frac{1}{\tau_{\varepsilon_x}} + \frac{1}{\tau_{\varepsilon_y}} + \frac{1}{\tau_{\Delta E}} = 2 + \frac{d\ln E'}{d\ln E}. \tag{29}$$

5.3.3. *A new direction?*

Two years ago, we learned that the E167 experimenters [36] using particles at 28 GeV from the SLAC linac had observed doubling of the energy of

some of the bunch as a result of the plasma wakefield process in a 90 cm plasma column. This result is shown in Fig. 10. There have been great hopes attached to plasma research for more than the half-century since the initiation of the controlled fusion programs of the 1950s. Now, not only is the power generation direction progressing as exemplified by ITER, but we have a splendid example of acceleration within a plasma.

In the plasma wakefield process, an electron bunch initiates a strong plasma wave that can accelerate particles behind the head of the bunch or particles in a trailing bunch. An order-of-magnitude estimate of the energy gain may be made as follows. Represent the electric field in the wave by

$$E(x,t) = E_0 \cos[k(x - vt)], \tag{30}$$

where x is in the direction of motion of the bunch. From Maxwell's equation $\mathrm{div}\,E = \rho/\epsilon_0$, we have

$$E_0 = \frac{ne}{\epsilon_0 k} = \frac{nec}{\epsilon_0 \omega_p}, \tag{31}$$

where n is the electron density in the plasma, ω_p is the plasma frequency, and c (over)estimates the phase velocity of the wave. For $n = 10^{23}/\mathrm{m}^3$, the result is about $30\,\mathrm{GeV/m}$.

This is a big step, but it does not solve all of our problems just yet. We still have Eq. (27)

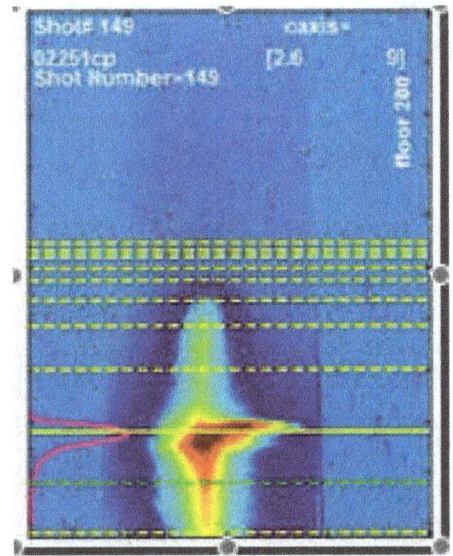

Fig. 10. The beam energy spectrum of a 28.5 GeV electron beam after interaction with an accelerating plasma, showing evidence of a high accelerating gradient in the plasma channel. The solid yellow line is at 30 GeV, and the dotted lines above are at 10 GeV intervals [36].

to contend with. There is still the power problem. But why not recapture the beam power after collision? That is the mission of the Energy Recovery Linac investigations [37], and then the power in the equation may just represent the inefficiency of the recovery process. How emittance can be preserved within the plasma environment is another interesting question.

But the point we wish to make is that there is evidence of a new departure. It is much too early to guess how this may play out. This plasma-related physics is alluring, and may point a way to relieving the cost-and-scale difficulties that hinder further progress in the familiar directions of the past. Again, a direction stimulated by interest in HEP may find application in the near term as relatively low cost room-sized x-ray sources for medicine [38].

6. Concluding Remarks

The radio frequency accelerating technology initially adopted by Wideroe, Lawrence, and successors has provided an eight-order-of-magnitude increase in cms energy. A similar eight-order-of-magnitude luminosity increase has been achieved since the introduction of colliders. Early in this process, the accelerator community, of necessity, relied on technology developed for other purposes; radar and the steel industry are examples.

As accelerator projects became larger in scope, influence on technology development became possible. Superconducting materials for magnets and rf accelerating structures are the obvious examples. HEP applications have taken the lead in conductor development. Today, an MRI examination involves one's insertion within a superconducting magnet, astonishingly enough. Earlier, detector advances found a role in medical diagnostics with CAT scanners.

The interplay between accelerators and information technology has not been a subject of this article because our interests have been elsewhere in the construction process, and so any remark on this subject is made from afar. Computers are a pervasive presence throughout life today, and that circumstance has been of major benefit to our science. Computers are cheap — per pound comparable in price with food — and that has been of great benefit to such accelerator systems as controls and operator interfaces, in addition to the design tools already

noted. Innovation, such as the World Wide Web, stimulated in the HEP environment is an important consequence of this circumstance.

Initial results are coming in from Auger [39], and again we are reminded that protons with energy up to the 10 J level enter the upper atmosphere. A constant aggravation is the recognition that Someone Else knows how to do it better.

References

[1] E. M. McMillan, *Phys. Rev.* **68**, 143 (1945).
[2] V. Veksler, *J. Phys. (Sov.)* **9**, 153 (1945).
[3] E. Courant, M. S. Livingston and H. Snyder, *Phys. Rev.* **88**, 1190 (1952).
[4] J. Schwinger, *Phys. Rev.* **75**, 1912 (1948).
[5] K. Robinson, *Phys. Rev.* **111**, 373 (1958).
[6] L. Alvarez, *Phys. Rev.* **70**, 799 (1964).
[7] G. K. O'Neill, *Phys. Rev.* **102**, 1418 (1956).
[8] C. Bernardini *et al.*, *Phys. Rev. Lett.* **9**, 407 (1963).
[9] *Handbook of Accelerator Physics and Engineering*, eds. A. Chao and M. Tigner (World Scientific, 2006), 3rd printing.
[10] M. A. Furman and M. S. Zisman, *Handbook of Accelerator Physics and Engineering*, op. cit., p. 277.
[11] Two examples are D. A. Edwards and M. J. Syphers, *Introduction to the Physics of High Energy Accelerators* (Wiley, 1993) and H. Wiedemann, *Particle Accelerator Physics* (Springer, 1999).
[12] The lectures organized under the CERN Accelerator School (CAS) and the United States Particle Accelerator School (USPAS) are readily available resources. The former may be accessed at http://cas.web.cern.ch and the latter at http://uspas.fnal.gov
[13] K. Johnsen, *Proc. Nat. Acad. Sci. U.S.A.* **70**(2), 619 (1973).
[14] M. Livingston, *The Future of Electron Synchrotrons*, available at http://epaper.kek.jp/p65/PDF
[15] G. Cocconi, *Phys. Rev.* **111**, 1699 (1958).
[16] The article by B. Richter on the history of the SLAC accelerators at http://www-conf.slac.stanford.edu/40years/histories/BL-SI7-1184.pdf makes good reading.
[17] D. Möhl, G. Petrucci, L. Thorndahl and S. van der Meer, *Phys. Rep.* **58**(2), (1980).
[18] http://www.bnl.gov/RHIC
[19] H. T. Edwards, *Ann. Rev. Nucl. Part. Sci.* **35**, 605 (1985).
[20] S. Nagaitsev *et al.*, *Phys. Rev. Lett.* **96**, 044801 (2006).
[21] *A Proposal to Construct and Test Prototype Superconducting RF Structures for Linear Colliders*, ed. H. Lengeler (Apr. 1992 TTF Collaboration, DESY).
[22] M. Dohlus, P. Schmüser and J. Rossbach, *Introduction to Ultraviolet and X-Ray Free Electron Lasers* (Springer, 2008).

[23] P. Oddone in *Proc. UCLA Workshop: Linear Collider B\bar{B} Factory Conceptual Design*, ed. D. Stork (1987).

[24] The two reports treating the PEP-II design may be obtained through SPIRES as slac-r-352 and slac-r-353. The KEKB report is at http://www-acc.kek.jp/kekb/publication/KEKB_design_report

[25] https://edms.cern.ch/file/445829/4/ Vol_1_Chapter_1.pdf and following files.

[26] http://superb.kek.jp

[27] http://www.pi.infn.it/SuperB

[28] U.S. Linear Collider Technology Options Study, Mar. 1, 2004. A summary may be found at http://www.interactions.org/pdf/ITRPexec.pdf

[29] http://linearcollider.org

[30] http://tesla.desy.de/new_pages, TDR_CD/PartII/accel.html

[31] D. Edwards *et al.*, *Proc. XX Int. Linac Conference* (Monterey, CA, 2000).

[32] Advanced Accelerator Research Annual Report 2005 (DOE Office of High Energy Physics).

[33] https://dms.uslarp.org

[34] http://www.fnal.gov/projects/muon_collider

[35] http://www.ihep.ac.cn/data/ichep04/c_paper/9-0457.pdf

[36] Material on this series of experiments may be found at http://slac.stanford.edu/grp/arb/siemann.pdf

[37] http://epaper.kek.jp/p07/ PAPERS/MOYKI03.PDF

[38] http://epaper.kek.jp/p07/ PAPERS/THPMN114.PDF

[39] http://www.auger.org/

Helen and Donald Edwards met at Cornell a half-century ago, when she was a graduate student and he a faculty member. Since then, they have collaborated in accelerator science and technology activities at Cornell, Fermilab, SSC Laboratory, and DESY. Helen is a MacArthur Fellow, a member of the American Academy of Arts and Sciences, and a member of the National Academy of Engineering. Both are American Physical Society Fellows. For recreation, they head toward the Arctic whenever possible.

Reviews of Accelerator Science and Technology
Vol. 1 (2008) 121–141
© World Scientific Publishing Company

World Scientific
www.worldscientific.com

Synchrotron Radiation

A. Hofmann

Former Physicist at CERN
CH-1211 Geneva 23, Switzerland
albert.hofmann@cern.ch

The physics of synchrotron radiation, undulator radiation, and free electron lasers is reviewed with an emphasis on the underlying physical principles and the experimental observables, such as the radiation spectrum, angular distribution, and radiation polarization.

Keywords: Synchrotron radiation; undulator; wiggler; storage ring; FEL (free electron laser).

1. Introduction

Synchrotron radiation is emitted by a charge undergoing a transverse acceleration while moving with relativistic velocity toward an observer. The resulting Doppler effect produces a high frequency spectrum. Two time scales enter into the treatment of this radiation. The motion of the particle and the creation of the radiation are described by an emission time t', while its reception by an observer is given by a delayed and compressed observation time, t. Many properties of synchrotron radiation can be understood in a qualitative way, either by going into a frame moving with the source, or by comparing the radiation in the two time scales. These methods reveal the small vertical opening angle, the high frequency spectrum, and the polarization properties.

For quantitative description the electric and magnetic potentials, created by the charge, are first given in emission time and then expressed in observer time. The corresponding fields are given by the Liénard–Wiechert equation, which describes all relevant classical properties of synchrotron radiation. A magnetic field is used to provide the transverse acceleration — either as a long homogeneous magnet, emitting ordinary synchrotron radiation, or as a spatially periodic magnet, giving quasi-monochromatic undulator radiation. The latter is widely used since its radiation has a small opening angle in both planes and its spectrum, as well as its polarization properties, can be modified.

The optical properties of synchrotron radiation are dominated by the small natural opening angle and the resulting diffraction limitations of the image resolution. The radiation emittance is the product of angular and spatial beam dimensions. It approaches the theoretical limit if emitted by an ideal electron beam. However, the circulating electrons have themselves a finite angular and spatial distribution due to the energy change during the photon emission. This quantum excitation leads to a finite emittance and energy spread of the electron beam which has to be convoluted with that of the photons. Ideally this effect should be small and not dilute the photon emittance; in this case the radiation is called diffraction-limited or spatially coherent.

The energy loss of the electrons is on average replaced by a high frequency longitudinal voltage, called rf voltage. It focuses the electrons in energy and longitudinal position and collects them in so-called bunches having a nominal energy and revolution frequency around which they oscillate. This longitudinal modulation affects the motion of an individual circulating electron in so far as it is no longer periodic and results for most cases in a continuous and not in a line spectrum. Some attempts were made to obtain very short bunches having only a small longitudinal modulation. This would not only give a line spectrum but also temporal coherence with all electrons in a bunch radiating in phase such that their fields add up and give a manifold increase in power. But this can be done for rather long wavelengths only.

However, undulator radiation can itself create a short period modulation of the electron energies in a bunch. Through the energy dependence of the path length in a deflecting magnet, this creates also a longitudinal intensity modulation resulting in temporal coherent radiation of small spectral width and high intensity. Such devices have been built mainly for linear accelerators and are called free electron lasers (FELs), which are becoming more and more important as radiation sources.

2. Radiation from a Relativistic Particle

We start with the general case of radiation emitted by an accelerated relativistic charge. It is the base for more specialized sources, like ordinary synchrotron radiation emitted from a circular orbit or undulator radiation from a spatially periodic magnet. For this reason we give in this section some derivations of expressions which can later be adapted to special sources. For each case we rely on some qualitative treatments to illustrate the underlying physics and to estimate some radiation properties.

Synchrotron radiation is of interest only if it is emitted by a relativistic particle and we assume that $v = \beta c$ with $\beta \approx 1$, $\gamma = 1/\sqrt{1 - \beta^2} \gg 1$, and make the corresponding approximations. In some derivations the exact value of β has to be carried along for a while, because it might later be subtracted from unity, e.g. $1 - \beta \approx 1/2\gamma^2$. In the figures used to illustrate synchrotron radiation we have to choose smaller values for β and γ, otherwise some properties, like the opening angle, become invisible.

There are several dedicated books on synchrotron radiation [1–4] and also some on electrodynamics which have a chapter on this subject [5–8]. Some of the early work is reported in Refs. 9–13. A history of the development of the field can be found in Ref. 14. The use of synchrotron radiation for different fields in science has expanded significantly over the past few decades. By now there are more than 50 storage rings operating as sources in many laboratories around the world.

2.1. *Estimation of the opening angle*

The basic physics and some radiation properties can be understood from qualitative arguments. With the

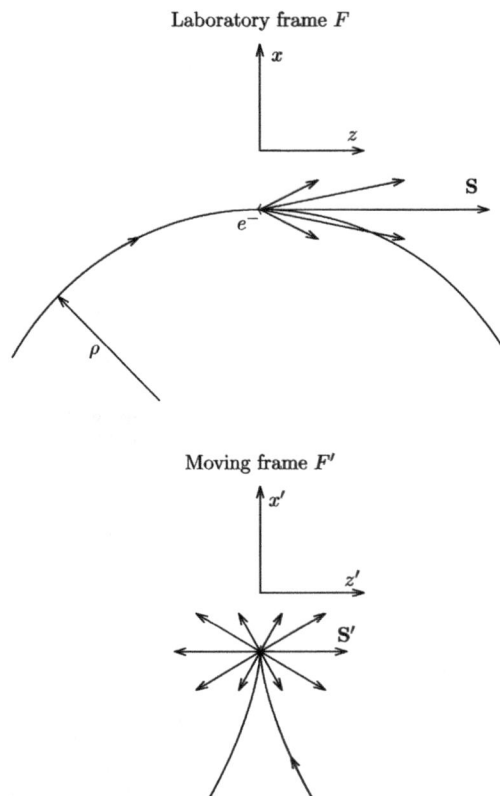

Fig. 1. Opening angle estimated by going into a moving frame; from Ref. 4, modified.

help of Fig. 1 we can obtain the typical opening angle. It shows an electron moving in a laboratory frame F on a circular orbit of radius ρ and emitting synchrotron radiation. In an inertial frame F' which moves with a constant velocity $\mathbf{v} = \boldsymbol{\beta}c$, being at one instant the same as that of the electron, the trajectory has the form of a cycloid. It has a cusp where the electron undergoes an acceleration in the $-x'$ direction and emits radiation with a distribution being approximately uniform in F'. Going back to the laboratory frame F, by applying a Lorentz transformation, this radiation becomes peaked in the forward direction, with a photon emitted along the x' axis in F' appearing at an angle $1/\gamma$ in F. The typical opening angle of synchrotron radiation is therefore of order $1/\gamma$. For ultrarelativistic electrons with $\gamma \gg 1$ the radiation is confined within very small angles around the direction of the electron motion.

2.2. *Relevant motion*

For a quantitative treatment of synchrotron radiation we have to distinguish between two time

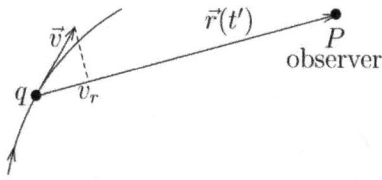

Fig. 2. Particle trajectory and radiation geometry.

scales: the emission time t', in which we describe the location and motion of a charge q and the radiation it emits; and the observer time t, when this is detected at a distance $\mathbf{r}$, as illustrated in Fig. 2. The light, propagating with velocity c over the distance r, causes a delay r/c for the observation. For a charge which moves with velocity $\mathbf{v}$, having a component $v_r = \mathbf{n} \cdot \mathbf{v}$ along the unit vector $\mathbf{n}$ pointing toward the observer, the distance r changes by $dr = -v_r dt' = -\mathbf{n} \cdot \mathbf{v} \, dt'$ and correspondingly also the delay

$$t = t' + \frac{r(t')}{c}, \quad \frac{dt}{dt'} = 1 + \frac{1}{c}\frac{dr}{dt'} = 1 - \mathbf{n} \cdot \boldsymbol{\beta}. \quad (1)$$

This simple relation is made complicated by the fact that the distance $r(t')$ is a function of the emission time t' which is difficult to calculate for a given observation time t.

Synchrotron radiation is usually observed in the forward direction, in which the angle between the particle motion $\boldsymbol{\beta}$ and the observation direction $\mathbf{n}$ is small, $\mathbf{n} \cdot \boldsymbol{\beta} \approx \beta$. For the assumed ultrarelativistic motion we have $\beta \approx 1$ and $dt/dt' \approx 1 - \beta \approx 1/2\gamma^2$. The observer time interval is strongly compressed compared to the emission time which leads to the high frequency spectrum of synchrotron radiation.

2.3. *Retarded potentials*

For a charge distribution with density $\eta(x, y, z)$ at a distance $r(x, y, z)$ from an observer, the electric potential V is given by Coulomb's law:

$$V = \frac{1}{4\pi\epsilon_0} \int \frac{\eta(x, y, z)}{r} dx \, dy \, dz.$$

In the stationary case this potential is evaluated directly from the given charge density. However, for moving charges these quantities have to be evaluated at the earlier emission times t'_i of each charge q_i, which corresponds to a chosen observation time

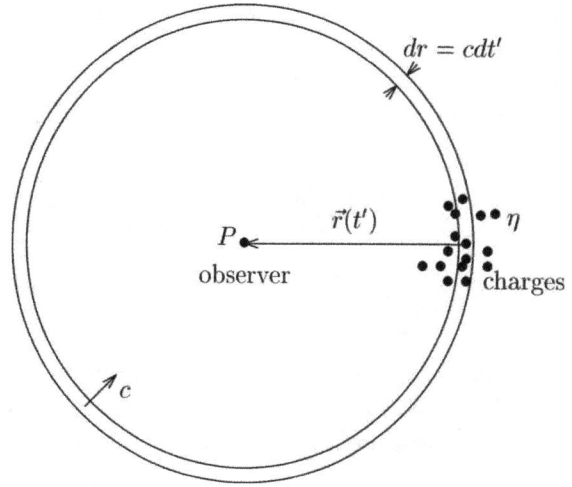

Fig. 3. Collapsing sphere representing the integration over charges.

t. The integration over the charge density $\eta(t')$ represents a thin sphere of the radius r collapsing with the speed of light c toward the observer P, while counting all the charges on its way, as illustrated in Fig. 3. In this process the charges moving toward the observer contribute for a longer time, and therefore more, to the potential than charges moving away.

A single charge of finite radius b shown in Fig. 4 contributes in this process to the potential $V(t)$ for a time $\Delta t'_0 = 2b/c$ if it is at rest, but for a moving charge the time of contribution is

$$\Delta t'_v = \frac{2b}{c - v_r} = \frac{2b}{c(1 - \mathbf{n} \cdot \boldsymbol{\beta})} = \frac{\Delta t'_0}{1 - \mathbf{n} \cdot \boldsymbol{\beta}}.$$

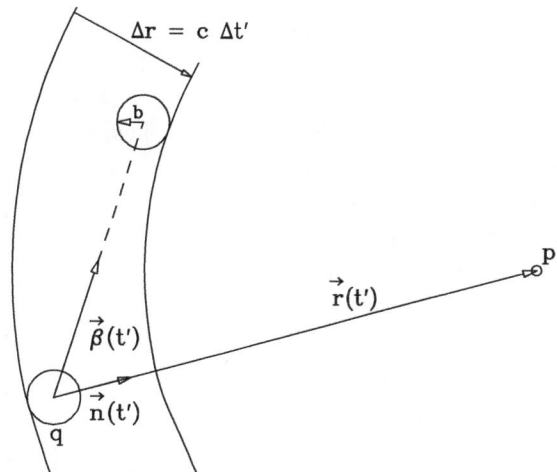

Fig. 4. Contribution of a moving charge to the potential.

Since the ratio between the two is independent of the particle radius, we can let b go to zero and get for the scalar electric potential $V(t)$ and also for the related magnetic vector potential $\mathbf{A}(t)$ of a moving charge

$$V(t) = \frac{q}{4\pi\epsilon_0} \left(\frac{1}{r\left(1 - \mathbf{n}\cdot\boldsymbol{\beta}\right)} \right)_{\mathrm{ret}},$$

$$\mathbf{A}(t) = \frac{\mu_0 q}{4\pi} \left(\frac{\mathbf{v}}{r\left(1 - \mathbf{n}\cdot\boldsymbol{\beta}\right)} \right)_{\mathrm{ret}}.$$

In the above expressions the part inside the parentheses has to be evaluated at the emission time t' but expressed in the later observer time $t = t' + r(t')/c$. This will be emphasized in the future with the index "ret" for larger expressions but not always for a single variable. The quantities $V(t)$ and $\mathbf{A}(t)$ are the retarded potentials of a moving charge, also called Liénard–Wiechert potentials. To obtain them we used the "duration" of the contribution by a moving charge to the potential at the observer, which looks a little simpleminded. However, in quantum mechanics the electromagnetic interaction is described as an exchange of virtual photons. In this picture the rate of photons received increases as the charge moves toward the observer due to the Doppler effect, which augments the strength of the interaction.

2.4. *Fields of a moving charge*

From the potentials $V(t)$ and $\mathbf{A}(t)$ we obtain the corresponding fields $\mathbf{E}(t)$ and $\mathbf{B}(t)$ of a moving charge by using the relations

$$\mathbf{E} = -\operatorname{grad} V - \frac{\partial \mathbf{A}}{\partial t}, \quad \mathbf{B} = \operatorname{curl} \mathbf{A}$$

with the Lorentz convention $\nabla\cdot\mathbf{A} - \dot{V}/c^2 = 0$. This looks straightforward; however, the difficulty in carrying out this operation lies in the fact that the relation between potentials and fields requires differentiation with respect to time t and the position $\mathbf{r}_p$ of the observer. Any change Δt or $\Delta\mathbf{r}_p$ in these coordinates changes the time t' the radiation was emitted through the relation $t = t' + r(t')/c$, and in turn also the quantities $\mathbf{r}(t')$, $\mathbf{R}(t')$, and $\mathbf{n}(t')$ in a rather complicated way. We give here only the final result and refer for the derivation to the corresponding literature. The electromagnetic fields of a moving and accelerated charge are given by the Liénard–Wiechert equation

$$\mathbf{E}(t) = \frac{q}{4\pi\epsilon_0} \left(\frac{(1-\beta^2)(\mathbf{n}-\boldsymbol{\beta})}{r^2(1-\mathbf{n}\cdot\boldsymbol{\beta})^3} \right.$$
$$\left. + \frac{[\mathbf{n}\times[(\mathbf{n}-\boldsymbol{\beta})\times\dot{\boldsymbol{\beta}}]]}{cr(1-\mathbf{n}\cdot\boldsymbol{\beta})^3} \right)_{\mathrm{ret}}, \quad (2)$$

$$\mathbf{B}(t) = \frac{\mathbf{n}_{\mathrm{ret}}\times\mathbf{E}}{c}.$$

It exhibits the following properties:

- The fields $\mathbf{E}$ and $\mathbf{B}$ are perpendicular to each other.
- For a stationary charge, we get Coulomb's law.
- For vanishing acceleration $\dot{\boldsymbol{\beta}} = 0$ we get the field of a charge moving with constant velocity in a slightly unusual form. It can be reduced to Coulomb's law by a Lorentz transformation, which indicates that there is no energy radiated in this case.
- The general field expression has two terms one being proportional to $1/r^2$, called the near field, and the other being proportional to $1/r$, called the far field, or radiation field. At large distances the second term dominates and the first is often neglected. This is justified if one wants to calculate the radiated power to which the near field does not contribute. However, the far field alone does not satisfy Maxwell's equation, in particular div $\mathbf{E} \neq 0$.
- For the far field the electric and magnetic components, $\mathbf{E}$ and $\mathbf{B}$, are perpendicular to each other and both are perpendicular to the direction of propagation given by the unit vector $\mathbf{n}$, which points from the charge to the observer.

3. Radiation Field and Power

In nearly all applications of synchrotron radiation one does not observe the field directly but only its power or photon flux to which the near field does not contribute. Therefore, we will omit the first term of (2) and use only the second part giving the radiation field. Furthermore, we concentrate here on the radiation from a single elementary charge, $q = e = 1.602\cdot10^{-19}$ C, and consider effects of many charges later. As mentioned before, the evaluation of this equation is in general difficult due to the complicated relation between the time t', in which the charge motion and photon emission are described, and the time t, used by the observer.

However, one is sometimes interested only in the instantaneous emitted power and its angular distribution, while the time development of the field received by an observer is less important. We calculate now the field radiated at a given emission time $t' = 0$, with the observation time at a distance r being just $t = r/c$.

In most experimental applications the time evolution of the received field is not observable but only its spectral distribution. It is advantageous to form directly the Fourier transform of the field (2). This still involves the relation between the time scales but in a way which is easier to handle. Furthermore, the response of matter to electromagnetic radiation is usually given as a function of frequency.

3.1. *Emitted power and its angular distribution*

The power seen by the observer is described by the Poynting vector **S**, which represents an energy element dU passing per time element over a solid angle $d\Omega$:

$$\mathbf{S} = \frac{1}{\mu_0}[\mathbf{E} \times \mathbf{B}] = \frac{E^2}{\mu_0 c}\mathbf{n} = \frac{d^2U}{r^2 d\Omega dt}\mathbf{n},$$

where we use only the farfield, for which $\mathbf{n} \cdot \mathbf{E} = 0$. We have to specify which time element we mean. An energy dU is emitted by the charge in an emission time dt' but received in a compressed observer time $dt = (1 - \mathbf{n} \cdot \boldsymbol{\beta})dt'$. The Poynting vector contains the field $E(t)$ given in observer time and relates therefore to the received power. However, the power radiated by the charge is more relevant and represents some kind of averaged received power. We give here its angular distribution using the time relation (1):

$$\frac{dP}{d\Omega} = \frac{d^2U}{d\Omega dt'} = \frac{d^2U}{d\Omega dt}\frac{dt}{dt'} = \frac{r^2 E^2}{\mu_0 c}(1 - \mathbf{n} \cdot \boldsymbol{\beta}). \quad (3)$$

3.2. *Instantaneous emitted field and power*

We calculate the field radiated at a fixed emission time $t' = 0$ and choose a coordinate system (x, y, z) with a charge e at the origin moving in the z direction with $\mathbf{v} = \boldsymbol{\beta}c$ and undergoing a normalized transverse acceleration $\dot{\boldsymbol{\beta}}$ in the x direction, usually provided by a magnetic field **B** in the $-y$ direction for $e > 0$, as shown in Fig. 5. We also use the angles θ and ϕ of the corresponding spherical coordinate system as well as

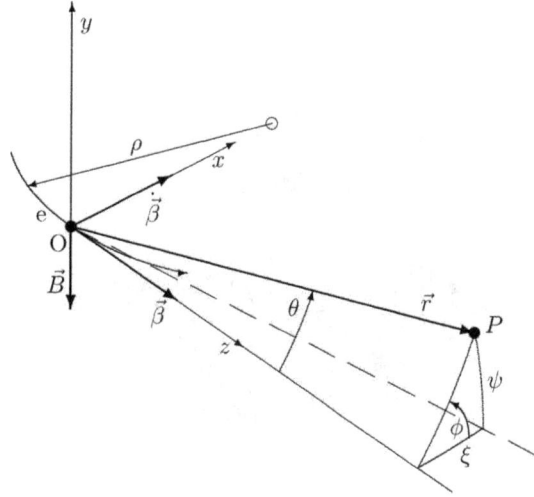

Fig. 5. Geometry of instantaneous radiation [4].

their projections $\xi = \theta \cos\phi$ and $\psi = \theta \sin\phi$ on the horizontal and vertical planes. With the unit vector **n** in the **r** direction, the normalized velocity vector $\boldsymbol{\beta}$ and acceleration $\dot{\boldsymbol{\beta}}$ expressed in this coordinate system, and noting the approximation $1 - \mathbf{n} \cdot \boldsymbol{\beta} \approx \frac{1+\gamma^2\theta^2}{2\gamma^2}$, we obtain from (2) the electric radiation field

$$\hat{\mathbf{E}} = -\frac{e\gamma^4[1 - \gamma^2\theta^2\cos(2\phi), -\gamma^2\theta^2\sin(2\phi)]}{\pi\epsilon_0\rho r(1 + \gamma^2\theta^2)^3}.$$

Equation (3) gives the angular distribution of the radiated power

$$\frac{dP}{d\Omega} = P_0 F_u(\theta, \phi) = P_0[F_{u\sigma}(\theta, \phi) + F_{u\pi}(\theta, \phi)],$$

$$F_u = \frac{3\gamma^2}{\pi}\frac{[1 - \gamma^2\theta^2\cos(2\phi)]^2 + [\gamma^2\theta^2\sin(2\phi)]^2}{(1 + \gamma^2\theta^2)^5},$$

$$P_0 = \frac{2r_e c m_0 c^2 \gamma^4}{3\rho^2} = \frac{2r_e c^3 e^2 E_e^2 B^2}{3(m_0 c^2)^3}. \quad (4)$$

The total instantaneous radiated power P_0 is obtained by integrating over the solid angle and is expressed here with the classical electron radius $r_e = e^2/4\pi\epsilon_0 m_0 c^2 = 2.818 \cdot 10^{-15}$ m and the energy $E_e = m_0 c^2 \gamma$. The power is presented as a normalized function F_u being the sum of contributions obtained from the two field components. They represent the horizontal and vertical polarization modes. The total angular power distribution is shown in Fig. 6 as a function of the horizontal and vertical coordinates x and y and of the polar angles θ and ϕ. The radiation

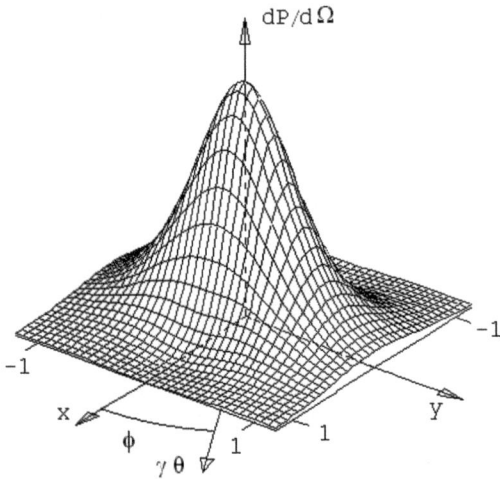

Fig. 6. Instantaneous angular power distribution of radiation [4].

has a small opening angle with the variances

$$\langle\theta^2\rangle = \frac{1}{\gamma^2}, \quad \langle\xi^2\rangle = \frac{3}{8\gamma^2}, \quad \langle\psi^2\rangle = \frac{5}{8\gamma^2}.$$

Longitudinal acceleration emits considerably less power for the same force and is of no practical interest.

3.3. Radiation field in frequency domain

The difficulty with the relation between the time scales t' and t is reduced by calculating directly the Fourier transform of the field

$$\tilde{\mathbf{E}}(\omega) = \frac{1}{\sqrt{2\pi}} \int_{-\infty}^{\infty} \mathbf{E}(t)e^{-i\omega t}dt.$$

This involves t since we are interested in the spectrum seen by the observer. However, a formal transformation of the integration variable $t = t' + r(t')/c$ results in an integral over t' which can be carried out using some approximations. We would like to give not only the field but also the power as a function of frequency. This is expressed as the angular spectral energy distribution, i.e. the energy dU radiated into a solid angle $d\Omega$ and frequency element $d\omega$:

$$\frac{dU}{d\Omega} = \frac{2r^2}{\mu_0 c} \int_0^{\infty} |\tilde{E}(\omega)|^2 d\omega,$$

$$\frac{d^2U}{d\Omega d\omega} = \frac{2r^2|\tilde{E}(\omega)|^2}{\mu_0 c}.$$

(5)

The factor 2 comes from the fact the spectral energy density is taken at positive frequencies only, contrary to the field given for both signs. This is common

practice since power can be measured directly but not the sign of the frequency. The field, however, is rarely accessible to direct measurements.

In the frequency domain we give the distribution of the radiated energy rather than that of the power used in the time domain. The Fourier transform involves an integral of the field $E(t)$ over time which relates it closer to the emitted energy. From the instantaneous power it is impossible to obtain information about the spectrum. However, in cases where the power is constant over a certain time, like for a charge circulating in a homogeneous magnet, we use an average power by dividing the energy radiated in one turn by the revolution time or, in the case of an undulator, by dividing the radiated energy by the traversal time.

4. Synchrotron Radiation

4.1. Geometry

We consider the radiation emitted by a charge which moves over a certain length on a circular trajectory of bending radius ρ with constant linear and angular velocities $v = \beta c$, $\omega_0 = \beta c/\rho$, as shown in Fig. 7. This is the case of ordinary synchrotron radiation emitted by a charged particle in a long bending magnet.

The instantaneous distribution of the emitted radiation is given by the expression (4) and illustrated in Fig. 6. However, this distribution is now swept with angular velocity ω_0 in the horizontal direction and cannot be observed but is received as a very short flash in time t which determines the radiation spectrum.

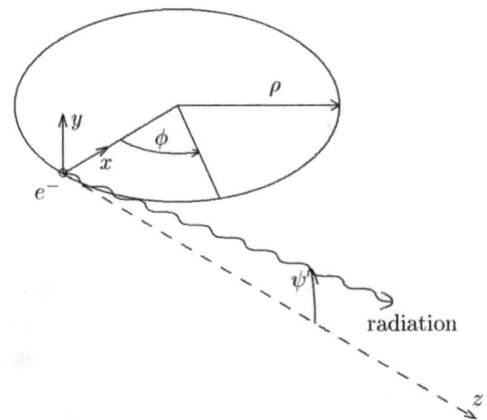

Fig. 7. Geometry of synchrotron radiation.

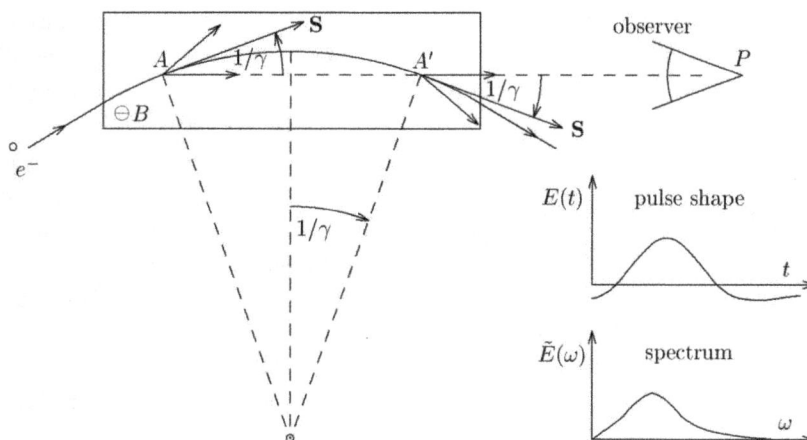

Fig. 8. Spectrum of the radiation [4].

4.2. *Qualitative spectrum*

To estimate the typical frequency of synchrotron radiation, we consider an electron going through a long magnet, where it emits radiation which reaches an observer P; see Fig. 8. We ask ourselves how long the observed pulse of radiation lasts. Due to the small opening angle of $\approx 1/\gamma$ the observer receives the light for a limited time only. The radiation seen first was emitted at the point A, where the electron trajectory has an angle of $1/\gamma$ with respect to the direction pointing toward the observer, while the one seen last originated at the point A', where this angle is $-1/\gamma$. Since both the electron and the photon move toward the observer, the length of the received radiation pulse is just the difference $\Delta t = \Delta t_e - \Delta t_\gamma$ in travel times of the electron and the photon going from A to A':

$$\Delta t = \frac{2\rho}{\beta\gamma c} - \frac{2\rho\sin(1/\gamma)}{c} \approx \frac{4}{3}\frac{\rho}{c\gamma^3}. \qquad (6)$$

The typical frequency, $\omega_{\text{typ}} \approx \frac{2\pi}{\Delta t}$, is $\propto \gamma^3$. A factor of γ^2 is due to the difference in velocities and a factor of γ is due to the different trajectory lengths of the electron and the photon in the magnet.

This illustration contains some interesting information concerning the formation length of synchrotron radiation, i.e. the length of the interaction between electron and photon. The observed light originates approximately from the trajectory arc between the points A and A' which has the length $\ell \approx 2\rho/\gamma$. We assumed that the magnetic field, and therefore the curvature, is constant over this length. By introducing variations within this, we

can modulate the emission and alter the spectrum. This is done in an undulator.

4.3. *Qualitative polarization*

Since the acceleration of the electron is radial (usually horizontal), we expect the electric field to lie in this same plane, resulting in a radiation with predominantly horizontal polarization. For observation in the orbit plane this is entirely the case; however, above and below this plane, a vertical or elliptical polarization component is also observed. This is illustrated in Fig. 9, where a circulating electron in a storage ring is observed from above or below the orbit plane and seen as a left- or right-handed elliptical motion. The electric field vector of the radiation makes a similar motion, resulting in elliptical polarization. It has one sign above, becomes completely

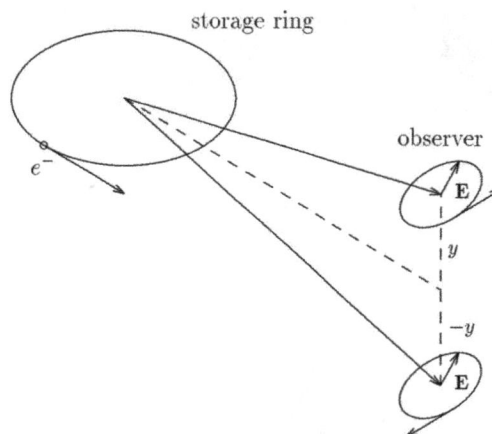

Fig. 9. Elliptic polarization [4].

horizontal in the middle, and has the opposite sign below the median plane. For undulator radiation we expect no elliptical but only horizontal and vertical components.

4.4. *Angular distribution*

For an observer the instantaneous distribution is swept horizontally with angular velocity ω_0. To get the average vertical distribution we express the function F_u in (4) with the projected angles ξ and ψ, use $d\xi = \omega_0 dt'$ and integrate over it, giving

$$\frac{dU/U_0}{d\psi} = \frac{21\gamma}{32(1 + \gamma^2\theta^2)^{5/2}} \left[1 + \frac{5}{7} \frac{\gamma^2\theta^2}{1 + \gamma^2\theta^2} \right],$$

which is normalized by U_0, the total energy radiated, and is shown in Fig. 10. The two terms in the square brackets correspond to the horizontal (σ mode) and vertical (π mode) polarization modes. Their vertical opening angles have the variances

$$\langle\gamma^2\psi^2\rangle_\sigma = \frac{1}{2}, \quad \langle\gamma^2\psi^2\rangle_\pi = \frac{3}{2}, \quad \langle\gamma^2\psi^2\rangle = \frac{5}{8}.$$

4.5. *Spectrum*

The sweeping beacon of radiation determines the spectrum as seen by a stationary observer. It is characterized by the critical frequency

$$\omega_c = \frac{3c\gamma^3}{2\rho},$$

which is identical to the typical frequency we estimated from a qualitative treatment. Since the spectrum itself is given by a relatively complicated

expression involving either modified Bessel functions of fractional order or Airy functions, we only illustrate the spectrum here with figures. First, we show the electric field seen by an observer in the median plane $\psi = 0$ in Fig. 11. A linear plot of the spectrum is shown on top as a function of ω/ω_c. It has a maximum close to the critical frequency. The time domain field is shown at the bottom as a function of $\omega_c t_p$, where $t_p = t - r_p/c$ is the reduced observation time where an uninteresting delay factor has been omitted. Although of no practical interest, this time function illustrates some of the radiation properties. It represents a very short pulse of typical length $2/\omega_c$ and its average value vanishes in accordance with $\tilde{E}(\omega) \to 0$ for $\omega \to 0$.

The spectral power density

$$\frac{dU/U_0}{d\omega} = \frac{1}{\omega_c} \left[S_\sigma\left(\frac{\omega}{\omega_c}\right) + S_\pi\left(\frac{\omega}{\omega_c}\right) \right],$$
$$\int_0^\infty S\left(\frac{\omega}{\omega_c}\right) d\left(\frac{\omega}{\omega_c}\right) = \frac{7}{8} + \frac{1}{8} = 1, \tag{7}$$

is given by a normalized distribution function S presented as a sum of the two polarization contributions which are plotted in Fig. 12 in a double-logarithmic scale. All decrease slowly toward small frequencies and fast toward high frequencies, with a maximum close to ω_c, which divides the spectrum into two

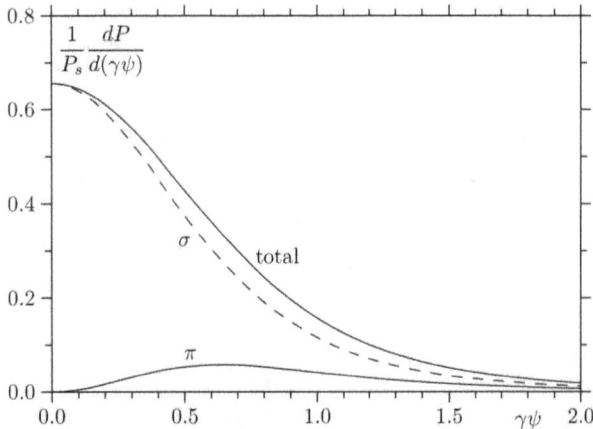

Fig. 10.   Frequency-integrated angular distribution.

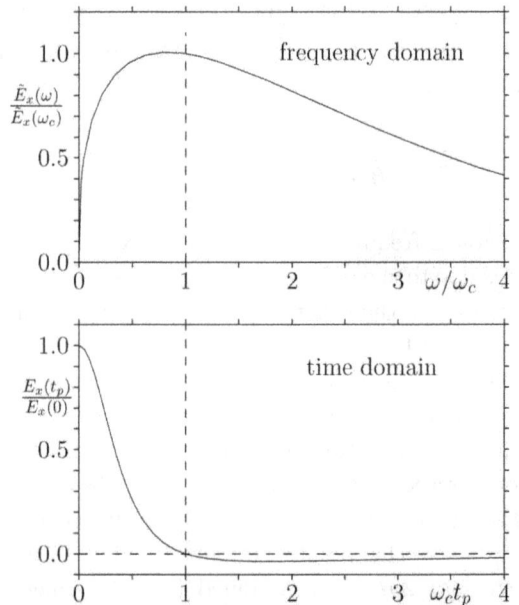

Fig. 11.   Median plane field in frequency and time [4].

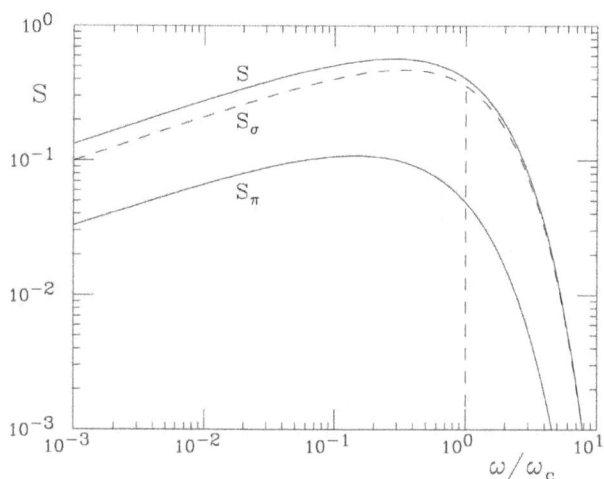

Fig. 12. Normalized power spectrum.

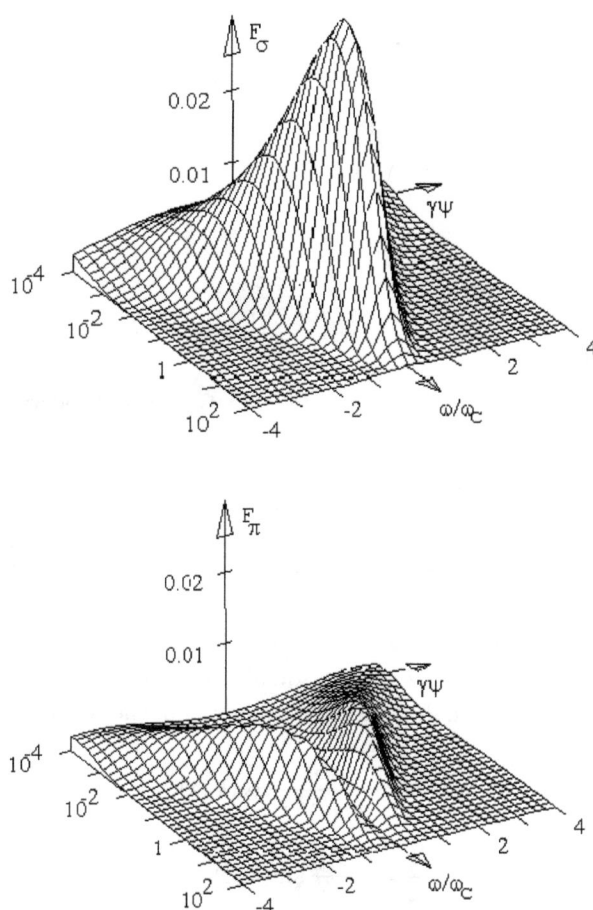

Fig. 13. Angular spectral power density of horizontal (top) and vertical (bottom) polarization [4].

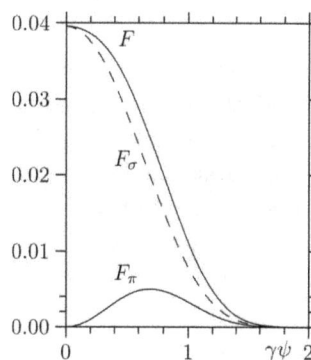

Fig. 14. Angular distribution at ω_c.

parts of equal total power. The horizontal polarization becomes more dominant toward high frequencies where the vertical opening is smaller. Integrating over frequency gives the total power, with the fractions 7/8 and 1/8 for the two polarization modes.

4.6. *Angular spectral energy distribution*

In Fig. 10 we show the angular distribution and in Fig. 12 the spectral distribution. The first one includes all frequencies and the second one all angles. However, there is a correlation between the two which is not visible in these presentations. We give now the angular spectral distribution which expresses the energy radiated into a solid angle and frequency element:

$$\frac{d^2 U/U_0}{d\Omega d\omega} = \frac{\gamma}{\omega_c} F(\omega, \psi) = \frac{\gamma}{\omega_c} [F_\sigma + F_\pi].$$

Here, we use again a normalized function that is the sum of the contributions by the two polarization modes. They are plotted in Fig. 13. The horizontal polarization (top) is concentrated around the median plane $\psi = 0$, while the vertical one (bottom) vanishes there. For both, the angular distribution is wide at low frequencies and narrow at high frequencies and the amplitude maximizes around the critical frequency. As a further illustration, the angular distribution at the frequency ω_c is shown in Fig. 14.

5. Weak Undulator Radiation

An undulator is a spatially periodic magnetic structure designed to emit quasi-monochromatic radiation from relativistic electrons [15, 16]. It has become one of the main sources at synchrotron radiation facilities.

We start with a plane harmonic undulator with period length λ_u, or wave number $k_u = 2\pi/\lambda_u$, as shown in Fig. 15. It has in the median plane ($y = 0$)

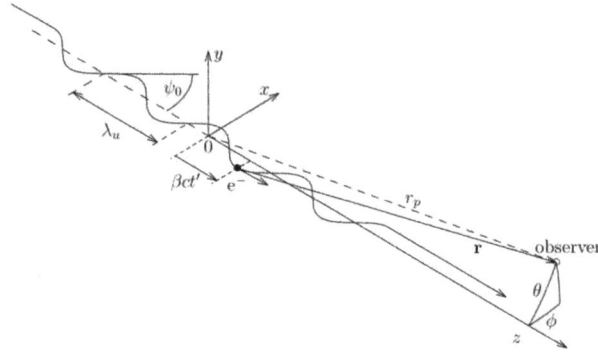

Fig. 15.   Geometry of undulator radiation; from Ref. 4, modified.

a magnetic field B or curvature $1/\rho$ of the form

$$B(z) = B_y(z) = B_0 \cos(k_u z) \propto \frac{1}{\rho}.$$

For a relatively weak magnetic field a particle moves along the z axis on a sinusoidal trajectory with $z \approx \beta c t'$ and a small transverse excursion of the form

$$x(z) = \frac{eB_0}{m_0 c \gamma k_u^2} \cos(k_u z),$$

$$\frac{dx}{dz} = -\frac{K_u}{\gamma} \sin(k_u z). \tag{8}$$

We introduce the undulator parameter

$$K_u = \frac{eB_0}{m_0 c k_u} = \gamma \psi_0,$$

which gives the ratio between the maximum deflecting angle ψ_0 and the natural opening angle $1/\gamma$ of the radiation. We assume here that $K_u < 1$, $\psi_0 < 1/\gamma$, in which case the observer receives light having a small opening angle of $\approx 1/\gamma$ in both planes, with a smaller wiggle angle ψ_0 and a weak intensity modulation resulting in quasi-monochromatic radiation.

5.1. Qualitative treatment

There are different ways to understand undulator radiation. One is based on the interference between the radiation from the different periods of the undulator. Each emits radiation toward an observer at an angle θ, as shown in Fig. 16. These contributions interfere with one an other. We get maximum intensity at a wavelength λ for which they are in phase.

The time interval between the arrival of adjacent contributions is $\Delta T = \frac{\lambda_u}{\beta c} - \frac{\lambda_u \cos\theta}{c}$ and the corresponding frequency ω_1 is

$$\omega_1 = \frac{2\pi}{\Delta T} \approx \frac{2\gamma^2 k_u \beta c}{1 + \gamma^2 \theta^2} = \frac{2\gamma^2 \Omega_u}{1 + \gamma^2 \theta^2}. \tag{9}$$

Constructive interference is obtained for the frequency ω_1.

Undulator radiation can probably best be understood by going into a frame which moves with the average electron velocity, as shown in Fig. 17. In the laboratory frame the electron moves on a sinusoidal trajectory, consisting for a weak undulator of a drift velocity βc in the z direction and a transverse oscillation with frequency $\Omega_u = k_u \beta c$. The latter is small for $K_u < 1$, has a negligible influence on the longitudinal motion, and is nonrelativistic. Going into the moving frame, the electron has no longitudinal motion but executes a harmonic oscillation in the x^* direction with an increased frequency $\Omega_u^* = \gamma \Omega_u$, due to the Lorentz contraction of the undulator period $\lambda_u^* = \lambda_u/\gamma$. The oscillating electron emits monochromatic dipole radiation with frequency Ω_u^* mainly perpendicular to the x^* axis but with a large opening angle. Going back into the laboratory frame, this radiation becomes peaked

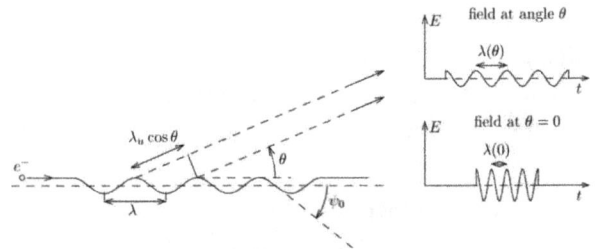

Fig. 16.   Undulator spectrum from interference.

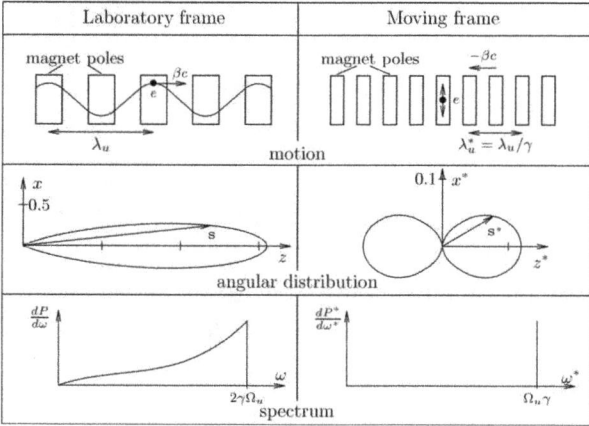

Fig. 17. Weak undulator radiation in a moving and a stationary frame [4].

forward with a typical opening angle of $\theta \approx 1/\gamma$ around the z axis. The frequency becomes Doppler-shifted:

$$\omega_1 = \frac{\Omega_u^*}{\gamma(1 - \beta\cos\theta)} \approx \frac{2\gamma^2\Omega_u}{1 + \gamma^2\theta^2} = \frac{\omega_{10}}{1 + \gamma^2\theta^2}. \quad (10)$$

As a consequence the radiation is no longer monochromatic but depends on the angle θ with respect to the z axis.

5.2. Angular distribution

The geometry we use for weak undulator radiation is the same as that of the instantaneous emission shown in Fig. 5. Since the transverse motion is small and results only in an angular deflection $\psi_0 \ll 1/\gamma$ being smaller than the natural opening angle, we observe actually the instantaneous angular distribution expressed by (4) and shown in Fig. 6 for the total radiation. We separate this now for the two polarization modes and plot the normalized distributions in Fig. 18 as a 3D plot and in Fig. 19 as cuts through the horizontal and vertical planes. The σ mode peaks along the axis while the π mode vanishes in the xz plane, as we expected earlier from Fig. 9; it also vanishes in the yz plane.

The total power emitted over all angles is, according to (4), proportional to the square of the magnetic undulator field B^2 and is therefore modulated $\propto \cos^2(k_u z)$, yielding half the value for the radiated energy U_u and the average power P_u compared to a constant field, but gives the same form of

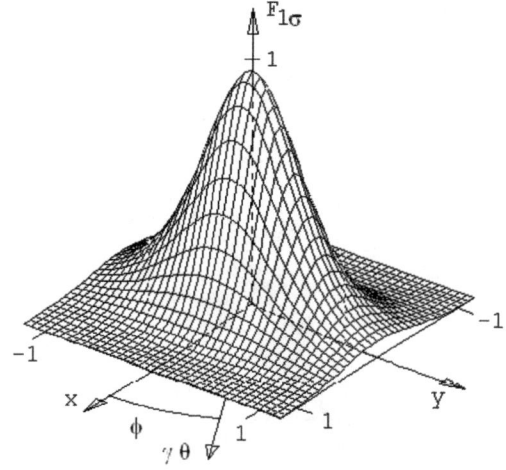

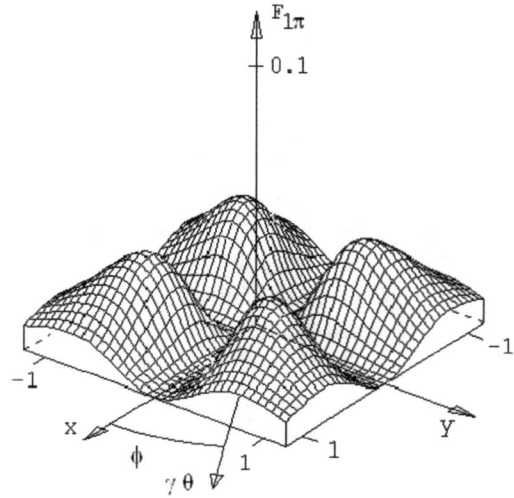

Fig. 18. Normalized angular power density of the horizontal (upper) and vertical (lower) polarization for weak undulator radiation [4].

the angular distribution

$$\frac{dU_u}{U_u d\Omega} = \frac{dP_u}{P_u d\Omega} = \gamma^2[F_{u\sigma}(\theta,\phi) + F_{u\pi}(\theta,\phi)],$$

$$U_u = \frac{r_e c^2 e^2 E_e^2 B_0^2 L_u}{3(m_0 c^2)^3}, \quad P_u = \frac{r_e c^3 e^2 E_e^2 B_0^2}{3(m_0 c^2)^2}.$$

5.3. Angular spectral power density

According to (4) the radiation field is proportional to the transverse acceleration or curvature $1/\rho$, which is modulated for an electron in the undulator as $1/\rho = \cos(\Omega_u t')/\rho_0$, resulting in a corresponding modulation for the emitted field

$$\mathbf{E}(t_p) = -\hat{\mathbf{E}}\cos(\Omega_u t') = -\hat{\mathbf{E}}\cos(\omega_1 t_p). \quad (11)$$

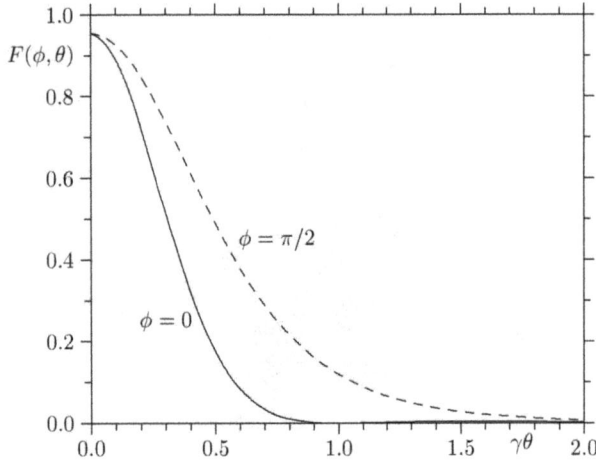

Fig. 19. Cuts through the angular distribution of undulator radiation at the $\phi = 0$ and $\phi = \pi/2$ planes.

Here we have used the frequency (10) and the fact that the phase is the same in the two time scales. The minus sign is due to the opposite acceleration at the origin compared to (4).

To Fourier-transform this field we integrate between $t' = \pm\pi N_u/\Omega_u$ in emission and $t_p = \pm\pi N_u/\omega_1$ in observation time, and get the well-known Fourier transform of a truncated cosine function being $\propto \sin v/v$. Applying this to the radiated power, we get for its distribution in the angle and frequency

$$\frac{d^2 U_u}{d\Omega d\omega} = U_u \gamma^2 \left(F_{u\sigma}(\theta, \phi) + F_{u\pi}(\theta, \phi) \right) f_N(\Delta\omega),$$

$$f_N = \frac{N_u}{\omega_1} \left(\frac{\sin\left(\pi N_u \Delta\omega/\omega_1\right)}{\pi N_u \Delta\omega/\omega_1} \right)^2,$$

where $\Delta\omega = \omega - \omega_1$ and f_N is a normalized spectral function.

5.4. *Spectral power density*

The spectrum of undulator radiation is governed by two effects. First, the frequency ω_1 depends on the angle θ of observation with respect to the axis; second, for a given θ the frequency has a distribution around ω_1 of finite width given by the function $f_N \propto \sin^2 v/v$. The larger the number (N_u) of periods is, the narrower this distribution becomes, turning in the limit into a δ function containing a single frequency, ω_1. For this limit we first integrate the distribution over the azimuthal angle ϕ to obtain the power distribution with respect to θ and translate

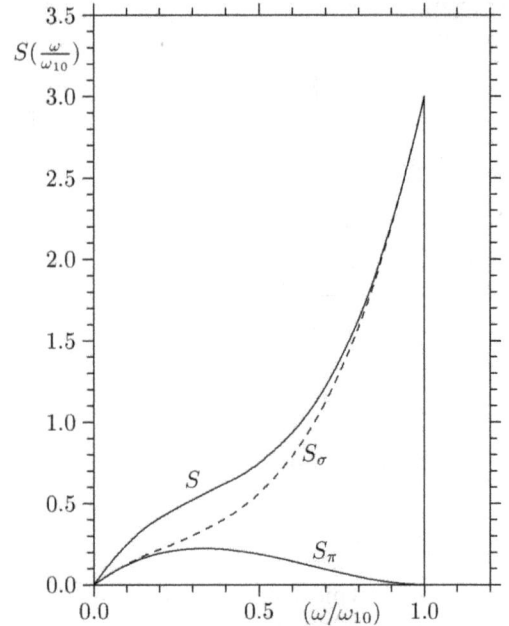

Fig. 20. Normalized spectral power density $S = (dP/P_0)/(d\omega/\omega_{10})$ of undulator radiation.

this into the frequency ω_1. The spectrum obtained this way is shown in Fig. 20 for the two polarization modes and the total radiation. For a finite number of periods the sharp edge at $\omega = \omega_{10}$ will be smoothed out.

6. Strong Undulators and Wiggler Magnets

The majority of existing undulators operate with a parameter $K_u > 1$, and several of our approximations have to be modified; however, the basic physics is the same [17]. We still assume a plane harmonic magnetic field and calculate the electron trajectory more accurately. For a magnetic deflection the square of the velocity is constant, $\dot{x}^2 + \dot{z}^2 = \beta^2 c^2$, and there must be some influence of the transverse motion on the longitudinal one, which we no longer neglect. Apart from a modulation it results in a reduced average longitudinal velocity $\langle \dot{z} \rangle = \beta^* c$:

$$\beta^* \approx \beta \left(1 - \frac{K_u^2}{4\gamma^2} \right), \quad \gamma^{*2} = \frac{1}{1 - \beta^{*2}} \approx \frac{\gamma^2}{1 + \frac{K_u^2}{2}}.$$

We have introduced here a normalized longitudinal drift velocity β^* and the corresponding Lorentz factor γ^*, which allows one to express some strong undulator parameters in the same way as we did for a weak one.

This reduction of the longitudinal drift velocity gives a transverse oscillation frequency $\Omega_u = k_u\beta^*c$ and influences the spectrum we estimated from the constructive interference between emissions from adjacent undulator periods in (9) and illustrated in Fig. 16. We now get for the frequency at the angle θ

$$\omega_1 = \frac{k_u\beta^*c}{1 - \beta^*\cos\theta} \approx \frac{2\gamma^2 k_u c}{1 + \gamma^2\theta^2 + K_u^2/2}. \qquad (12)$$

The frequency ω_1, which is seen at a given angle θ, depends now on the strength of the undulator field B_0, i.e. on the undulator parameter K_u. This offers a way to adjust the undulator spectrum by changing the field strength.

Some further understanding of strong undulator radiation can be obtained by comparing the emission in a laboratory frame F and a system F', moving with the average longitudinal velocity $\langle\dot{z}\rangle = \beta^*c$, as shown in Fig. 21. Since the absolute velocity of the electron is fixed, the transverse motion modulates the longitudinal velocity component. While the electron traverses the axis, its transverse velocity is large and the longitudinal one must be smaller, going actually backward in the moving system. At the maximum excursion the transverse velocity vanishes and the longitudinal one is larger than average, going forward in the moving system. As a result the trajectory in F' has the form of a figure eight. This represents a non-linear motion with an x^* component of frequency Ω_u^* and odd harmonics of it, and a z^* component of

frequency $2\Omega_u^*$ and harmonics of this. The electron radiates at these frequencies mainly perpendicular to the x^* axis for the odd, and perpendicular to the z^* axis for the even harmonics of Ω^*. In the laboratory this radiation is Doppler-shifted and peaked forward along the z axis for the odd, and around it for the even harmonics.

For large values of K_u and higher harmonics, the radiation pattern becomes very complicated. In Fig. 22 the spectrum up to the third harmonics is shown for the two polarization modes and their sum. The radiation emitted on the z axis creates the sharp peaks for the odd harmonics which are missing for the even harmonics being distributed around this axis. In Fig. 23 the angular distribution is shown for the two polarization modes of the third harmonic. It shows the complicated structure with a maximum along the axis for the horizontal polarization and vanishing intensity for the vertical one on the x and y axes.

For strong undulators with few periods, the interference effects between the emission from different periods are less pronounced and the spectrum resembles more the one obtained from dipole magnets. Such devices are called wiggler magnets, a distinction which is not very sharp. The same device can be an undulator at small fields with a spectrum containing few harmonics, but can emit at high fields many overlapping harmonics typical of a wiggler.

In the extreme case a wiggler will consist of a single strong dipole acting as a source of radiation, with two weak ones to bring the electron trajectory

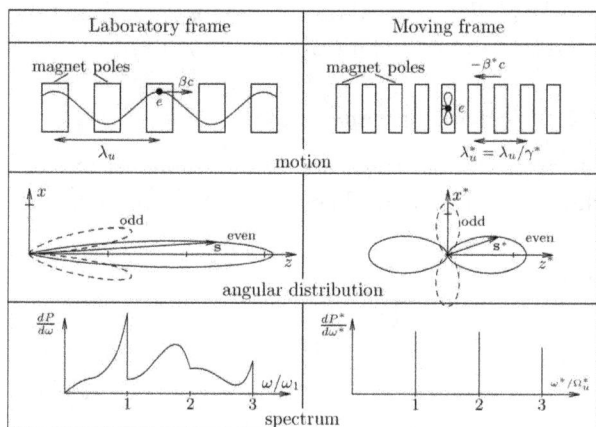

Fig. 21. Strong undulator radiation in a moving and a stationary frame [4].

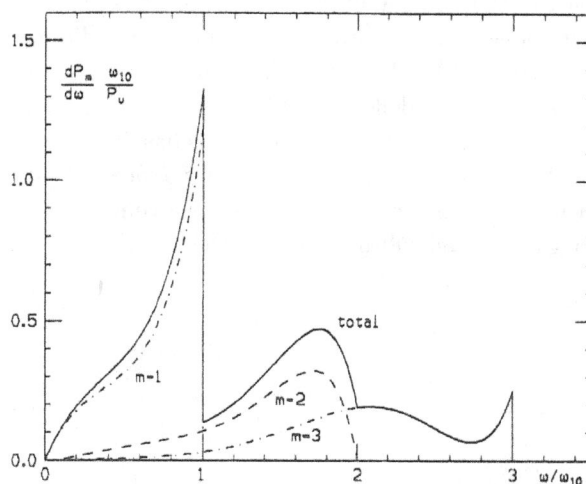

Fig. 22. Strong undulator spectrum.

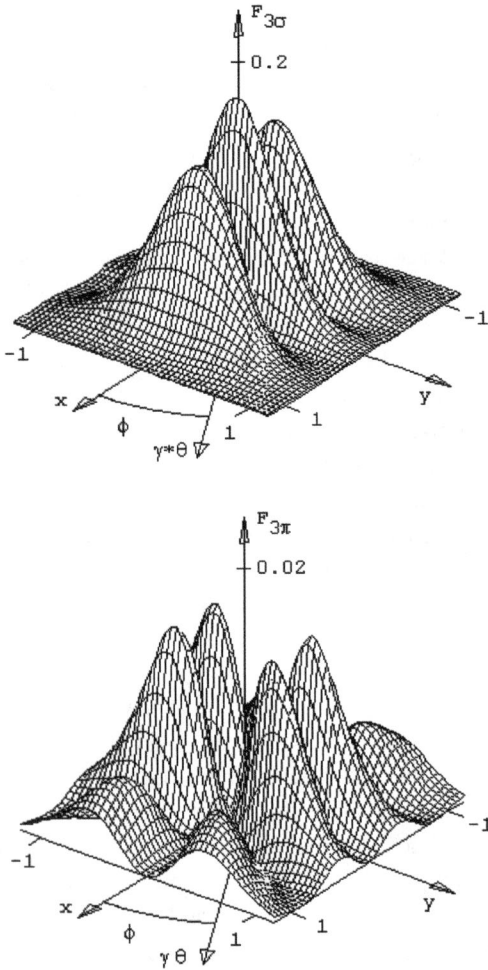

Fig. 23. Angular power density of the horizontal (top) and vertical (bottom) polarization modes for third harmonics of strong undulator radiation.

back to the nominal orbit; see Fig. 24. Its spectrum is basically that of ordinary synchrotron radiation characterized by the critical frequency ω_c. However, since the overall deflection of the three magnets vanishes, a change of their field strengths has little effect on the rest of the machine and allows a local adjustment of ω_c, and therefore of the spectrum. Such a wiggler is often called a wavelength shifter.

7. Application

7.1. *Optics of synchrotron radiation*

One of the interesting properties of synchrotron radiation is the natural small rms opening angle of order $\sigma'_\gamma = \sqrt{\langle \theta^2 \rangle} \approx 1/\gamma$. It has important effects on the optics of this radiation, which becomes dominated by diffraction. This is illustrated in Fig. 25, where a lens is used to form a 1:1 image of the radiation source. Such measurements are often carried out to determine the cross section of the electron beam emitting the radiation. Due to the small natural opening angle, only a central part of the lens is illuminated. The wave nature of light leads to diffraction resulting in a finite image size. In optics books this is treated for a circular opening of diameter D giving an image of rms radius

$$\sigma_{\mathcal{R}} \approx 1.22 \frac{R\lambda}{D} \approx 0.3 \frac{\lambda}{\sigma_\gamma} \approx 0.3\lambda\gamma,$$

where we have taken $D \approx 4R\sigma'_\gamma$. This can limit the resolution of the image severely, especially for $\gamma \gg 1$. It is interesting to calculate the product of image size and opening angle, which is the natural emittance of a round photon beam and its horizontal and vertical components,

$$\epsilon_\gamma = \sigma_{\mathcal{R}}\sigma'_\gamma = \epsilon_{\gamma x} + \epsilon_{\gamma y} \approx 0.3\lambda \geq \frac{\lambda}{2\pi}.$$

This depends only on the wavelength of the radiation. A detailed calculation gives a lower limit of the radial emittance obtained for a Gaussian angular distribution. For undulator radiation the distribution is not Gaussian, giving about twice this limit. For synchrotron radiation the horizontal distribution cannot be observed since the beacon of light sweeps in that direction. The vertical distribution is approximately Gaussian and gives an emittance $\epsilon_{\gamma y} \approx \lambda/4\pi$.

7.2. *Electron storage rings*

Electron storage rings are used as sources of synchrotron radiation. They have an assembly of

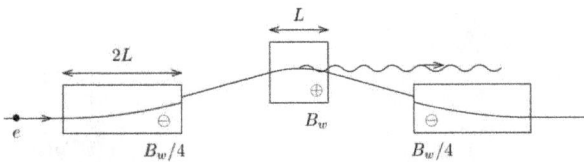

Fig. 24. Wiggler acting as a wavelength shifter [4].

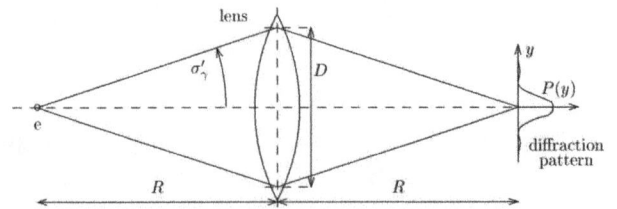

Fig. 25. Imaging with synchrotron radiation.

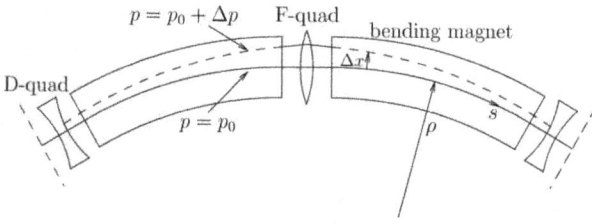

Fig. 26. Magnet lattice consisting of bending and quadrupole magnets.

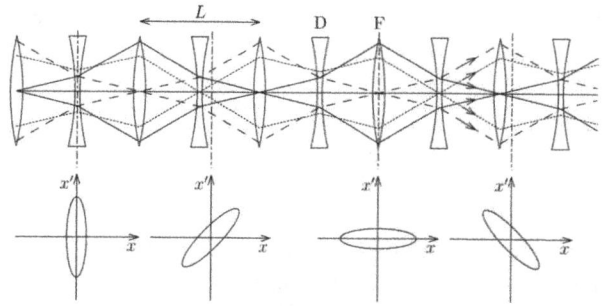

Fig. 27. Electron trajectory over many turns [4].

deflecting dipole and focusing quadrupole magnets, called lattice, a section of which is shown in Fig. 26. The bending magnets determine a closed central orbit of curvature $1/\rho$ and circumference C_0 for electrons with a nominal momentum $p_0 = m_0 c \beta \gamma$. The quadrupoles provide transverse focusing and keep electrons with slight deviations from the nominal momentum, position, or angle close to the central orbit. Each of them focuses in one plane and defocuses in the other, but together they provide focusing in both planes. They are referred to as F quads of D quads, according to their effect in the horizontal plane.

The focusing is illustrated in Fig. 27, where the trajectory of an electron is shown in successive turns but disregarding the curvature in the dipoles located between the quadrupoles. At a fixed location s the angle x'_k and the position x_k of an electron are observed in each turn k. If the two are plotted against each other, the points lie on an ellipse which is narrow and high close to a defocusing lens, a D quad, where the angular deviation is large and the spatial one small. Correspondingly, close to an F quad the angles are small and the deviations large, leading to a wide and low ellipse. In between there is a correlation between angle and position leading to a tilted ellipse. The area A of these ellipses is constant around the ring and proportional to the electron emittance ϵ_{ei}. The ratio between the maxima of the two is also of interest and expressed with the help of the so-called beta functions $\beta_x(s)$ and $\beta_y(x)$. At the center of the quadrupoles the ellipse is upright, giving simple expressions for the area and the beta function,

$$ A = \pi \hat{x}'_k \hat{x}_k = \pi \epsilon_{ei}, \quad \beta_x(s) = \frac{\hat{x}}{\hat{x}'}, $$

while elsewhere the relations are a little more complicated.

For a beam of many electrons we define an average emittance $\epsilon_e = \langle \epsilon_{ei} \rangle$. The electron optics can be optimized for special requirements by choosing a certain arrangement of quadrupoles with different strengths and distances. For some experiments one likes that the synchrotron radiation comes from a small spot which is best obtained where the beta function is small. For undulator radiation the frequency depends on the angle θ of observation $\omega_1 = \omega_{10}/(1 + \gamma^2 \theta^2)$ and its spread can be kept small by locating the undulator where the beta function is large and this angle small.

The trajectory of the electron shown in Fig. 27 represents an oscillation around the nominal orbit, called betatron oscillation. The number of such oscillations per turn is called the tune, Q_x or Q_y; it is a measure of the overall focusing strength. This oscillation around the ring is obviously not harmonic. However, plotting the excursion x_k against turn k at a fixed location s results in points lying on a harmonic curve. Since the oscillation is sampled only once per turn, it does not give the tune Q but only its fractional part as sidebands of the revolution frequency harmonics.

A particle with a momentum deviation Δp has a different curvature in the bending magnets but finds, thanks to focusing, a new closed orbit displaced by $\Delta x(s) = D_x(s)\Delta p/p_0$, with $D(s)$ being called dispersion; see Fig. 26. It is a function of the longitudinal position s around the ring. Such a particle has also a different circumference, $\Delta C = \alpha_c \Delta p/p_0$, with a proportionality factor α_c, called momentum compaction. For strong focusing the off-momentum particle gets deflected quickly back toward the nominal orbit and stays close to it. In this case dispersion and momentum compaction are both small. An electron of nominal momentum has a revolution frequency $2\pi\beta c/C_0$. It changes with momentum deviation Δp, for two reasons: first, the circumference changes; and second, the electron velocity varies, $\Delta\beta/\beta = (\Delta p/p_0)/\gamma^2$, giving

$$\frac{\Delta\omega_r}{\omega_r} = -\left(\alpha_c - \frac{1}{\gamma^2}\right)\frac{\Delta p}{p} = -\eta_c\frac{\Delta p}{p}.$$

The quantity η_c is sometimes called the "slippage factor" and is positive for most rings, which means that an electron with an excess of momentum has a lower revolution frequency and takes longer to make one turn.

Storage rings have at least one cavity which contains an electromagnetic field and voltage oscillating with frequency $\omega_{\rm rf} = h\omega_r$ being a harmonic h of the nominal revolution frequency ω_r. Its longitudinal electric field component accelerates the particle and can provide an energy change, as shown in Fig. 28. A nominal particle with correct energy and revolution frequency passes each turn through the cavity at the synchronous time t_s and receives an energy $\hat{V}e\sin(\omega_{\rm rf}t_s) = U_s$, which compensates for its energy loss due to synchrotron radiation. An electron coming earlier, smaller t, receives more energy and takes longer for the next turn, which corrects some of its timing error. Correspondingly, an electron arriving late gets less energy and arrives earlier next turn. Furthermore, a particle with excess energy takes longer for the next turn and arrives late when the rf-voltage is lower, and vice versa. As a consequence this rf-voltage compensates on average for the energy loss due to synchrotron radiation and provides focusing for timing and energy deviations. In general each particle executes an oscillation around its synchronous time and nominal energy, called synchrotron or energy oscillation. There are

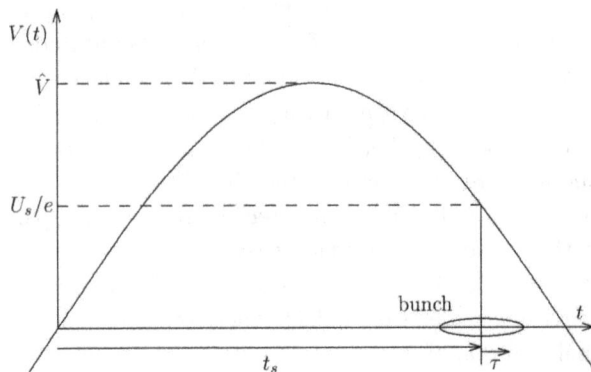

Fig. 28.   Energy replacement and focusing by an rf-field.

h such synchronous times t_s and the oscillating electrons form h so-called bunches with distributions of rms values σ_τ, σ_E, σ_x, and σ_y in time, energy, and transverse excursions.

7.3. *Quantum effects*

Being a rather macroscopic device, a storage ring is at first glance not expected to show much quantum effect. However, the emission of synchrotron radiation occurs in quanta which have an effect on the electrons and can modify the spectrum.

The circular electron orbit resembles a little bit a giant atom. It is far from the lowest energy state; on the contrary, its quantum number is extremely high, in which case a classical treatment can be used as far as the electron orbit is concerned.

The energy of the emitted radiation has to be provided by the electron and is replaced on average by the rf-voltage. A small fraction of the electron energy is contained in the betatron and synchrotron oscillations, and it also gets reduced by the radiation emission while the cavity provides only a longitudinal acceleration. This leads to damping of these oscillations. For the energy oscillation this damping is even more evident. The synchrotron radiation power is a strong function of the electron energy and takes more energy away during the part of the oscillation when the energy is high, and less when it is low. These damping effects are very important for the operation of electron storage rings as they correct errors coming from electronic noise, scattering on the residual gas in the vacuum chamber, and misaligned injection.

The radiation is emitted in quanta of energy $E_\gamma = \hbar\omega$, with $\hbar$ being Planck's constant divided by 2π. Sometimes we like to give the number, $\dot{n}(\epsilon_\gamma)$, of photons with a certain energy ϵ_γ being emitted per unit time instead of the radiation power. With the above relation we can convert the spectral power expression into

$$\frac{dP}{d\omega} = \hbar\frac{d\dot{n}}{dE_\gamma/E_\gamma}.$$

The power per absolute frequency interval is proportional to the photon flux per relative energy interval.

With each emitted photon the electron suffers a sudden energy loss, which will excite an energy oscillation. This and the damping effect lead to a Gaussian equilibrium energy distribution with an rms value σ_E. If this emission happens at a location of finite horizontal dispersion D_x, the electron has afterward a different equilibrium orbit corresponding to its new energy and will make a betatron oscillation around it, as shown in Fig. 29. Together with the damping this leads to a finite horizontal equilibrium emittance of the electron beam. Ideally, there is no dispersion in the vertical plane, but the finite opening angle of the photon emission leads to a very small vertical emittance. However, due to alignment errors there is in praxis always a small vertical dispersion that is the dominant contribution to the vertical emittance. It is amazing that for a large device like a storage ring, quantum effects are important, leading to a beam size of the order of mm which can be measured easily by imaging.

In some cases quantum effects can change the spectrum. For normal synchrotron radiation the calculated spectrum decays fast at high frequency but has no limit. There will be a few photons which have a higher energy than the electron, which is obviously impossible. The reason for this is the neglect of the recoil of the electron at the moment of emission. A proper calculation gives a spectrum having a limited upper frequency. For most practical cases this is of no concern unless the emitted photon carries a significant fraction of the electron energy, which could happen in the high fields at the interaction point of a future linear collider.

7.4. *Radiation emitted by many particles*

The n_e different electrons in the beam usually have no fixed phase relation between them, and the total radiation power is just the sum of the individual contributions $P_{\text{tot}} = \sum P_0$, with P_0 being the power per electron (4).

The size and angular spread, as well as their product (the emittance), of the photon beam are of interest. It is a convolution of the natural properties of the emitted photons and that of the electrons. For the first one we get at best $\epsilon_\gamma = \lambda/4\pi$ in each plane. Ideally the emittance of the electron beam should be smaller or at least comparable in order to avoid dilution of the emittance. In the second case the two should also be matched, i.e. having the same ratio between size and angle. In this case the final photon beam has the smallest possible emittance; it is called diffraction-limited or spatially coherent at a given wavelength λ. The transverse electron beam dimensions can in this case not be resolved using radiation of this wavelength to image the cross section. An electron optics with strong focusing and weak bending can give a small dispersion and, according to Fig. 29, small quantum excitation of horizontal betatron oscillation. A lot of work has been done to develop small emittance rings. However, only for rather long wavelengths is it possible to reach the diffraction limit.

It is of interest to achieve temporal coherence by having a bunch length shorter than the emitted wavelength $c\sigma_\tau < \lambda$. In this case the radiation contributions of all electrons are in phase and their fields have to be added, as illustrated in Fig. 30. With the field being proportional to the number (n_e) of electrons in the bunch, the resulting power becomes proportional to n_e^2, giving an important enhancement. Bunches that are shorter than the wavelength represent a periodic current and emit a spectrum consisting of lines at harmonics of the revolution frequency. Attempts have been made to operate a storage ring in this mode, but one succeeded only for relatively long wavelengths that were of limited interest for applications. However, it is possible to achieve temporal

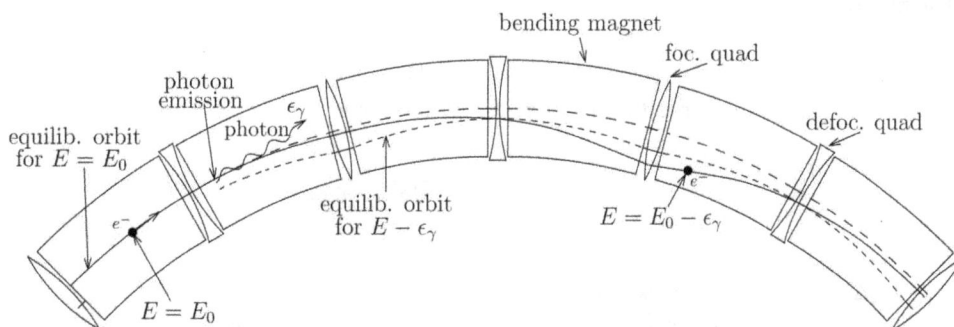

Fig. 29. Quantum excitation of horizontal betatron oscillation at finite dispersion.

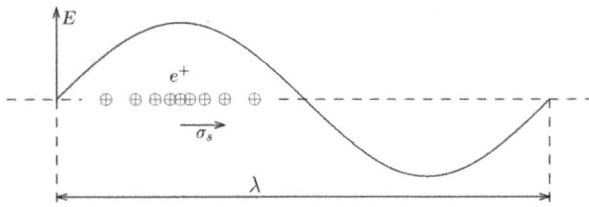

Fig. 30. Temporal coherence for a bunch shorter than the wavelength [4].

coherence in another device, called the free electron laser (FEL).

A certain amount of temporal coherence is also obtained by filtering a narrow band of the undulator spectrum. This results in a wave packet having a determined phase relation over a large distance, called coherence length. But no enhancement of the total radiated power is obtained this way.

8. Free Electron Laser

Temporal coherence can be obtained by modulating the beam intensity at a period equal to the desired wavelength. This is done in an FEL, where the emitted radiation itself is used to create this modulation [18]. In Fig. 31 the functioning of such a device is illustrated by separating the three main actions — energy modulation, conversion into intensity modulation, and temporal coherent emission — while in most cases they are superimposed.

The electron beam traverses the first undulator, where it emits radiation at wavelength λ_{10} on the axis, called spontaneous radiation. This emitted electromagnetic wave has a certain amount of coherence at this wavelength which will interact with the cotraversing electron beam. Since the electric field is perpendicular to the main propagation direction of the electrons, it will not change their energy except at locations where the particles make an angle due to the undulator motion.

This interaction between the electron beam and an electromagnetic wave of wavelength $\lambda_{10} = \lambda_u(1 + K_u^2/2)/2\gamma^2$, while both are going through the first

undulator, is illustrated in Fig. 32. Here, an electromagnetic wave with horizontal field E_x enters an undulator at time $t' = 0$. At the same time an electron enters, shown as •, and traverses the undulator with longitudinal velocity $v_z = \beta^* c < c$. Its trajectory has the form of a sine function shown as a solid curve. The electric field, being perpendicular to the z axis, does not change the electron energy except where this makes an angle and has a transverse

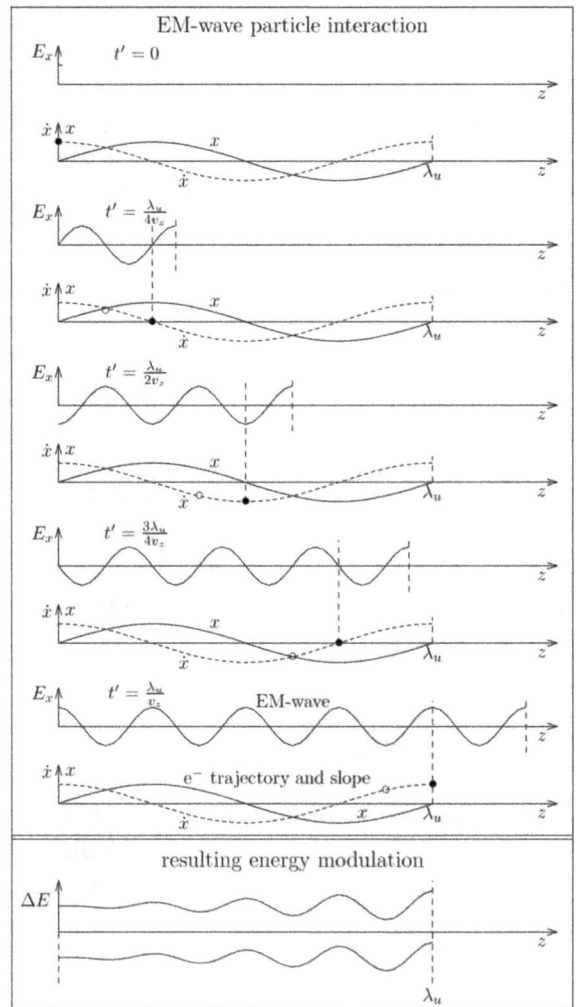

Fig. 32. Interaction of an EM wave in an undulator with electrons of $\beta^* = \frac{4}{5}$ and resulting energy modulation.

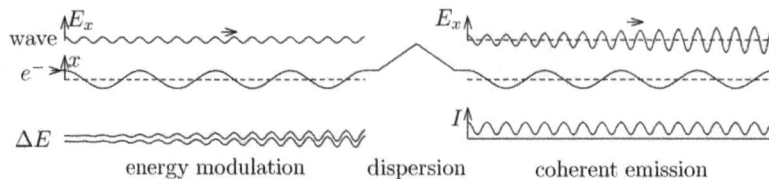

Fig. 31. Basic function of an FEL.

velocity component $\dot{x}$, which is shown in the figure as a dashed curve. This results in a power exchange $P = eE_x\dot{x}$. We watch the interaction between the two at different times during the traversal.

At $t' = 0$ the field E_x and the particle velocity are both positive, resulting in an energy increase of the electron. Later, $t' = \lambda_u/4v_z$, the electron has traversed a quarter of the undulator period while the faster wave has gone a little further. In this position the field and the transverse velocity vanish and no energy is exchanged. At $t' = \lambda_u/2v_z$ the transverse field and velocity are both negative, leading again to an increase in the electron energy. At the next point there is again no interaction but at $t' = \lambda_u/v_z$ the field and the transverse velocity are again both positive, resulting in an energy increase for the electron. It should be noted that by this time the wave has advanced by one wavelength compared to the electron.

A second particle, ∘, being longitudinally displaced by $\lambda_{10}/2$ compared to the first one, finds just the opposite relation between E_x and $\dot{x}$ than the first one and suffers a loss of energy. A string of many particles with different longitudinal positions will get an energy modulation with a period length of $\approx \lambda_{10}$.

At the end of the first undulator, the beam enters a dispersive section. It has three bending magnets which create a distortion of the electron trajectory by deflecting it first by a positive angle, followed by twice the opposite angle back toward the axis, and finally a positive angle again to bring the electron beam back on-axis. The action of the dispersion section is shown in Fig. 33, which gives the trajectories of particles with nominal energy and with positive and negative deviations. Due to the momentum dependence of the curvature in the magnetic field, the higher energy particle has a shorter path length, and vice versa. The electron beam enters the section at "a" with an energy modulation. By the time it reaches the center "b," the particles with higher energies have moved to an earlier relative time and those with less energy to a later one. At the exit "c" the path length differences have moved particles of maximum positive and negative energy deviations to the same longitudinal position. We have now a beam which has no longer a modulation of its central energy but one of its energy spread. As a consequence also the particle density, or current I, has a longitudinal modulation with period λ_{10}. This beam will now

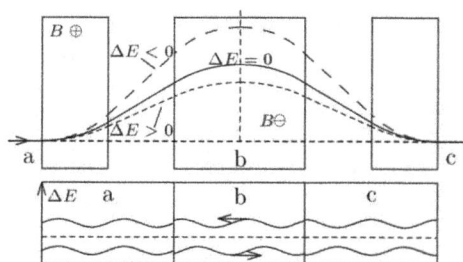

Fig. 33. The dispersion section converts an energy modulation into one of intensity.

radiate with temporal coherence at the frequency ω_{10}.

In our example shown in Fig. 31, we have separated the undulator into one for energy modulation and a second one for coherent emission as well as a dispersion section in between, to illustrate the different steps. Actually, there are FELs of this type, usually called optical klystrons. In most cases, however, the path length in a strong undulator depends on energy and acts like the dispersion section. As a result the coherent emission already starts somewhere at the beginning of the undulator. Most FELs consist of a single strong undulator, as shown in Fig. 34, in which the energy modulation and its conversion into an intensity modulation as well as the coherent emission occur. In some cases one uses an external light source to start this process, called seeding, and the FEL represents an amplifier with a certain gain. In other cases one uses the spontaneous radiation and the FEL acts like an oscillator. The FEL can operate in a single traversal and the radiation builds up exponentially along the path. To get a sufficient gain the undulator might have to be very long. Much effort is presently put in such FELs to produce coherent x-rays. For visible and ultraviolet light there are mirrors available. In this case the FEL can be placed between two focusing mirrors

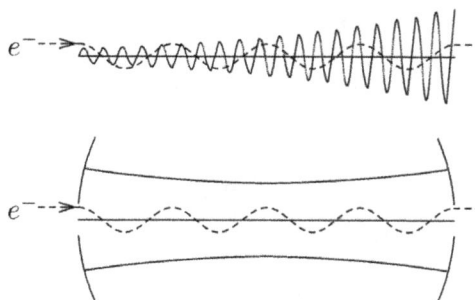

Fig. 34. Single- and multiple-passage FELs.

forming an optical cavity. The radiation goes many times through the undulator and its intensity builds up if the gain is larger than the relative losses on the mirrors. It should be pointed out that the electrons entering the undulator are always new and not yet energy- or intensity-modulated.

9. Synchrotron Radiation Facilities

By now there are many synchrotron radiation centers operating in different parts of the world. Research in a variety of fields like biology, physics, chemistry, and even archeology is carried out in these laboratories using one or several storage rings as the main instrument(s). The actual photon sources are bending magnets, wigglers, and undulators. Bending magnets deliver a fan of radiation with a horizontal width determined by collimation. The energy of the storage ring is chosen to get the desired spectrum from these bending magnets, resulting in ring energies of 1–2 GeV for ultraviolet light and several GeV for x-ray research. The spectrum from individual wiggler magnets can be adjusted by changing the magnetic field strength without affecting other beam lines. With available wiggler fields of several teslas, even a low energy ring can be used for x-ray work. Undulators have a spectrum which is determined by the period length λ_u, the adjustable parameter K_u, and the electron energy. Most of these devices use permanent magnets with a variable gap to adjust the field. Since the period length λ_u should be at least twice the gap size to produce a sufficient field strength and the aperture for the beam cannot be too small, the desired undulator spectrum also influences the choice of the beam energy. Undulator radiation has some narrow peaks which can be adjusted to limit the total power deposited on mirrors for a given intensity at the desired part of the spectrum.

A list of operating sources can be found on the website www.lightsources.org [19] and in the booklet Ref. 20. It contains parameters of existing facilities and of others being constructed or planned as well as interesting general information.

References

[1] A. A. Sokolov and I. M. Ternov, *Radiation from Relativistic Electrons*, translated by S. Chomet and edited by C. W. Kilmister American Institute of Physics translation series (New York, 1986).

[2] V. A. Bordovitsyn, *Synchrotron Radiation Theory and Its Development* (World Scientific, Singapore, New Jersey, London, Hong Kong, 1999).

[3] H. Wiedemann, *Synchrotron Radiation* (Springer-Verlag, Berlin, Heidelberg, 2002).

[4] A. Hofmann, *The Physics of Synchrotron Radiation* (Cambridge University Press, 2004).

[5] J. D. Jackson, *Classical Electrodynamics* (John Wiley, New York, l962, 1974, 1998).

[6] M. Schwartz, *Principles of Electrodynamics* (McGraw-Hill, New York, 1972).

[7] L. Eyges, *The Classical Electromagnetic Field* (Dover, New York, 1980).

[8] J. Schwinger, L. L. DeRaad Jr., K. A. Milton and W.-Y. Tsai, *Classical Electrodynamics* (Perseus/Westview, 1998).

[9] A. Liénard, *L'Éclairage Électrique* **16**, 5 (1898).

[10] E. Wiechert, *Archives Neerlandaises* **5**, 549 (1900).

[11] G. A. Schott, *Annalen der Physik* **24**, 635 (1907).

[12] G. A. Schott, *Philos. Mag.* **13**, 194 (1907).

[13] J. Blewett, *Phys. Rev.* **69**, 87 (1946).

[14] J. P. Blewett, *Nucl. Instrum. Methods* **A266**, 1 (1988).

[15] V. L. Ginzburg, *Izv. Akad. Nauk SSSR, Ser. Fiz.* **11**, 165 (1947).

[16] H. Motz, *J. Appl. Phys.* **22**, 527 (1951).

[17] D. F. Alferov, Yu. A. Bashmakov and E. G. Bessonov *Synchrotron Radiation*, ed. N. G. Basov (New York Consultants Bureau, 1976).

[18] L. R. Elias, W. M. Fairbanks, J. M. J. Madey, H. A. Schwettmann and T. J. Smith, *Phys. Rev. Lett.* **36**, 717 (1976).

[19] http://www.lightsources.org (→Facility Information → Light Source Facilities).

[20] J. Murphy, *Synchrotron Light Source Data Book*, BNL 42333 (NSLS Brookhaven National Laboratory).

Albert Hofmann worked on a range of different accelerators. Starting at a low energy Cockcroft-Walton device, he worked at Harvard/MIT on an electron-positron collider CEA, at CERN on the proton-proton collider ISR, at SLAC on the linear collider SLC, the rings SPEAR and PEP, as well as on accelerator issues of synchrotron radiation, and at CERN on the commissioning and beam physics of LEP. Since his retirement up to now he spent short periods at SLAC on PEP-II, the Brazilian light source LNLS, and at Lyncean Technologies, a light source company. He was awarded the American Physical Society's Wilson Prize in 1996.

Reviews of Accelerator Science and Technology
Vol. 1 (2008) 143–161
© World Scientific Publishing Company

World Scientific
www.worldscientific.com

Medical Applications of Accelerators

Hartmut Eickhoff

Gesellschaft für Schwerionenforschung GmbH (GSI),
Darmstadt, Germany
H.Eickhoff@gsi.de

Ute Linz

Forschungszentrum Jülich,
Jülich, Germany
u.linz@fz-juelich.de

Particle accelerators play an essential role in the field of medical applications. A large variety of systems is in use for diagnostic purposes, such as the production of radioactive tracers for imaging or x-ray radiography. The dominant application, however, is related to the treatment of cancer patients. This article puts emphasis on cancer treatment, presenting the status and developments of the corresponding technical systems, and gives a brief overview of the biophysical properties and medical aspects of these treatments.

Keywords: Ion therapy; medical accelerators; gantry.

1. General Aspects of Radiotherapy

The concept of all radiation therapy is to achieve severe cell damage inside the tumor volume by applying a required high dose with photon or particle beams and to minimize the damage in the healthy tissue outside the tumor. The effects of radiation are quantified according to the dose, which is defined as the energy deposited per mass unit and measured in grays ($1\,\mathrm{Gy} = 1\,\mathrm{J/kg}$).

The difference between tumor control and unacceptable damage to nearby healthy tissue requires dose control on the percent level. Essential in this is the accurate definition of the tumor volume now made possible by CT (computerized x-ray tomography) and MRI (magnetic resonance imaging) methods. The challenge is to conform the delivery precisely to the irregular 3D shape of the tumor.

The following forms of radiation have been used for tumor therapy:

- Electromagnetic radiation (x- and γ-rays)
- Ions ($Z \leq 18$)
- Neutrons
- π mesons (pions)

X-rays still make up the major share. The role of ion beams is gradually increasing. Gamma rays are nowadays restricted to the application in the gamma knife, which is gradually being replaced by linac-based radiosugery systems with a stereotactic frame. Neutrons have also lost much of their relevance and pions are no longer applied.

For each of these kinds of radiation therapy, specific accelerator types are in use. In addition, accelerators are employed for the production of radioactive isotopes, for both diagnostic and therapeutic purposes.

1.1. Radiation properties

The energy loss per unit distance of ionizing particles in matter or linear energy transfer (LET) has long been viewed as a major parameter for describing the qualitative differences of the biological effects of various kinds of radiation [1]. Numerically, LET is identical with the stopping power of a particle.

At first approximation, $20\,\mathrm{eV/nm}$ can be considered the lower limit of the so-called "high-LET" radiation [2]. This amount of energy deposited in a cell nucleus causes two DNA double-strand breaks, on the average.

The local energy deposition of photons does not reach this limit. Similarly, more than 95% of the dose of a 200 MeV proton beam comes from particles depositing less than 10 eV/nm. The 20 eV/nm limit is reached only within the last 50 μm of the trajectory.

A carbon ion beam traveling a similar distance as a 200 MeV proton beam has high-LET properties in the last 50 mm range, i.e. including the distal part of the plateau region.

Kempe *et al.* have recently shown that the lateral penumbra (FWHM of the dose peak) and the longitudinal standard deviation of the energy deposition show a similar dependence on the number of nucleons [2]. These clinically interesting parameters differ most dramatically from protons to helium. The differences are approximately twice as high as between helium and carbon. For the heavier ions up to neon, they are further reduced.

Figure 1 shows the depth–dose distribution for photons and ions: whereas photons (and similarly neutrons) have a maximum dose close to the surface and an exponential decay of the dose with increasing depth, ion beams deposit the dose maximum (Bragg peak) near the end of their range, which can be adjusted with a proper selection of the particle energy.

Robert Wilson was the first to recognize the therapeutic advantage of the "inverse dose–depth profile" of ions [3]. He described the unique conditions for the treatment of deeply seated tumors, with a minimum effect on the tissue in the entrance channel. The wider proton curve arises from the higher multiple scattering and range-straggling of protons; the tail of the carbon curve comes from nuclear breakup of the projectile into lighter (longer-ranged) fragments. Ionization density for charged particles varies with Z^2. Therefore, heavier ions — in addition to having sharper stopping points — are more lethal to malignant cells.

Besides the "physical dose", d_P, determined by the number and energy of the particles, stopped in a defined volume, the "biologically effective dose", d_E ($= d_P$ * RBE; RBE = relative biological effectiveness), is essential for the treatment, taking into account the different biological efficiencies of the various radiations in tissue. The RBE describes the biological effect of a radiation source of interest in comparison to a reference radiation dose — mostly 250 kV x-rays or ^{60}Co gamma rays. For protons, the RBE is close to 1.

The advantage of C ions is their selectively enhanced RBE in the Bragg peak region. The RBE factor for other particle species such as neutrons or ions with an atomic mass > 16 is also high, but uncontrolled secondary effects like fragmentation play a dominant role leading to side effects.

1.2. *Treatment modalities*

The development of multileaf collimators to be integrated into clinical x-ray linacs and the implementation of inverse treatment planning have helped to actualize intensity-modulated radiation therapy with x-rays (IMXT). IMXT has improved tumor conformity in comparison to conventional wide-field x-ray irradiation. But, in return, IMXT increases the integral dose of normal tissue outside the target volume (see Sec. 5).

The well-defined stopping point of protons and other ions makes it easier to conform the radiation dose to an irregularly shaped tumor by independently varying the energy of the stopping particles. Pencil beam scanning systems have been designed for this purpose, but the required precision of control and response time are challenging.

But even with less sophisticated delivery systems, dose distributions of proton and ion beams

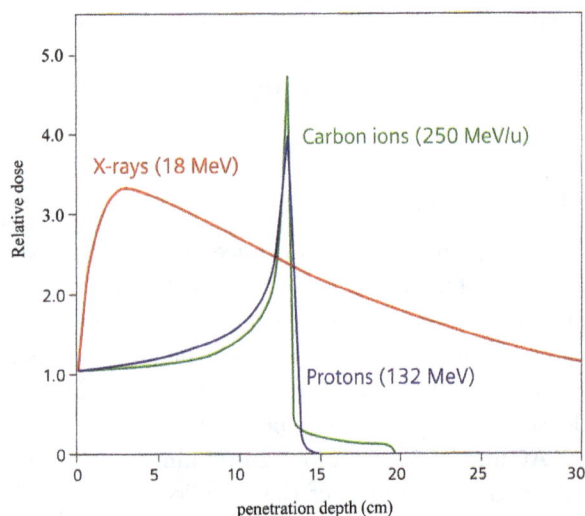

Fig. 1. Depth–dose distribution for photons, protons and ions.

can compete with those of the best x-ray systems, and successful clinical programs with these beams have been going on for almost 50 years (for recent reviews see. e.g., Refs. 4 and 5). The implementation of active beam scanning has further improved the dose distribution of ion beams. Their physical properties as compared to photons mean that therapeutic ion beams will always show a more favorable dose distribution, with a less integral dose outside the target volume.

The treatment modalities for ion beam therapy (IBT) can be distinguished as "passive" versus "active" [6]. For the passive modality the accelerator delivers a beam with nearly constant or only slowly varying properties (e.g. energy). By means of various mechanical components the beam is shaped in both the transverse and the longitudinal direction to achieve an appropriate conformity at the location of the tumor (see Fig. 2). After beam extraction from the cyclotron or synchrotron, the beam is spread out by means of scattering foils or wobbler magnets. A fine degrader gives the possibility of adjusting the mean energy, while a ridge filter provides spreading-out of the Bragg peak. The bolus,

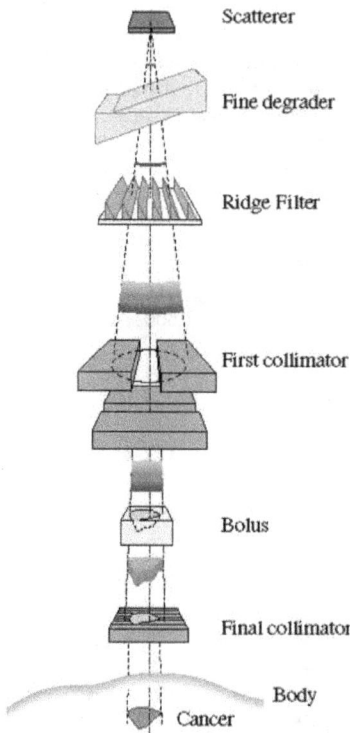

Fig. 3. Schematic of the "active" treatment modality (intensity-controlled rasterscan).

individually fabricated for each patient, has the 3D shape of the tumor surface and adapts the distal edge of the treatment volume to the tumor. In addition, collimators are used to shape the beam. Early proton therapy was performed with fixed-energy synchrocyclotrons (e.g. Harvard, Uppsala); their "passive" delivery systems were decoupled from the accelerator, requiring only a fast and reliable beam cutoff system. In these first facilities, treatment ports were changed by moving the patient.

The concept of the "active" treatment modality is to avoid all beam-shaping components in the beam line and to provide the appropriate beam properties from the accelerator. In Fig. 3 the principle of the "intensity-controlled" rasterscan procedure is shown; this was developed and applied at GSI during the last 10 years to about 400 patients, with very good results [5].

Within this treatment method the tumor volume is dissected into slices ("isoenergy slices") of different depths. These slices are irradiated with ions of specific energies, correlated to the required penetration depth. By sequentially treating the slices with adequate intensities, the required dose profile for the tumor volume is achieved.

To cover the lateral dimensions of the tumor, the ion beam passes two fast scanner magnets that deflect the ions in both the horizontal and the vertical direction after being accelerated to the required energy in a synchrotron and slowly extracted.

The rasterscan control system determines the excitation of the scanning magnets to deposit the required dose profile, measuring the number of ions

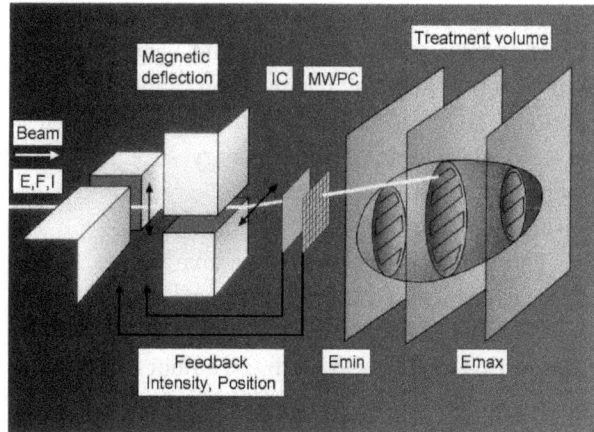

Fig. 2. Beam-shaping devices for the "passive" treatment modality.

at a specific irradiation point by means of ionization chambers, and the position and beam width at
each scanning point by means of fast multiwire proportional counters in front of the patient. When a
required dose limit of an isoenergy slice has been
reached, the beam extraction is interrupted very
quickly ($< 0.5\,\mathrm{ms}$).

The requirements of the "passive" modality for
the accelerator systems are rather low, as nearly constant beam parameters are required; the disadvantages of this modality are reduced tumor conformity,
as compared to the "active" mode and enhanced
fragmentation due to scattering processes at the collimators and boli. The "active" method, on the other
hand, demands fast energy variation at the accelerator level to provide beams of different penetration
depths. Apart from energy variation, intensity and
beam spot variations on a pulse-to-pulse basis can be
required at the treatment location in order to minimize the treatment time.

At present only few of the existing IBT facilities
are equipped with the rasterscan treatment modality
(PSI and GSI), but for the major part of the planned
facilities "active" treatment systems are foreseen [7].

2. Demand of Radiotherapy Facilities

As part of their planning process, four European
groups (HIT, MEDAUSTRON, ETOILE, CNAO)
have evaluated treatment data of cancer patients to
set up a list of potential candidates for IBT [8].

Even though the approaches were quite different (one-day or three-month national survey, single-
clinic experience or review of epidemiological data),
the suggested indications are quite similar. According to this report, approximately 5–6% of all newly
diagnosed cancer patients or 13–16% of all cancer
patients receiving irradiation should benefit significantly from IBT.

Numberwise, the 10 most frequent indications
do not at all include those for which the longest
IBT experience exists, i.e. uveal melanomas, chor-
domas and chondrosarcomas. It is the malignancies which cause the high societal tumor burden,
such as non-small-cell lung cancer (NSCLC), gastric,
prostate, liver, or head and neck cancer (Table 1).
Even though less than half of these tumors are considered candidates for IBT, they are so common that
they make up the major fraction of all anticipated
indications.

Table 1. Cancer types, their incidences and the possible fraction to profit from IBT. The incidences are from various public sources, with emphasis on European numbers. The fraction sizes are rounded values from the consensus report of the ENLIGHT (European Network for Light Ion Hadron Therapy) working group. The list is sorted according to the expected numbers of patients for a population of 10 million.

Indication	Incidence (no./10^5 people)	% for IBT	Annual cases
NSCLC	54	20	1080
Gastric cancer	16	45	720
Prostate cancer	80*	25	575
Liver cancer	10	50	500
Head & neck cancer	13	25	325
Non-Hodgkin lymphoma	14	20	280
Rectal cancer	11.4	20	228
Pancreatic cancer	9.8	20	196
Bladder cancer	10.1	15	152
Cervical cancer	6.7	20	134
Uveal melanoma	0.6	100	60
Brain tumors	2.3	25	58
Bile duct cancer	2.7	20	54
Soft tissue sarcoma	1	40	40
Hodgkin lymphoma	2	20	40
Anaplastic thyroid cancer	0.7	45	32
Salivary gland cancer	0.5	45	23
Pediatric malignancies	0.6 ($3.6/10^5$ children)	15	9
Chordoma	< 0.1 (0.06)	100	6
Chondrosarcoma	< 0.1 (0.03)	100	3
Sum			**4515**

*Related to 10^5 men.

For a population of 10 million people, a cautious estimate prognosticates approximately 4500 patients with primary malignancies, annually, for IBT. In addition, approximately 40% of the local recurrences are likely to benefit from IBT as well.

A somewhat different categorization was established by the Italian TERA project [9]. They concluded that a minimum of 1% of the patients receiving x-ray radiation therapy today should definitely profit from proton irradiation. These would be indications for which clinical data existed already, e.g. chordoma or chondrosarcoma of the base of the skull, and uveal melanoma. 12% of the patients receiving x-ray treatment should be likely to benefit from a treatment with protons, too, but for these indications clinical studies are still lacking.

Finally, 3% of the patients could profit from high-LET ions, such as carbon. For these, clinical data are missing, as well. But a clinical rationale for these indications could be derived from the experience with fast neutrons.

3. Accelerator Systems for Radiation Therapy

3.1. *Accelerators for x-ray therapy*

X-ray therapy is the most widely used radiation treatment. The x-ray depth–dose relation is exponential (Fig. 1). Hence, treating a tumor located, for example, 25 cm inside a patient involves significant doses up- and downstream of the treatment field. These doses can be mitigated by multiport treatments, where beams are brought in from several angles overlapping at the tumor. Restricting the beam cross-section to the different projections of the tumor along the different beams is accomplished with sophisticated collimators.

X-rays are produced by electrons striking a heavy-metal target. 5–30 MeV S-band electron linacs are the mainstay of radiation therapy today ($\approx$ 5000 worldwide; principal manufacturers Varian, Siemens, General Electric, Mitsubishi and Toshiba).

The very broad bremsstrahlung spectrum is "hardened" by using absorbers to filter out contributions from lower energies. Powered by either a magnetron (< 6 MeV) or a klystron (> 6 MeV), these accelerators operate at repetition rates up to 1 kHz.

S-band linacs are a highly successful spinoff from high-energy and nuclear physics programs. The low

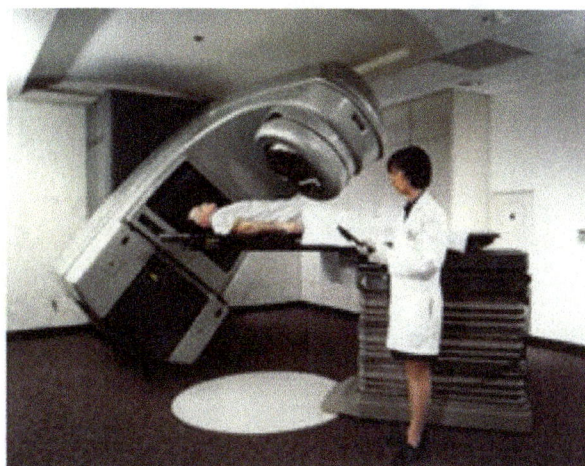

Fig. 4. Modern electron linac for photon treatment.

rigidity of the electron beams and very efficient packaging of the accelerator and beam transport systems have permitted overall lengths of 1–2 m. This compactness paired with efficiency and reliability has been the key to their acceptance in clinical applications. The development of the isocentric gantry configuration (the patient lies stationary in the center of a circle on which the x-ray beam can be rotated) has allowed multiple treatment ports.

3.2. *Accelerators for IBT*

The dose–depth behavior, shown in Fig. 1, reveals the advantage of using ions rather than x-rays for therapy.

The first use of ion beams, or more precisely of protons, in humans was for pituitary hormone suppression in the treatment of patients with metastatic breast carcinoma by John H. Lawrence [11]. This was in 1954 at Berkeley, where the brother of Dr. Lawrence (Ernest O.) had developed the first cyclotron.

Since then more than 60,000 patients have been treated with ions. The largest fraction, nearly 90%, is attributed to proton treatments, followed by C and He ions. Nearly 30 IBT facilities are at present in operation, mainly in Europe, the USA and Japan. With more than 11,000 treatments, the Loma Linda University Medical Center (LLUMC) in the USA has treated the highest number of patients (see Table 2). Inaugurated in 1990, it was the first IBT center in the world, exclusively devoted to clinical applications.

Table 2.   Worldwide IBT facilities (status in 2007) [12].

Place	Country	Particle	Start of treatment	Total patient treated
ITEP, Moscow	Russia	p	1969	4024
St. Petersburg	Russia	p	1975	1327
PSI, Villingen	Switzerland	p	1984	4875
Dubna	Russia	p	1999	402
Uppsala	Sweden	p	1989	840
Clatterbridge	England	p	1989	1701
Loma Linda, California	USA	p	1990	11,414
Nice	France	p	1991	3129
Orsay	France	p	1991	4143
iThemba Labs	South Africa	p	1993	500
MPRI (2), Indiana	USA	p	1993	379
UCSF, California	USA	p	1994	920
HIMAC, Chiba	Japan	ion	1994	3795
TRIUMF, Vancouver	Canada	p	1995	130
PSI, Villingen	Switzerland	p	1996	320
GSI, Darmstadt	Germany	ion	1997	384
HMI, Berlin	Germany	p	1998	1014
NCC, Kashiwa	Japan	p	1998	552
HIBMC, Hyogo	Japan	p	2001	1658
HIBMC, Hyogo	Japan	ion	2002	271
PMRC (2), Tsukuba	Japan	p	2001	1188
NPTC, MGH, Boston	USA	p	2001	2710
INFN-LNS, Catania	Italy	p	2002	151
Shizuoka Wakasa	Japan	p	2003	570
WERC, Tsuruga	Japan	p	2002	49
WPTC, Zibo	China	p	2004	537
MD Anderson Cancer Center, Houston, Texas	USA	p	2006	527
FPTI, Jacksonville, Florida	USA	p	2006	360
NCC, Ilsan	South Korea	p	2007	155

In contrast to the very compact electron linacs, used for x-ray therapy, systems for IBT are huge and much more expensive. Acceleration is by cyclotrons or synchrotrons. A separate beam delivery section distributes the beam through fixed-angle or rotating beam lines to various treatment rooms.

A proton energy of 250 MeV allows penetration of 30 cm tissue. An average beam current of a few nanoamperes is adequate for treatment times of ~1–2 min for all but the largest therapy fields.

The first hospital-based proton therapy accelerator was the 250 MeV synchrotron at Loma Linda, built by Fermilab. It has a weak-focusing lattice, injected by a 2 MeV RFQ with a single-turn kicker. Operating on a 2 s cycle, the half-integer resonant extraction provides reasonably flat spills with a 25% duty factor at any desired energy up to 250 MeV. The beam is transported to two fixed-beam rooms and three gantry rooms.

Synchrotron-based facilities (Hitachi, Mitsubishi) have come on line in Tsukuba, Shizuoka and Wakasa Bay, Japan.

Cyclotron-based facilities (IBA, Sumitomo) are operating in Boston (Massachusetts General Hospital) and Kashiwa (National Cancer Center East). The fixed-energy (235 MeV) proton beam is degraded, energy-selected and collimated right after the cyclotron, with the beam current being increased to provide constant emittance and brightness of beam delivered to the gantries. Intensity control is achieved via the internal proton ion source.

For a carbon beam to penetrate 30 cm in tissue, an energy of approximately 430 MeV/u, corresponding to a magnetic rigidity of 6.6 Tm, is required. Isocentric delivery, therefore, presents a formidable challenge (see Subsec. 4.1).

Treatments with helium beams began in the mid 1950's at Berkeley's 184″ synchrocyclotron. Between 1978 and 1992, therapy trials with heavier ions, e.g.

neon or argon, took place at Berkeley's Bevalac, a combination of the Bevatron and the linear accelerator HILAC.

HIMAC, in Chiba, Japan, has operated since 1994 with carbon beams. Based on two 16 Tm synchrotrons (which produce 30 cm range ion beams for $Z \leq 14$), it is injected by a 6 MeV/u RFQ-Alvarez linac and two separate ion source platforms. It has three treatment rooms; one with a horizontal, one with a vertical and one with both horizontal and vertical beam delivery ports [13].

A medical synchrotron (HARIMAC) for proton and carbon treatments is operating at Harima Science Garden City in Hyogo prefecture, Japan [14].

3.3. *IBT facilities planned or under construction*

The good results of IBT led to major activities worldwide to design and construct respective facilities. Table 3 gives a summary of the present situation, indicating that about 15 new IBT facilities will probably come into operation within the next four years. Additional facilities like the ETOILE project in Lyon, France or the MAASTRO project in Maastricht, the Netherlands are in the design stage. Nearly all these systems are integrated into hospital environments, rather than nuclear research institutes, as was the case in the past. Whereas technical developments were dominant in the beginning, at present a significant focus is put on the medical aspects of patient treatment and follow-up procedures, which permit high patient throughput (> 1000 patients/year).

It should also be emphasized that a drastically increased commercial interest is seen in such facilities, as the cost per patient amounts to approximately €20,000. At present, several companies (e.g. SIEMENS, IBA and VARIAN) offer turnkey facilities and partly also run the operation.

The planned facilities deliver either protons or protons and light ions, preferably C ions. Although C ions are to some extent superior to proton beams, for certain treatments the "low-LET" radiation of protons is preferred or combined with "high-LET" C irradiation.

With regard to the accelerator-technical aspects, it should be noted that, according to the difference in the maximum magnetic rigidity (2.5 Tm for protons vs. 6.6 Tm for C for a range of 30 cm), the

application of different technologies has to be considered. Concerning economical aspects for proton beams, both cyclotrons and synchrotrons are appropriate for the acceleration to the desired energy. For C beams, because of their larger magnetic rigidity, only synchrotrons are at present cost-effective. This offers the additional advantage of active beam modulation.

Cyclotron-based facilities accelerate the beam to the maximum energy required and reduce it at the beginning of the high energy beam line by means of degraders and energy-analyzing systems followed by collimators to shape the transverse beam dimensions to the acceptance of the beam delivery system. Due to this procedure a significant part of the primary beam intensity is destroyed during delivery. This leads to activation and necessitates a higher beam intensity to compensate for the losses.

Facilities based on synchrotrons offer the possibility of changing the extraction energy in time intervals of a few seconds. Degraders and collimation systems can thus be avoided, providing maximum use of the accelerated beam intensity.

However, in comparison to a cyclotron with a 100% duty factor, a synchrotron has a duty factor of only about 25–40%. The beam intensities required for cancer treatment (2–3 Gy/l/min) are in the order of a few times 10^8/s for light ions or some 10^{10}/s for protons. The short synchrotron injection time relative to its "cycle time" leads to enhanced pulse intensities of several mA for protons, required from the injection system, especially if a single-turn injection is used. All synchrotron-based systems are equipped with slow extraction systems, applying half or third integer resonant extraction procedure. These systems make use of either tune variation by means of changing the excitation of dedicated quadrupoles during the extraction time, or momentum-changing systems like betatron cores, or transverse beam excitations by means of applying a transverse electric rf field to drive particles over the resonance.

Within the Proton Ion Medical Machine Study (PIMMS) group [15], the characteristics of the slow extraction method were investigated in detail and published together with a technical design of a medical synchrotron and a beam delivery system, evaluated from an optimized beam extraction with respect to the beam properties (lateral beam dimensions and spill structure).

Table 3.   Facilities planned or under construction (status in 2007).

Place	Country	Particle	Max clinical energy (MeV)	Beam direction	Number of treatment rooms	Start of treatment planned
RPTC, Munich	Germany	p	250 sc(*) cyclotron	4 gantries with scanning, 1horiz.	5	2008
PSI, Villigen	Switzerland	p	250 sc(*) cyclotron	2 gantries, 2D parallel scanning, 1 horiz.	3	2007/08 (OPTIS2/ GANTRY2)
U Penn, Philadelphia	USA	p	230 cyclotron	4 gantries, 1horiz.	5	2009
Gunma	Japan	p, ion	synchrotron			2009
MedAustron, Wiener Neustadt	Austria	p, ion	synchrotron	2 gantries 1–2 horiz.	3,4(?)	2012
Trento	Italy	p	? Cyclotron	1 gantry 1 horiz.	2	2010(?)
CNAO, Pavia	Italy	p, ion	430/u synchrotron	1 gantry(?) 3 horiz., 1 vert	3, 4	2009(?)
HIT, Heidelberg	Germany	p, ion	430/u synchrotron	1 gantry, 2 horiz.	3	2008
iThemba	South Africa	p	230 cyclotron	1 gantry, 2 horiz.	3	?
RPTC, Cologne	Germany	p	250 sc(*) cyclotron	4 gantries, 1 horiz.	5	?
WPE, Essen	Germany	p	230 cyclotron	3 gantries, 1 horiz.	4	2009
CPO, Orsay	France	p	230 cyclotron	1 gantry 4 fixed	3	2010(?)
PTC, Marburg	Germany	p, ion	430/u synchrotron	145° fixed. 3 horiz.	4	2010
North Illinois PT, Chicago	USA	p	250	2, 3 gantries, 1, 2 horiz.	4	2011
Kiel	Germany	p, ion	430/u synchrotron	90°, 90° + 45°, 90° + 0°, fixed	3	2012
ETOILE, Lyon	France	p, ion	430/u synchrotron	1 gantry, 2 horiz.	3	2012

* sc = superconducting.

In Europe, two compact, synchrotron-based light ion therapy facilities within a hospital environment are at present in the commissioning phase: in Germany the HIT facility [16] at the University Hospital of Heidelberg (Fig. 5), and in Italy the CNAO facility in Pavia [17]. The basic specifications of HIT may be summarized as follows:

- Ion species : H, He, C, O
- Ion range (in water) : 20–300 mm
- Ion energy [*]: 50–430 MeV/u
- Extraction time : 1–10 s

- Beam diameter : 4–10 mm (h/v)
- Intensity (ions/spill)*: $1 \cdot 10^6$ to $4 \cdot 10^{10}$
- Fast change of ion species
- Three treatment sites for patients
- Integration of an isocentric gantry (* dependent upon ion species)

The CNAO specifications are similar; the installation of a gantry is foreseen as an upgrade.

Both HIT and CNAO have a nearly identical concept for the source and linac section (Fig 6): a double ECR source with an analyzing system is used

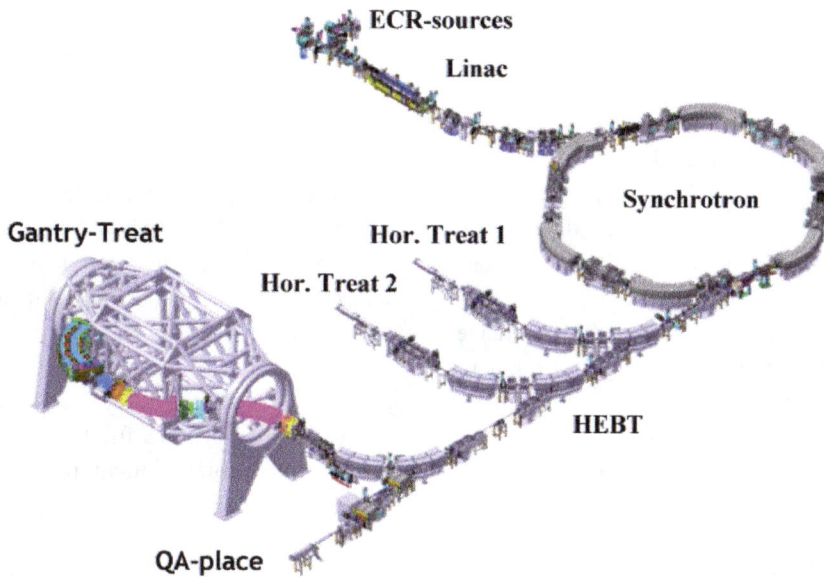

Fig. 5. Overview of the HIT accelerator and beam line systems.

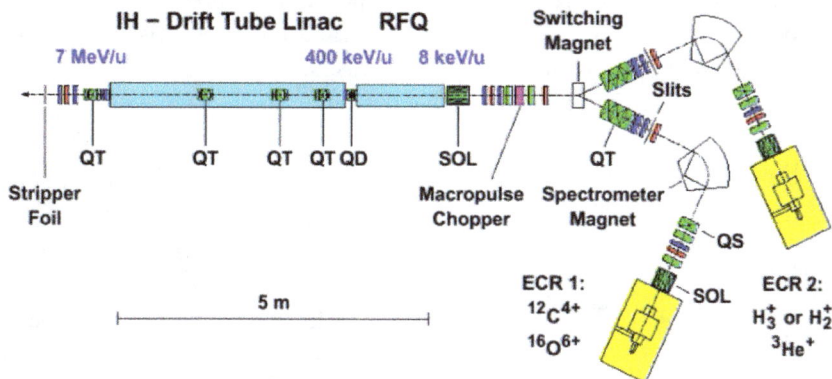

Fig. 6. Ion sources and linac section of the HIT facility with its main components.

in order to provide the possibility of a fast ion change, e.g. protons and carbon ions; the linac section consists of a compact RFQ and IH-DTL linac for an acceleration to 7 MeV/u, followed by a medium beam transport system with a stripper system and ion optical elements to match the beam to the synchrotron injection.

The accelerator design of HIT (Fig. 5) was developed at GSI. The layout of the CNAO synchrotron is based on the PIMMS design of a CERN study group.

Three other European initiatives (MedAustron in Wiener Neustadt, ETOILE in Lyon and the planned IBT facility in Stockholm) will also use this design, optimized for pixel scanning and isocentric delivery of carbon beams.

All facilities will have a capacity of more than thousand patients per year.

3.4. *Other forms of accelerator-based medical applications*

3.4.1. *Accelerators for neutron therapy*

Fast neutrons (14–70 MeV) have been used for therapy for over 50 years. D-T generators were prolific sources of 14 MeV neutrons for many years. Cyclotrons producing neutrons via {p-Be} or {d-Be} reactions in the 60–70 MeV range were employed in the 1970s and 80s [18]. Despite considerable technological efforts, including a compact, gantry-mounted, superconducting cyclotron as generator

[19] or multileaf collimators to deliver the beam, fast neutron therapy has not lived up to the early expectations. With depth–dose distributions similar to lower energy x-rays, localization of the dose into a well-defined volume is difficult. Only shallow tumors, such as salivary gland tumors and adenoid cystic carcinomas of the head and neck, profited from the high ionization density of neutrons without serious toxicity [20].

Neutron therapy remains much more expensive than conventional x-ray therapy. In comparison to IBT, tumor conformity is less favorable and it is more hazardous to clinical staff [18].

In the 1970s, approximately 20 fast neutron facilities were active. Today this number is down to less than 5 centers. The University of Washington and the Harper University Hospital in Detroit are those with the longest experience. The Fermilab program was transferred to Northern Illinois University.

Since the 1950s, boron–neutron capture therapy (BNCT) has attracted a lot of attention. Without question, BNCT exploits a unique principle — the high capture cross section of boron 10 for thermal neutrons.

The reaction (^{10}B[n,^{4}He]^{7}Li) occurs when the stable ^{10}B is irradiated with low-energy or thermal neutrons yielding high-energetic ^{4}He nuclei of $1.47\,$MeV and recoiling ^{7}Li ions of $0.84\,$MeV. With a combined path length of approximately 12 microns the radiation effect should be restricted to those cells, which have taken up a sufficient amount of ^{10}B atoms. The idea came up to use low energy accelerators rather than nuclear reactors to produce the thermal ions. However, it is not the lack of adequate neutron sources which impairs the development of this treatment modality.

Critical to success is the tumor specificity of the pharmaceutical. These problems are of a very basic nature. It is, therefore, unlikely that BNCT will be turned into an effective therapy [21].

3.4.2. π-meson therapy

Radiobiological experiments in the 1960s were in favor of π mesons (pions) as probably the most promising particles for radiation therapy with a superior physical dose distribution and an increased biological effectiveness [22].

Three meson factories — at Los Alamos, USA (LAMPF); Villigen, Switzerland (PSI); and Vancouver, Canada (TRIUMF) — conducted medical trials with π mesons. They treated many hundreds of patients in 1970–80 with innovative pion transport systems.

In contrast to the working hypothesis, however, the radiation quality for normal tissue was not any different from that of the targeted tumor cells. Low dose rates resulted in long irradiation times of an hour or more. This is not only impractical in the clinical routine. It is often not tolerated by the patients. The inherent contamination of the primary pion beam with electrons, muons and other particles complicated the treatment further. As most of the contaminants exhibit a longer penetration depth than the pions themselves, tumor conformity was additionally impaired [23].

Still, for some large, irregularly shaped soft tissue and bone sarcomas, encouraging results could be obtained [24]. But these rare entities did not justify a specific pion program. In the meantime, all three programs have been discontinued or replaced by IBT projects.

3.4.3. Radioisotope production

Radioactive isotopes, are widely used in both diagnostic and therapeutic applications. Isotopes, either alone or attached to physiologically relevant molecules, are used as tracers for structural or functional imaging in order to detect concentrations or activities in tissues. Short-lived positron emitters, (such as ^{11}C, ^{15}O and ^{18}F) for positron emission tomography (PET) or gamma emitters (e.g. ^{67}Ga, ^{111}In and ^{201}Tl) for single photon emission computed tomography (SPECT) are typical examples. PET isotopes are produced with small single- or dual-particle (H or H/D) cyclotrons ($\leq 20\,$MeV) close to the end-user in the clinic.

With the exception of ^{99}Tc, which is a reactor nuclide, the same accelerator types can be used for the production of SPECT isotopes, as well. However, level IV accelerators ($\leq 40\,$MeV) are often preferred [25].

Commercial production is concentrated in a few centers with elaborate distribution networks to provide rapid delivery of short-lived isotopes. Research isotopes are also produced at higher-energy accelerators (200–800 MeV protons) in TRIUMF, Vancouver, BLIP, Brookhaven, and, until recently, LAMPF, Los Alamos.

While gamma-emitting isotopes are best suited for diagnostics, alpha- and beta-emitters are preferred for therapeutic applications, limiting the dose to the immediate vicinity of the isotopic application. The radioactive dose is applied either as open source (such as iodine in thyroid treatments) or surgically implanted incapsulated material (referred to as interstitial or brachytherapy).

About 200 accelerators are used worldwide to produce radioisotopes. PET isotopes are most commonly produced by (p, n) reactions with low-energy (11–15 MeV) cyclotron beams. The commercially available cyclotrons are compact, self-shielded, highly reliable, and totally automated. Targetry and autochemistry units are usually included, providing complete hands-off preparation of isotopes in a form ready for administration. Manufacturers include CTI (Knoxville, Tennessee, USA), IBA (Louvain-la Neuve, Belgium) and Ebco (Vancouver, Canada).

Small cyclotron technology has been revolutionized by the development of high-quality H^- ion sources, which solved the thermal, mechanical and activation problems associated with beam extraction. For energies up to 30 MeV, magnetic fields in the cyclotron can be high, leading to compact structures, but for higher energies the magnetic field must be reduced to avoid Lorentz stripping of the H^- ions. Beam currents for the PET isotope systems are modest (e.g. 50 μA). The higher-energy cyclotrons used for production of longer-lived isotopes push the limits of current for this technology up to 1 mA. New technologies are being explored for isotope production using proton- and ion-based linac systems (SAIC-FNAL, AccSys).

3.4.4. *Radiography*

With the advent of high-flux, high-quality x-rays from synchrotron radiation sources, new opportunities were seen for monochromatic x-rays in diagnostics. A notable example is the coronary angiography program, which started at the Stanford Synchrotron Radiation Laboratory (SSRL) already nearly 30 years ago [26].

The idea was to use two sets of x-rays of narrow bandwidth to depict the coronary vessels. With x-rays just above and below the K edge of a contrast agent, the subtraction images achieved a 10,000-fold better contrast enhancement.

Despite such achievements, digital subtraction coronary angiography (DCA), which was also pursued at other synchrotron centers (e.g. DESY, Hamburg; KEK, Tsukuba; NSLS, Brookhaven; and ESRF, Grenoble), has not found much support among clinicians. This is in part due to problems with artefacts and impaired representation of certain sections of the heart vessels. The main issue is, however, the lack of small, cost-effective synchrotron sources.

Relevant x-ray energies are in the 30 keV range. Storage rings with e^- or e^+ beam energies ≥ 2.5 GeV are adequate for this purpose. For the lower-energy rings, wigglers with high magnetic fields are required.

Electron-beam-computed tomography or MRI is not necessarily superior as far as imaging quality is concerned. But, in contrast to the synchrotron source for DCA, CT and MRT are versatile diagnostic instruments applicable for all kinds of imaging purposes.

4. New Developments for IBT Facilities

New technological developments are driven both by the requirements of the oncologists to optimize the treatment technique and by the commercial side to design more compact and/or more cost-effective solutions for the accelerator systems.

With respect to the patient treatment the following topics are in the focus of recent R&D: optimization of quality assurance by improved imaging, application of "in-beam PET," optimization of treatment planning programs, treatment of moving organs and optimal three-dimensional conformation of the tumor volume. In particular, the last two topics have direct implications for the technology of accelerator systems.

4.1. *Gantries*

For an optimal 3D conformation of the tumor volume, an irradiation with a horizontal beam is not sufficient, even if the additional horizontal degrees of freedom of the patient couch are considered. An optimal solution is the installation of an "isocentric gantry" with 360° rotation perpendicular to the rotation axis of the patient couch.

Today's conventional radiation therapy is inconceivable without a rotating isocentric gantry which

Fig. 7. Isocentric gantry at the NPTC (now Francis Burr) proton treatment facility of the Massachusetts General Hospital, Boston, USA.

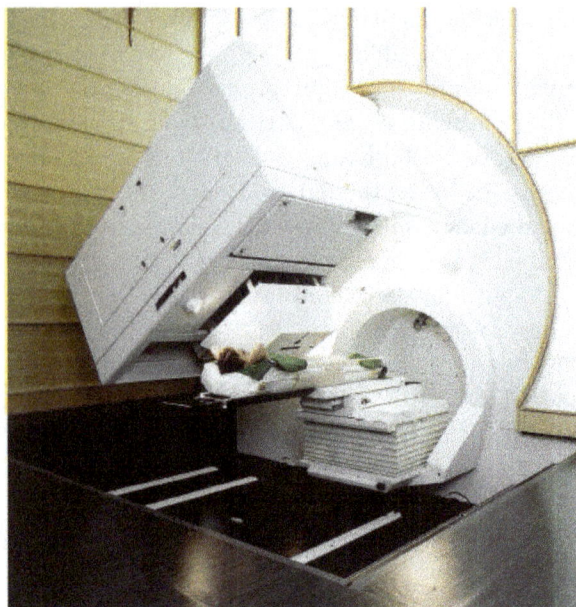

Fig. 8. "Eccentric" gantry for proton treatment, installed at PSI.

permits treatment from the medically most desirable angle(s).

The proton therapy centers which have been opened during the last five years have at least one treatment room with gantry; for example, Loma Linda has three and the Munich Rinecker Proton Therapy Center has four gantries installed.

The magnetic rigidity of a 250 MeV proton beam is approximately 2.5 Tm. This determines the minimum size of a gantry. In addition, a sufficient flight path must be provided from the last magnet to the patient to allow for beam spreading, dosimetry and field definition. In the case of the Loma Linda gantry with a diameter of approximately 13 m, the latter effect contributes roughly 6 m to the size.

Alternatives are imaginable, as will be exemplified in the following. However, the examples show, as well, that the medical field is rather conservative in accepting technology which does not fit into standard diagnostic or hospital routines. Medical technology systems which comprise new features but do not impair the existing workflow are more likely to be accepted in the clinic.

A compact "eccentric" gantry system is in operation at the Paul Scherrer Institute, Villigen, Switzerland, in a proton therapy line attached to the 600 MeV cyclotron. The patient table and the massive bending-magnet system counterrotate around a common center, reducing the diameter to approximately 4 m. In addition, the treatment field is obtained by means of an active spot-scanning system

allowing the patient to be placed much closer to the last magnet. This concept, however, has not been pursued despite its compactness [27]. The fact that the patient rather than the gantry moves is considered too high a risk which may impair the access to the patient in an emergency. As a consequence PSI has built a second gantry, which is of the conventional isocentric type.

Since 1990, a gantry-mounted 60 MeV superconducting deuteron cyclotron for neutron therapy has been in operation at the US Harper Grace hospital [19]; an extension to a 250 MeV gantry-mounted superconducting proton synchrocyclotron with a weight of less than 35 tons and fields higher than 8 T is under development [28].

For carbon ions, various gantry alternatives have been considered, but due to the high magnetic rigidity of these particles the gantries are quite massive and expensive [29, 30].

At the Lawrence Berkeley Laboratory, Berkeley, USA, patients were treated with light ions, sitting in front of the beam, with the patient support rotating around the vertical axis. To enable adequate treatment planning, a custom-made CT was purchased, which permitted imaging in the seated position. However, this was not only a costly solution. Positioning errors are more likely to occur in the seated than in the supine position. This would not

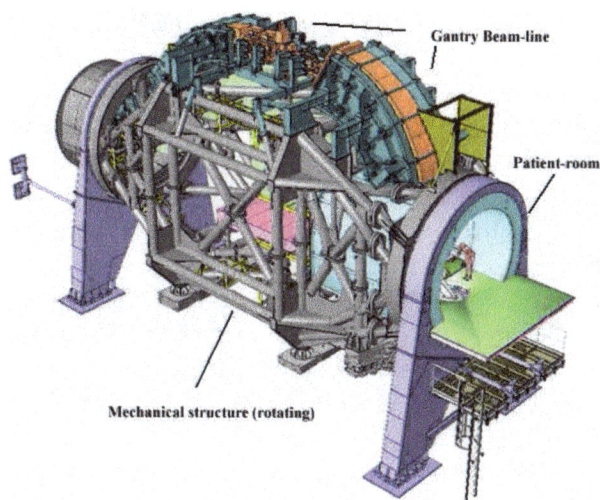

Fig. 9. Light ion gantry (HIT).

only be an issue for the single imaging session as the basis for the treatment plan but for all the irradiation fractions and hence the precision of the whole radiation therapy. The installation of an additional vertical beam line, as foreseen for the CNAO project in Italy, the combination of a vertical and a horizontal beam line as in HIMAC, Chiba, Japan, or tilted beam lines, as realized at the IBT center in Hyogo, can only be considered a compromise. For cost and weight aspects, one could also think of a sector gantry covering the horizontal position $\pm 45°$ to $60°$ But the gold standard will be the isocentric gantry permitting the greatest flexibility in the irradiation angle.

The worldwide first isocentric $360°$ gantry for light ions is presently installed at the HIT facility in Heidelberg [16]. Due to the large maximum magnetic rigidity of 6.6 Tm (carbon ions of 430 MeV/u), the magnet components, which are normal-conducting, are rather heavy, corresponding to a mechanical structure with a total moving weight of about 200 tons and a total weight of 600 tons (including all beam line components and shielding counterweights).

A recent design study by Furukawa *et al.* [31] suggested a gantry for carbon ions with limited field size ($15 \times 15\,\text{cm}^2$) and range limitation (25 cm). This slim version of the Heidelberg gantry should weigh only about half of the original. However, in view of the increasing problem of obesity — at least in the western societies — this is probably only a temporary solution.

Major investigations have been performed to provide the required beam properties to the patient independent of the rotation angle of the gantry system in order to ease the operation of such a device. Especially for slowly extracted beams from a synchrotron, this is not trivial as the beam emittances in both planes are different and matching conditions of the beam have to be provided.

The PIMMS concept proposes a mechanical "rotator" concept of the high-energy beam transport line to fulfil the condition of rotation-independent gantry settings. In this study also a new gantry concept ("Riesenrad-gantry") was proposed with a rotating patient treatment cabin and no rotating beam line components except for a $90°$ bending magnet.

Some investigations are in progress to study the possibility of light ion gantries with superconducting magnets. In addition the FFAG concept is proposed in studies for new generations of gantries with applications of compact superconducting combined-function bending magnets [28, 32].

Even though it is difficult at the moment to estimate the percentage of suboptimal treatments due to restricted irradiation angles, it is highly recommended to address the engineering challenge of a less bulky ion gantry. But whatever the final successful design will be, it will likely have the patient table at rest and the patient in the supine position.

4.2. *Use of radioactive beams*

The occurrence of ^{11}C fragment nuclei in ^{12}C ion beams has been turned into a helpful diagnostic tool. Measuring the annihilation radiation of ^{11}C positrons provides a noninvasive means of monitoring the range and even dose of the treatment beam *in situ* [33, 34]. It improves the particle range data derived from the Hounsfield units of the planning CT. They can comprise inaccuracies due to positioning errors, newly developing variations in tissue structure (e.g. edema, scar formation, mucus deposits) or hardening of the x-rays [35].

Enghardt *et al.* have demonstrated that in-beam PET can be fully integrated into a therapeutic beam line (see Subsec. 4.4 for more details) without disturbing the irradiation and with only minor influence on the treatment time. They have been able to detect and alter positioning errors or density variations [36].

At NIRS, Chiba, a radioactive ^{11}C beam is produced and used for PET measurements prior to the actual treatment. Applying a thin (1 mm), pure particle beam provides more accurate range measurements. As compared to in-line PET which uses the broadened peak of the built-up secondaries, the signal intensity (approximately 7×10^6 pps) with a radioactive ion beam can be about an order of magnitude higher [37]. However, the yield of secondary positron-emitting ions is at best 1% of the primary beam intensity. This causes activation of equipment and requires additional radiation safety measures. Finally, the patient receives an unnecessary additional radiation dose in the range of approximately 5% [38]. The low yield of a pure ^{11}C beam, the cost for the extra shielding, and the increased exposure of patients and possibly staff favor in-line PET over a radioactive carbon beam for treatment. But, even for diagnostics, the advantage of a higher signal intensity has to be weighed against additional cost and effort for a second beam course.

4.3. *Alternative accelerator structures*

At present, tumor irradiation of moving organs (for example due to breathing) is performed at few facilities (e.g. HIMAC) by "gating" procedures: the actual position of the organ is detected and the beam delivery takes place only when the position is within a predefined tolerance, which means that beam extraction is gated with a signal detecting the position of the organ.

Respiration gating requires fast beam switch features (such as by the transverse knockout technique for slow extraction from the synchrotron). For an optimized treatment and an effective usage of the treatment time, an active 3D position control of the beam is desirable. The lateral beam adjustment can be performed with fast scanner magnets, used for the rasterscan irradiation. To cover the longitudinal or range deviations due to organ movement, a fast energy variation in the millisecond range of the beam is required. An active energy variation of the corresponding time scale combined with the accelerator would be desirable. One proposed solution for this demand is the fixed-field, alternating-gradient (FFAG) accelerator concept [32, 39].

Developed about 50 years ago, FFAG has recently been reconsidered. In the "nonscaling"

FFAG design, fixed-field, combined-function bending magnets are foreseen; this concept allows fast beam acceleration with compact machines without ramping the magnets. In predesign studies, it is anticipated to accelerate carbon beams up to 400 MeV/u, with a repetition rate of 200 Hz, using compact and cost-effective normal- or superconducting structures. Such a system would allow for the fast energy variation required for the treatment of moving organs.

Figure 10 shows an example for the design of a three-stage FFAG accelerator. It should be noted that a multistage concept is required, as linear field FFAGs limit the momentum range, which can be accepted, to a factor of approximately 3; hybrid FFAG designs are under investigation which extend the momentum range to a factor of 6 [32, 39].

One of the initiatives of the TERA foundation in Italy was the design of a linac booster (LIBO) for proton therapy. LIBO should operate at the very high frequency of 3 GHz, and be able to accelerate protons up to 250 MeV. Fast energy variations would also be featured [40]. The facility further offers the possibility of dual use, i.e. to produce isotopes during the times when no treatments take place. CABOTO (carbon booster for therapy in oncology) is an evolution of this concept suited to accelerate carbon ions [7].

The "dielectric wall linac accelerator" (DWA), which is under development at ANL, LLNL and UC

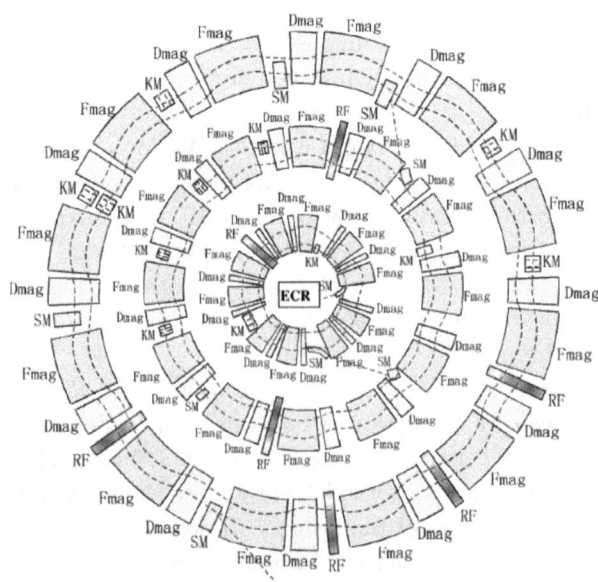

Fig. 10.   Triple-cascade radial sector FFAG system [38].

Davis is an approach to enhancing the accelerating gradients from about 10 MeV/m to 100 MeV/m, using new dielectrics capable of allowing very high voltages. Such a device could be small enough to be mounted on a gantry [41].

The use of both superconducting (K250 by Varian/ACCEL) and conventional cyclotrons (C235 by IBA) has meanwhile been consolidated at proton therapy centers. At LNS in Catania, Italy, a compact, superconducting cyclotron able to accelerate both protons to 250 MeV and carbon ions to 300 MeV/u has recently been proposed [42]. This cyclotron should operate with an average magnetic field between 3.2 T and 4.2 T at the extraction radius. A size of $5 \times 3 \, m^2$ is considered, and a weight of about 350 tons.

As a long-term vision toward compact treatment facilities, development of ion acceleration by means of high-power lasers is under investigation at different laboratories (see e.g., Refs. 43–45). However, at present the beam properties of laser-accelerated ions are not at all what is required for therapy (high intensity, well-defined energy) and many technical challenges are waiting to be resolved [46].

In addition to new accelerator concepts, the application of antiprotons was proposed [47]. The arguments are the high RBE-factor. At present, such a proposal seems far-fetched, not the least because of the enormous investment costs of such a facility.

4.4. Treatment technique

4.4.1. In-line PET diagnostics

In order to get a clear picture of the dose distribution during the irradiation process, an "in-line" PET diagnostic system [7, 34] has been implemented at GSI during the experimental therapy program. This PET camera allows one to detect the position of gamma rays of decaying positron emitters produced by either projectile or target fragmentation during irradiation. The most relevant positron-emitting nuclei are ^{11}C (half life 20 min) and ^{15}O (half life 2 min).

4.4.2. Advanced positioning — and diagnostic systems

For the HIT facility a sophisticated combined positioning and diagnostic system has been developed and commissioned by Siemens. This system, based

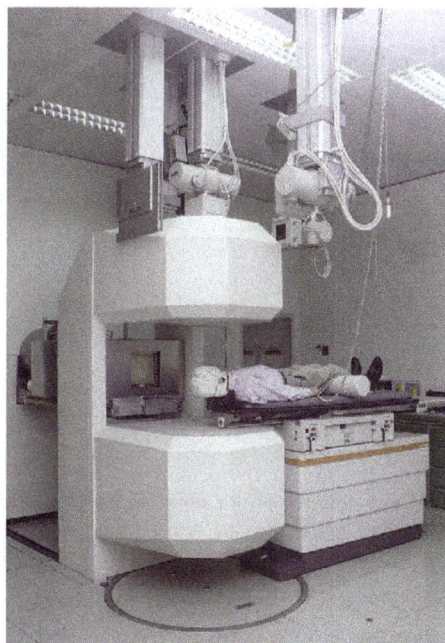

Fig. 11. Positron emission tomography (PET) camera at GSI.

on robotic technology (Fig. 12), allows an optimized workflow of the patient treatments, and accurate and flexible usage of patient positioning and position verification.

As an alternative to respiration gating, GSI is developing a 3D online motion compensation method [48]. In contrast to the former procedure, the GSI development can actively compensate for 3D variations of a moving tumor volume (Fig. 13). This is achieved with the scanning magnets, which perform corrections in both lateral planes. In addition,

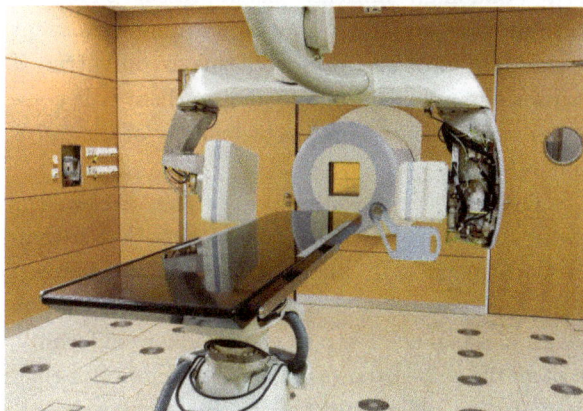

Fig. 12. Combined robotic system at the HIT treatment place.

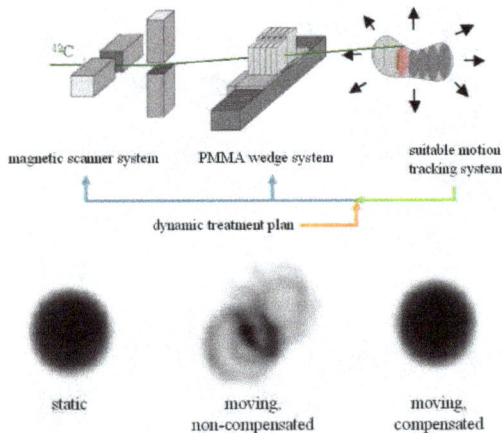

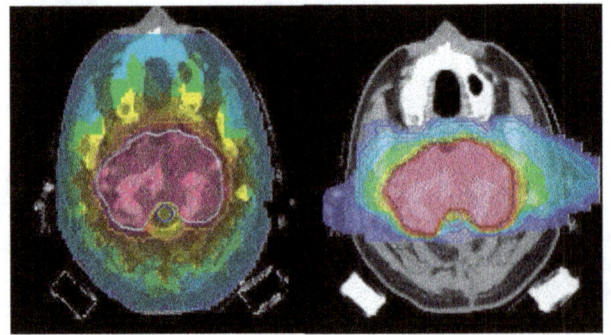

Fig. 14. Dose distribution for nine field IMRT photon treatment (left) and two field carbon treatment (right).

Fig. 13. 3D online motion compensation method, schematic (*upper part*) and test results for non-compensated and compensated irradiation (*lower part*).

changes in the third dimension are enabled (within certain limits) using a computer-controlled PMMA wedge system to cover variations of the penetration depth.

5. Outlook from a Medical Perspective

As the requirements of therapy facilities are predominantly defined by the demands of oncologists, medical aspects have to be considered in addition to technical issues. This section will address a few of these medical perspectives relevant for future development of accelerator-based radiotherapy.

5.1. *Photons vs. ions*

Since the introduction of 3D treatment planning and intensity-modulated radiation therapy (IMRT) for x-rays, one might have had the impression that photons provide equally tumor-conforming treatment schemes as protons. However, this is only true if state-of-the-art megavoltage photon therapy is compared with outdated proton technology of the 1970s and '80s.

The physical depth–dose profile of protons ensures that proton therapy performs better than the treatment with photons, if equally advanced techniques are compared. Even if modern photon techniques can achieve reasonable tumor conformity, it is at the expense of a higher exposure of normal tissue to the radiation dose [49, 50].

Several studies have estimated the risk of developing late toxicity or a second tumor as a function of the normal tissue dose [51, 52]. For patients likely to live longer than 10 years, the risk of a second malignancy after IMRT is said to be close to 2%, or nearly twice as high as after conventional radiation therapy. This limits application of IMRT, particularly, in children. Proton therapy, on the other hand, is considered the external radiation delivery technique with the lowest risk of second malignancies [53]. For health-economic considerations, this will be a relevant issue. In general, it will be necessary to demonstrate that the improved dose distribution of ion beams can be translated into significant clinical benefits for the patient.

5.2. *The optimum ion*

Protons are still the major ion type used for IBT, but other ions were studied for their biological effects from early on.

During the 1980s, when interest in medical applications of heavy-ion beams was revived, mainly in Europe and Japan, carbon ions were considered the ions of choice, because they demonstrated high-LET qualities within the Bragg peak and low-LET behavior in the entrance channel of their trajectory. Their advantageous biological effects have in the meantime been realized in several thousand successfully treated patients (for reviews see e.g. Ref. 5). However, systematic experimental studies to find the optimum ion for clinical use have not been pursued. As relativistic nuclei fragment when they interact with matter, they produce a "tail" dose of secondary particles. These are not only lighter isotopes of the initial ion but nuclides of all the elements in the periodic table to the left of the mother particle. This extends the area of increased RBE from the Bragg peak region, where it is desired, to the range beyond the peak where a steep dose falloff is preferred. In proton beams, secondary particles — mainly knocked-out protons — contribute less than 20% of the total dose [2]. They raise the threshold dose level over the whole trajectory path of the beam. The major advantage of proton therapy comes, therefore, from their improved physical selectivity, i.e. Bragg peak and magnetic deflectability, which they share with all other ions. In He ion beams, protons are also the main secondary particles. Because of the mass difference, however, they penetrate significantly deeper into the target, causing a small tail dose (approximately 3% of the total dose) behind the Bragg peak. This unwanted dose contribution from secondary particles increases further for the heavier ions. But, whereas dose contributions of protons and ^{4}He ions remain relatively constant, ^{11}B, ^{10}B and other heavier fragment nuclei are responsible for a high-LET component in the Bragg peak tail of carbon ions [54].

Ions with $Z > 6$ are, thus, unlikely to undergo a clinical revival. They display too high a LET in the plateau region and the fragmentation tail vitiates the gain of the Bragg peak. But ions with an atomic number between 1 and 6, in particular He or Li, could be interesting alternatives to carbon. They display high-LET effects only in the final Bragg peak region; in the plateau it is negligible. The tail dose is not yet significant and, not least, their production should be more economical. Clinical studies in this direction are, therefore, highly recommended. The new clinical ion beam centers in Heidelberg and Pavia will offer the technical conditions for performing these tests with different ion sources.

5.3. *Number of fractions*

Fractionation of radiation, which means dividing a radiation treatment into several individual sessions or "fractions" (e.g. a 60 Gy total dose delivered in 30 fractions, of 2 Gy each), facilitates selective recovery of normal tissue. With a reduced level of the normal tissue dose, the irradiation with ions can either be escalated to the tolerance level of the normal tissue or condensed to fewer fractions. In any case, the chances are higher to kill the tumor cells in the target region. Fewer fractions means, in addition, a positive effect on the economic side of IBT.

Reduction of the number of fractions has been a major research topic at the National Institute

of Radiological Sciences (NIRS) in Chiba, Japan. Preclinical studies predicted an increase in the therapeutic ratio of carbon ion beam irradiation for hypofractionated regimens using higher doses per fraction [55, 56].

Indeed, excellent clinical results have been obtained and long-term follow-up has not revealed serious late toxicity.

For selected patients with hepatocellular carcinoma, for example, three-year local control rates of up to 95% were achieved. These patients had received four fractions of 13.2 GyE each within one week. Inoperable stage I non-small-cell lung cancer (NSCLC) has been treated with similar success with only four fractions, and a dose escalation study is underway to treat this malignancy with a single dose of up to 44 GyE [57]. This extreme reduction of the treatment scheme has raised a couple of issues.

For single fractions, the risk of microscopic cold and hot spots, due to the narrow track structure of the high-LET radiation, is highest. Fortunately, complications attributable to such heterogeneities have not been reported. But it is important to keep the risk in mind. A combination treatment of ions and x-rays could possibly be an alternative.

It has also been questioned whether the beam intensity is sufficient for such high-dose irradiation fractions. For cyclotrons with their high beam output, the intensity is clearly sufficient. But even for the HIMAC synchrotron with approximately 10^9 carbon ions/second, these high doses can be supplied under the required conditions.

5.4. *Cost aspects*

IBT is presently about 2–3 times as expensive as IMRT [58]. The total investment cost for an IBT facility, including the building, accelerator, beam delivery systems and treatment installations, is at present in the range of €100–150 million. Besides these investment costs, the operation costs are quite high, as both operation utilities (e.g. electric power) and increased costs for technical personnel have to be considered. To justify such a price gap requires more than clinical superiority. In countries with a public health insurance system, as is the case in most European countries, the healthcare providers will tolerate such differences at best temporarily. When a certain number of institutions offer a new technology, a prototype status can no longer be claimed as an excuse for higher costs. To be financed, subsequent facilities might be obligated to guarantee lower prices.

In the health market, cost-effectiveness analyses are gaining more and more importance. Often seen with great skepticism, in particular by affected patients, such studies can be — if carefully designed and performed — a helpful means of putting competing technology into perspective.

In 2005, a group at the Karolinska Institute published a cost–benefit assessment for four different cancers (prostate, breast, head and neck, and medulloblastoma) comparing proton therapy with conventional radiation therapy. They concluded that for appropriately chosen risk groups, proton therapy could be cost-competitive. In the case of pediatric medulloblastoma, they even proposed a significant cost saving of €23,600 and an additional 0.7 quality-adjusted life-year (QALY) per pediatric patient, if treated by proton therapy.

We are now approaching a situation where we have clinical cancer centers which provide modern IBT and high-level photon technology alike. They can offer the radiation modality which is considered most effective for a certain tumor entity. In the case of radioresistant base-of-skull tumors, for example, the gold standard might be IBT. In many other cases we can only extrapolate from physical and radiobiological considerations.

The new facilities provide the means of treating hundreds of patients by IBT, annually. This assures the numerical basis for randomized, controlled clinical trials. These are indispensable for finally deciding whether the increased tumor dose and/or reduced normal tissue exposure of an ion beam treatment plan translates into higher survival and lower toxicity as compared to the most advanced photon radiation techniques. The outcome of such clinical studies can then serve as a solid basis for further cost–benefit assessments.

References

[1] R. E. Zirkle, The radiobiological importance of linear energy transfer, in *Radiation Biology*, ed. A. Hollaender (McGraw-Hill, New York, 1954), Vol 1, pp. 315–350

[2] J. Kempe *et al.*, *Med. Phys.* **34**, 183 (2007).

[3] R. R. Wilson, *Radiobiology* **47**, 487 (1946).

[4] J. D. Slater, *Technol. Cancer Res. Treat.* **5**, 81 (2006).

[5] D. Schulz-Ertner *et al.*, *J. Clin. Oncol.* **25**, 953 (2007).

[6] G. Kraft, *Prog. Part. Nucl. Phys.* **45**, S473 (2000).

[7] U. Amaldi *et al.*, *J. Radiat Res.* **48** (Suppl.), A27 (2007).

[8] R. Mayer *et al.*, *Radiother Oncol.* **73**, S24 (2004).

[9] U. Amaldi, *Radiother Oncol.* **73**, S191 (2004).

[10] G. Gademann, Socio-economic aspects of hadrontherapy, *Proc. 1st Int. Sympo. Hadrontherapy* (Como, Italy; Octo. 18–21, 1993), p. 59.

[11] J. H. Lawrence, *Cancer* **10**, 795 (1957).

[12] M. Jermann, Hadron therapy patient statistics, http://ptcog.web.psi.ch/Archive/Patient_data_Dec-07.pdf (Mar. 2008).

[13] K. Sato *et al.*, *Nucl. Phys. A* **588**, 229 (1995).

[14] A. Itano, *Natl. Inst. Radiol. Sci. M* **156**, 6 (2002).

[15] PIMMS Parts I and II, CERN/PS 1999-010 DI and CERN/PS 2000-007 DR, (Genova).

[16] H. Eickhoff, HICAT: The German Hospital-Based Light Ion Cancer Therapy Project (EPAC 2004, Lucerne).

[17] S. Rossi, Developments in proton and light-ion therapy (EPAC 2006, Edinburgh).

[18] H. Blattmann *et al.*, NUPECC (Dec. 27–41, 1994).

[19] R. Maughan *et al.*, Status Report for the Harper Hospital Superconducting Cyclotron Neutron Therapy Facility, CP600, *Proc. Cyclotrons & Applications* 18 (2001).

[20] F. J. Prott *et al.*, *Bull. Cancer. Radiother.* **83**, 115s (1996).

[21] U. Linz, *Technol Cancer Res. Treat.* **7**, 83 (2008).

[22] P. H. Fowler *et al.*, *Nature* **189**, 524 (1961).

[23] H. Blattmann, New developments at PSI, in *Ion Beams in Tumor Therapy*, ed. U. Linz (Chapman and Hall, Weinheim, 1995), pp. 333–340.

[24] G. Schmitt *et al.*, *Radiat. Res. Suppl.* **8**, S272 (1985).

[25] S. M. Qaim, *Radiochim. Acta* **89**, 223 (2001).

[26] E. B. Hughes *et al.*, *Nucl. Instrum. Methods. Phys. Res. A* **246**, 719 (1986).

[27] E. Pedroni *et al.*, *Z. Med. Phys.* **14**, 25 (2004).

[28] J. B. Flanz, *Nucl. Instrum. Methods. Phys. Res. B* **261**, 768 (2007).

[29] U. Weinrich, Gantry design for proton and hadrontherapy facilities, in *EPAC 2006* (Edinburgh).

[30] D. Trbojevic *et al.*, A dramatically reduced size in the gantry design for the proton-carbon therapy, in *EPAC 2006* (Edinburgh).

[31] T. Furukawa *et al.*, Design study of scanning system and rotating gantry for HIMAC new treatment facility, in *Proc. NIRS–CNAO Joint Symp. Carbon Ion Radiotherapy* (2006), pp. 142–153.

[32] C. Johnstone *et al.* in New nonscaling FFAG for medical applications, in *PAC 2007*.

[33] W. Enghardt, *Phys. Med. Biol.* **37**, 2127 (1992).

[34] W. Enghardt *et al.*, *Nucl. Instrum. Methods. Phys. Res. A* **525**, 284 (2004).

[35] M. Krämer *et al.*, *Phys. Med. Biol.* **45**, 3299 (2000).

[36] W. Enghardt *et al.*, *Radiother. Oncol.* **73**, S96 (2004).

[37] M. Kanazawa *et al.*, Present status of HIMAC at NIRS, in *Proc. 1999 Particle Accelerator Conference* (New York, 1999), pp. 599–602.

[38] M. Kanazawa *et al.*, *Nucl. Phys. A* **701**, 244c (2002).

[39] T. Misu *et al.*, *Phys. Rev. ST — Accel. Beams* 7, 094701, 2004(E-Publ).

[40] U. Amaldi *et al.*, *Nucl. Instrum. Methods Phys. Res. A* **521**, 512 (2004).

[41] Y.-J. Chen *et al.*, Compact proton accelerators for cancer therapy, TUPAS059, in *PAC 2007* (Albuquerque).

[42] M. Maggiore *et al.*, Design studies of the 300 AMeV Superconducting Cyclotron for Hadrontherapy, in *PAC 2007* (Albuquerque).

[43] A. Noda *et al.*, Ion production with a high-power short-pulse laser for application to cancer therapy, in *EPAC 2002* (Paris), pp. 2748–2750.

[44] V. Malka *et al.*, *Med. Phys.* **31**, 1587 (2004).

[45] M. Borghesi *et al.*, *J. Phys. Conf. Ser.* **58**, 74 (2007).

[46] U. Linz *et al.*, *Phys. Rev. ST–Accel. Beams.* 10, 094801, 2007 (E-Publ.).

[47] M. H. Holzscheiter *et al.*, *Radiother. Oncol.* **81**, 233 (2006).

[48] S. Grötzinger *et al.*, *Phys. Med. Biol.* **51**, 3517 (2006).

[49] R. Lin *et al.*, *Int. J. Radiat. Oncol. Biol. Phys.* **48**, 1219 (2000).

[50] R. Miralbell *et al.*, *Int. J. Rad. Oncol. Biol, Phys.* **54**, 824 (2002).

[51] E. Hall *et al.*, *Int. J. Radiat. Oncol. Biol. Phys.* **56**, 83 (2003).

[52] U. Schneider *et al.*, *Phys. Med.* 17, S97 (2001).

[53] E. B. Hug, *Radiother. Oncol.* 73, S35 (2004).

[54] N. Matsufuji *et al.*, *Phys. Med. Biol.* **48**, 1605 (2003).

[55] S. Koike *et al.*, *Radiat. Prot. Dos.* **99**, 405 (2002).

[56] K. Ando *et al.*, *J. Radiat. Res.* **46**, 51 (2005).

[57] H. Tsujii *et al.*, *J. Radiat. Res.* **48** (Suppl), A1 (2007).

[58] M. Goitein *et al.*, *Clin. Oncol.* **15**, S37 (2003).

[59] J. Lundkvist *et al.*, *Acta. Oncol.* **44**, 850 (2005).

Hartmut Eickhoff is a physicist at the Gesellschaft für Schwerionenforschung mbH (GSI), Darmstadt. In the mid 1990s his contact to medical applications of accelerators started, when he became responsible for the accelerator modifications at GSI for the Therapy Pilot Project, applying the intensity controlled rasterscan for patient treatment with carbonions. Until 2005 he was the GSI project leader of the accelerator facility of the Heavy Ion Therapy complex HIT at the Clinics in Heidelberg. He is the head of the GSI accelerator department.

Ute Linz is a chemist, biologist and approbated physician specializing in radiooncology. In her present position at the Jülich Research Center she has mainly worked on therapeutic applications of ion beams and on the treatment of malignant gliomas. As a sideline she provides biomedical consultancy services to business partners active in the medical field and supports tumor patients in her private practice.

Reviews of Accelerator Science and Technology
Vol. 1 (2008) 163–184
© World Scientific Publishing Company

World Scientific
www.worldscientific.com

Industrial Accelerators

Robert W. Hamm

R&M Technical Enterprises, Inc.,
4549 Mirador Drive,
Pleasanton, CA 94566, USA
bob.hamm@comcast.net

About half of the particle accelerators produced worldwide are used for industrial applications. These commercial systems utilize a wide range of accelerator technologies and cover numerous applications over a broad range of business segments. While this is not a high profile business, these "industrial accelerators" have a significant impact on people's lives and the world's economy, as many products contain parts that have been processed by charged particle beams. Wide scale adoption of many of these processing tools has resulted in the rapid growth of the business of producing and selling them. This paper is a review of the current status of industrial accelerators worldwide, including the technologies, the applications, the vendors and the sizes of the markets.

Keywords: Industrial accelerators; accelerator applications; particle beam processing; particle beam applications; particle beam irradiators; radiation processing; commercial accelerators.

1. Introduction

"Industrial accelerators" are defined for purposes of this review to be all charged particle accelerators that generate external beams for use in a beam process other than for medical treatment or basic physics research. Those devices that use low energy charged particles internally, such as cathode ray tubes, x-ray tubes, radio frequency and microwave tubes, and electron microscopes, are not included. For the most part, the accelerators covered in this review are all readily available for purchase from commercial companies worldwide or from other for-profit enterprises such as government-backed companies and government laboratories outside the United States.

In the author's experience, most users or buyers of industrial accelerators care more about the beam being able to satisfy the requirements of their application than the specific technology used to generate that beam. Hence, this paper will only briefly review the origin of the science and technology used in these systems, and will then describe in more detail the many applications in which they are employed and the parameters of the systems used for each application.

The worldwide market and the present commercial vendors for the accelerators used in each application will be discussed. The list of vendors in the manufacturing of accelerators is in a continuous state of change, as new vendors appear and others are purchased by competitors or by other businesses in their application area. As a particular application becomes a mature business, the size of each vendor increases but their numbers dwindle due to the competitive nature of the business. Also, due to this competitive nature, some of the details of the construction of specialized accelerators from some companies are not revealed. The same is true for sales data from a particular company. However, the basic accelerator technology used in their equipment is quite evident and will be outlined, and the sales data available from all the sources will be combined to indicate the general size of the markets.

The present review will therefore attempt to cover the current status of the very diverse but well-established industrial accelerator applications from the perspective of not only the technology and its uses but also the commercial aspects and the impact on society. The summary of these accelerators and applications will necessarily be brief, but the reader can obtain more details of particular systems or applications from books and review papers that have previously been published. A number of short review papers [1–6] have been written over the past 35 years, some focusing on specific areas of these industrial accelerator applications, and many books and book chapters have been published about these systems [7–10]. The International Atomic Energy Agency (IAEA) sponsors a number of ongoing accelerator applications projects [11] and has held a series of symposia on these applications [12–14]. Several international conferences devoted to accelerator applications [15–17] are also held periodically.

2. Origin of Industrial Accelerators

Most industrial applications of accelerators evolved from basic or applied science programs using research accelerators installed at large universities and national laboratories. The accelerators subsequently developed for these industrial uses have evolved into high quality products, with each one specifically tailored to a specific application.

Much of the science and technology used in modern industrial accelerators was initially developed in the 1930's for physics research, as is elegantly described in a recent book by Sessler and Wilson [18]. The techniques developed in that period included use of direct voltages (high voltage multipliers and electrostatic charging), rf linear acceleration and circular acceleration using magnets (cyclotrons, betatrons, and synchrotrons). The accelerators that now employ these techniques for industrial applications include both electron beam and ion beam systems.

2.1. *Direct voltage accelerators*

The use of direct voltages to produce high energy charged particle beams for bombarding targets dates back to the pioneering work of Cockcroft and Walton [19] and Van de Graaff [20] in the early 1930's. Over the next three decades a number of different accelerators were developed using this technique, which basically involves putting charged particles through a voltage drop to impart energy to them. While each accelerator employs a somewhat different technology, the beam acceleration is accomplished using some type of high voltage power supply. These include the charge-carrying belt or "chain" of the Van de Graaff type accelerator, the voltage multiplier of the Dynamitron and the Cockcroft–Walton generator, and the transformer charging of the Insulating Core Transformer (ICT). The HV power supply is usually built as part of the accelerator for the high energy beam

(> 300 keV) systems, but is an external component connected to the accelerator through a high voltage cable for the simpler low energy systems. The capabilities of each type of direct voltage accelerator are listed below:

- Cockcroft–Walton accelerators are used for either electrons or ions at voltages up to 5 MeV and currents up to 100 mA.
- Dynamitrons are used for either electrons or ions at energies from 0.5 to 5 MeV and at beam powers up to 300 kW for electrons or 60 kW for ions.
- Van de Graaff accelerators are used for either electrons or ions (only ions for tandem configurations) at voltages from 1 to 15 MeV. The voltage is precisely variable and currents range from a few nA to as much as 1 mA.
- ICT accelerators provide electron beams at voltages from 0.3 to 3 MeV and currents up to 50 mA.

2.2. *Radio frequency linear accelerators*

The use of radio frequency (rf) cavities to accelerate ions was first proposed by Ising [21] in 1924, and initially developed by Wideroe [22] in 1928 and Sloan and Lawrence [23] in 1930. Significant advances in these resonant cavity accelerator concepts were made by Alvarez [24] in 1948 and by Knapp and others [25] in 1964. New linear accelerator (linac) structures continue to be developed and now these devices cover a wide range of operating frequencies and output charged particle energies. All of these structures use rf-generated voltage to accelerate "bunches" of charged particles in synchronization with the rf frequency. Due to the large difference in the masses and velocities of the electron and ions, industrial linacs used for electrons are usually different than those used for ions.

Most industrial electron linacs use the side-coupled cavity developed in 1964 by Knapp and Nagle at Los Alamos National Laboratory. Although the traveling wave structure originally proposed by Sloan [26] for the first electron linac is still in use, the stability of the standing wave structure has made it the structure of choice for industrial systems. These systems require electron-gun injection at low energies (10–30 keV) and operate at rf frequencies from 800 MHz (L-band) to 9 GHz (X-band), although the majority use the 3 GHz (S-band) structure that dominates the cancer therapy electron linac market. For industrial applications, these accelerators are used to produce electrons at energies from 1 to 15 MeV, with output beam powers from 1 to 100 kW, depending on the application.

Ion linacs are the newest member of the industrial accelerator family. Although the linac systems originally developed in the 1930s accelerated ions for physics research, it was not until the development of the radio frequency quadrupole (RFQ) structure in the early 1970s by two Russian scientists, Kapchinskii and Teplyakov [27], that the size of these systems became practical for industrial applications. The importance of the RFQ to the development of industrial ion linacs has previously been described by several authors [28–30]. All present industrial ion linac systems use the RFQ as the first (or only) accelerating structure and operate at rf frequencies from 100 to 600 MHz.

2.3. *Circular (magnetic) accelerators*

The major categories of circular accelerators used for industrial applications are cyclotrons, betatrons, and synchrotrons, and were developed or conceived roughly in that order. First developed by Lawrence and Livingston [31] at the same time as the Sloan and Lawrence linac in 1930, the cyclotron concept was quickly scaled up from the prototype energy of 80 keV to higher ion beam energies. While the use of cyclotrons rapidly expanded as a research tool, they also quickly became an instrument for the use of ion beams in practical applications, with many of the ideas driven by the early pioneers of the technology.

Cyclotrons employ a magnetic field to contain the ions in a circular spiral path as they are accelerated by an rf voltage applied to a gap between two D-shaped electrodes placed within the magnetic field. They employ both rf and magnetic focusing of the beam to keep it contained within the acceleration region. Modern industrial cyclotrons employ many new features, such as the acceleration of negative ions to make the extraction of the beam from the magnetic field easier and magnets with stronger focusing and larger gaps to allow good vacuum pumping. The details of this accelerator technology and its development history are thoroughly presented in another paper of this issue [32].

Because the acceleration of electrons in the cyclotron was impractical due to the electrons quickly becoming relativistic, the betatron was developed in 1940 by Kerst and Serber to make cyclic acceleration of electrons practical [33]. They used the concepts of magnetic induction that had been proposed separately by Slepian and Wideroe in the 1920's to produce the first working unit at 2.3 MeV. This design was later scaled up to higher energy electrons and by the 1950's and 1960's a number of these accelerators were used for the industrial application of nondestructive testing, as well as for medical irradiation, with more than 200 built. However, soon after the development of the present standing wave linac these devices became less practical, but they still find limited industrial use even today in nondestructive testing.

A more recent development of an accelerator that uses a magnetic field in the acceleration of electrons is the Rhodotron™, originally proposed by Pottier [34] and now commercially built by Ion Beam Applications (IBA) in Belgium. This accelerator uses a coaxial resonator at a frequency of 107.5 MHz to produce an energy gain of 1 MeV for an electron transiting through it. Magnets placed around the circumference of the cavity are used to obtain multiple passes through the cavity to produce higher electron energies. The cavity can be excited in a continuous wave (CW) mode as in the cyclotron, allowing it to produce high power output beams up to 10 MeV.

The synchrotron produces the highest energy electron beams used in industrial applications. This accelerator concept was first developed in the 1940's by Goward in England and General Electric in the US based on the ideas of Oliphant [35]. A ring of magnets with a changing magnetic field is used to maintain an injected beam of particles in a constant radius orbit as they are accelerated in an rf cavity by tracking the frequency of the rf accelerating voltage as the particles are accelerated. The original General Electric system accelerated electrons to 300 MeV and was the first to produce the synchrotron radiation for which these systems are used extensively today [36]. Modern synchrotrons are employed for the acceleration of both ions and electrons, but the only industrial application of these systems is for synchrotron radiation from electron synchrotrons.

3. Industrial Accelerator Applications

What should be obvious from the above summary of the origins of the science and technology of industrial accelerators is that all of them were originally developed for areas of research in physics and other fields. However, as soon as these devices became available for use, scientists began to find practical applications for them based on the interactions that they were observing of the energetic beams with the materials being used in their experiments.

These early physics accelerators have since evolved for use in a very wide range of industrial applications. The broad categories of applications and the individual applications within each group are listed below:

- Materials processing — encompasses the largest group of industrial accelerators, including ion implantation of semiconductors and metals, irradiation of plastics, food and other products by ions, electrons and x-rays to promote chemical changes and sterilization, and electron beam welding and cutting of ceramics and metals.
- Materials analysis — unique to ion accelerators and includes all of the ion beam analysis techniques used for trace element detection and quantification, element profiling and precision determination of elemental ratios, including accelerator based mass spectrometry.
- Nondestructive analysis — includes accelerators used to generate x-rays or neutron beams to examine materials for flaws and hidden features; also includes the use of x-rays and neutrons to detect contraband (security inspections) and the use of neutrons for mineral detection.
- Radioisotope production — includes all electron and ion accelerators employed for the production of radioactive materials for use in medical procedures and industrial applications.

Most of these applications are quite mature and have previously been reviewed in detail by many others. The present review only briefly talks about the science behind the applications and focuses more on the accelerator technology, the current vendors for the equipment used in each field and the approximate annual market size for each application.

The market for industrial accelerators has grown to be a significant international business. Table 1 lists the number of accelerator systems sold to date for each application, as estimated by the author from available literature and consultations with vendors. These quantities include a number of accelerators originally purchased for physics research, but now

Table 1. Total number of industrial accelerators sold worldwide (not including medical accelerators).

Application	Total (2007)
Ion implantation	∼ 9500
Electron cutting and welding	∼ 4500
Electron beam & x-ray irradiators	∼ 2000
Ion beam analysis (including AMS)	∼ 200
Radioisotope production (including PET)	∼ 900
Nondestructive testing (including security)	∼ 650
Neutron generators (including sealed tubes)	∼ 1000
Synchrotron radiation	50
Total	18,700

used mainly for industrial applications. Note that the quantities are larger than reported by other studies that only counted currently operating systems. The total number sold will be larger because some of these industrial systems are no longer in use, many having been replaced by newer models with improved performance and reliability.

It should also be noted that the impact on the world's economy is much larger than just the sales of these accelerator systems would indicate, as has been discussed by Nunan [37]. Products and processes produced have a much larger monetary value than the equipment itself, usually by a factor of 100–1000 times the initial capital costs.

3.1. Ion implantation accelerators

This application represents by far the largest market for industrial accelerators. With ∼ 8000 units sold from 1980 to 2006 [38], this market is about the same size as the market for medical accelerators, which was reported recently to have a worldwide installed base of 9100 systems [39]. Ion implantation was first carried out in the 1940's by physics researchers for the production of solid targets containing deuterium, tritium and helium-3 atoms. Ion implantation of silicon using accelerators for doping was reported as early as 1950 [40] but the first "industrial" ion implantation system was delivered in 1965 to Fairchild Semiconductor by the High Voltage Engineering Corporation. Several other companies began commercial production of these systems in the same time frame [41].

By the early 1980's, the development of high current machines (>10 mA) made ion implanters the primary tool in the semiconductor production business of doping integrated circuits. Improved control over the dose and depth distribution of dopants provided by ion implantation, coupled with the efficient production coming from the use of photoresistant patterned layers as dopant masks, resulted in the rapid growth of complementary metal-oxide semiconductor (CMOS) device production. This then allowed CMOS devices to replace older bipolar transistor designs. Ion implantation applications continued to be developed by the semiconductor industry and the accelerators now used cover a wide range of ion species, beam energies and beam currents. Ions from protons to antimony are used for implantation in silicon at energies from hundreds of electron volts to almost 10 MeV. The industry is most interested in the implanted dose, i.e. the integrated charge deposited in each unit area of the material being bombarded, so the beam currents are typically as high as possible within the cooling requirements of the material being implanted. The various semiconductor implantation applications are shown in Fig. 1 within the "phase space" of the ion energy and implanted dose.

While the most common ion implantation application is the doping of semiconductor materials as discussed above, Fig. 1 illustrates that there are several other important applications in this field. These include "mesotaxy", the technique of forming a buried implanted layer within another material, and the fabrication of silicon on insulator (SOI) substrates by the direct implantation of oxygen into the silicon substrate to form a buried oxide layer.

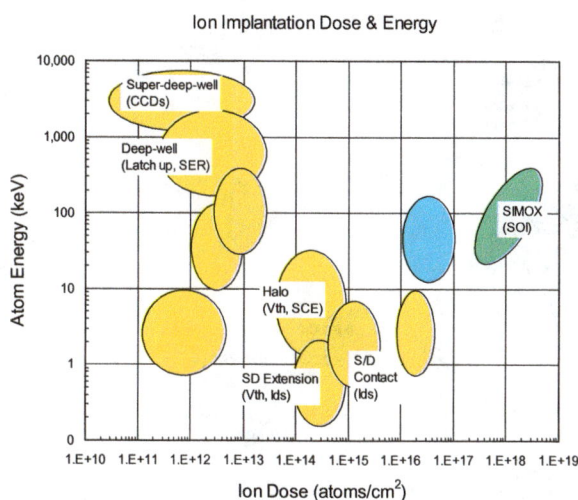

Ion Implantation Dose & Energy

Fig. 1. Ion energy and implanted dose ranges for doping of the key device parameters in CMOS transistors and for the fabrication of SOI and other heterogeneous layer wafers [42].

A competing approach for the fabrication of SOI and related wafer types is the implantation of hydrogen ions within a silicon substrate to allow the "cleaving" of a very thin layer of material for the production of laminated transistors and for other electronic, photonic and microelectronic mechanical systems (MEMS) applications.

Ion implantation is also widely used to modify the surface properties (hardness, stress, adhesion, friction, dielectric, and resistance to corrosion and chemicals) of metals, ceramics and glasses without changing the bulk properties. Common uses for metals include cutting tools and artificial human joints that last longer after ion implantation. For ceramics and glasses, ion implantation is used to harden them and change their optical properties.

It should be noted that a recently developed technique known as "plasma immersion implantation" [43] has become increasingly used for many of these applications at ion energies up to 100 keV. These devices are used when ultrahigh implantation doses are required. This technique uses the material to be implanted as one of the electrodes in a pulsed plasma discharge which is technically not an accelerator and thus will not be included in this review, although these systems are included in the market data.

The ion implantation accelerators used in the semiconductor industry were initially cascade generators with voltages up to several hundred kilovolts. Both the applications and the accelerators have since gone through many generations of development, with a wide range of accelerator types now being employed. It is easiest to classify them by their output beam energy and current, as detailed below.

Low energy / high current systems are usually referred to as "high current implanters" and cover the energy range from a few hundred eV to tens of keV. These variable energy machines employ single gap acceleration and produce beam currents up to 50 mA. Figure 2 shows the layout of a low energy, high current ion implanter that uses a horizontal ribbon beam and vertical scanning of the wafer.

Medium energy / medium current systems were the original accelerators used for ion implantation, with variable beam energies in the 50–300 keV range and output currents in the 0.01–2 mA range. They are typically used for irradiating one silicon wafer at a time either by scanning the beam over the

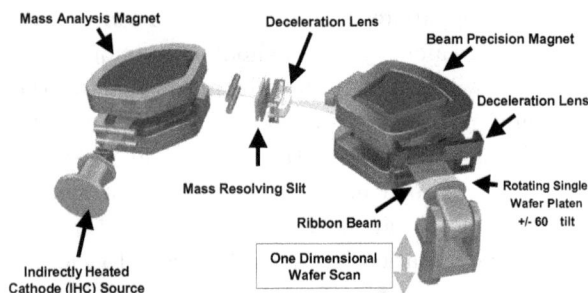

Fig. 2. Layout of a conventional low energy, high current ion implanter (Varian Model HCP, 0.2–0.6 keV).

wafer or by moving the wafer under the beam. These accelerators are usually multigap direct voltage units powered by a small voltage-multiplier high voltage power supply. The layout of a modern medium current system that uses a horizontal magnetically scanned ion beam and vertical wafer motion is shown in Fig. 3. The power magnets used in the beam line also allow implantation of massive molecular ions, such as $B_{10}H_{14}$ and $B_{18}H_{22}$, to provide high throughput doping at boron equivalent energies below 0.5 keV.

High energy/low current systems were driven by the need to reduce "latch-up" and "soft error rates" in CMOS and DRAM devices. This led to the development of ion implanters capable of accelerating the ions used as dopants to energies greater than 1 MeV. These variable energy accelerators cover the

energy range from 1 to 10 MeV at low currents up to hundreds of microamperes. Beam energies from 3 to 10 MeV are typically used for fabrication of advanced CMOS imagers and similar devices. These high energy systems can be either linacs or tandem charge-exchange columns, with both accelerator types generating high-charge-state ions for the upper energy ranges. The linacs are constructed from a number of MHz rf resonant cavities, as originally proposed by Glavish [44]. A typical high energy ion implanter accelerating structure is shown in Fig. 4.

The cavities in the Glavish structure are individually controlled for amplitude and rf phase to allow the acceleration of a wide range of ion species. A complete high energy ion implanter employing such a linac is shown in Fig. 5. This system uses a multiple number of linac stages (each stage operating at a maximum voltage of 80 kV), with interleaved quadrupole focusing elements to provide a wide range of output energies.

The economics and production requirements of IC manufacturing with smaller and smaller details have imposed severe constraints on the ion implanter equipment industry. They require stringent regulation of ion beam current and energy, with minimal contaminants and high uniformity, while being very reliable under 24/7 operation. Moreover, the operational requirements for IC production have become increasingly demanding. Production of an

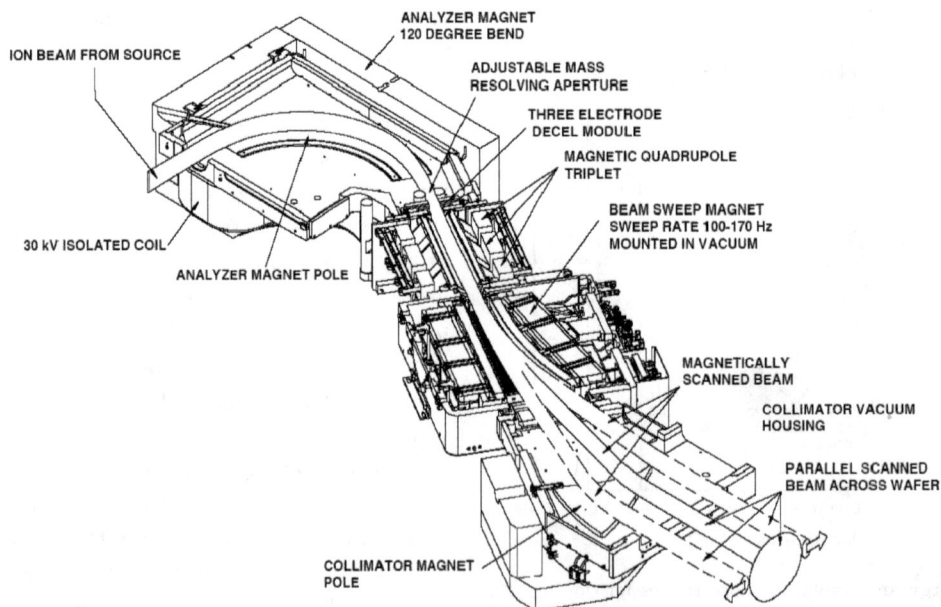

Fig. 3. Layout of modern medium current ion implanter system (Nissin Ion Model 9600A, 3–320 kV).

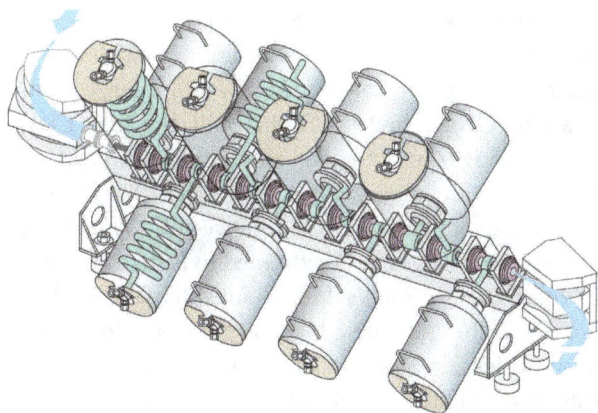

Fig. 4. Glavish high energy ion implantation linac structure. (*Courtesy Axcelis Technology*).

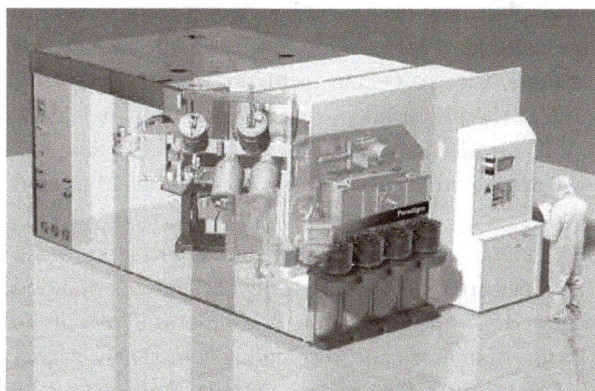

Fig. 5. High energy ion implantation system (Axcelis Model Paradigm XE, 10 keV to 4 MeV).

IC in the 1980's might have required 3–7 major implants, while today a modern IC can require 30 or more. To meet these demands, commercial implanters have evolved into expensive, high quality equipment, which has in turn raised the economic and technical barriers to entry in this accelerator market. As a consequence there are only a few major ion implantation equipment vendors worldwide.

Three large vendors today account for most of the implant system sales worldwide for semiconductor applications: Varian Semiconductor Equipment (USA), Axcelis Technology (USA) and Nissin Ion Equipment Company (Japan). Axcelis and Sumitomo Heavy Industries also operate SEN Corporation, a joint venture in Japan. A fourth major vendor, Applied Materials Inc. (USA), announced in early 2007 that it was exiting the ion implanter business.

The applications of ion implantation outside of the mainstream semiconductor production industry present opportunities for smaller commercial operations to innovate and serve specialty markets, including ion implanters for metal implantation. While the mainstream implanters often can be modified for these specialty applications, there are a number of vendors that sell to these niche markets. These include Ulvac Technologies (Japan), IHI Corporation (Japan), China Electronics Technology Group (China), Ibis Technology (USA), and Advanced Ion Beam Technology (Taiwan), which is aiming to move into the void left by the exit of Applied Materials. This group also includes the electrostatic accelerator manufacturers High Voltage Engineering Europa B.V. (Netherlands), National Electrostatic Corporation (USA), Danfysik (Denmark) and Applied Energetics (formerly North Star Research Corp., USA).

A number of accelerators in research facilities are also being used by industrial customers for ion implantation. An example is the production of microfilters using high energy heavy ions [45]. While this product and process is an industrial application of accelerators, the accelerators used are usually large high energy research accelerators located at national laboratories. Thin polycarbonate sheets are bombarded with heavy ions (argon) at energies sufficient for the ions to penetrate through the material (usually several MeV/nucleon). The ion tracks left in the material can then be etched to produce uniform small holes that are used to filter viruses out of blood and to purify water of all contaminants.

Ion implantation systems have a sale price ranging from ~$1.5–5 million, depending on their size and material-handling capability. The current market shares for semiconductor systems are 17% for high energy systems, 30% for medium energy systems and 50% for high current systems, with the last being the fastest-growing segment. As illustrated in Fig. 6, the market for ion implanters over the last 25+ years has followed the well-known "boom and bust" cycle for semiconductor equipment sales. The major surges are associated with the construction of new IC fabrication facilities first in the US, then in Japan, Korea and Taiwan, and now in mainland China. The overall sales for ion implantation systems are presently ≈ 500 units per year, with sales totaling > US$1.4 billion in 2007.

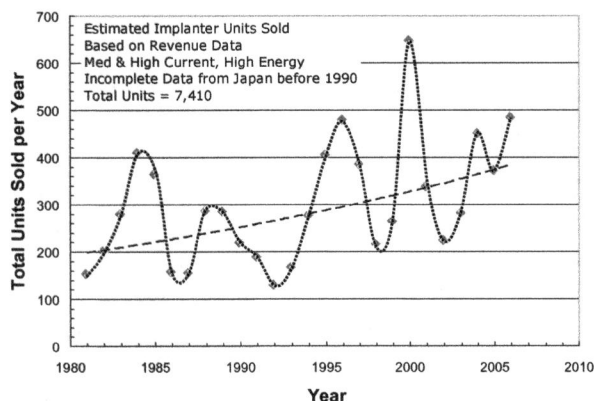

Fig. 6. Annual sales of ion implanter units since 1981 estimated from reported financial data. Actual sales are higher than shown in this compilation due to incomplete data for some years and some regions of the world [38].

3.2. *Electron beam processing accelerators*

Almost as large an industrial application as ion implantation, the uses of electron beam processing accelerators fall into three broad categories:

- Material treatment — includes welding, melting and hardening of metals, and cutting and drilling of metals and other materials such as ceramics.
- Radiation processing — includes polymer grafting and cross-linking, degradation and decomposition of plastics and waste gases, and curing of monomers and oligomers as well as epoxy-based composites.
- Sterilization and preservation — includes medical products and wastewater sterilization, and irradiation of food and feed products for disinfestation and preservation.

3.2.1. *Electron beam welding and cutting*

The first of these three categories is more isolated as a business market than the other two but has the longest history. The electron beam technology used in these systems dates back to 1869, when Hittorf [46] discovered electron beams. However, Pirani [47] was the first to use electron beams for fusion tests with metals in 1905. In 1948, Steigerwald, who was at that time developing more powerful electron microscopes, used electron beams as a thermal tool for drilling watch stones and for soldering, melting and welding in a vacuum [48]. In 1952 he built the first electron beam processing machine. At around

the same time in France, Stohr discovered electron beam welding during manipulation on x-ray tubes at the research laboratory of the Atomic Energy Commission (CEA). In 1956 Stohr built the first EBW machine in France to join zirconium alloys. Mario Sciaky was a classmate of Stohr in the 1930's, and a license agreement was signed between CEA and Sciaky S.A. which was later extended to Sciaky Brothers in the US, now Sciaky Inc., a subsidiary of Phillips Service Industries [49]. In 1958 Steigerwald welded 5-mm-thick zircaloy pieces together and discovered the "deep welding effect" of electron beams. In 1963 he founded the company Steigerwald Strahltechnik GmbH, now a member of the All Welding Group AG.

Outside of Germany and France work began elsewhere, particularly in the US and Great Britain, on developing new electron beam equipment. Some of the companies formed for this task are still in this business. Electron beam technology is now so widely used in the materials processing field that it is difficult to list all the different users and new applications that are constantly being developed for welding and cutting.

Typical applications for these electron beam machines are welding, annealing and cutting of metal parts. Holes in a 1.25-mm-thick metal sheet up to $125\,\mu$m diameter can be cut almost instantly with a taper of only a few degrees. Similarly, they can be used to cut very precise slots. Hence, this equipment is commonly used by the electronics industry to aid in the etching of circuits in microprocessors.

The main parts of a typical electron beam processing system are shown schematically in Fig. 7. The electron accelerator is the key part of the "beam head," which contains an insulator with the cathode, the bias cup and the anode of the electron gun integrated into it. An external high voltage power source is connected through a high voltage cable. This subassembly is placed in a vacuum system, with a valve separating the beam generating system from the beam optics channel and the "working" vacuum chamber. A viewing system is used to adjust the beam's focus point on the work piece in the chamber and to observe the welding process.

The insulator, electron gun design and high voltage power supply are the three critical components that determine the energy and power in the electron beam. The control circuitry is also designed to protect the system from discharges or sparking.

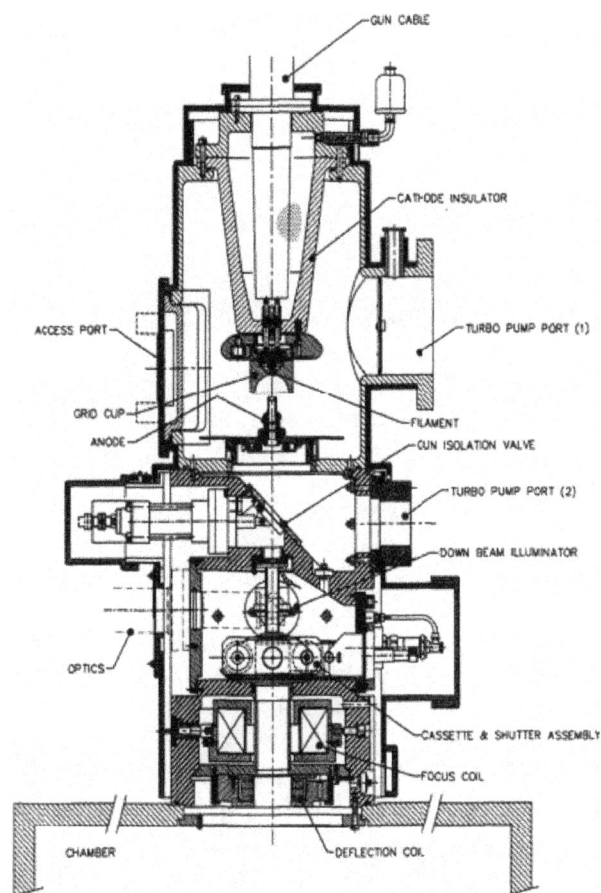

Fig. 7. Schematic layout of a typical electron beam system. (*Courtesy Cambridge Vacuum Engineering*)

In most electron beam processing machines, the electrons from a thermal emitter electron gun are accelerated to energies in the range of 60–200 keV. The machining or cutting process is usually performed in a working vacuum chamber to reduce the scattering of the electrons by gas molecules, although new systems can be used at atmospheric pressure. An electromagnetic lens is used to focus the beam onto the work piece, and a magnetic deflection system is used to locate and move the beam on the object being bombarded as required. The electron beam can thus be used to precisely bombard a location on the object, with the kinetic energy of the electrons converted into thermal energy to melt or vaporize the material, either fusing it or removing it to form holes or cuts. The electrons penetrate the object according to their initial energy, making it possible to weld much thicker pieces with high energy electrons than is possible with conventional welding

processes. Also, because the electron beam is tightly focused on the object, the total heat input is much lower than that of conventional welding. The objects cool quickly, which allows many dissimilar metals to be welded together. Almost all metals can be processed with electron beams — the most common are superalloys, refractory or reactive metals, and stainless steels.

In a high vacuum work chamber, materials as thick as 15 cm can be processed at distances from the electron gun of as much as 70 cm. For thinner work pieces, the welding can take place in a lower vacuum (100 mT) work chamber, allowing parts as thick as 5 cm to be processed at stand-off distances of about 40 cm. This reduces the time needed to pump out the chamber and get the process started. Electron beam processing can also be performed in air, but the parts must be within a few inches of the exit of the beam from the acceleration stage. While less common, this technique allows work pieces of any size to be processed.

A new development now being commercialized for electron beam welding is the "plasma window" developed at Brookhaven National Laboratory by Hershcovitch [50] and licensed to Acceleron Incorporated. This device, shown schematically in Fig. 8, uses a dense plasma discharge to separate the vacuum in the electron gun region from the atmosphere. The plasma's ionized atoms and molecules are much hotter than the air at room temperature. The higher thermal velocities make them collide more often with air molecules trying to enter the vacuum, thus stopping most of them. The electron beam passes through the plasma easily because of its high velocity, thus allowing it to be used for welding operations in air. While a drawback is the amount of power to maintain the plasma, this new system has expanded the applications of electron beam welding to large structures, such as in the production of airplanes and ships.

As has been the case in other mature accelerator business sectors, most of the major vendors of electron beam welding and cutting equipment have been consolidated into a few large international players. However, there are a number of smaller vendors that are pursuing business in a particular geographic region or niche market. The major vendors include Sciaky, Inc. (USA), All Welding Group AG, consisting of the PTR Group and Steigerwald

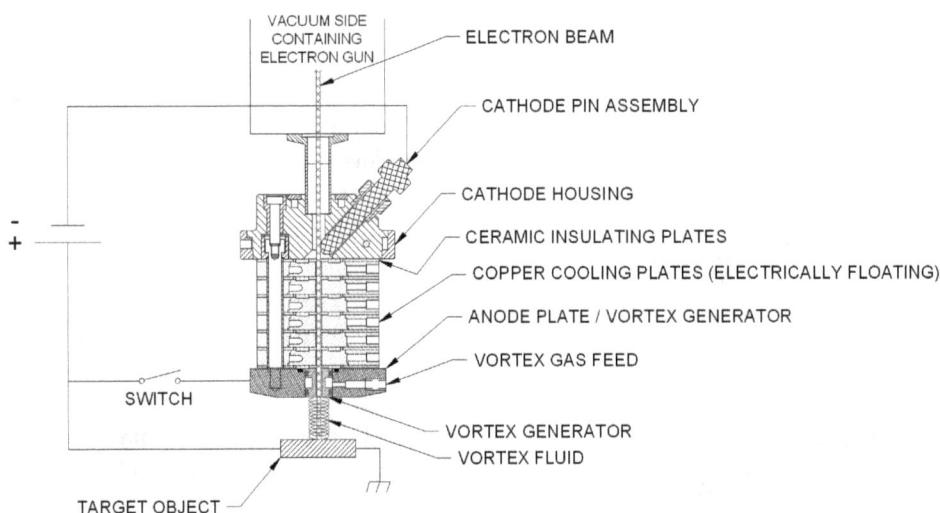

Fig. 8.   Plasma window for the electron beam welding system.

Strahltechnik (Germany), Cambridge Vacuum Engineering (England) and Bodycote Techmeta (France). Smaller vendors include Pro-beam (Germany), Orion (Russia), Mirero (Korea), Omegatron (Japan), NEC Corporation (Japan), Mitsubishi Electric Corporation (Japan) and Acceleron (USA).

The cost of a typical electron beam welding/cutting system is ~US$500,000–2,500,000 depending on the energy and power of the system, as well as its product-handling capability (i.e. working space inside the vacuum chamber). The installed base is growing at more than 100 systems per year, with the market for these systems estimated by the author to be ~US$150 million.

3.2.2. Electron beam irradiation accelerators

Accelerators used for the other two broad categories of electron beam processing applications can be combined and labeled simply as electron beam irradiation systems (also referred to as electron beam irradiators). The applications are also commonly referred to as radiation processing. Radiation processing began as a little-known technology used by a few companies in the 1950's and has now grown into a mature, reliable technology that is being used for a wide variety of processes in many sectors of the world economy. In radiation processing, energetic electrons serve as a source of ionizing energy that promotes a chemical process in a chosen material. They do this by generating ions and slow electrons in the bombarded material, which then modify the

bonds in the material's atoms and molecules through interactions with the free radicals. This can change both the chemical and the physical properties of the bulk material.

The electron beam energy needed is determined by the thickness and density of the material being treated. The current needed is determined by the desired processing speed, the radiation dose required to accomplish the process and the power dissipation capability of the material being processed. Because these processes cover a wide range of electron beam properties, it is useful to divide electron beam irradiators into four categories using output beam energy ranges.

- 100–300 keV — usually single-gap, self-shielded sheet beam systems with no beam scanning. These are used for curing inks and thin film coatings on sheet material and for cross-linking plastic laminates and insulation on single-strand wire. Output beam currents vary from 10 to 2000 mA and can treat a 1–3-m-wide strip of material.
- 450–1000 keV — larger dc systems usually employing scanned beams with self-shielding. They are mainly used for cross-linking, curing and polymerization processes in the tire, rubber, wire and plastics industries. Output beam currents vary from 25 to 250 mA and can treat a 0.5–2-m-wide strip of material.
- 1–5 MeV — scanned beam systems capable of producing 5–300 kW beam power. They are used for cross-linking, curing and polymerization of

thicker materials, and for sterilization of medical products. Scanned beam width can be up to ∼2 m.

- 5–10 MeV — high energy scanned beam systems capable of producing 5–700 kW beam power. They are used for medical product sterilization and for cross-linking, curing and polymerization of even thicker materials. They are also used for food irradiation, wastewater remediation, and gemstone color enhancement, particularly for topaz and diamonds. The higher power units are used as x-ray generators for medical product sterilization, food irradiation and curing of carbon-fiber-reinforced plastic composite materials.

The lowest energy systems are used for surface curing and drying, since the energies are typically less than 300 keV. Like the electron beam welding systems, they use an external high voltage power supply. The power is fed through a cable to an elongated cathode inside the "beam head." The electrons emitted from the wide cathode are accelerated to an anode containing a very thin foil window. The accelerated electron beam is transmitted through this window to irradiate the material moving past it, as shown schematically in Fig. 9. The product that is being treated enters and exits the irradiation area through openings in the shielding around the system. This self-shielding allows the systems to be used on a factory floor in an assembly line.

Low energy sheet beam systems are available in a variety of sizes and power levels, depending on the width of the material to be treated and the processing speed required. A typical low voltage system has a beam power of up to 50 kW, which is limited by the power dissipation in the beam exit window. It can treat a 2 m width of material moving at a speed of 15 m/s with a radiation dose of 1 kGy and a dose nonuniformity of less than ±5% [51]. The largest of these systems can have a beam width up to 3 m and a beam power of 100 kW.

The next-largest accelerators used for direct electron bombardment of materials fall into the two medium energy categories that cover an energy range from 0.45 to 5.0 MeV. These are usually direct voltage machines, such as the Cockcroft–Walton, ICT or Dynamitron, in which the high voltage generator is actually integrated into the accelerator. All of these systems use multigap graded accelerating columns. A typical Dynamitron accelerator is shown in Fig. 10 with the column removed from the gas tank. These medium energy units are some of the original electron beam irradiation systems employed in industry many years ago and still represent a large percentage of the systems being sold. Beam powers can be up to 200 kW and the beam is usually scanned over a width up to 2 m. The beam exits a "scanning horn" through a foil window, with the irradiated product moving past it in air. These units generate large amounts of radiation and ozone, so they are usually

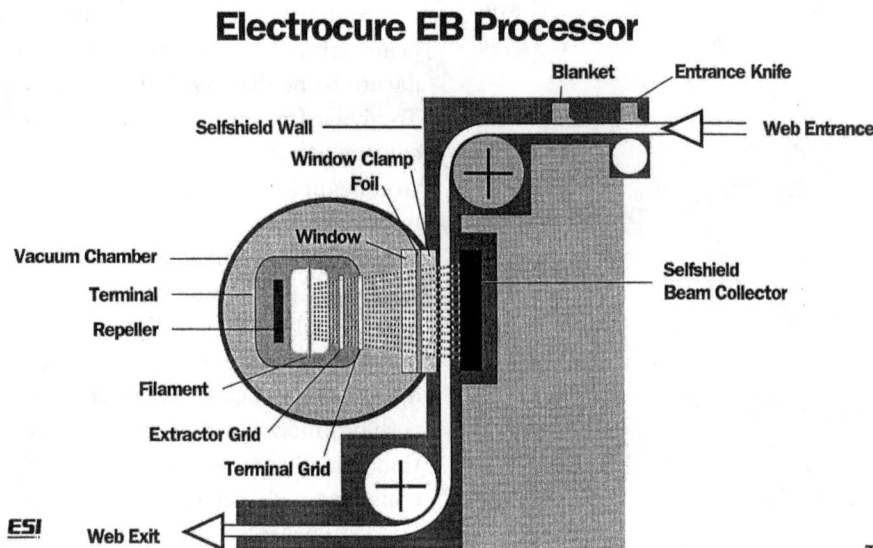

Electrocure EB Processor

Fig. 9. Schematic of a sheet beam irradiator for surface curing. (*Courtesy Energy Sciences Inc.*)

Fig. 10. Dynamitron accelerator removed from tank. (*Courtesy University of Albany Ion Beam Laboratory*)

located in a shielded vault with a special ventilation system.

The highest energy electron irradiators usually employ rf acceleration, either in rf linacs or other single-gap rf cavity systems. These systems have a maximum energy of 10 MeV to avoid the activation of the bombarded materials from nuclear reactions that occur above the bombarding energy. This category includes the most powerful of the electron beam irradiators, the Rhodotron™. The electron linacs in the energy range from 5 to 10 MeV use standing wave structures at 800–6000 MHz and have beam power outputs of 5–100 kW. Most of these systems are used for medical product sterilization. A typical layout is shown in Fig. 11. The single-cavity rf systems typically operate in the few-to-200 MHz range

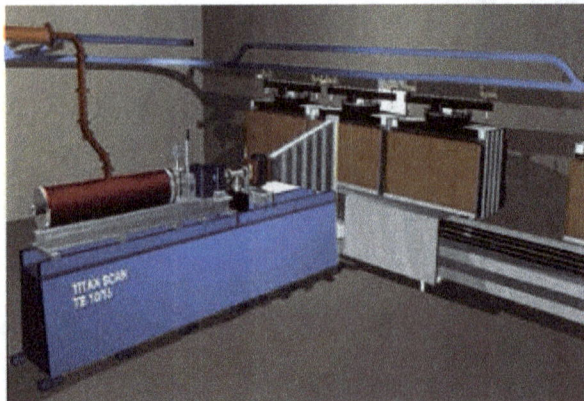

Fig. 11. Layout of a 10 MeV linear accelerator and process conveyor for metical sterilization. (*Courtesy L-3 Communications*)

Fig. 12. Rhodotron™ accelerator used for electron irradiation. (*Courtesy IBA Industrial*)

and cover the energy range from 1 to 5 MeV. The Rhodotron, shown in Fig. 12, can produce electron beams up to 10 MeV and output beam powers up to 700 kW. All of these high energy machines can be used with a water-cooled tantulum target to generate high energy x-rays for food irradiation, which has been approved by the US FDA for energies up to 7.5 MeV [52].

Radiation processing applications using electron beam irradiators are too numerous and varied in nature to be discussed in great detail in this review. To give a feel for how many common industrial and consumer products are directly processed using electron beam irradiation or contain radiation-processed materials, the applications can be divided into four subcategories: curing and drying of coatings and materials; production of cross-linked, scissored and grafted polymers; sterilization and food irradiation; and wastewater and gas treatment. There are other applications, such as advanced composite curing, viscose production and thermomechanical pulp production, which have been demonstrated but have not yet gone into widespread commercial use.

While electron beam irradiation systems for curing thin materials and surface coatings have been in use for several decades, the market continues to expand in spite of competition from other curing techniques, such as ultraviolet light. One advantage of these sheet beam systems is that the radiation curing process resulting from the molecular changes induced by the charged particles is virtually instantaneous, occurring as the material passes through the beam. Moreover, it can be done at room temperature. The beam current can also be computer-controlled to accurately provide the radiation dose required for the material and desired processing speed.

There is also a push to market these systems as a "green" technique [53]. Curing of coatings and adhesives on woods, metals and polymers can eliminate the use of catalysts or solvents and reduce pollution and the safety problems associated with these toxic and corrosive materials. For example, electron beam curing has recently been used to replace autoclave curing of composites. Electron beam curable epoxy resins have been commercially available since the mid-1990's and have been used successfully in the production and repair of some of the composite components in aircraft [54].

Cross-linking in polymers takes place when the free radicals released in a material bombarded by energetic electrons cause adjacent molecules to link by covalent bonding. The new material created can have a molecular structure that makes it stronger and more heat and chemical-resistant, or it can become "memory-sensitive," with an ability to stretch and shrink when heated above a certain temperature. The energetic electrons can also be used to "scissor" the molecular bonds in other materials or even cause bonding between different molecules (grafting).

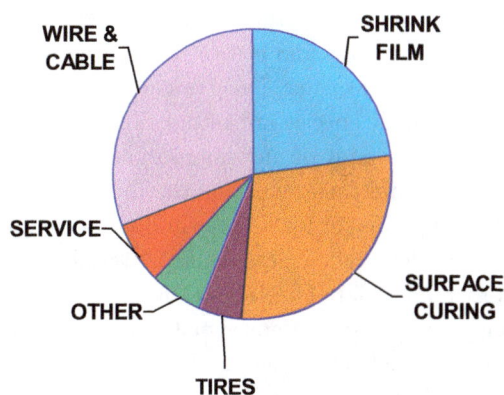

Fig. 13.   E-beam irradiator end-user markets [57].

Table 2, based on a presentation given by Machi [55] in 2005, summarizes the many types and uses of commercially produced polymers.

The cross-linking of polymers is the largest existing industrial application of radiation processing, occurring mainly in systems that are in factory production lines. Within this application the largest uses are in the wire and cable industry, the heat shrink film and tube industry and the tire industry, as shown in Fig. 13. The commercial value of these products exceeds US$ 50 billion per year [56].

The second-largest application of radiation processing is the sterilization of single-use disposable medical products and supplies such as surgical gowns, surgical gloves, syringes, sutures and the like. High energy electrons, typically between 3 and 10 MeV, are required for the sterilization of these products in order to penetrate the product packaging and provide a minimum dose throughout the package. As the product is passed through

Table 2.   Commercially produced cross-linked and grafted polymers.

Product	Use
Cross-linked polyethylene (PE) and PVC	Heat- and chemical-resistant wire insulation; pipes for heating systems
Cross-linked foam PE	Insulation, packing and flotation material
Cross-linked rubber sheet	High quality automobile tires
Cross-linked polyurethane	Cable insulation
Cross-linked nylon	Heat- and chemical-resistant auto parts
Heat-resistant SiC fibers	Metal and ceramic composites
Vulcanized rubber latex	Surgical gloves and finger cots
Cross-linked hydrogel	Wound dressings
Acrylic-acid-grafted PE film	Battery separators
Grafted PE fiber	Deodorants
Curing of paints and inks	Surface coating and printing

the sterilization system, the energetic electrons interact with its contents and create secondary energetic species such as electrons, ion pairs and free radicals. These secondary species are responsible for disrupting the DNA chains of any microorganisms, inactivating them and thus rendering the product sterile.

The requirement for a uniform dose in the package means that lighter, less dense materials tend to be most appropriate for electron beam sterilization. For larger packages or denser materials, the conversion of the electrons to energetic x-rays is required to get the desired penetration and dose distribution.

Although the technique was introduced in 1956 when Johnson & Johnson developed the first commercial electron beam sterilization process [58], the poor reliability of the early accelerator-based systems inhibited its early adoption. Instead, Cobalt 60 (gamma) irradiators were used due to their higher output and much higher reliability.

The reliability and performance of the critical accelerator components slowly improved over the next two decades at national high energy physics research laboratories. Industry's involvement in developing radiographic and oncology machines during this period also advanced the durability and reliability of electron accelerators. These improvements, along with the advent of automated computerized controls, caused the industry to re-evaluate the technology. Even though ethylene oxide (EtO) gas and ^{60}Co gamma radiation are still the most commonly used sterilization methods, installation of electron beam sterilization systems is increasing. This is partly fueled by present concerns about the world's supply of ^{60}Co and the problems associated with handling these large radioactive sources. In fact, medical products with a commercial sales value well in excess of US$10 billion/year [59] are sterilized today using electron beam irradiators. In addition, pharmaceutical companies are choosing to use electron beam sterilization of product ingredients.

A related application of high energy electron and x-ray irradiators is the disinfestation and preservation of food. Food irradiation is currently approved for selected products in more than 50 countries and is being performed at ~200 facilities worldwide, mostly with gamma radiation from radioactive sources [60]. However, these systems process only a small fraction of the food that could be irradiated. As this process becomes more widely accepted, this application

could become the largest use of electron beam irradiators. This is because of the growing concern over the use of large radioactive sources, the shortage of these sources, and the availability of new, more powerful high energy electron accelerators that are reliable and economical.

The major vendors in the electron beam irradiator equipment business are distinguished in general by the sizes of the systems they manufacture. There are many vendors that manufacture and sell the low energy sheet beam irradiators while there are only a few suppliers for the large cross-linking and sterilization systems. However, there are vendors in at least nine countries and the number of vendors in developing nations that are now developing or starting production of these systems is continuing to grow.

The vendors for low energy systems include Energy Sciences, Inc. (USA), PCT Product & Manufacturing, LLC, [formerly RPC Industries (USA)], IBA (Belgium), Electron Crosslinking AB (Sweden), Advanced Electron Beams (USA), Wasik Associates (USA) and Nissin High Voltage Corp. (Japan).

The price for a sheet beam irradiator system is US$200,000–1,000,000 but it has a lifetime of 15 years and can be used to process many millions of dollars' worth of products during this time. The present worldwide market for these self-shielded accelerator systems is estimated to be ~US$30 million per year and is increasing annually by more than 8%.

For the larger systems required for cross-linking of thicker or denser materials and for medical sterilization and food irradiation, the vendors are primarily IBA (Belgium), which also owns RDI in the US, Nissin High Voltage Corp. (Japan), Denki Kogyo Co., Ltd. (Japan), IHI Corporation (Japan), L-3 Communications Pulsed Sciences Division (USA), Vivirad (France), Mevex (Canada) and the Budker Institute of Nuclear Physics (BINP) in Novosibirsk Russia, which has sold more than 70 systems, mostly in Russia. EB TECH Co., Ltd. (Korea), with BINP assistance, has developed a large electron beam system for treating wastewater (1 MeV, 400 kW) and the Center for Advanced Technology in Indore, India is also developing, with assistance from BINP, a 2.5 MeV, 100 kW air-core transformer type electron beam accelerator for industrial use in India.

A typical high energy electron beam irradiator system (1–10 MeV) has a sales price in the range of US$1 to 8 million. This market is now estimated to be more than US$100 million per year and is expected to grow to well over US$1 billion per year in the next decade as the demand for "green" electron beam processing increases and new applications are developed.

3.3. *Ion beam analysis accelerators*

The use of ion beams for analysis of materials is a relatively new industrial application of electrostatic accelerators originally developed and used for nuclear physics research. From the 1930's through the late 1970's, these systems (commonly known as tandem Van de Graaffs) were primarily used for the investigation of the nuclear and atomic structure of elements. As more and more reactions were identified and cross-sections measured, a major shift was made to use the accelerators and the vast data base acquired through the basic research for materials analysis applications.

The precise energy resolution of the ion beams generated by electrostatic accelerators makes them ideal for analytical analysis techniques currently used in many fields, ranging from materials science and environmental science to the study of cultural heritage and biological samples. In particular, electrostatic systems are now widely used by the semiconductor industry for quality control and by environmental sciences for pollution monitoring [61]. A typical tandem Van de Graaff–based system for ion beam analysis is shown in Fig. 14.

Materials analysis techniques all rely on the vast database of knowledge that exists for almost all of the nuclei in the periodic table. Known reactions and cross-sections are used to determine the properties of materials, as well as the presence, quantity and

Fig. 14. Commercial accelerator system for ion beam analysis. (*Courtesy National Electrostatics Corporation*)

distribution of a trace element in a known material. Most of these applications use one of the following physics techniques:

- Rutherford backscattering (RBS)
- Elastic recoil detection analysis (ERDA)
- Nuclear reaction analysis (NRA)
- Particle-induced x-ray emission (PIXE)
- Particle-induced gamma ray emission (PIGE)
- Nuclear resonance reaction analysis (NRRA)
- Resonant scattering analysis (RSA)
- Charged particle activation analysis (CPAA)

A newer technique that has been rapidly growing in use is accelerator mass spectrometry (AMS). This differs from the above in that a small sample of the material to be analyzed is itself ionized to form the beam. AMS uses the capability of an accelerator and specially designed beam line combination to differentiate mass and charge in order to measure long-lived radioisotopes such as ^{14}C as markers or environmental and biomedical tracers with sensitivities as high as 1 part in 10^{15}.

Industrial applications of AMS include uses in the applied fields of geology, oceanography and biomedical science. In fact, the use of AMS in drug development is the fastest-growing industrial application of this technique. The high sensitivity of AMS permits the use of "microdosing" for the determination of the pharmacokinetics of new drugs. Drug development regulations permit drug doses labeled with ^{14}C to be administered to human subjects at 1% of the nominal dose for studies using AMS to determine the drug uptake in various organs.

The first demonstrations of AMS measurements were made in 1977 by teams at Rochester/Toronto and General Ionex Corporation [62] and soon after at Simon Fraser University and McMaster University. This sensitive analysis tool continues to be used by a number of research facilities, but industrial use is following the same evolution as the other ion beam analysis techniques.

All commercial AMS systems use electrostatic accelerators, primarily tandem Van de Graaffs. However, new users are looking for smaller systems that do not require physicists to operate them and have high reliability and availability. In response to this demand, compact single-stage electrostatic accelerator systems are now becoming available that are

Fig. 15. Compact single-stage ^{14}C AMS system. (*Courtesy National Electrostatics Corporation*)

smaller and easier to use. An example of a new single-stage AMS system is shown in Fig. 15.

The major vendors for turnkey AMS systems are National Electrostatics Corporation (USA) and High Voltage Engineering Europa (Netherlands). The sales from these and smaller companies in this business are still divided between research facilities and industrial users, but it appears that more than 20 systems per year are now sold for industrial use. Sales prices range from US\$400,000 to over US\$1.5 million and the current annual market is ∼US\$30 million.

3.4. *Radioisotope production accelerators*

Not long after the development of the first cyclotrons in the 1930's, these accelerators were used to produce radioisotopes. For many years these research systems continued to be used parasitically for production of radioisotopes for medical and industrial applications. It was not until the 1960's that several companies (Phillips, Allis Chalmers and The Cyclotron Corporation) started to sell small cyclotrons designed specifically for dedicated radioisotope production using the existing technology of the larger research systems [63]. These were positive ion cyclotrons with internal ion sources and usually used internal targets to increase the current available for bombarding the targets.

In the 1970's the next generation of cyclotrons (the negative ion cyclotrons) was commercially developed. These systems were able to deliver external ion beams using stripper foils to extract the beam with high efficiency, but required high vacuum and still used internal ion sources. In the 1980's the third generation of cyclotrons was commercially introduced. These were deep valley negative ion systems using external ion sources. They required less input power and were able to produce even higher beam currents (> 1 mA). It was also during this period that the "compact" cyclotrons for hospital-based production of the short-lived radioisotopes used in positron emission tomography (PET) were introduced.

The field of nuclear medicine uses a wide variety of radionuclides to both diagnose and treat cancer and other diseases. While the majority of the diagnostic procedures use radioisotopes produced in nuclear reactors, more than 10 million nuclear medicine diagnostic procedures are performed every year with accelerator-produced radioisotopes. While linacs are starting to be commercially available, the bulk of these isotopes are still produced using cyclotrons in the energy range from 10 to 30 MeV for protons. The energy range for commercially available systems is 7–70 MeV.

More than 50 radioisotopes are routinely produced by accelerators for medical and industrial use [64], including the most common traditional gamma ray emitters, such as ^{201}Th, ^{67}Ga, ^{123}I, ^{11}In and ^{103}Pd, and the short-lived positron emitters, ^{11}C, ^{13}N, ^{15}O and ^{18}F, used in PET. While the majority of the accelerator-produced radioisotopes are used for medical purposes, there are a variety of industrial applications that use radioisotopes, such as for measuring flow, determining moisture and thickness, and mutating plants.

Although most of the accelerator-based production of radioisotopes has been done using cyclotrons, proton linacs can also be used. The high energy physics research proton linacs at Brookhaven National Laboratory and Los Alamos National Laboratory have been used for more than 25 years to commercially produce a large variety of medical radionuclides. However, these older physics research linacs were too large and costly to reproduce for commercial production. But the ion linac development work at Los Alamos National Laboratory in the late 1970's significantly reduced the size and cost of the modern linac. The development of the

RFQ accelerating structure was the key to the practical implementation of compact proton linacs for this and many other applications [65]. This unique device allowed the use of higher frequency linac structures and lower energy ion injectors, decreasing the size and cost of the systems. Innovative modular approaches to the rf power systems have also made compact proton linacs practical as commercial radioisotope production systems, with several now in use.

Accelerators for medical radioisotope production fall into two major categories: those producing radioisotopes for single photon emission computed tomography (SPECT), industrial applications and therapy; and those for PET. The first category requires proton energies above 20 MeV, while the dedicated PET accelerators are all below 20 MeV.

Both cyclotron and linac-based systems are offered commercially for both applications. The larger systems for SPECT and other long-lived radionuclides range in proton energy from 22 to 70 MeV, but most of them are around 30 MeV. These dedicated turnkey systems are capable of producing time-averaged external beam currents of 1 mA or more and can be used to simultaneously bombard multiple targets.

Modern dedicated PET radioisotope production accelerators (both cyclotrons and linacs) are much smaller, covering the proton energy range from 7 to 18 MeV. An example of a compact PET cyclotron is shown in Fig. 16. Most of the compact cyclotrons and linacs operate with external beam currents of 100–200 µA and the smaller systems (7–11 MeV) include radiation shielding that eliminates the need for a vault. These small systems are designed for installation in a hospital. One company has even installed a 7 MeV linac in a trailer to make a mobile PET radioisotope production facility, as shown in Fig. 17 [66].

The largest radioisotope accelerator market is for the smaller dedicated PET systems. These short-lived radionuclides must be produced in close proximity (or within a two-hour delivery radius in the case of ^{18}F) to the site where they will be used. The market demand, which is large and growing at a robust pace, has spurred the continued development and innovation of these compact accelerators, with more than 500 systems now in use worldwide.

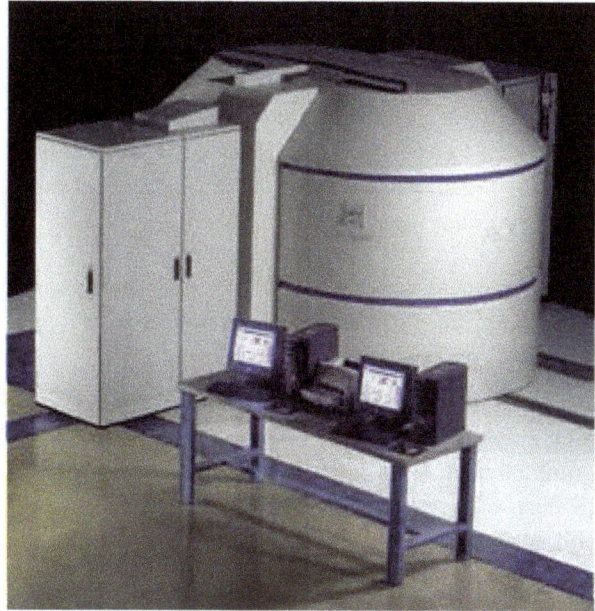

Fig. 16. Self-shielded compact cyclotron for PET radioisotope production (Siemens RDS Eclipse).

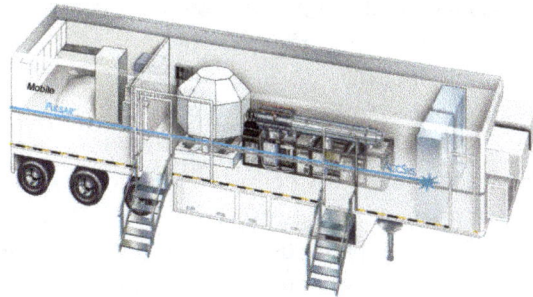

Fig. 17. Mobile PET radioisotope production system (AccSys Mobile PULSAR®7).

Because radioisotope production accelerators are a fast-growing market, there are a large number of vendors. Six are well established, with several more manufacturers now seeking to enter the market. The leading vendors for cyclotrons are GE Healthcare (Sweden), Siemens Medical Systems (USA), Ion Beam Applications SA (Belgium), Advanced Cyclotron Systems (Canada) and Sumitomo Heavy Industries (Japan). Two new cyclotron vendors entering the business are Samyoung Unitech Co. (Korea) and Thales GERAC (France). The only established linac vendor is AccSys Technology, Inc. (USA).

The selling price for dedicated radioisotope production accelerator systems (cyclotrons and linacs) ranges from ~US\$1–3 million for the smaller PET systems to US\$5–30 million for the larger SPECT systems. The PET accelerator market is by far the larger of the two, with more than 50 per year being sold. The total annual market for all of the dedicated radioisotope production accelerators is ~US\$70 million.

3.5. *Nondestructive testing accelerators*

This industrial application uses electron linacs in the energy range from 1 to 16 MeV to generate high energy x-rays from a tungsten target for radiographic inspection of thick materials that cannot be inspected using standard x-ray tubes at lower voltages. These electron linacs have been in use for almost 50 years for the inspection of large metal castings and welded joints to locate flaws and for the inspection of large solid fuel rocket motors. The high energy x-rays from these accelerators have sufficient energy to penetrate these thick parts without being stopped and can produce radiographs of their path through the material. The resolution of these radiographs necessary to locate small flaws in the material requires that the x-ray spot be as close to a point source as possible. This means that the electron beam needs to be focused onto the production target to a spot as small as possible without melting it.

Because the parts being inspected are often very large and heavy, early commercial units were designed to be mobile so that they could be moved around the part. With the advent of real-time detection systems, high energy x-ray inspection systems were developed using the computed tomography (CT) technique. These CT systems often use a stationary linac and stationary detectors, while the inspected part is rotated and translated. Also, the *in situ* inspection of parts in fixed installations, such as nuclear power plants and bridges, has led to the development of very compact portable linacs [67]. These employ 9 GHz resonant cavity structures instead of the 3 GHz microwave structures used in conventional electron linacs designed for this application and for medical therapy.

A more recent and much larger application of nondestructive testing linacs is the inspection of large cargo containers and semitrailers at border entry points. Although originally deployed to stop the entry of weapons and illicit materials, these systems are now also being used for screening of cargo to collect duties and taxes on shipments that have not been properly declared [68]. First deployed in 1983 at the Eurotunnel between Europe and the UK [69], this security application has become very widely adopted and has resulted in an exponentially expanding market demand for compact electron linacs packaged for this use. Border inspection systems using these high energy electron linacs are now deployed in more than 20 countries. These new systems, an example of which is shown in Fig. 18, are self-shielded and emit a collimated fan beam that passes through the cargo and is recorded on a line detector opposite it for the production of a radiograph of the truck's or cargo container's contents. The latest innovation in these inspection systems is the use of simultaneous dual energy beams that allow the differentiation of heavy elements, such as uranium or lead.

The two major accelerator vendors in this business are Varian Medical Security & Inspection Products (USA), and Nuctech (China). Other, smaller vendors include L&W Research, Inc. (USA), HESCO (USA), EuroMeV (France), MEVEX (Canada) and JME Ltd. (UK) All of these vendors produce electron linacs except for JME, which manufactures a small portable betatron for radiographic inspection.

More than 650 systems have been sold for this application during the past 50 years, but that number is rapidly increasing, with as many as 100 systems now being sold annually, predominantly for security applications. An NDT system costs from ~US\$300,000 to US\$2 million, depending on its output energy and configuration. The present annual

Fig. 18. New security inspection electron linac. (*Courtesy Varian Medical Corporation — Security and Inspection Products*)

market for these systems is ~US\$70 million, but it is expected to grow to more than US\$270 million per year within five years.

3.6. *Neutron-generating accelerators*

Neutron sources are generally categorized as utilizing either radioactive materials or induced nuclear reactions. The first category includes nuclear reactors and radioisotopic sources, and represents the bulk of the neutron sources currently used in industry. The second category includes all methods of accelerating charged particles to produce neutrons by induced nuclear reactions. These devices range from the electrostatic acceleration of particles in the classic "sealed tube" neutron generators, which produce monoenergetic neutrons via the $D(d, n)^3He$ or the $T(d, n)^4He$ reaction, to large accelerators (electrostatic tandems, cyclotrons and linacs).

Accelerator-based neutron sources can be separated into two classes: electron accelerators that produce neutrons from secondary nuclear reactions [e.g. $Be(\gamma, n)$ using bremsstrahlung x-rays produced by electron bombardment of a tungsten target], and ion accelerators that generate neutrons via direct nuclear reactions. For ion accelerators, a number of induced nuclear reactions can be used to generate neutrons, as illustrated by the data from Hawkesworth [70] shown in Fig. 19.

A large number of publications have been produced about the generation of neutrons using accelerators, including several review papers [71–73]. Most of the early work was done using electrostatic accelerators, but advances in modern compact ion linacs have made them viable neutron generators for industrial applications in nondestructive testing (NDT).

When used with a suitable moderator, these neutron generators can replace conventional radioactive sources in thermal neutron radiography applications ranging from corrosion detection in aircraft structures to detection of voids in munitions. They can also be used as neutron sources for fast neutron and resonant neutron radiography, as the neutrons can penetrate thick materials and detect hidden features. These NDT techniques have been demonstrated in applications as diverse as mineral assay and homeland security.

The principal commercial vendors for sealed tube neutron generators are Thermo Scientific

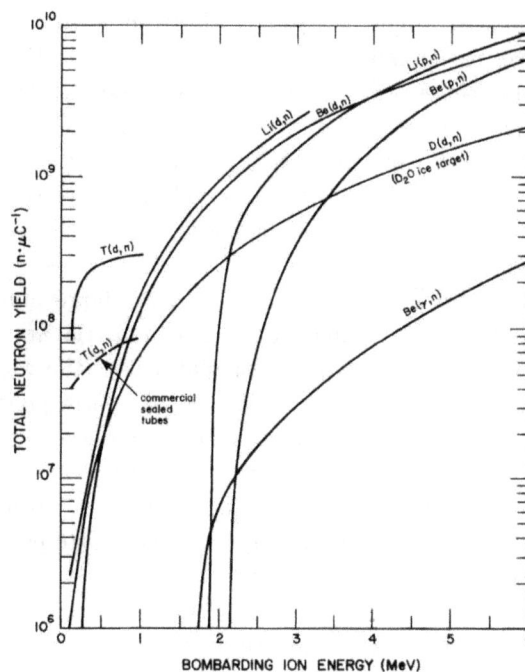

Fig. 19. Neutron yields for low energy particle beam reactions [70].

(USA), EADS Sodern (France) and All-Russia Research Institute of Automatics–VNIIA (Russia). However, three of the largest industrial producers in the US of these small devices — Halliburton Co., Schlumberger Well Services and Baker Atlas — build units for internal use in oil well logging tools.

Larger accelerator-based neutron generators are built by AccSys Technology, Inc. using its proton and deuteron linacs and by IBA using the Dynamitron. Cyclotrons from Sumitomo Heavy Industries and electrostatic accelerators from NEC and HVEE have also been sold for this application.

The market for these systems is dominated by the sealed tube units, with more than 50 systems per year now being sold. There are only a few ion accelerators sold per year for this application, but new security applications could dramatically increase sales in the future. The total market value for this business is very hard to obtain because of the larger number of vendors and the mix of products sold by each one, but is estimated by the author to be ~US\$30 million per year.

3.7. *Synchrotron radiation accelerators*

Soon after the discovery of synchrotron radiation from the General Electric 300 MeV electron accelerator, high energy synchrotrons built for physics research began to be used as sources of these intense x-rays for materials studies. These accelerators have gone through three generations of development and can now generate a wide spectrum of x-ray energies. While there are only a few of these large accelerators actually deployed in industrial settings, more than 40 systems installed in research centers worldwide are being used by a large number of industrial customers [74]. The industrial applications of synchrotron radiation all utilize the wide range of x-ray energies available and the corresponding wavelengths to probe materials on a microscopic scale to determine their structure. These x-ray beams are more intense than can be obtained in any other way. They can be absorbed, reflected, scattered or refracted as they pass through the materials. These processes can be used in conjunction with the ability to "tune" the wavelength of the x-rays to identify specific elements and the structure of materials, including biological molecules.

Most of the industrial uses of the current synchrotron radiation accelerators are in the semiconductor, chemical and biomedical fields. Semiconductor applications include lithography, studies of material interfaces and other production issues. Applications in the chemical industry include the study of the properties such as stress or texture of various materials produced, as well as the chemical reactions themselves. Applications in the biomedical field include protein crystallography, imaging molecular structures and studying molecular dynamics in tissue cells.

In the past decade there have been a number of electron synchrotrons built as turnkey systems by accelerator suppliers and it is expected that this trend will continue. The Oxford Instruments Accelerator Technology Group (UK) has built several superconducting systems for use in semiconductor lithography and Danfysik (Denmark) has recently built normal conducting systems in Canada and Australia. Sumitomo Heavy Industries (Japan) also provides compact synchrotrons for industrial use. This accelerator business is still in its early stages and is not expected to be a large market for at least a decade, after which the free electron laser — the next generation of synchrotron light source — is expected to become an industrial tool.

4. Future Industrial Accelerators and Applications

New accelerator technology is usually first developed for research applications and must be proven by research laboratories and universities before it is adopted by the industrial or medical community. Several technologies that are now approaching more general acceptance for industrial applications include the free electron laser (FEL), superconducting cyclotrons and linacs, and the new fixed-field, alternating-gradient ring (FFAG).

The FEL is the next generation of synchrotron light source that uses the electron beam from a linac (or electrostatic accelerator) and a permanent-magnet wiggler to create a tunable source of high energy radiation that can be used for many applications that must now be performed at a much larger electron synchrotron facility. Similarly, the new superconducting cyclotron and linac structures will allow an increase in the efficiency and a reduction in the size of these accelerators as the cryogenic technology becomes more proven and less costly as a result of its widespread use in large research and medical accelerators [75].

The FFAG is actually an accelerator concept that has been around for many years, having first been proposed in 1954 simultaneously by Symon in the US and Ohkawa in Japan [76]. It is currently being developed for use in high energy physics research at several national laboratories. However, it is also being developed as a neutron source for medical use and, if it proves to be useful in this application, will be adapted quickly for other applications of neutron beams.

There is no doubt that as advances are made in existing accelerator technology to improve performance and capability and as the new technologies mature into commercial products, the uses and applications of industrial accelerators will continue to grow. It is quite certain that most of the accelerator manufacturers and users are busy working on new uses and markets, but much of this effort is heavily guarded for competitive reasons.

5. Conclusion

The industrial applications detailed in this review utilize a wide variety of accelerators and, except for the electron synchrotron, there is not one accelerator that is unique to a single application. The important tool for each of these applications is not the accelerator itself but the charged particle (or secondary particle) beam produced. The accelerator's output must meet the beam specifications for any application before it is a useful tool in that business. In addition, the initial capital cost, the operating cost and the reliability of the entire system play an important role in these for-profit applications.

The industrial accelerator community is continuously seeking ways to lower production costs, so new accelerator technologies that can increase the return on investment (ROI) for a particular application are always being sought. However, a new system must be proven in an industrial setting for some period before it will find widespread acceptance. This is particularly the case with a new application for an industrial accelerator. This slow pace is usually due to the risk that the research equipment used to develop the application technique will not provide the standard of reliability, availability, and ease of operation required for economical commercial use.

Hence, the introduction of a new accelerator technology can require a number of years for widespread market penetration. However, as seen from the data presented here, the market for industrial accelerators is surprisingly large and is continuing to grow at an accelerated pace in many fields. The annual market for all industrial accelerators described in this review is estimated to presently exceed US\$2 billion and is growing at more than 10% per year. More than 18,000 industrial accelerator systems have been sold to date and the annual sales now exceed 900 units per year.

Acknowledgments

This review is a brief summary of work of the many experts in the individual applications of industrial accelerators. The author would also like to thank Michael I. Current, Marshall Cleland, David Schlyer, William Reed, Kenn Lachenberg and Greg Norton for their input and assistance in ensuring the accuracy of the review. As in any review of such a diverse subject, it is likely that there are commercial companies and new applications that may have been inadvertently missed and the author apologizes for any such omissions.

References

[1] I. L. Morgan, *IEEE Trans. Nucl. Sci.* **NS-20**(3), 36 (1973).

[2] J. L. Duggan, *IEEE Trans. Nucl. Sci.* **NS-30**(4), 3039 (1983).

[3] K. Bethge, 1988 *European Accelerator Conference* (World Scientific, 1989), p. 158.

[4] D. Lewis, *Proc. 5th European Particle Accelerator Conference* (Instituet Of Physics, 1996), p. 135.

[5] G. A. Norton and G. M. Klody, *Applications of Accelerators in Research and Industry* (AIP Press 1997), CP392, 1109.

[6] A. Denker, H. Homeyer, H. Kluge and J. Opitz-Coutureau, *Nucl. Instrum. Methods Phys. Res. B* **240**, 61 (2005).

[7] E. A. Abramyam, *Industrial Electron Accelerators and Applications* (Springer-Verlag, 1988).

[8] H. Schopper (ed.), *Advances of Accelerator Physics and Technologies* (World Scientific, 1993), Chap. 18.

[9] A. Vertes, S. Nagy and Z. Klencser (eds.), *Handbook of Nuclear Chemistry* (Kluwer, 1994).

[10] C. H. Sommers and X. Fan (eds.), *Food Irradiation* (Blackwell, 2006).

[11] G. Mank *et al.*, *Acc App' 07* (American Nuclear Society, 2007), p. 53.

[12] *Int. Symp. Utilization of Accelerators* (IAEA, Jun. 5–9, 2005, Croatia).

[13] *School of Ion Beam Analysis and Accelerator Applications* (IAEA; Mar. 13–24, 2006).

[14] *Emerging Applications of Radiation Processing*, IAEA-TECDOC-1386 (2004).

[15] *European Conference on Accelerators in Applied Research and Technology (ECAART)*, biannual.

[16] *Conference on Applications of Accelerators in Research and Industry (CAARI)*, biannual.

[17] *Nuclear Applications and Utilization of Accelerators (AccApp)*, biannual.

[18] A. Sessler and E. Wilson, *Engines of Discovery* (World Scientific, 2007).

[19] J. D. Cockcroft and E. T. S. Walton, *Proc. R. Soc. A* **136**, 619 (1932).

[20] R. J. Van de Graaff, *Phys. Rev.* **38**, 1919 (1931).

[21] G. Ising, *Ark. Mat. Fys.* **18**(30), 1 (1924).

[22] R. Wideroe, *Arch. Electrotechnik* **21**, 387 (1928).

[23] D. H. Sloan and E. O. Lawrence, *Phys. Rev.* **38**, 2022 (1931).

[24] L. W. Alvarez, *Phys. Rev.* **70**, 799 (1946).

[25] B. C. Knapp, E. A. Knapp, G. J. Lucas and J. M. Potter, *IEEE Trans. Nucl. Sci.* **NS-12**(3), 159 (1965).

[26] D. Sloan, US Pat. No. 2,398,169 (9 Apr. 1946).

[27] I. M. Kapchinskii and V. A. Teplyakov, *Prib. Tekh. Eksp.* **2**, 19 (1970).

[28] H. Klein, *IEEE Trans. Nucl. Sci.* **NS-30**(4), 3313 (1983).

[29] J. D. Schneider, *6th European Particle Accelerator Conference Institute of Physics* (1998), p. 128.

[30] L. Young, *Proc. 2003 Particel Accelerator Conference*, IEEE Catalog No. 03CH37423C, 60 (2003).

[31] E. O. Lawrence and M. S. Livingston, *Phys. Rev.* **37**, 1707 (1931).

[32] M. Craddock and K. Symon, this issue.

[33] D. W. Kerst, *Phys. Rev.* **58**, 841 (1940).

[34] J. Pottier, *Nucl. Instrum. Methods Phys. Res. B* **40/41**, 943 (1989).

[35] E. J. N. Wilson, *Proc. 5th European Particle Accelerator Conference*, (Institute of Physics, 1996), p. 135.

[36] H. C. Pollock, *Am. J. Phys.* **51**(3), 278 (1983).

[37] C. S. Nunan, *9th Fermilab Industrial Affiliates Roundtable on Applications of Accelerators*, SLAC eCONF C8905261 (1989), p. 55.

[38] M. I. Current, private communication.

[39] T. Guertin, Varian Medical Corporation Mid-Year Review, (May 2007).

[40] R. S. Ohl, *Bell System Tech. J.* **31**, 104 (1951).

[41] P. H. Rose, *Nucl. Instrum. Methods B* **6**, 1 (1985).

[42] M. I. Current and N. R. White, Health Physics Soc. mid-year mtg. (Jan. 2008), CD proc. (2008).

[43] J. R. Conrad, J. L. Radtke, R. A. Dodd and F. J. Worzala, *J. Appl. Phys.* **62**, 4591 (1987).

[44] H. F. Glavish, *Nucl. Instrum. Methods Phys. Res. B* **24/25**, 771 (1987).

[45] A. Denker *et al.*, *Proc. 17th Int. Conf. Cycl. & Appl.* (Part. Acc. Soc. of Japan, 2005), p. 10.

[46] M. Z. Widermann, *Electro-Chemie* **6**, 12 (1895).

[47] M. von Pirani, US Pat. No. 848,600 (26 Mar. 1907).

[48] H. B. Cary and S. C. Helzer, *Modern Welding Techn.* (Pearson Education, 2005), p. 202.

[49] H. A. James, *J. Vac. Sci. & Tech.* **7**, 539 (1970).

[50] A. Hershcovitch, *J. Appl. Phys.* **78** (1995).

[51] B. Laurell and E. Foll, *RadTech Europe 2007* (Nov. 2007), www.crosslinking.com

[52] M. R. Cleland and F. Stichelbaut, *Proc. AccApp '07* (American Nuclear Society, 2007), p. 648.

[53] M. Laksin and J. Epstein, *RadTech Report* (Mar./Apr., 2007), p. 15.

[54] C. Günthard and D. W. Lee, *Rad. Phys. Chem.* **57**, 641 (2000).

[55] S. Machi, Emerging applications of radiation processing. IAEA-TECDOC-1386 (2004), p. 5.

[56] M. Cleland, private communication.

[57] A. J. Berejka, *Proc. AccApp' 07* (American Nuclear Society, 2007), p. 661.

[58] L. R. Calhoun *et al.*, *Med. Plastics Biomater.* **26** (Jul. 1997).

[59] A. J. Berejka, An assessment of the United States measurement system: addressing measurement barriers to accelerate innovation, NIST Spec. Publ. 1048, App. B (2007), p. 164.

[60] A. G. Chmielewski and A. J. Berejka, Trends in radiation sterilization. IEEE Consultants meeting (May, 2005), p. 41.

[61] G. A. Norton and G. M. Klody, *Proc. Appl. of Acc. in Res. & Ind.* CP392 (AIP Press, 1997), p. 1109.

[62] C. Tuniz and G. Norton, *Nucl. Instrum. Methods B* (2008), to be published.

[63] D. M. Lewis, *Proc. 4th European Particle Accelerator Conference* (World Scientific, 1995), p. 357.

[64] D. J. Schlyer, *Handbook of Radiopharmaceuticals* (John Wiley & Sons, 2003), p. 1.

[65] R. W. Hamm, *The World & I* (Jan. 1990), p. 362.

[66] R. W. Hamm and M. E. Hamm, *EPAC 2006* (Edinburgh, Scotland, 2006), p. 1911.

[67] A. V. Mishin, *Proc. 2005 Particle Accelerator Conference*, IEEE Cat. No. 05CH37623, (2005), p. 240.

[68] P. Bjorkholm and L. D. Boeh, Jr., *Port Technol. Int.* **31**, 2 (2006).

[69] P. J. Bjorkholm and J. Johnson, *Cargo Screening Int.* **49** (Jun. 2004), p. 49.

[70] M. R. Hawkesworth, *Atomic Energy Rev.* **15**, 169 (1977).

[71] R. W. Hamm, *Proc. Third World Conf. Neutron Radiography* (Kluwer, 1990), p. 231.

[72] N. Colonna *et al.*, *Proc. 15th Int. CAARI*, AIP Conf. Proc. **475**, 1045 (1998).

[73] R. W. Hamm, *SPIE Conf. Proc.* **4142**, 6 (2000).

[74] G. R. Neil, *Proc. AccApp '07* (American Nuclear Society, 2007), p. 3.

[75] E. J. Minehara, *Proc. AccApp '07* (American Nuclear Society, 2007), p. 168.

[76] F. E. Mills, *Cycl. & Their Appl. 2001, 16th Int. Conf.* (AIP, 2001), cp600, p. 195

Robert W. Hamm is an industrial accelerator physicist who has designed and built many types of particle accelerators throughout his 40-year career. He recently retired as CEO and President of AccSys Technology, Inc., a successful linear accelerator manufacturing company he co-founded with three other physicists in 1985. He enjoys scuba diving, fishing and big game hunting in the U.S. and abroad.

Reviews of Accelerator Science and Technology
Vol. 1 (2008) 185–210
© World Scientific Publishing Company

World Scientific
www.worldscientific.com

The Development of Superconducting Magnets for Use in Particle Accelerators: From the Tevatron to the LHC

Alvin Tollestrup

Fermil National Accelerator Laboratory,
P. O. Box 500, Batavia, Illinois 60510, USA
alvin@fnal.gov

Ezio Todesco

CERN, Accelerator Technology Department,
Geneva, 1211 Switzerland
ezio.todesco@cern.ch

Superconducting magnets have played a key role in advancing the energy reach of proton synchrotrons and enabling them to play a major role in defining the Standard Model. The problems encountered and solved at the Tevatron are described and used as an introduction to the many challenges posed by the use of this technology. The LHC is being prepared to answer the many questions beyond the Standard Model and in itself is at the cutting edge of technology. A description of its magnets and their properties is given to illustrate the advances that have been made in the use of superconducting magnets over the past 30 years.

Keywords: LHC; Tevatron; superconducting magnets; hadron colliders.

1. Introduction

Superconductivity has played a key role in the development of magnets for the accelerators used in high energy physics. And yet, as we shall see, it is at best an unholy alliance! The challenge for the last 60 years has always been to push accelerators to higher energy and in general this has paid off with exciting and often unexpected results.

Table 1 lists the accelerators that we will consider in this article along with the beam energy, magnetic field strength and machine circumference. The Tevatron was the first successful synchrotron using superconducting magnets and will be used to illustrate the many difficult problems that had to be overcome. A detailed description of the Large Hadron Collider (LHC) will be used to show the

state-of-the-art technology. There is an inevitable change in style brought about by this approach and the authors hope that the reader will either enjoy the approach or forgive the authors.

To understand why superconductivity has played such a pivotal role in accelerator development, consider the landscape in the 1970s. The Standard Model was beginning to emerge but important pieces were missing. The Fermilab Main Ring was the largest operating proton synchrotron with a radius of 1 km, a peak field of 2 T, a power consumption of more than 50 MW and an operating energy of 400 GeV. Type II superconductors were becoming available that offered the prospect of operating at fields of more than 4 T with no resistive losses. The possibility of doubling the energy of the Fermilab accelerator and at the same time reducing the power consumption was irresistible to Robert R. Wilson, the founding director of the laboratory. The original contract for the laboratory did not specifically define the machine that was to be built, only the total cost. He was in the enviable position of having constructed the laboratory and its accelerator under budget by about US$30 million and his plan was to use the excess

Table 1. Four accelerators using superconducting magnets and their major parameters.

	E (GeV)	B (T)	Length (m)	First beam
Tevatron	980	4.3	6280	7 1983
HERA	920	5.0	6336	4 1991
RHIC	100/n	3.5	3834	6 2000
LHC	7000	8.3	26659	9 2008

money to build a second superconducting ring with 1 TeV energy. A feeling for the environment in which the Tevatron was conceived can be gleaned from the following paragraph from Ref. 1.

"The design process, and if carried out, the construction of the Doubler, builds upon our experience at NAL. We have not proceeded on the basis of deciding what is readily practicable, designing to that, adding up the cost and attempting the result. Instead, we have set a cost goal and keep designing, redesigning, haggling and improving until we have done what we set out to do. Occasionally, we are forced to admit that we are not clever enough to achieve our cost goal and admit defeat, but not without a struggle." A nice contrast with the present process of building a new machine! (Note: In most of this article we will refer to the Tevatron even though in its early stages it was called the Doubler or the Energy Saver.)

During this period there was active investigation for using this new technology at many other high energy physics laboratories. A 4 GeV experimental ring, ESCAR [2], was under construction at LBL. Brookhaven was developing a 400 × 400 GeV pp collider called ISABELLE [3], and Rutherford Lab [4] did some crucial development of superconducting cable while studying the possibility of building the SPS with superconducting magnets, an effort that was discarded in favor of conventional (i.e. warm) magnet technology.

The Tevatron was first commissioned in 1983 as a fixed-target machine that accelerated protons to peak energy and extracted them to a target where they produced secondary beams of particles [5]. However, the spectacular success at CERN with its pbar-proton collider shifted the emphasis to using the large rings in the collider mode. This was first achieved at the Tevatron in 1986 and all of the other rings in Table 1 have been initially designed for use in this mode. The availability of superconducting cable has had a dramatic impact on accelerator design and that will now be explored.

2. Superconducting Accelerator Magnets

Normal magnets use iron to shape the field and water-cooled copper coils to supply the ampere turns. The current density in practical magnets is of the order of 5 A/mm^2 or less due to the

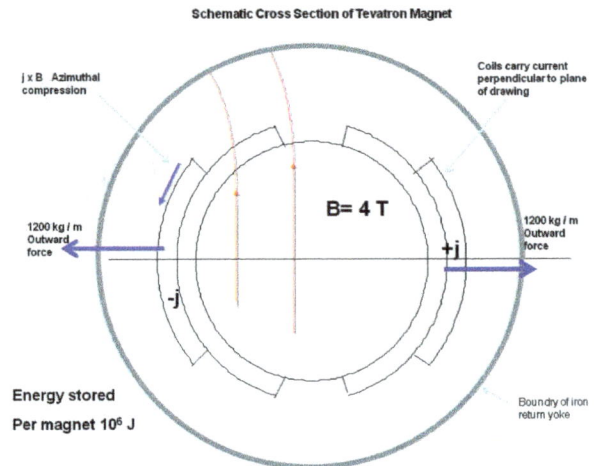

Fig. 1. Schematic drawing of the Tevatron dipole magnet transverse cross-section, showing the configuration of the coil and iron return yoke as well as the direction of some of the forces acting on the coil.

difficulty of removing the joule heating. In contrast, superconducting cable can operate at current densities more than 100 times greater. This offers the possibility of designing magnets whose field shape is governed by the geometry of the current-carrying conductors and the iron plays a secondary role of providing a flux return and shielding for the external space [6]. Figure 1 shows a very abstract schematic of the Tevatron dipole cross section.

The design is based on the superposition of two simple solutions to Maxwell's equations. If the field is expanded in cylindrical coordinates, it is easy to show that a current sheet with the current flowing in the longitudinal direction and in which the density varies as the cosine of the angle in the transverse plane (i.e. the so-called $\cos\theta$ layout [4]) will produce a uniform dipole field inside. If this current sheet is inserted in a coaxial iron return yoke, the field induced in the iron produces a uniform field within the cylindrical hole and adds to the uniform field within the current sheet. The current density is picked to give the desired field inside the coil and the radius of the hole in the yoke is picked to give a peak field at the pole that is less than saturation. In the Tevatron, the iron provides about 18% of the central field.

A number of choices were made that were not necessary in later accelerators. The diameter of the beam tube is 76.2 mm and the diameter of the hole in the yoke is about 250 mm and the current density is such that the field in the center is about 4 T.

By keeping the field in the iron below saturation, the field is proportional to the current. In the Tevatron the quadrupoles and dipoles are all in series and so it is important to have the fields track each other during acceleration. In addition, the yoke is at room temperature and the coil must be held at liquid helium temperature, and so there is a cryostat that fits in the space between the coil and the yoke. The yoke is thus not available to support the large magnetically induced forces. We will discuss these choices later but now we must discuss some properties of superconductors.

2.1. *Superconductors*

The gross properties of a superconducting cable are contained in a graph of current density versus critical field. If the current density exceeds the critical value, the superconducting state is destroyed and one says that the conductor "goes normal" or "quenches." This is shown in Fig. 2 for Nb–Ti at 1.8 K and 4.6 K.

On this same plot one can show the load line of the magnet which is the relation in the magnet between current density in the winding and the high field point at the conductor. The case shown is for a magnet with an operating high field point of 4 T and operated at 4.6 K. If the current is raised past the quench point, or if the ambient temperature should increase, the magnet would quench. The term "short sample limit" is used to characterize the quench point shown in the above figure and derives

its meaning from the measurements made on short samples of the conductor in a test rig where various current densities, field strengths and temperatures can be applied. The figure also indicates that the operating field could be increased by lowering the temperature to 1.8 K, a solution that was chosen at the LHC.

2.1.1. *Superconductor cable development*

One of the great success stories for the HEP community was the commercialization of Type II superconductor alloys into useful cable [7, 8]. The collaboration of magnet builders, materials scientists and industry produced a spectacular advance. The graph in Fig. 2 shows in a grossly oversimplified manner the conditions that must be met. The alloy we will be concerned with is 46.5% wt Nb–Ti alloy. However, to make a useful conductor this material must be in filamentary form and surrounded by copper. Figure 3 shows a 0.5 mm copper strand with over 2000 imbedded Nb–Ti filaments that have a diameter of about 8 μm.

There are several reasons for this structure [4]. The first is that in a Type II superconductor, the flux penetrates the filament in small jumps as the field is increased. These flux jumps release a small amount of heat. The copper carries this heat away and keeps the filament below the transition temperature. If the filament should pass into the normal state, its resistance is very high and the joule heating will start to

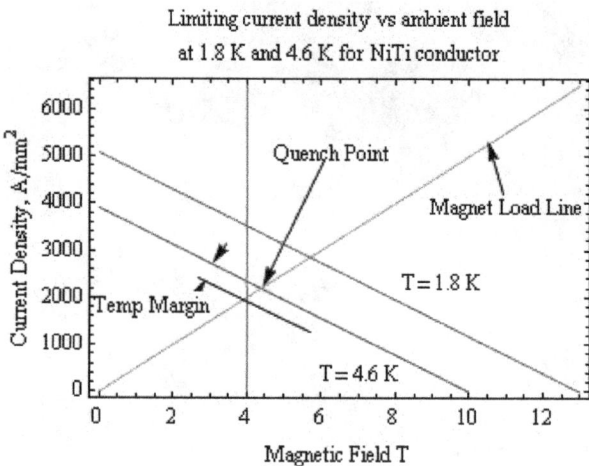

Fig. 2. Critical current for Nb–Ti at two temperatures versus the ambient field. The magnet load line is determined by the magnet design and is the high field point in the winding versus the winding current density.

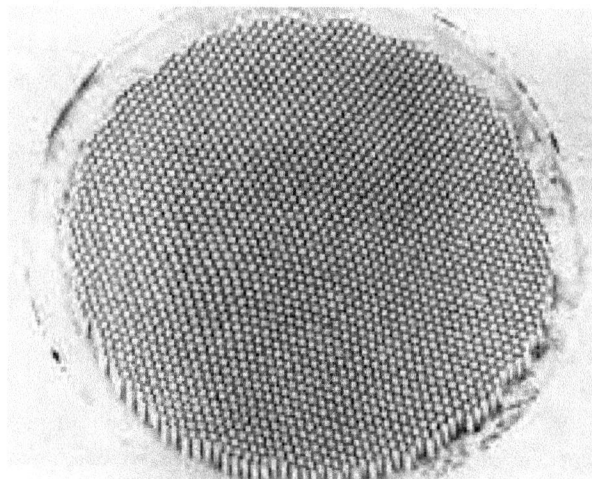

Fig. 3. Photomicrograph of a 0.5 mm strand. The copper has been etched away to show the individual Nb–Ti filaments. This strand will carry about 200 A at 4.6 K in a field of 4 T.

turn the whole strand normal and unless the current is turned off the strand will destroy itself. The second reason for the copper is to provide an excellent conductor for the current should the strand start to go normal. The copper used is very pure and at low temperature has only 1% of its room temperature resistance and can carry the full current long enough for protective action to be taken.

The strand is produced from rods of Nb–Ti about 3 mm in diameter and 65 cm long. Wilson originally purchased enough alloy for 1/6 of the ring and had it formed into these rods by (Teledyne) Wah Chang. This material was then distributed to manufacturers of superconducting strands, where it was loaded into hexagonal copper rods with a 3 mm coaxial hole. Next, 2000 of these rods were loaded into a close-packed array in an ~ 250-mm-diameter copper cylinder and extruded under heat and high pressure into a cylinder ~ 75 mm in diameter. At this point the cylinder was heat-treated and drawn down into the 0.5 mm strands. The exact details of this industrial process had large effects on the ultimate current-carrying capacity of the strand. There was intense competition among the participating companies to produce the best results and win the largest orders.

But the strand is still not suitable for use in a magnet because the magnet must be pulsed for use in an accelerator and the energy stored in it may exceed 1 MJ. In order for this to happen in tens of seconds, the inductance of the magnet must be kept low, which implies a small number of turns, which in turn requires very large currents. Thus in the Tevatron the conductor must carry about 4000 A, which requires 23 of the strands shown in Fig. 3. But this causes more problems.

Consider two of the strands in the cable that are necessarily joined tightly at the ends. If there is a net flux linkage in this loop when the magnet is pulsed, a large loop current will flow and the strands will not share the cable current equally. And a variation on this theme is seen within the strand where there can be loops enclosing flux between the filaments themselves. This latter problem was solved by twisting the strand through 360° every few inches during fabrication. The first problem was solved using an idea originated at Rutherford Lab [4]. This cable can be visualized by considering the 23 strands as being wound in a helix around a small coaxial cylinder and

then flattening the resulting cylinder into a flat ribbon cable, as shown in Fig. 4. Magnets made with this cable showed large heat loads due to the eddy currents flowing through the loops generated by the top and bottom strands crossing each other. This problem was empirically fixed by coating every other strand with copper oxide, which is an insulator, and that effectively broke open the loops. Finally, as the cable must be positioned around a cylinder, as can be seen from Fig. 4, it is processed through a rolling mill that forms its cross section into a trapezoid.

The cable is next spiral-wrapped with an overlapping film of 25-μm-thick Kapton to provide electrical insulation. Finally, there is a spiral wrap of epoxy-impregnated glass cloth tape that when the coil is wound and cured holds the mass together (see Fig. 4).

The development of a successful conductor was a major accomplishment and was carried out by a close collaboration between Fermilab and the other national labs, materials scientists and industry. The initial purchase of a large amount of alloy allowed the distribution of identical raw materials to industry for the exploration and optimization of the many parameters that affect the ultimate current density and stability of the finished conductor.

2.1.2. *Fabrication of the coil package*

The coil is wound on a form and then placed in a precision mold and heated under high pressure to cure

Fig. 4. The cable developed for the Superconducting Super Collider, showing the epoxy-impregnated glass tape, the Kapton insulation and the filaments that have been exposed by etching.

the epoxy and produce an object that can be handled easily. Here again there was much to be learned. As we shall see, the dimensions of the coil had to be controlled to about 0.025 mm in order to produce the correct field and to keep the conductors from moving under the large magnet forces present during excitation.

Originally the mold was machined out of a solid piece of steel the length of the magnet. This was possible but slow for the $\sim$ 300-mm-long model magnets, and challenged the available machines to maintain a high accuracy over a length of $\sim$ 7 m. This led to the invention of laminated tooling, which was a major innovation [9]. The requirements on the coil are that its local cross section be the same over the length of the magnet. It turns out that industry has a well-developed capacity to make precision stampings out of sheet stock. And, even better, industry can quickly produce the dies necessary for stamping to an accuracy of a few μm. Thus, by forming the mold from steel stampings and stacking them together on a precision flat bed, one can produce a very accurate mold quickly and change it if necessary. Figure 5 shows such a mold with the hydraulic press in the rear.

The fabrication of the coils required a lot of research and trial. Too much epoxy could completely seal off access for the liquid helium cooling. The

tolerance on the Kapton film thickness and the glass tape had to be carefully monitored to prevent the accumulation of changes in their dimension from causing large changes in the coil package dimensions.

2.1.3. *Constraining the forces*

Constraining the forces and maintaining the geometry of the coil package were two central problems that had to be solved when making a transition away from normal iron magnets, where the iron controls the field shape. The Tevatron coil has a 76.2 mm diameter and the magnetic field pressure in the bore is about 600 N/cm². A 1 m length of cable in the median plane of the coil package has a force of about 16,000 N forcing it outward. In addition, as can be seen from Fig. 1, the cable near the poles exerts a large azimuthal force that tends to compress the winding toward the median plane. The azimuthal force decreases to zero at the median plane, but the sum of all of the turns is large. The question of field accuracy will be discussed later, but the result of calculations shows that changes of any of the dimensions should be constrained to the order of 0.025 mm.

Early in the program, a novel method for containing the forces was tried. The coils were assembled on a series of spaced-out titanium rings and then overwrapped with stainless banding. There were two layers of banding under high tension wrapped in opposite directions to balance the torque. Figure 6 shows a one-foot model magnet using this technology.

It was soon found that the structure was not rigid enough to contain the forces, and the technique was abandoned. During the early part of the program many experiments were made on short magnets which could be quickly constructed and tested in an open cryostat filled with liquid helium.

Fig. 5. A mold section made from lamination. The coil fits in the bottom and a second piece of the mold comes down from above. The pipes carry hot liquid to heat the mold and cure the epoxy.

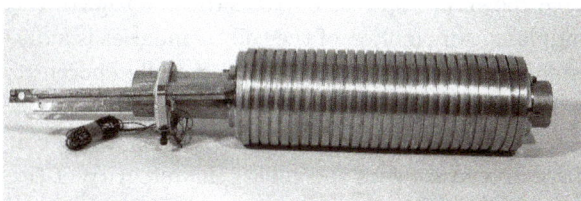

Fig. 6. A very early one-foot model magnet using spiral-wound stainless steel banding in two oppositely wound layers.

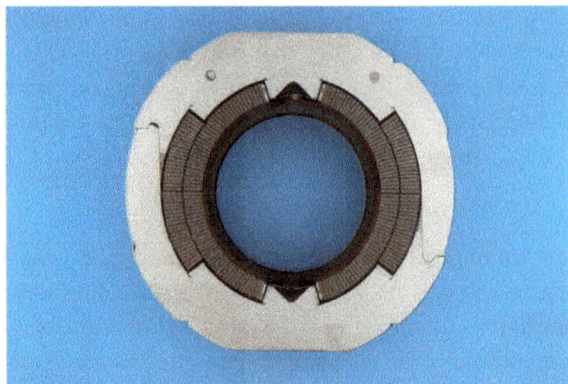

Fig. 7. A model magnet was constructed and, after testing, was sliced apart. This slice shows the details of the coil and stainless steel collars.

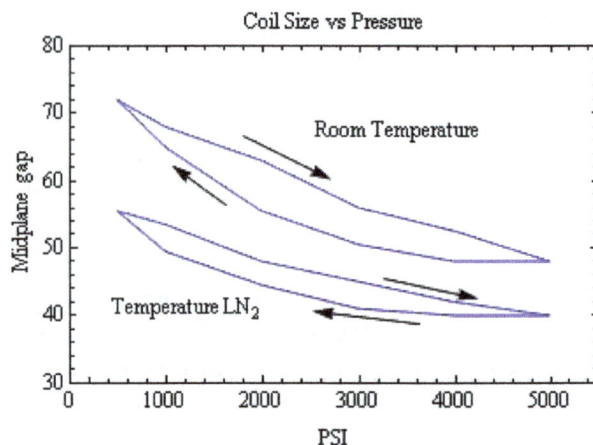

Fig. 8. Compressibility of the coil package at room temperature and at that of liquid nitrogen.

The solution found not only worked but has been an integral part of all subsequently constructed accelerator magnets. It was the original use of the technology mentioned above and shown in Fig. 5. Stainless steel stampings were made in a form that created a steel jacket around the outside of the coil.

The collars are split asymmetrically and the next layer down would have the joint between the two pieces reversed. The stainless steel is 1.5 mm thick and the layers are joined by axial welds along the outside. The whole package is later impregnated with epoxy to strengthen the structure. Additional rigidity is obtained by small dimples pressed into the steel that mesh with the underlying layer, and can be seen clearly near the top in Fig. 7.

The collars are first assembled into short packs, which are then placed around the coil package. The whole assembly is placed in a hydraulic press that closes the collars, and an automatic welding machine makes the axial welds along the outside that lock the structure together. There are two key points for maintaining the field accuracy. The first is that the collars must be thick enough to resist the tremendous horizontal force mentioned above. The second is a much more difficult problem to solve. The azimuthal force shown in Fig. 1 compresses the coil toward the midplane. Any motion of the coil boundaries is a disaster. The coil angles have been carefully chosen to make a uniform field and they are set by the collars, as can be seen in the above figure. As long as the coil stays in contact with the collar, the integrity of the field is assured.

So the key to the problem is to collar the coil with a press that compresses the coil package enough

so that the elastic forces are always greater than any magnet force during excitation. A number of problems had to be solved and it was not clear that a solution existed. The main obstacle was that the coil, when cooled, shrank more than the stainless steel collars releasing some of the strain. Figure 8 shows the compressibility of the coil package at room temperature and at LN_2 temperature. This data was obtained by using a slice of a collared magnet, as shown in Fig. 7. The collars were cut along the midplane on one side, which released the collaring pressure. The force required to close the collar back to its original size gave a direct measure of the prestress in the package. The same measurement made at LN_2 temperature gave a measure of the force lost when the coil was cold. It was necessary that this force was great enough to ensure that the coil package didn't pull away from the collar during excitation.

One might think that the solution was to apply a high-enough collaring pressure to compensate for the differential shrinkage, but there was a limiting pressure that the insulation could stand before turn-to-turn shorts developed. The solution found involved very careful control of all of the dimensions, but would not have worked for higher field magnets, and alternative techniques have been developed and will be described later.

2.1.4. Field errors

This is an appropriate place to discuss the question of field errors. As mentioned before, since the field is primarily determined by the geometry of the currents, any error in the coil shape will show up as deviations in the field. Since the bore is a current-free region, a harmonic expansion of the field can be made and the coefficients determined from the known current distribution. The information in the expansion is generally displayed as follows [5]:

$$B_y(x,y) = \text{Re}\left[B_1 \sum_1^\infty (b_n + ia_n)\left(\frac{x+iy}{R_{\text{ref}}}\right)^{n-1}\right],$$

$$B_x(x,y) = \text{Im}\left[B_1 \sum_1^\infty (b_n + ia_n)\left(\frac{x+iy}{R_{\text{ref}}}\right)^{n-1}\right].$$

It is customary to quote the value of the multipole fields at a radius $R_{\text{ref}} \sim 2/3$ of the aperture divided by the central dipole value. Accelerators require that these error fields be of the order of 10^{-4} of the central bending field for stable beam behavior, and thus of the order of a few gauss. Therefore, both b_n and a_n are expressed in units, i.e. the actual value is multiplied by 10^4 to get a number of the order of 1.

The b_n are the so-called normal harmonics and are all driven by current distributions that are symmetrical between the top and bottom halves of the coil. The a_n are called the skew harmonics and are driven by left–right-symmetrical current distributions.

It is interesting to look at the Tevatron coil in Fig. 9 as a simple example. Assume that there is perfect right–left and top–bottom symmetry. Then all of the a's and all of the even b's are zero. It is easily shown in the case of "thin" coils, b_3 and b_5 can also be set to zero by adjusting the two coil angles. Thus the first term that comes into the expansion is the 14th pole, which varies like x^7, and evaluation of this term shows that it is negligibly small over the inner 2/3 of the aperture.

Table 2 gives the rms values measured for the 870 dipoles initially installed in the Tevatron [10, 11]. The data on the quadrupoles is available in Ref. 12 and a description of the measurement facility in Ref. 13. The values shown are integral values averaged over the length of the dipole. Control of the normal and skew quadrupole was the most difficult. However, it was possible to correct for an error in the coil by offsetting the center of the coil from the axis of the hole in the iron yoke, because the first effect of such an offset in the iron is to produce a small induced quadrupole. This correction will be described below.

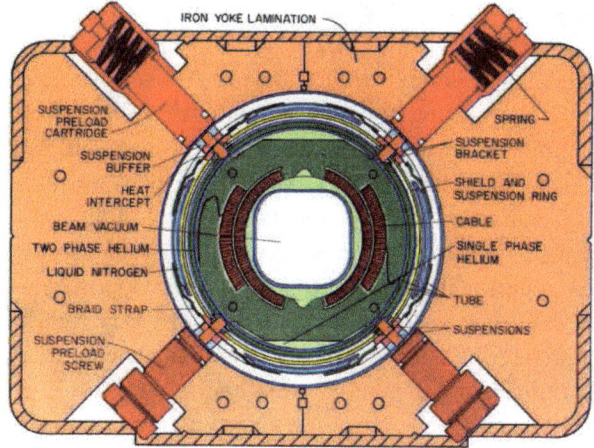

Fig. 9. Cross section of a Tevatron dipole.

Table 2. The rms values of the various multipole components of the Tevatron dipoles. The measurement was made at 4000 A. The coefficients are in units of 10^4 of the central field at a radius of 25.4 mm.

n	b_n	b_n	b_n	a_n	a_n
	Design	Mean	RMS	Mean	RMS
2		0.09	0.48	0.17	0.50
3	0.04	0.95	3.12	0.10	1.16
4		−0.23	0.77	−0.07	1.46
5	1.04	−0.57	1.32	−0.10	0.46
6		−0.07	0.32	−0.07	0.55
7	4.44	5.48	0.54	−0.07	0.29
8		0.04	0.17	0.22	0.26
9	−12.09	−12.52	0.33	−0.07	0.41
10		0.02	0.23	0.28	0.38
11	3.63	3.70	0.26	0.08	0.25
12		−0.01	0.20	−0.24	0.25
13	−0.82	−0.80	0.19	−0.05	0.22

There is a second systematic effect that must be corrected. The ends of the coil where the cable reverses direction cause a small change in the effective length of the magnet that varies like x^2, the same as the variation of the sextupole in the body of the coil. These two effects can be made to cancel. As the path moves off center the magnet becomes shorter, but the sextupole moment of the coil can increase the field so that the field integral through the dipole is constant.

2.1.5. *Persistent current effects*

Superconducting cable comes with a small problem in that it produces small systematic error fields. According to the Bean model [14], the currents that flow are always at the critical limit and the total current is the sum of the transport current and the shielding current, which is trying to keep the flux from penetrating the metal. A cylindrical filament in a uniform field has an induced dipole moment due to the shielding currents, and since there are several million of these filaments in the winding, they can coherently produce a residual field that can lag the ambient field and produce an open loop magnetization curve. Since the windings are very nearly symmetrical, only the spatially symmetric normal harmonics are generated and their magnitude is of the order of tens of gauss, a magnitude that is negligible for the guide field but has big implications for the dynamics of the focused beam. The biggest effect comes from the sextupole moment which controls the chromaticity of the machine. If it were static, it could be easily compensated for by the correction elements in the machine.

However, the trouble comes because these persistent currents decay logarithmically from their dynamic value if the magnet excitation is suddenly held constant [15]. The magnet is cycled from injection to high field and then finally back to low field for another cycle, where it may remain for a long period while particles are injected. For the Tevatron, this can be of the order of 30 min as both the protons and the antiprotons must be injected. While injection is taking place, the persistent sextupole field decay is a relatively slow process, but it turns out that at the start of acceleration, the magnetic moment returns to its dynamic value ("snap-back") with only a very small change of field. This sudden change in chromaticity must be carefully corrected for stable operation.

The fact that there is an open loop means that there is also an energy loss when a magnet is cycled around a loop and this heat must be carried away by the cryogenic system [16]. Both the heating and the magnetization can be reduced by making the filament smaller in the conducting strand. In the Tevatron the filaments are about $8\,\mu$m in diameter and the hysteresis loss is about 200 J per magnet cycle.

3. The Tevatron Magnet Development

As indicated in the introduction, the development of the Tevatron was an experiment. Never before had over 1000 superconducting magnets been produced.

Thus constructing a factory as part of the experiment was essential to the process of learning how to make magnets. There were two lines of attacking this problem. The first was to start a vigorous model program that was based on 300-mm-long model magnets. These could be produced rapidly; in some cases a model with a different parameter could be produced in a week. This program was the basis for understanding how to fabricate and insulate the coils and was also crucial for the cable development program.

At the same time tooling was being developed for full length magnets based on what was being learned from the model tests. In the end almost 200 full length coils were fabricated and tested before the actual construction of the machine began. Some of these magnets were later used in beam lines where the field quality was not so important. A crucial piece of this program involved the development of instrumentation that could measure the field of a coil package at room temperature [17]. After the coil was collared, it was taken to a measurement stand where a full length probe consisting of a set of parallel stretched wires was inserted. The coil was excited with a 10 A sinusoidal wave form. A phase-locked voltage integrator measured the flux through the parallel loops and derived the multipoles. The system was able to monitor the lower harmonics up through the sextupole terms and was thus able to close a feedback loop around the factory. If the field components started to drift, the cause could be looked for and corrected. But, more important, systematic effects due to small systematic changes in the components could be corrected by placing small shims to slightly change the angle subtended by the coils. This measurement was very cost-effective, in that it insured that a coil placed in a cryostat and yoke at considerable cost and labor would not have serious defects when it was measured as a completed package.

3.1. The cryostat and yoke

A cross section of the completed magnet is shown in Fig. 9. The yoke construction followed the normal procedure of using stamped iron laminations stacked together in two sections split symmetrically about the vertical center line [18]. The only special feature was that the magnets were long enough so that, if straight, the curve of the beam through the magnet would have had a sagitta of 5 mm consuming ±2.5 mm of aperture. This problem was successfully solved by curving the yoke and forcing the cryostat and coil, which were constructed as straight elements, into the curved cavity. Welds were then made to lock the structure in place.

The cryostat must fit in the space between the outside of the collars and the inside of the iron yoke. The iron is at room temperature — a real challenge and an ingenious piece of engineering!

The main part of the cryostat comprises three stainless cylinders consisting of the beam tube and two closely spaced cylinders around the outside of the collars. The region between the beam tube and the first stainless cylinder is filled with liquid helium under pressure and forms the supply pipe. After going through a string of eight magnets, it expands through a Joule–Thomson valve and passes back along the string in a space between the first and second stainless steel cylinders. It is in intimate contact with the inner volume and absorbs heat from the coil assembly, boils, and passes back to the liquefier as a two-phase liquid. This keeps the inner coil at almost a constant temperature. There is an intermediate shield and then a liquid nitrogen shield before coming to the outside cylinder which closes off the cryostat insulating vacuum.

We will leave the cryogenics system [19] here, but one should appreciate the tremendous success of these enormously dispersed systems. An obvious challenge was making the thousands of leak tight welds necessary to form the cryostat and the difficulty of leak-checking the completed system. The LHC was an even bigger challenge.

3.2. The coil support system

The Tevatron had to solve one problem that was not pertinent to later machines. Since the iron was warm, there had to be a coil support system devised that had limited heat leak and yet held the coil package firmly in place. When the coil is cooled it shrinks both axially and radially. It is firmly anchored in the axial direction at its center. There are nine support points along the axis and the ones on either side of the center must allow for axial slipping.

However, more troublesome is the shrinkage in the radial direction, which is almost 1 mm. As can be seen in Fig. 9, each of the nine stations has four

support pillars at 45° which are points on the collars that do not change radius as the coil is excited and becomes slightly elliptical. The bottom two are fixed at a position that will center the cold coil properly, but are also adjustable. The top two are spring-loaded and allow motion of the support point as the coil is cooled. The springs must be strong enough so that if the coil moves off center their restoring force is larger than the magnetic force, pulling the coil in the direction it is offset.

It turned out to be very useful to be able to move the coil slightly off center, because that allowed the correction of the intrinsic quadrupole errors in the coil package. The induced field from the iron due to an offset coil is a quadrupole whose strength is proportional to the offset. Each magnet was measured cold in the Magnet Test Facility, and quadrupole error field was determined. Shims were then placed under the external bolts that shifted the coil package by an amount to set the total moment to zero.

3.3. Quench protection

A superconducting magnet is intrinsically unstable. If some piece of the conductor is forced into the normal state, the subsequent joule heating may drive more of the conductor normal and the normal zone will propagate. Consider a small length, δz, receiving a small pulse of heat that is sufficient to drive it into the normal state. The current will immediately transfer to the copper, which has about 1% of its room temperature resistivity. If δz is sufficiently small, the heat capacity of the surrounding medium may be enough to overcome the joule heating in the copper, and the cable returns to the superconducting state. However, the magnet is most vulnerable when it is at full field and operating very near its upper limit, and there will be some δz that is long enough so that the joule heating will overcome the cooling and the quench wave will propagate along the cable with a velocity between 1 and 10 m per second. As more of the cable heats up, it will heat adjacent turns and the quench will propagate through the winding. If the current is not turned off, the cable will destroy itself and the insulation.

To calculate the limits that must be met, consider the following expression for the increase in temperature of a small section of cable with resistance $R(T)$ and heat capacity $c_p(T)$ and carrying a current $i(t)$:

$$dT = \frac{i(t)^2 R(T) dt}{c_p(T)}.$$

Collecting the temperature-varying terms together, we can write

$$\int_0^t i^2(t) dt = \int_{T_1}^{T_2} \frac{c_p(T) dT}{R(T)}.$$

The right hand side can be evaluated. The resistance is essentially determined by the copper, which is very pure and cold at the start. The heat capacity is small for the metal because initially it is at low T where the heat capacity varies as the third power of T, but the liquid helium contained in the strands provides the large initial heat sink. T_2 is the upper temperature that the coil will survive. Putting in these numbers, one obtains a limit for the integral over time of the square of the current. For the Tevatron this number was of the order of 7×10^6 A^{2}s for a temperature increase to about 200°C, and exceeding 12×10^6 A^{2}s will damage the conductor. Since the magnet current is of the order of 4000 A, something must be done in a fraction of 1 s if the magnet is not to destroy itself.

A Tevatron magnet contains about 300 kJ and it is not practical to extract this energy in a short time. The solution was to short the offending magnet which bypasses the bus current around it and at the same time turn the whole magnet normal and absorb the energy in the magnet as heat in the whole winding, which then contains the temperature increase to reasonable values [20]. In the meantime, the energy from the rest of the machine is extracted in the normal manner.

The solution involved building into the magnet some stainless steel foils that acted as heaters. When a quench was detected, a capacitor was discharged into the foils, which provided heat to a large section of the coil package and which spread the quench uniformly throughout the winding.

The signal to detect a quenching magnet is the resistive voltage developed across the normal zone. We will not discuss this more here, but the quench detection and protection system represents a major portion of the control system.

A quench of an accelerator magnet system is a violent event. Not only must the magnets be protected electrically, but there are also very large

mechanical forces due to the very rapid vaporization of the liquid helium. This requires very careful design of the cryostat to survive these events without opening up leaks in the myriad of welds, as well as a gas-handling system that minimizes the loss of helium.

3.4. *Progress at other machines*

The first operation of the Tevatron was on July 7, 1983, when it accelerated beam to 512 GeV [5]. Subsequently it has been used in both the fixed target mode and as a pbar-p collider for almost 25 years. The above description outlines the problems that were solved and indeed the cross section of any of the subsequent accelerator magnets shows the heritage of this early work. It is interesting to examine some of the initial choices that were made. For instance, the decision to keep the iron warm had a major influence on the magnet design. The Tevatron covered a larger area than any existing cryogenic installation, and cooling both the coil and the iron yoke would have needed either an enormous plant or a very long time. There was great concern that if the magnets were not reliable and required replacement, the downtime would not be acceptable. Early operational experience verified that this was a wise decision, as many changes and corrections were necessary. However, the experience now is that perhaps one magnet a year needs replacement and the time to replace a magnet and resume operation has turned out to be about one week. The big advantage of having cold iron is that it can help support the large magnetic forces within the magnet. HERA [21] was the first to take advantage of this, and the reliability of superconducting magnets has been quite sufficient to justify this choice in the design. The HERA magnets were also longer, 8.824 m vs. 6.4 m, and worked at a higher field, 4.68 T vs. 4 T, for the Tevatron.

An additional advance at HERA was the use of aluminum collars. It was known at the time the Tevatron magnets were being developed that the loss of compression in the collared coil due to the greater thermal contraction of the coil compared to the stainless steel collars could be alleviated by using aluminum collars. HERA developed an elegant system for controlling this differential contraction. An additional improvement was to use spacers within the coil block to enhance the approximation to a $\cos\theta$

current distribution and reduce the lower harmonics in the field.

Finally, HERA was the first to have the magnets produced industrially. This involved transferring the design to industry and carefully monitoring the product through the production cycle [22]. The construction started in 1984 and was operational in 1990.

During the period from 1985 to 1995 there was a great deal of intense work on superconducting magnets at other locations. The Relativistic Heavy Ion Collider (RHIC), at Brookhaven National Laboratory, developed a very simple single-shell coil that used the iron collars directly to contain the winding.

A big challenge for this magnet was careful control of the iron properties and the successful modeling and control of the effects of saturation on the field quality. It was the first magnet that employed saturated iron in the yoke [23].

The Superconducting Super Collider (SSC) story is a modern tragedy, but it did involve all of the US national laboratories in the magnet development program, and the technology of magnet construction was refined. However, the present state of the art is exemplified by the LHC, which has successfully overcome the complex mechanical problem of having two magnets in one yoke and cryostat while increasing the field up to 8.3 T. In addition, the fields in the two apertures must track each other precisely in magnitude and shape from injection at 450 GeV, where the persistent currents generate a large sextupole to 7 TeV, where loss of beam can cause enormous damage to the machine. Another bold innovation is the use of superfluid helium for cooling the magnets. A detailed description of the magnet production gives an excellent overview of the present state of the technology.

4. The Main Dipole in the Large Hadron Collider

4.1. *Design*

The development of models to prove the feasibility of superconducting dipoles with a 10 T magnetic field, i.e. significantly higher than that obtained in previous accelerators magnets, started nearly 20 years ago [24]. These field values could be achieved either by lowering the operational temperature of the Nb–Ti to around 2 K, or by using a material with a larger critical field and critical current, such as Nb_3Sn [25].

Lowering the operational temperature of the Nb–Ti had the disadvantage of making the magnet more sensitive to any heat deposition in the coil, either from the beam or from the magnet itself (mechanical movements, flux jumps). On the other hand, Nb_3Sn presented worse mechanical properties, strain sensitivity, difficulties in manufacturing and, last but not least for such a large project, higher costs. At the beginning of the 1990s, the Nb–Ti option was retained.

Nearly 10 years of short models and long prototypes (10–15 m) resulted in three generations of dipoles. The first one [26] was based on a two-layer coil with 17-mm-width cable, aluminum collars, a 10 m length and a 50 mm aperture. In the second one the length and the aperture were increased to 15 m and to 56 mm respectively, the cable width was reduced to 15 mm, and the cryogenic line was placed outside the dipole cryostat [27]. In the third generation, the collar material has been changed to stainless steel, with a revised yoke design that gives better support and with a gap between the two iron yoke halves that is closed at room temperature [28].

In the final design, the LHC dipole has a short sample field of 9.7 T and an operational field of 8.3 T in a 56 mm aperture bore using Nb–Ti cables cooled at 1.9 K [28, 29]. Compared to the previous accelerator dipoles, the LHC cable has a larger width (~15 mm), and a larger strand diameter of the inner cable (~1 mm; see Fig. 10).

The main strand parameters are listed in Table 3. The critical current specifications are given at 1.9 K, and correspond to having a current density in the superconductor of ~2100 A/mm^2 at 9 T or ~1500 A/mm^2 at 10 T. These current densities are about 30% larger with respect to that specified 25 years earlier for the Tevatron magnets [30].

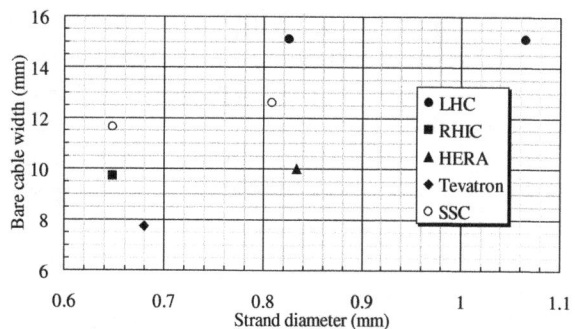

Fig. 10. Strand diameter and cable width used in five dipoles.

Table 3.   Main parameters of the LHC dipole strand.

	Inner	Outer
Filament diameter (mm)	0.007	0.006
Number of filaments	~ 8900	~ 6500
Strand diameter (mm)	1.065 ± 0.0025	0.825 ± 0.0025
Copper–noncopper ratio	1.65 ± 0.05	1.95 ± 0.05
Critical current (A) at 10 T	≥ 515	
Critical current (A) at 9 T		≥ 380
RRR	≥ 150	≥ 150

Table 4.   Main parameters of the LHC dipole cable.

	Inner	Outer
Number of strands	28	36
Mid-thickness (mm) at 50 MPa	1.900 ± 0.006	1.480 ± 0.006
Keystone angle (°)	1.25 ± 0.05	0.90 ± 0.05
Transposition pitch (mm)	115 ± 5	100 ± 5
MIITS [300 K] (MA^{2}s)	45 [8 T]	30 [6 T]
Critical current (A) at 10 T	≥ 13750	
Critical current (A) at 9 T		≥ 12960
Interstrand cc resistance ($\mu\Omega$)	≥ 15	≥ 40
RRR	≥ 70	≥ 70

The cable parameters are summarized in Table 4; the values for the critical current correspond to assuming 5% degradation of the strand performance.

The cable insulation is made up of two polymide layers 50.8 μm thick, with 50% superposition, plus a third adhesive layer 68.6 μm thick wrapped in such a way that it leaves a 2 mm gap to ease the superfluid helium penetration between cable turns. The polymide aims at withstanding a turn-to-turn voltage of around 100 V [31].

After some iterations on the coil design, a six-block, two-layer $\cos\theta$ layout was selected [32], not far from the SSC dipole layout, with an $\sim 10\%$ larger aperture and $\sim 20\%$ larger cable (see Fig. 11). Due to the large coil width and collar width, the iron contribution to the field at the operational current is limited to 18%, as in the Tevatron dipoles (by comparison, it is 57% in the RHIC dipoles). Notwithstanding the large field, iron saturation in the LHC dipoles is much less critical than in the RHIC dipoles: at collision energy, it decreases the transfer function by about 0.6% in the LHC dipoles, and ten times more (7%) in the RHIC dipoles. This is due to the fact that in the LHC dipoles both the coil and the collars are very thick, and therefore the iron is far from the aperture.

The iron also has a limited beneficial effect on the LHC dipole short sample field, increasing it by

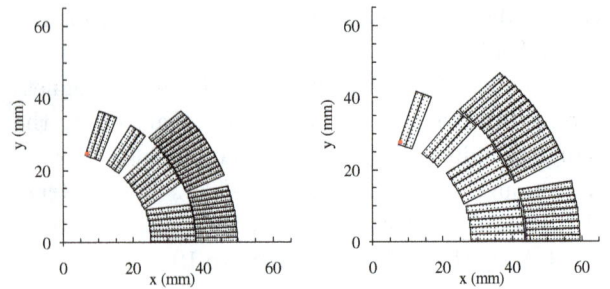

Fig. 11.   Coil layouts of the SSC (left) and the LHC (right) main dipole.

3.5%. Multipoles up to b_{11} have been optimized [32] at the level of $\sim 10^{-4}$ times the main component (i.e. one unit).

The curing cycle of the coils reaches a maximum temperature of 190°C and a maximum pressure of 100 MPa, for a few hours [33]. The curing is used to activate the glue of the third insulation layer and to give the correct dimensions to the coil. The cables from the two layers are joined with a ramp splice and the coils are powered in series; since the outer cable is smaller, this provides a larger current density in the outer layer of 23% (it was 30% in the SSC dipoles). The larger cable width, the improved cable properties, and (especially) the lower operational temperature (1.9 K instead of 4.2 K) allow the LHC dipoles to reach the unprecedented short sample field of 9.7 T and an operational field of 8.3 T, i.e. 86% of the short sample limit (see Figs. 12 and 13), and with a temperature margin of 1.5–1.6 K (inner/outer layer).

The electromagnetic forces acting in the azimuthal direction on the coil mid-plane at 8.3 T are ~ 450 kN/m, corresponding to a stress of ~ 60 MPa.

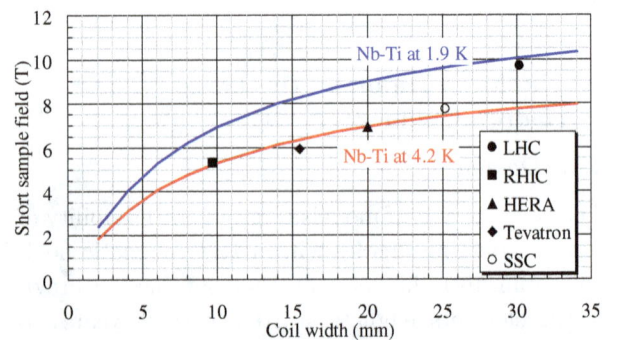

Fig. 12.   Short sample field versus coil width for five dipoles (markers), and estimate for an ironless $\cos\theta$ coil with a 0.3 filling factor. The same Nb–Ti cable properties have been assumed for all dipoles.

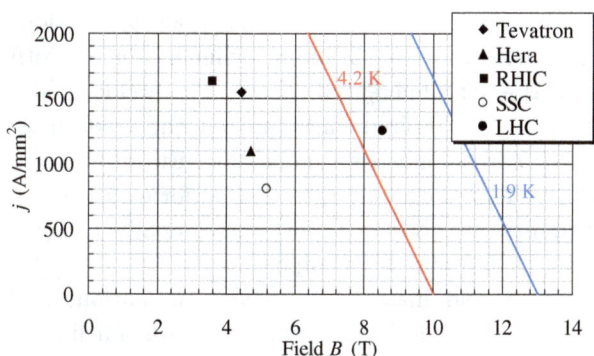

Fig. 13. Operational values of current density in the super-conductor and magnetic field in the bore for five dipoles (markers), and typical values of the critical surface of Nb–Ti at 4.2 K (red line) and at 1.9 K (blue line).

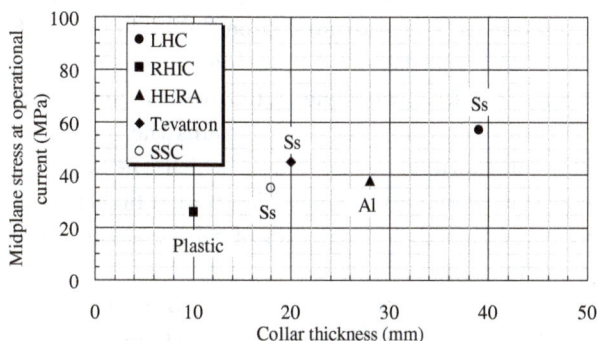

Fig. 14. Average stress in the mid-plane in operational conditions versus collar thickness and material for five dipoles.

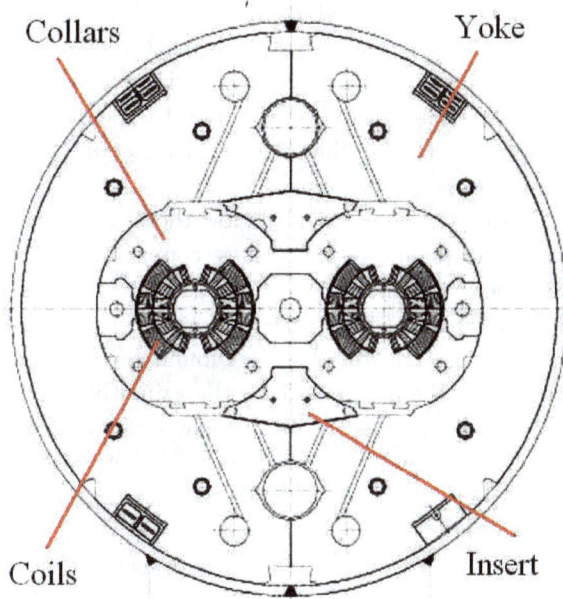

Fig. 15. Cross-section of the LHC dipole cold mass.

This stress is 30% larger than what was reached in the Tevatron dipoles (see Fig. 14, where the collar material is also reported). We point out that whereas larger forces can always be contained by an appropriate mechanical structure, large stresses induce strain in the coil which can give an ultimate limit to performance. For this reason we believe that a comparison of the mechanical challenges in superconducting magnets should be given in terms of stress and not of force.

The LHC accelerates and collides two counter-rotating proton beams. Contrary to the SSC, where one had two separate single-aperture dipoles, in the LHC a two-in-one dipole concept was developed: the two apertures share the same cryostat and the iron shielding to fit the limited space available in the tunnel. During the dipole prototype phase, both separate collar [24] and twin collar [26] options were considered. The final design [28, 29] presents a twin

stainless steel collar structure retaining both apertures in a common yoke (see Fig. 15). This option was taken to save costs, under the correct assumption that the resulting mechanical and magnetic coupling between the apertures would not have affected the performance. The large thickness of the collars (40 mm; see Figs. 14 and 15) is due to a design originally foreseen for aluminum collars; the switch to stainless steel collars was carried out in the final phase of the full-size prototypes [34]. As a result, the collars bear most of the electromagnetic forces at nominal field.

Collars are made up of 3-mm-thick laminations and with stringent specification on the permeability. Nevertheless, their protrusions (which are usually called "noses") give a nonnegligible contribution to the field quality, which is compensated for via the coil geometry. The iron yoke is made up of 5.8-mm-thick low-carbon-steel laminations. It is vertically split and the two halves are in contact (closed gap) both after assembly and after cool-down. Forces are transmitted from the yoke to the two-in-one collars also through an inclined iron insert (see Fig. 15).

During the assembly one usually aims at reaching a sufficient coil compression (prestress) to avoid coil movements in operational conditions. For the LHC dipoles, the target prestress at room temperature after collaring was fixed at 70–75 MPa

[33]. Collars are locked using rods, which results in a large spring back after collaring ($\sim 60\%$). For this reason the prestress needed during collaring is ~ 130 MPa; during the model and prototype phase it was carefully verified that the insulation could withstand these large pressures.

During the cool-down, the low thermal contraction of the stainless steel collars coupled with the large thermal contraction of the superconducting coils contributes to a significant prestress loss (see Fig. 16), similarly to what was found for the SSC dipoles [35]: the final target for the azimuthal stress at 1.9 K is ~ 30 MPa. The mechanism of the prestress loss is due not only to the differential thermal contraction but also to the hysteresis in the mechanical behavior of the coil [36]. Short models and prototypes were instrumented with capacitive gauges to measure the prestress level in the coil [37]. Short models having an unloaded coil at full energy did not show worse quench performances [38], and even though the series magnets had no capacitive gauges, there is evidence that part of the LHC dipoles have unloaded coils at full energy. The minimal required level of stress in superconducting magnets is still an open issue in the literature.

The sensitivity of the prestress on the azimuthal coil size has been measured with dedicated experiments [39]; a 0.1 mm larger coil gives a larger stress of ~ 12 MPa at room temperature and of ~ 6 MPa at 1.9 K. The tolerance on the prestress at room temperature after assembly has been set at ± 15 MPa, thus giving a ± 0.12 mm window on the coil size. Pole shims of variable thickness have been foreseen [33] to keep the prestress under control in the case of coil sizes outside tolerances.

The collared coil and the yoke laminations are enclosed by a shrinking cylinder welded with 150 MPa circumferential stress. The welding also imposes the desired curvature on the cold mass, which corresponds to a nominal sagitta of 9.14 mm [40]. The main parameters and performance of the cold mass are given in Table 5 and compared to Tevatron, HERA and RHIC dipoles.

The cold mass (coil, collars, laminations and shrinking cylinder) contains a static bath of helium II at atmospheric pressure. The iron yoke laminations contain the heat exchanger tube, which extracts the heat during the operation at 1.9 K and during cooling from 4.2 to 1.9 K. It is a seamless, round, oxygen-free copper tube with an outside diameter of 58 mm, and a thickness of 2 mm. In operation it carries a two-phase flow of saturated superfluid helium at 16 mbar [41]. The helium is provided by an external cryogenic distribution line (QRL) which runs parallel to the magnets. The line contains the helium in different thermodynamic states which are used for intercepting the applied heat loads at a higher temperature, thus saving the cost of refrigeration of the whole machine [41]. In the arcs, the QRL feeds the machine through the main quadrupoles. Temperature levels range from shields at ~ 70 K in the cryostat, to ~ 20 K for the cooling of the beam screen, to 1.9 K for the main dipole cold masses.

The cryostat is a low carbon steel cylindrical vacuum vessel of 914 mm diameter [42]. The cold mass lies on three support posts made of glass fiber-reinforced epoxy, with two heat-intercepting plates at 4–10 K and 50–65 K. According to the design, the posts at the extremities are free to move longitudinally, and the central one is free to move radially;

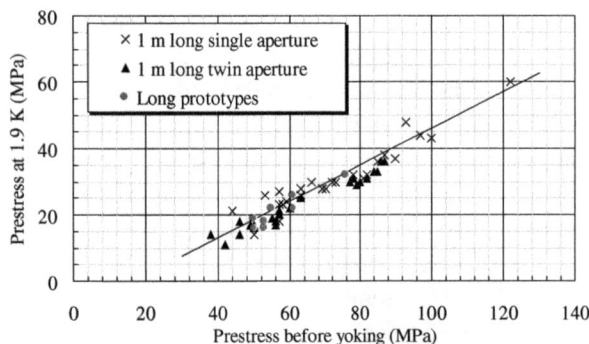

Fig. 16. Azimuthal prestress at room temperature after assembly versus azimuthal prestress at 1.9 K measured for LHC dipole short models and prototypes.

Table 5. Parameters of four superconducting dipoles; field, current, inductance and energy are given at nominal values (collision energy).

	LHC	Tevatron	HERA	RHIC
Field (T)	8.3	4.3	4.7	3.5
Current (kA)	11.8	4.3	5.0	5.05
Inductance (mH)	98.7	32	58	28
Energy (MJ)	6.93	0.30	0.73	0.35
Magnetic length (m)	14.3	6.4	8.8	9.45
Cold mass weight (t)	27.5	NA	NA	3.6
Nominal/injection field	15.5	6.5	20.3	8.62
Temperature (K)	1.9	4.2	4.5	4.3–4.6
Coil diameter (mm)	56	76	75	80
Number	1276	774	416	264

this has been done to avoid stress on the post during cool down and warm-up. During the production, it has been decided to block the central post to better control the dipole shape [43]. Contrary to the case of the RHIC dipoles, there was no need to use external welding on the cold mass to control the dipole shape.

As in the HERA machine, dipoles are equipped with correctors to compensate for sextupole, octupole and decapole components. Sextupolar correctors (usually called "spool pieces" in the specialized literature) are mounted on each dipole, whereas octupolar and decapolar correctors are mounted every second dipole.

The machine is divided into octants which are individually powered, i.e. 154 dipoles are powered in series. The dipole protection is ensured by quench heaters placed between the outer layer and the collars. The firing time after quench detection is of the order of 20 ms, which assures that the temperature of the hottest spot is less than 300 K. Each dipole is equipped with a cold diode that during a quench bypasses the high current (up to 12 kA) with a decay time of around 100 s [42].

4.2. Production and quality control of components and assembly

For such a large project as the LHC, cost is a major issue and costs relative to the magnetic system are strongly related not only to raw material price, but also to tolerances on components and assembly procedures. LHC large production offered a unique opportunity to judge the soundness of the required tolerances with good statistics. Here we will briefly review the main results, showing that in most cases the tolerances were not far from what was just needed. The other important aspect is that one faulty magnet is enough for the LHC not to work. For this reason, the techniques for the quality control and test are extremely important, since one has to avoid having even a single faulty magnet installed in the ring. In this section we will also review the strategies used in the quality control of such a large production.

The cable production was shared between two manufacturers for the inner layer and five for the outer layer. The cable critical current was measured on a sample of $\sim 25\%$ (see Fig. 17). The specifications on critical current have been met, and average values $\sim 10\%$ larger than specification have been obtained [45]. Some outliers as shown in Fig. 17 have been

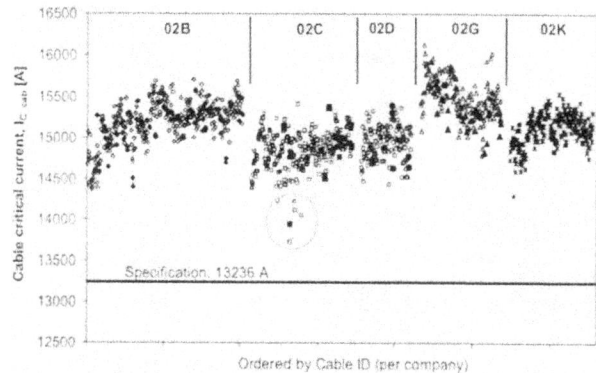

Fig. 17. Critical current measurements of the outer layer cable [45].

traced back to production features. The specification required the absence of cold welds within the length of superconducting cable used for winding each pole. During the production, a bunch of cold welds localized in the same section of the cable, and limiting the dipole performance to 60% of the short sample limit, were found in one case.

The field quality at injection is affected by the persistent currents in the cable. To monitor this effect, the magnetization of the strands was measured. A hysteresis cycle up to 1 T is performed, and the width of the loop at 0.5 T has to meet specification (30 and 23 mT for the outer and the inner layer, respectively) within 4.5% [46]. In order to meet this goal, strands with extreme values of magnetization were sorted during cabling. For the inner layer, an $\sim 13\%$ difference in the magnetization between the two producers was found; this was traced back to manufacturing procedures, and its impact on the performance was judged as acceptable. Notwithstanding the large number of manufacturers, a good homogeneity of the global production was obtained. This allowed the mixing of cable from different manufacturers in the same octant, contrary to the original baseline of the installation scheme (see also Subsec. 4.5). The measured cable dimension has been within the tight tolerances of ± 0.006 mm (see Fig. 18).

Three manufacturers (Alstom, Ansaldo Superconduttori and Babcock-Noell) have assembled the 1232 dipole cold masses plus 46 spares. The production steps involved the coil winding and curing, the assembly of collars and the collaring, the assembly of the iron yoke and the welding of the shrinking cylinder. The initial plan foresaw each octant to

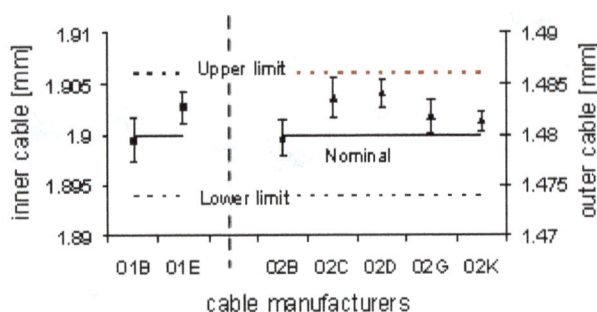

Fig. 18. Dimension of the cable during the production.

be produced by the same manufacturer; after the results of the first magnetic measurements, showing a remarkable homogeneity between the firms (see Subsec. 4.3), this scheme was abandoned and one third of the production was allocated to each manufacturer.

Copper wedges of 3.6 m length were produced by Outokumpu. Approximately 1% of the production has undergone several types of tests: physical and chemical tests, and dimensional tests. The transverse dimensions are critical for the correct position of the winding, i.e. for the field quality. The tight tolerance of ±0.030 mm has been kept through all the production [47]. A negligible influence on the field quality has been observed. On the other hand, a batch out of tolerance by around 0.050 mm used in the early part of the production (see Fig. 19) has shown a visible impact on b_3 [47].

The azimuthal size of the coils was measured at 70 MPa in 25–50% of the production, depending on the cold mass assembler. The curing is instrumental in giving the correct size to the coils: note that the pile-up of the tolerances on the cable (±0.006 mm) and on the polymide (±3%) would give a coil precision of ±0.3–0.4 mm. The initial tolerance of ±0.025 mm on the average coil size [33] has been shown to be not realistic: measured coil sizes [48] along the production are in general within ±0.2 mm, and for most of the production within ±0.1 mm (see Fig. 20 for the production in firm 1). This permitted the use of nominal pole shims for 94% of the whole production. Nonnominal shims of up to ±0.1 mm have been used in 6% of the coils, and of up to ±0.2 mm in four magnets; in this case, the expected change in the allowed harmonics $b_3, b_5, b_7, \ldots$ has been observed.

Stainless steel collars have been produced by two firms (5/8 by Malvestiti and 3/8 by FSG); tight tolerances of ±0.020 to ±0.030 mm (depending on the

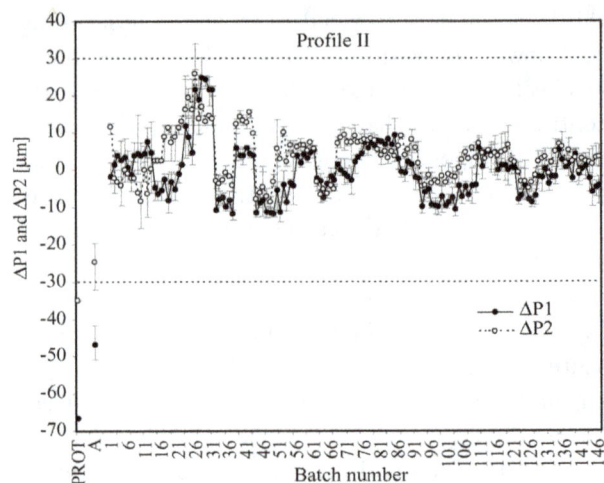

Fig. 19. Dimension of the copper wedge II along the production (difference with respect to nominal values), and two batches used at the beginning of the production.

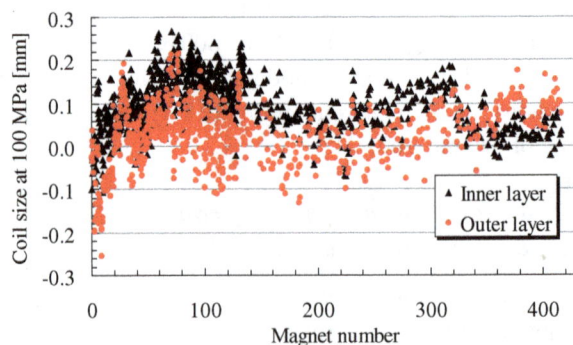

Fig. 20. Measured azimuthal coil size (difference with respect to nominal values, average left–right) in one of the dipole manufacturers.

position in the profile) on the complicated two-in-one shape have been set. Approximately 0.07% of the collars (three per magnet) were measured in around 100 points for quality control. In general the collars have been found to be precise within ±0.040 mm; the two manufacturers have shown similar dimensions. Despite the fact that the collars did not meet the original specifications, the impact on field quality has not been relevant [49]. Collar permeability was specified at 1.003 ± 0.002, and was kept all along the production.

Iron yoke laminations have been produced by two manufacturers. The tolerance on the stacking factor has been fixed to 98.50 ± 0.25%; since the iron contributes about 18% of the magnetic field, this tolerance gives a window of ±5 units on the transfer function. The situation here is less critical than in the case of RHIC, where more than 50% of the field was due to the iron and a stricter control of the iron mass was required.

The cold mass shape has been established using a laser tracker Leica^{TM} [40]. The tight tolerances on the longitudinal shape have been fixed to $\pm 1\,$mm along the reference orbit, and to $\pm 0.3\,$mm at the end of the magnet, where the multipolar correctors (spool pieces) are located. This tighter tolerance on the ends has been set to avoid harmonics feed-down from the correctors, which could have been critical for the beam. The sagitta of the produced magnets has a mean of $9.4 \pm 0.9\,$mm (one sigma), i.e. within the tolerances (see Fig. 21, left). Around 80 magnets from one of the assemblers have a systematic shift of the corrector position with respect to tolerances of about 0.4 mm; this nonconformity has been judged as acceptable for the beam dynamics (see Fig. 21, right).

All the information and tests relative to each cold mass have been stored in a Manufacturing Test Folder (MTF). Nonconformities that occured during the fabrication process have been recorded; when judged as not affecting the magnet performance, the anomaly has been registered and the assembly has been continued. On the other hand, when the nonconformity has been judged as nonacceptable (for instance a fault in the insulation), corrective actions have been taken. In each firm, two resident CERN inspectors followed the assembly, the documentation, and have provided the link to two CERN project engineers in charge of the production follow-up.

During the cold mass assembly, room temperature magnetic measurements were used as a diagnostic tool, as has been done for both the Tevatron [17] and RHIC [50]. Over the whole production, this

Fig. 22. A case of bad coil curing giving an inner radial movement of two turns of the inner layer, upper pole, of about 1 mm, found through anomalies in room temperature magnetic measurements.

technique allowed rescuing 19 faulty magnets (1.5% of the production) at the level of the collared coil: 5 wrong assembly cases (as double or a missing component), 2 wrong components, 8 cases of wrong procedure (a bad coil curing giving detachment of the last block of cables during collaring; see Fig. 22), and 4 other cases [51]. Magnetic measurements have also been used to successfully locate the position of electrical shorts during the assembly in 18 cases [52].

Thirty-one magnets (2.4% of the production) have been returned to the manufacturer and rebuilt after tests at 1.9 K at CERN: 14 of them for insufficient quench performance, 10 for electrical shorts or insulation faults and 7 for other reasons.

4.3. *Field quality performance*

4.3.1. *Test strategy*

As in the RHIC production, all magnets were measured at room temperature, and a sample has been measured at 1.9 K. Magnetic measurements at room temperature were carried out at two stages of the assembly, i.e. after the collaring (collared coil) and after the welding of the shrinking cylinder (cold mass). At the beginning of the production, all magnets were measured at 1.9 K with a static loadline and with a standard machine cycle. Once solid warm–cold correlations had been established, a sampling of 7–15% was carried out over the five year production. A total of 200 magnets, corresponding

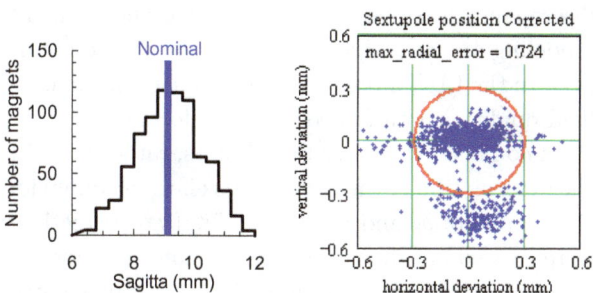

Fig. 21. Sagitta of the cold masses (left) and position of the correctors (right).

to 16% of the production, have been measured at 1.9 K.

4.3.2. *Systematic values*

The main novelties of the field quality control in the LHC dipoles with respect to previous machines are the even normal multipoles (quadrupole, octupole) which become allowed multipoles in the twin collar geometry. The iron yoke has been carefully shaped to minimize the impact of saturation [53] using numerical simulations; this optimization procedure has defined the position and the size of the holes in the iron, similarly to what was done in RHIC. The final prototypes showed a systematic quadrupole and octupole outside the targets, and a fine-tuning of both multipoles was carried out by reshaping the ferromagnetic insert between the collars and the yoke (see Fig. 15), using both simulations and a dedicated experiment [54]. The final insert chosen for the production has been shown to give systematic values within targets and no further corrective actions have been necessary during the production (see Figs. 23 and 24).

Concerns about the presence of a nonzero systematic b_4 inducing strong detuning triggered the insertion of octupolar correctors in half of the dipole cold masses; nevertheless, the production has shown very stable values close to zero (see Fig. 24).

The steering of the odd normal components (b_3 sextupole, b_5 decapole, b_7 14th pole; see Figs. 26 and 27) has also required corrective actions [55], as for the RHIC production [56]. Here the situation is more complicated since any change affects at the

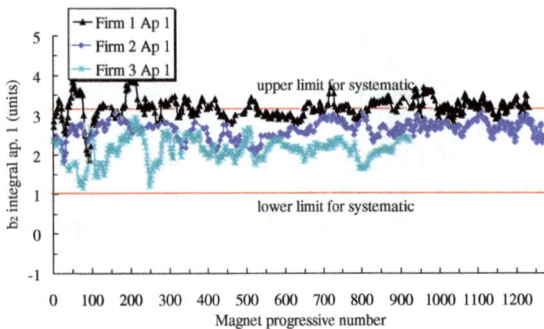

Fig. 24. Evolution of b_4 measured in aperture 1 during the production of 1276 dipoles, separated by the assembler (warm magnetic measurements in the cold mass, running average of five magnets per firm). The target limits for the global average are shown in red.

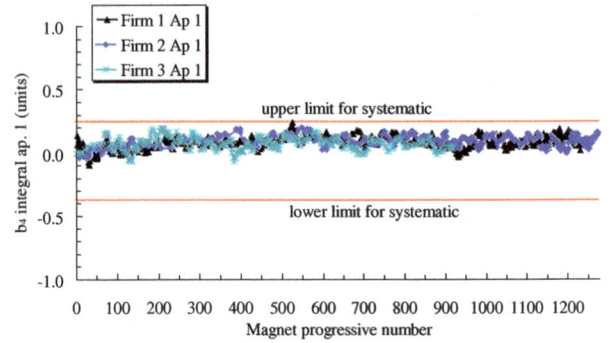

Fig. 25. Evolution of b_3 measured in aperture 1 during the production of 1276 dipoles, separated by the assembler (warm magnetic measurements in the cold mass, running average of five magnets per firm). The target limits for the global average are shown in red.

same time three multipoles and the main field, i.e. one has to control four variables at the same time.

At the beginning of the production both b_3 and b_5 were outside the target values by a sizeable value (see Figs. 25 and 26), giving an unacceptably large chromaticity at collision energy, beyond the capability of the sextupole correctors. The origin of this discrepancy was due both to neglected effects in the modeling and to a change in the beam dynamic targets. In the LHC dipole, the effect of deformations on field quality is dominated by the deformation of the thick coil inside the very rigid collars, rather than by the deformation of the collar. This effect accounts for three units of b_3 and one of b_5 [57]. Two corrections were carried out during the production, the first one after the measurement of nine dipoles (implemented in dipole 33) by modifying the shape of the inner layer copper wedges but keeping the same coil size

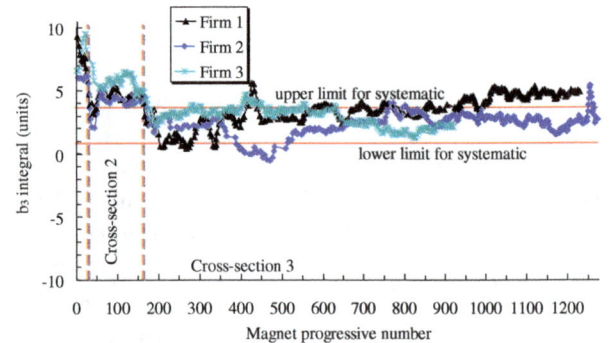

Fig. 23. Evolution of b_2 measured in aperture 1 during the production of 1276 dipoles, separated by the assembler (warm magnetic measurements in the cold mass, running average of five magnets per firm). The target limits for the global average are shown in red.

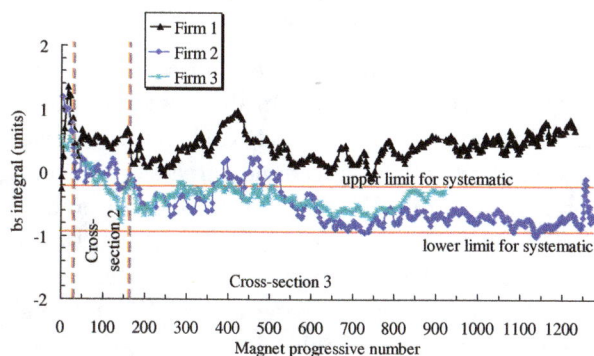

Fig. 26. Evolution of b_5 measured in aperture 1 during the production of 1276 dipoles, separated by the assembler (warm magnetic measurements in the cold mass, running average of five magnets per firm). The target limits for the global average are shown in red.

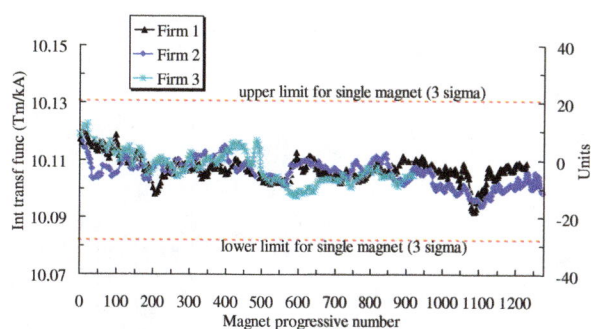

Fig. 27. Evolution of the integrated transfer function during the production, divided by the assembler.

to avoid changes in the tooling [55]; the second one after the completion of one octant (154 dipoles) consisted in adding a mid-plane shim, as done for the RHIC dipoles, to further lower b_3 and b_5 [55].

The agreement between the expected value of the corrections and their actual value is in general within 20%. Both corrections had an unexpected effect on b_7, which was driven out of the target range by 0.2 units, but it was considered acceptable. The final agreement between model and measurements is three units in b_3, one unit in b_5 and 0.5 units in b_7.

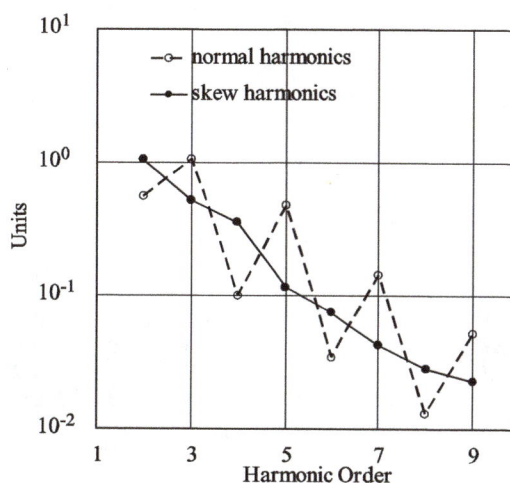

Fig. 28. Random components for LHC dipoles.

4.3.3. *Random components at room temperature*

The variability of the field quality from dipole to dipole, i.e. the so-called random component, is due to the component and assembly tolerances. This variability sets the ultimate limit on the precision available for steering the average values toward the design values over the whole production run. Moreover, random components of multipoles excite resonances which can limit the stability of the beam.

One of the critical field quality parameters is the reproducibility of the integrated transfer function. For the LHC dipoles this quantity has been remarkably stable along all the production (see Fig. 27), and, contrary to the HERA case [58], has shown negligible differences between the cold mass assemblers. The final spread over the entire production at room temperature is ~ 6 units. The option of reducing the transfer function spread via a control of the magnetic length (i.e. changing the longitudinal length of

the ferromagnetic laminations) has been used only for a few magnets during the production, just to test the validity of the method.

The random component of b_3 at room temperature is about 1 unit for the ~ 1100 magnets with the same cross section. The spread over all magnets is 1.5 units, compared to a target of 1.4 units. A global view of the random errors at room temperature is given in the semilogarithmic plot shown in Fig. 28. The random components correspond on average to a random positioning of the coil blocks of 0.025 mm rms. This is the lowest bound of what is usually considered in simulations [59] for estimating the random errors based on a random movement of the blocks (0.050–0.025 mm). The normal and skew components of the same order have different random components; this "saw-tooth" feature, which was already observed in Tevatron dipoles [60], is somewhat anomalous for the LHC since the skew harmonics show a rather flat

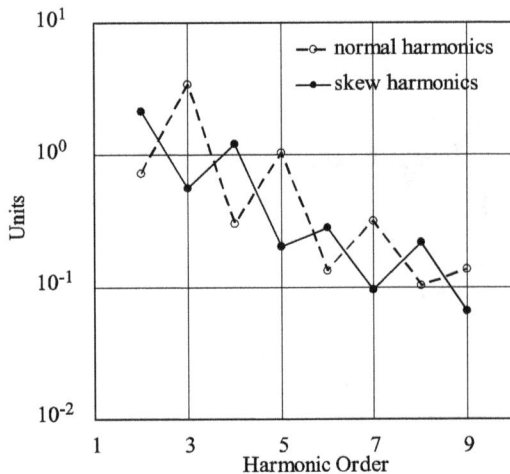

Fig. 29. Random components for HERA dipoles.

Table 6. Amplitude of the random block movements (mm) giving the measured random components.

	Tevatron	HERA	RHIC	LHC	LHC Firm 1
b odd	0.128	0.122	0.052	0.054	0.038
b even	0.052	0.020	0.006	0.012	0.010
a odd	0.070	0.024	0.008	0.018	0.012
a even	0.052	0.058	0.032	0.026	0.022
All	0.065	0.041	0.016	0.025	0.018

4.3.4. *Warm-cold correlations*

Warm–cold correlations were established during the early part of the production. In all cases the spread of the offset between "warm" and "cold" measurements was smaller than the spread measured at room temperature (see Table 7). In other words, the random part of the field quality is mainly determined by the room temperature measurements. For instance, the spread on b_3 at room temperature is 1–1.5 units and the spread of the warm–cold offset at high field (given by the coil geometry and the iron saturation) is ~ 0.2 units (see Table 7). The offset from room temperature to injection measurement, which is mainly due to the persistent current, is ~ 7 units, and its spread is ~ 0.5 units (see Figs. 30 and 31); this reflects the very good control of cable magnetization. The measured field hysteresis in the magnet is well explained by a model that relies on the magnetization measurements (see Fig. 32 for the b_3 case [63]).

The warm–cold offsets were monitored during the production, where a continuous sampling was taken; they have been remarkably stable (see Fig. 31). This has permitted steering the field quality using room temperature measurements. A negligible impact of the cold mass assembler on the warm–cold offsets was found in most cases, as expected.

behavior in the semilog plot (see the HERA results for comparison in Fig. 29). The amplitude of the random movements corresponding to each family of multipoles [61] for the four machines is given in Table 6. The exceptional results obtained by RHIC in the assembly precision are probably due to the thinner, one-layer coil, and to the simpler design.

The field quality of the LHC dipoles showed a very weak dependence on the cold mass assembler. It has already been pointed out that no signature of the assembler was visible on the transfer function. Some differences have been observed only in two allowed multipoles, b_5 and b_7, and in the skew component a_3. For this reason, the spread inside magnets assembled by the same firm is similar to the global spread for even multipoles, and is $\sim 30\%$ lower for the odd ones (see Table 6). A few trends [62] were observed in the production; some trends in b_3 and a_4 have been traced back to the position of the upper block in the inner coil, close to the pole, as also shown in Fig. 22.

Table 7. Spread of the multipoles (1 σ) ar room temperature (RT), and of the offsets between the multipoles measured at 1.9 K at injection energy (inj.) of collision energy (coll.) and room temperature.

	b_2	a_2	b_3	a_3	b_4	a_4	b_5
R.T.	0.59	0.89	1.51	0.39	0.08	0.28	0.54
Inj.–R.T.	0.32	0.39	0.40	0.09	0.04	0.09	0.10
Coll.–R.T.	0.33	0.18	0.20	0.10	0.04	0.04	0.07

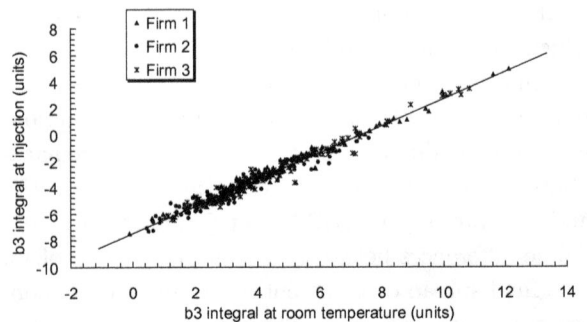

Fig. 30. Warm–cold correlation for b_3 between injection energy and room temperature measurements.

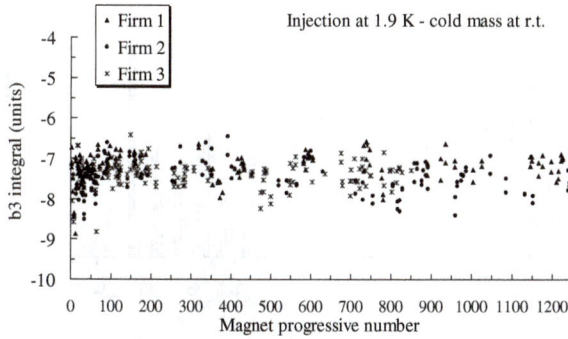

Fig. 31. Evolution of the difference between b_3 measured at injection energy at 1.9 K and at room temperature during the production, divided by the assembler.

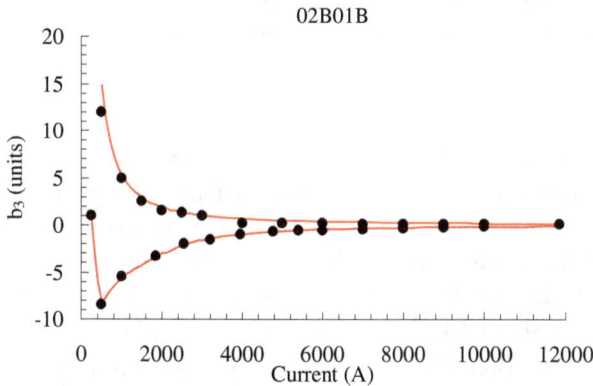

Fig. 32. Measured hysteresis in an LHC dipole (markers) and model based on the magnetization measurements (solid line) [65].

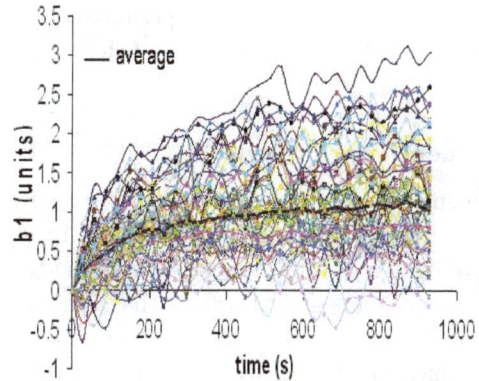

Fig. 33. Measured decay of the main field during the injection plateau measured in 35 LHC dipoles manufactured with cable of producer B [65].

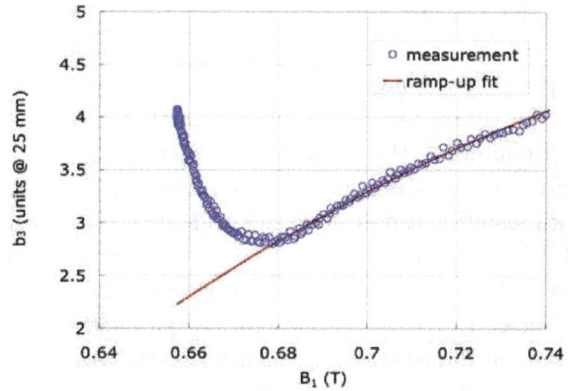

Fig. 34. Evolution of b3 versus main field during the beginning of the ramp (snapback) measured in one LHC dipole [65].

4.3.5. *Dynamic effects*

The foreseen duration of the injection plateau for the LHC at nominal operation is around 20 min. During this time the persistent currents decay about $+1.5$ units (transfer function, see Fig. 33), $+2$ units (b_3) and -0.35 units (b_5). These values have been measured for ~ 200 dipoles at 1.9 K. When the ramp is started, the decay suddenly disappears and the field harmonics snap back to their original value in a few seconds (see Fig. 34). For the LHC, special probes have been developed to measure these fast phenomena [63], and have been used also for measuring the Tevatron magnets during RUN II [64].

The Tevatron experience showed that the magnetic behavior of superconducting elements gives rise to time-dependent phenomena and loss of reproducibility, which can severely affect machine operation [15]. For the LHC dipoles, a few of them

underwent special tests to work out the dependence of the decay on the previous cycle parameters. Results are in qualitative agreement with the experience acquired on previous machines, i.e. that the amplitude of the decay is roughly proportional to the flat top field of the precycle, is smaller for a longer duration of the flat top in the precycle, and gets larger for a shorter preinjection porch.

Different empirical fits have been used for the decay (logarithm, single or double exponential) in the four machines. A similar situation holds for the snapback dependence on the current. A significant result linking the snapback amplitude and its decay constant in the case of an exponential fit was found during the first years of the LHC dipole production; this scaling law has been proven both for the LHC and Tevatron dipoles [64]. A major effort was carried out to build an LHC field model [66, 67] using all the relevant information and permitting the programming of the circuit magnets before day 1 of commissioning.

4.4. Quench performance

The quench performance has been measured on 12 test benches at CERN for all dipoles at 1.9 K [68]. The LHC magnets show little or no training to reach the nominal field, and therefore the standard test used for the early part of the production consisted of two quenches to get to 8.4 T; if this condition was not satisfied, the ultimate value of 9 T had to be reached within nine quenches. Otherwise, the magnet was tested again after a thermal cycle (warm-up and cool-down). These criteria have been updated for the mature phase of the production, replacing the two-quench–8.4 T with a three-quench–8.6 T requirement. They have been established according to: (i) optimizing the time necessary for testing, (ii) minimizing the expected quenches in the tunnel during commissioning, and (iii) detecting magnets with insufficient performance [68]. Fourteen magnets (1.2% of the production) have been returned to the assemblers for insufficient quench performance, and repaired. In the initial phase of the production, one dipole sustained severe damage during the test which provoked the destruction of the coil.

80% of the dipoles reached 8.3 T without or with one quench, and 17% with two quenches. 11% of the dipoles were tested after a thermal cycle: 76% reached 8.3 T without quenches, 19% with one quench, and 4% with two quenches. The histogram of the value of the first and the second quench, and of the first quench after the thermal cycle, is shown in Fig. 35.

The dipoles tested after the thermal cycle had some detraining, i.e. the value of the first quench after the thermal cycle was lower then the value of the last quench before warm-up. 20% of the dipoles had no detraining, 33% had a lower field of 0–0.3 T, 33% of 0.3–0.6 T, 12% of 0.6–1.0 T, and 2% of up to 1.4 T. These results must be read taking into account that only the "bad" dipoles were tested after a thermal cycle, and therefore the statistics could be biased.

Quench longitudinal localization through a quench antenna has been carried out for $\sim 1/6$ of the dipoles. Results show that 85–90% of the quenches originated in the coil ends. This proves that the level

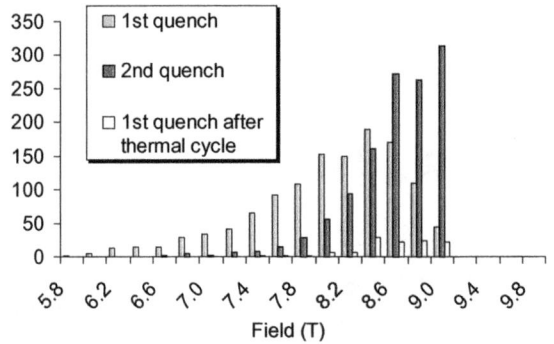

Fig. 35. First and second quench levels (all LHC dipoles), and first quench after the thermal cycle (a sample of 130 dipoles).

of prestress on the straight part of the coil was not the main limitation on the quench performance.

4.5. Installation strategy

Contrary to the original baseline, installation started when a large stock of measured magnets was available. Moreover, the magnetic measurements at room temperature provided the main field quality features of the dipoles, some months in advance of their arrival at CERN. For these reasons the Magnet Evaluation Board [69], in charge of the allocation of the magnets in the ring, has faced the "dream" situation of being able to sort practically all the dipoles in the ring to maximize the machine performance. To be more precise, the dipoles were produced in four families (i.e. according to the corrector package and the diode polarity) and therefore the sorting had to be done within these families, and not all over the machine.

The first two types of cross sections were installed in the first octant to minimize the spread on b_3. Both a local compensation and a compensation at a π or 2π betatron phase advance have been used. This possibility has also been used in the other sectors to further reduce the spread of b_3 and a_2 with respect to the original targets, with a gain of the order of a factor 2–3 on the required orbit corrector strength, coupling resonance and vertical dispersion, and third order resonance driving terms.

A more significant improvement in the machine performance has been obtained by sorting the dipoles according to the measured shape to maximize the mechanical aperture. Dipoles with shapes above tolerances have been allocated to slots where the beam envelope given by the optical functions is smaller, as the mid-cell positions. This action eliminated any aperture limitations on the main dipoles. No sorting has been done on the measured quench performance.

5. Outlook for Future Accelerator Magnets

The 9.7 T short sample field of the LHC dipoles is close to the ultimate limit of what can be done with Nb–Ti, which has been the "workhorse" superconducting material for accelerator magnets in the past 25 years. The record for a Nb–Ti dipole belongs to the 88 mm aperture dipole used in the cable test station at CERN [70], which has a two-layer $\cos\theta$ design with a 16.7 mm cable width, and a short sample field of 10.15 T. It reached 10.09 T after a short training period and is routinely used to measure the cable critical current at 10 T. Similar performances were obtained by the D19 model in LBL, having a maximum quench field of 10.06 T with a strongly graded two-layer coil of 2 ∗ 12 mm thickness [71].

A 10% increase to get an 11 T short sample limit would require doubling the coil width with respect to the LHC dipoles (see Fig. 36). Since ∼ 14 T is the critical field at 0 K and at zero current density for the Nb–Ti, the only possibility of achieving higher fields is to use materials with higher performance in terms of critical field and current density (see Fig. 37).

At the beginning of the 1990s, when the Nb_3Sn prototype for the LHC [25] was built, the critical current density at 10 T was only marginally larger than in Nb–Ti. Now, after 20 years of R&D in

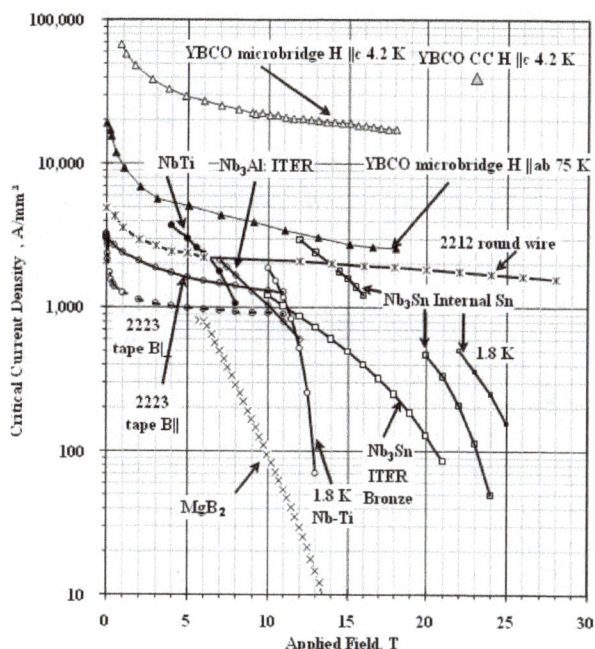

Fig. 37. Critical current of several types of superconductors versus magnetic field [78].

superconducting materials, a conductor bearing up to 3000 A/mm^2 at 12 T and 4.2 K is available, with a filament size of the order of 50 μm [72]. Using this cable, the fields reachable in a $\cos\theta$ dipole are shown in Fig. 36: a coil width of 30–45 mm can provide short sample fields in the range of 14–16 T.

The main issues for Nb_3Sn are its brittleness, the degradation induced by strain, and instabilities related to flux jumps [73]. They can be resolved by optimizing the design and the treatment of the strand, and by a clever mechanical structure. Besides the collars, invented for the Tevatron dipoles and used in all accelerator dipoles except RHIC, an alternative structure based on an aluminum shell pretensioned with bladders and keys has been proposed. This structure allows the highest peak stress to be reached at the end of cool-down and not during collaring. It has been successfully implemented in Nb_3Sn dipoles and quadrupoles [74, 75]. Another innovation in the coil layout is a design where non-keystoned Rutherford cables are arranged in rectangular blocks [76, 77].

Up to now, Nb_3Sn has been routinely used in solenoids to reach fields in the range of 10–20 T [79], but Nb_3Sn magnets have never been used in accelerators. Nevertheless, short models with accelerator-like field quality were built during the 1990s, attaining

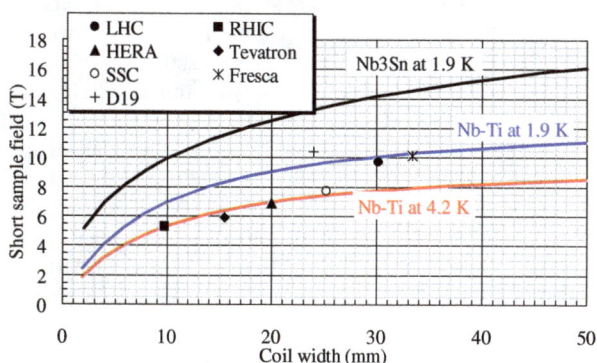

Fig. 36. Short sample field versus coil width for five dipoles (markers), and estimate for an ironless $\cos\theta$ coil at 4.2 K (red line) and at 1.9 K (blue line), with a filling factor of 0.3. The same Nb-Ti cable properties have been assumed for all dipoles. The $\cos\theta$ estimate is also given for the best Nb3Sn conductor available today.

fields in the range of 11–13 T [80, 81]. This R&D has rapidly evolved in the last few years. Simple designs such as racetrack dipole coils, without field quality, have been able to reach a bore field of 16 T [75] in a 1 m model. Recent results have also shown that 4-m-long Nb_3Sn quadrupole coils with a peak field of 12 T can be successfully manufactured [82]. FNAL [83] and the Lawrence Berkeley Laboratory [77] are today aiming at building Nb_3Sn dipoles with ~ 40 mm apertures with a coil width of 30–45 mm, and a short sample field in the range of 12–15 T. Nb_3Sn quadrupoles with a peak field in the range of 12–15 T could be used for the upgrade of the quadrupoles in the LHC interaction regions, and this would provide the first test of the reliability of Nb_3Sn magnets in an accelerator.

The discovery that high temperature superconductors such as YBCO and Bi2212 have critical fields well beyond 50 T when operated at LHe temperature opens up new possibilities for accelerator uses. The difficulties in fabricating coils with the material are very similar to Nb_3Sn, and going to fields in the range of 20–50 T poses really challenging problems for the magnet designer. But the fact that these materials can have high radiation resistance, can work at high temperature or can achieve very high fields may offer new applications such as low beta quads for interaction regions in accelerators or for providing intense cold muon beams for neutrino factories or muon colliders. Ever since Zeeman's early work, higher magnet fields have led the way to new physics, so the promise is still high!

Acknowledgments

A. T. would like to acknowledge the excitement and fun of working with Bob Wilson and the contributions of the many extremely talented people who worked on the Tevatron. We would both like to thank all the colleagues who were involved in the LHC dipole models, prototypes and production. A. T. acknowledges support from the Fermi Research Alliance, LLC under Contract No. DE-AC02-07CH11359 with the US Department of Energy.

References

[1] R. R. Wilson, E. L. Goldwasser, D. A. Edwards, W. B. Fowler, B. P. Strauss, G. Biallas, W. B. Hanson, D. F. Sutter, P. J. Reardon and P. C. Vander Arend, The Energy Doubler Design Study: a progress report, *Fermilab Phys. Notes* **FN-263**, 4 (1974).

[2] T. Elioff, W. Gilbert, G. Lamberston and R. Meuser, *IEEE Trans. Magn.* **11**, 447 (1975).

[3] J. R. Sanford, Isabelle, a proton–proton colliding beam facility, *BNL Report* **50718** (1978).

[4] M. N. Wilson, *Superconducting Magnets* (Clarendon, Oxford, 1983); K. H. Mess, P. Schmuser and R. Wolff, *Superconducting Accelerator Magnets* (World Scientific, Singapore, 1996).

[5] H. E. Edwards, *Ann. Rev. Nucl. Part. Sci.* **35**, 605 (1985).

[6] R. Palmer and A. V. Tollestrup, *Ann. Rev. Nucl. Part. Sci.* **34**, 247 (1984).

[7] B. P. Strauss, R. H. Remsbottom, P. J. Reardon, C. W. Curtist and W. K. McDonald, *IEEE Trans. Magn.* **13**, 487 (1977).

[8] D. C. Larbalestier, *IEEE Trans. Nucl. Sci.* **30**, 3299 (1983).

[9] K. Koepke, G. Kalbfleisch, W. Hanson, A. Tollestrup, J. O'Meara and J. Saarivirta, *IEEE Trans. Magnetics* **15**, 658 (1979).

[10] R. Hanft, B. C. Brown, W. E. Cooper, D. A. Gross, L. P. Michelotti, E. E. Schmidt and F. Turkot, *IEEE Trans. Nucl. Sci.* **30**, 3381 (1983).

[11] M. Wake, D. A. Gross, M. Kumada, D. Blatchley and A. V. Tollestrup, *IEEE Trans. Nucl. Sci.* **26**, 3894 (1983).

[12] E. E. Schmidt, B. C. Brown, W. E. Cooper, H. E. Fisk, D. A. Gross, R. Hanft, S. Ohnuma and F. T. Turkot, *IEEE Trans. Nucl. Sci.* **30**, 3384 (1983).

[13] B. C. Brown, W. E. Cooper, J. D. Garvey, D. A. Gross, R. Hanft, K. P. Kaczar, J. E. Pachnik, C. W. Schmidt, E. E. Schmidt and F. Turkot, *IEEE Trans. Nucl. Sci.* **30**, 3608 (1983).

[14] C. P. Bean, *Rev. Mod. Phys.* **36**, 31 (1964).

[15] G. V. Velev, G. Annala, P. Bauer, R. Carcagno, J. Di Marco, H. Glass, R. Hanft and R. Kephart, in *Particle Accelerator Conference* (2003), p. 1972.

[16] M. Wake, D. Gross, R. Yamada and D. Blatchley, *IEEE Trans. Magne.* **15**, 141 (1979).

[17] B. Peters, L. Harris, J. Saarivirta and A. Tollestrup, *IEEE Trans. Magne.* **15**, 134 (1979).

[18] G. Biallas, J. Finks, B. Strauss, M. Kuchnir, W. Hanson, E. Kneip, H. Hinterberger, D. Dewitt and R. Powere, *IEEE Trans. Magne.* **15**, 131 (1979).

[19] P. C. Vander Arend and W. B. Fowler, *IEEE Trans. Nucl. Sci.* **20**, 119 (1973).

[20] R. Stiening, R. Flora, R. Lauckner and G. Tool, *IEEE Trans. Magne.* **15**, 670 (1979).

[21] G. A. Voss and B. H. Wiik, *Ann. Rev. Nucl. Part. Sci.* **44**, 413 (1994).

[22] E. Apostolescu, R. Bandelmann, H. Boettcher, I. Borchardt, G. Deppe, K. Escherich, H. Kaiser, M. Leenen, O. Peters, H. Poggensee, S. L. Wipf

and S. Wolff, *IEEE Trans. Magne.* **28**, 689 (1992).

[23] M. A. Harrison, The RHIC Project, *BNL Rep.* **49794** (1994).

[24] R. Perin, D. Leroy and G. Spigo, *IEEE Trans. Magne.* **25**, 1632 (1989).

[25] H. ten Kate, A. den Ouden, D. ter Avest, S. Wessel, R. Dubbeldam, W. van Emden, C. Daum, M. Bona and R. Perin, *IEEE Trans. Magne.* **27**, 1996 (1991).

[26] M. Bona, R. Perin, E. Acerbi and L. Rossi, *IEEE Trans. Magne.* **32**, 2051 (1996).

[27] J. Billan, M. Bona, L. Bottura, D. Leroy, O. Pagano, R. Perin, D. Perini, F. Savary, A. Siemko, P. Sievers, G. Spigo, J. Vlogaert, L. Walckiers and C. Wyss, *IEEE Trans. Appl. Supercond.* **9**, 1039 (1999).

[28] L. Rossi, *IEEE Trans. Appl. Supercond.* **12**, 219 (2002).

[29] O. Bruning, P. Collier, P. Lebrun, S. Myers, R. Ostojic, J. Poole and P. Proudlock, *CERN Rep.* **2004–003** (2004).

[30] R. Palmer and A. Tollestrup, *Fermilab* **TM-1251**, 67 (1984).

[31] D. Tommasini and D. Richter, in *European Particle Accelerator Conference* (2008), p. 2467.

[32] S. Russenschuck, *CERN Rep.* **99-01**, 82 (1999).

[33] Technical specification for the LHC dipoles, IT-2997/LHC/LHC (2001).

[34] P. Fessia, D. Perini, R. Vuillerment and C. Wyss, *LHC Proj. Note* **288** (2002).

[35] A. Devred, T. Bush, R. Coombes, J. DiMarco, C. Goodzeit, J. Kuzminski, M. Puglisi, P. Radusewicz, P. Sanger, R. Schermer, G. Spigo, J. Thompkins, J. Turner, Z. Wolf, Y. Yu and H. Zheng, *AIP Conf. Proc.* **249**, 1309 (1992).

[36] P. Ferracin, W. Scandale, E. Todesco and D. Tommasini, *IEEE Trans. Appl. Supercond.* **12**, 1705 (2002).

[37] N. Siegel, D. Tommasini and I. Vanenkov, *LHC Proj. Rep.* **173** (1998).

[38] N. Andreev, K. Artoos, E. Casarejos, T. Kurtyka, C. Rathjen, D. Perini, N. Siegel, D. Tommasini, I. Vanenkov, *LHC Proj. Rep.* **179** (1998).

[39] P. Ferracin, W. Scandale, E. Todesco and D. Tommasini, *Phys. Rev. STAB* **5**, 062401 (2002).

[40] M. Bajko, R. Chamizo, C. Charrondiere, A. Kuzmin and F. Savary, *IEEE Trans. Appl. Supercond.* **16**, 429 (2006).

[41] Ph. Lebrun, *IEEE Trans. Appl. Supercond.* **10**, 1500 (2000).

[42] N. Bourcey, O. Capatina, V. Parma, A. Poncet, P. Rohmig, L. Serio, B. Skoczen, J.-P. Tock and L. R. Williams, *AIP Conf. Proc.* **710**, 487 (2004).

[43] F. Seyvet, J.-B. Jeanneret, A. Poncet, D. Tommasini, J. Beauquis, E. D. Fernandez Cano and E. Wildner, in *Particle Accelerator Conference* (2005), p. 2675.

[44] F. Rodríguez-Mateos, K. Dahlerup-Petersen, R. Denz Milani and F. Tegenfeldt, *IEEE Trans. Appl. Supercond.* **14**, 251 (2004).

[45] A. Verweij and A. Ghosh, *IEEE Trans. Appl. Supercond.* **17**, 1454 (2007).

[46] S. Le Naour, L. Oberli, R. Wolf, R. Puzniak, A. Szewczyk, A. Wisniewski, H. Fikis, M. Foitl and H. Kirchmayr, *IEEE Trans. Appl. Supercond.* **9**, 1763 (1999).

[47] B. Bellesia, F. Bertinelli, P. Fessia, G. Gubello, C. Lanza, W. Scandale and E. Todesco, *CERN LHC Proj. Rep.* **630** (2003).

[48] I. Vanenkov and C. Vollinger, in *Particle Accelerator Conference* (2003), p. 1951.

[49] B. Bellesia, F. Bertinelli, C. Santoni and E. Todesco, *IEEE Trans. Appl. Supercond.* **16**, 196 (2006).

[50] R. Gupta, M. Anerella, J. Cozzolino, D. Fisher, A. Ghosh, A. Jain, W. Sampson, J. Schmalzle, P. Thompson, P. Wanderer and E. Willen, *Proc. MT-15 Conference*.

[51] C. Vollinger and E. Todesco, *IEEE Trans. Appl. Supercond.* **16**, 204 (2006).

[52] B. Bellesia, G. Molinari, C. Santoni, W. Scandale and E. Todesco, *IEEE Trans. Appl. Supercond.* **16**, 208 (2006).

[53] C. Vollinger, *CERN Rep.* **99-01**, 93 (1999).

[54] L. Bottura, S. Redaelli, V. Remondino, S. Sanfilippo, F. Savary, W. Scandale and E. Todesco, in *Eighth European Particle Accelerator Conference* (2002), p. 2427.

[55] E. Todesco, B. Bellesia, L. Bottura, A. Devred, V. Remondino, S. Pauletta, S. Sanfilippo, W. Scandale, C. Vollinger and E. Wildner, *IEEE Trans. Appl. Supercond.* **14**, 177 (2004).

[56] R. Gupta, M. Anerella, J. Cozzolino, B. Erickson, A. Greene, A. Jain, S. Kahn, E. Kelly, G. Morgan, P. Thompson, P. Wanderer and E. Willen, *IEEE Trans. Magne.* **32**, 2069 (1996).

[57] P. Ferracin, O. Pagano, V. Remondino, W. Scandale, E.Todesco and D. Tommasini, *IEEE Trans. Appl. Supercond.* **12**, 1727 (2002).

[58] R. Meinke, *IEEE Trans. Magn.* **27**, 1728 (1990).

[59] R. Gupta, *Part. Accel.* **54**, 379 (1996).

[60] J. Herrera, R. Hogue, A. Prodell, P. Wanderer and E. Willen, in *Particle Accelerator Conference* (1985), p. 3689.

[61] B. Bellesia, E. Todesco, in *European Particle Accelerator Conference* (2006), p. 2601.

[62] E. Todesco, B. Bellesia, P. Hagen, C. Vollinger, *IEEE Trans. Appl. Supercond.* **16**, 419 (2006).

[63] T. Pieloni, S. Sanfilippo, L. Bottura, M. Haverkamp, A. Tikhov, E. Effinger, E. Benedico and N. Smirnov, *IEEE Trans. Appl. Supercond.* **14**, 1822 (2004).

[64] G. Ambrosio, P. Bauer, L. Bottura, M. Haverkamp, T. Pieloni, S. Sanfilippo and G. Velev, *IEEE Trans. Appl. Supercond.* **15**, 1217 (2005).

[65] S. Sanfilippo, talks given at field quality working group, http://www.cern.ch/fqwg

[66] N. Sammut, L. Bottura and J. Micallef, *Phys. Rev. STAB* **9**, 012402 (2006).

[67] N. Sammut, L. Bottura, P. Bauer, G. Velev, T. Pieloni and J. Micallef, *Phys. Rev. STAB* **10**, 082802 (2007).

[68] A. Siemko and P. Pugnat, *IEEE Trans. Appl. Supercond.* **17**, 1091 (2007); V. Chohan *et al.*, *Particle Accelerator Conference* (2007), p. 3742.

[69] P. Bestmann, L. Bottura, N. Catalan-Lasheras, S. Fartoukh, S. Gilardoni, M. Giovannozzi, J.-B. Jeanneret, M. Karppinen, A. Lombardi, K.-H. Meß, D. Missiaen, M. Modena, R. Ostojic, Y. Papaphilippou, P. Pugnat, S. Ramberger, S. Sanfilippo, W. Scandale, F. Schmidt, N. Siegel and A. Siemko, in *Particle Accelerator Conference* (2007), p. 3739.

[70] D. Leroy, G. Spigo, A. Verweij, H. Boschman, R. Dubbeldam and J. Pelayo, *IEEE Trans. Appl. Supercond.* **10**, 178 (2000).

[71] D. Dell'Orco, S. Caspi, J. O'Neill, A. Lietzke, R. Scanlan, C. B. Taylor and A. Wandersforde, *IEEE Trans. Appl. Supercond.* **3**, 637 (1993).

[72] D. Dietderich, E. Barzi, A. Ghosh, N. Liggins and H. Higley, *IEEE Trans. Appl. Supercond.* **17**, 1481 (2007).

[73] A. Zlobin, V. Kashikhin and E. Barzi, *IEEE Trans. Appl. Supercond.* **16**, 1308 (2006).

[74] S. Caspi, D. Dietderich, P. Ferracin, S. Gourlay, A. Hafalia, R. Hannaford, A. Lietzke, A. McInturff, G. Sabbi, A. Ghosh, A. Andreev, E. Barzi, R. Bossert, V.; Kashikhin, I. Novitski, G. Whitson and A. Zlobin, *IEEE Trans. Appl. Supercond.* **17**, 1122 (2007).

[75] A. Lietzke, A. Bartlett, P. Bish, S. Caspi, L. Chiesa, D. Dietderich, P. Ferracin, S. Gourlay, M. Goli, R. Hafalia, H. Higley, R. Hannaford, W. Lau, N. Liggens, S. Mattafirri, A. McInturff, M. Nyman, G. Sabbi, R. Scanlan and J. Swanson, *IEEE Trans. Appl. Supercond.* **14**, 345 (2004).

[76] G. Ambrosio, N. Andreev, E. Barzi, P. Bauer, D. Chichili, K. Ewald, L. Imbasciati, V. Kashikhin, S. Kim, P. Limon, I. Novitski, J. P. Ozelis, R. Scanlan, G. Sabbi and A. Zlobin, *IEEE Trans. Appl. Supercond.* **11**, 2172 (2001).

[77] P. Ferracin, S. Bartlett, S. Caspi, D. Dietderich, S. Gourlay, A. Hafalia, C. Hannaford, A. Lietzke, S. Mattafirri, A. Mcinturff, G. Sabbi, *IEEE Trans. Appl. Supercond.* **16**, 378 (2006).

[78] P. Lee, http://magnet.fsu.edu/~lee/plot/plot.htm

[79] T. Kiyoshi, S. Matsumoto, M. Kosuge, M. Yuyama, H. Nagai, F. Matsumoto and H. Wada, *IEEE Trans. Magn.* **12**, 470 (2002).

[80] H. J. ten Kate *et al.*, *IEEE Trans. Magn.* **27**, 1996 (1991).

[81] R. Benjegerdes *et al.*, *Particle Accelerator Conference* (1999), p. 3233.

[82] P. Ferracin, G. Ambrosio, M. Anerella, E. Barzi, S. Caspi, D. Cheng, D. Dietderich, S. Gourlay, A. Hafalia, C. Hannaford, A. Lietzke, A. Nobrega, G. Sabbi, J. Schmalzle, P. Wanderer and A. Zlobin, *IEEE Trans. Appl. Supercond.* **17**, 1023 (2007).

[83] F. Nobrega, A. V. Zlobin, G. Ambrosio, N. Andreev, E. Barzi, R. Bossert, R. Carcagno, S. Feher, V. S. Kashikhin, V. V. Kashikhin, M. J. Lamm, I. Novitski, Yu. Pischalnikov, C. Sylvester, M. Tartaglia, D. Turrioni and R. Yamada, *IEEE Trans. Appl. Supercond.* **17**, 1031 (2007).

Alvin Tollestrup came to Fermilab during the initial design of the Tevatron and helped develop the first superconducting accelerator magnets. He was co-spokesman of the CDF detector during its construction and initial operation and participated in the discovery of the top quark. Lately he helped with the design study for a muon collider. He is grateful for having a career that witnessed the discovery of the pi meson and helped with the discovery of the top!

Ezio Todesco has been working at the University of Bologna (Italy) on beam dynamics in particle accelerators for 10 years, mainly on issues related to perturbative theory of resonances, chaotic motion and long-term stability. Since 1998 he has been working at CERN in the design and construction of the superconducting magnets for the Large Hadron Collider (LHC).

Reviews of Accelerator Science and Technology
Vol. 1 (2008) 211–235
© World Scientific Publishing Company

World Scientific
www.worldscientific.com

Development of Superconducting RF Technology

Takaaki Furuya

KEK–High Energy Accelerator Research Organization,
1-1 Oho, Tsukuba, Ibaraki, 305-0801 Japan
takaaki.furuya@kek.jp

The development of superconducting RF cavity technology started in the 1960s, aiming for continuous wave operation of high gradient linacs with excellent electric power efficiency. After continuous efforts for more than four decades, the initial achievable field gradient of only 2 MV/m was gradually improved and approached the theoretical upper limit of 50 MV/m on Nb cavities. At each stage of this development, many applications have been extended in various fields. This article will review those advances and applications of the superconducting RF cavities.

Keywords: Superconducting cavity; microwave resonator; high gradient linac; high intensity storage ring; crab cavity.

1. Introduction

The development of superconducting RF cavity technology started in the 1960s. Due to the sufficiently low surface resistance of a superconductor, continuous wave (CW) operation at a high acceleration gradient was considered in superconducting (SC) RF linacs, where a high accelerating gradient of more than 30 MV/m was expected by choosing the cavity material, because the achievable accelerating gradient should be determined by the SC critical magnetic field as well as the operating temperature. Hence, the development of SC RF cavities has mainly been carriedout in the field of particle accelerators.

Extremely small surface resistance of a few tens of nano-ohms was, however, easily perturbed and additional residual resistance due to the various surface phenomena, such as surface defects, electron multipacting and field emission, which will be described later, dominated all the characteristics of the cavity performance. As a result, the achievable accelerating gradient was far below the theoretical one and the application of the SC cavities was restricted to a small area of accelerators.

The first effective use an SC cavity was in heavy ion accelerators, where the stable CW field was essential for obtaining the precise ion beams. The required field gradient was not so high in this application but the accelerating structures of various β $(= v/c)$ had to be developed to accelerate slow ion beams. ATLAS at ANL is a typical example of this application which contains 62 resonators of 7 different designs and accepts a range of β from 0.01 to 0.15. The ATLAS complex was commissioned in 1978 [1]. Since then, it has been delivering ion beams for nuclear and atomic research.

In the 1980s, a powerful RF system was required in storage rings for high energy lepton colliders, where a CW operation was needed to compensate for an energy loss due to synchrotron radiation. The worldwide research on a multicell SC cavity improved the gradient to 5 MV/m. From the viewpoint of saving AC power and space for the RF section, this gradient of 5 MV/m was worthy compared to that of the normal conducting (NC) cavity, which has a field limitation of 1–2 MV/m in CW operation due to the RF power loss in the cavity wall.

In 1988, KEK commissioned the TRISTAN SC system of 509 MHz five-cell cavities made of pure Nb sheets, which was the first large scale application of SC cavities [2]. The system had a total effective length of 48 m and provided an accelerating voltage of 200 MV to the electron and positron beams of 16 mA, upgrading the TRISTAN energy from 28 to 32 GeV. The 500 MHz storage ring RF of HERA (DESY) and the 1500 MHz recirculating linac of CEBAF (JLAB) were also commissioned, in 1990 and 1996 respectively [3, 4]. The SC system of HERA including 16 four-cell cavities provided a voltage of 50 MV, delivering a power of 1 MW to the electron beam. The recirculating linac, CEBAF, consists of two SC linacs in which 340 five-cell cavities are used in total. The original gradient of 5 MV/m was improved to 7.5 MV/m and the beam energy upgraded from 4 to 6 GeV, which will be improved to 12 GeV by adding high gradient RF units. The largest SC storage ring RF was commissioned for LEP at CERN in 1995 [5]. The 352 MHz four-cell structure made of Nb-sputtered copper cavities was developed to save the total amount of Nb materials. The SC cavity system, 600 m in length, provided a CW acceleration voltage of 3.5 GV in 1999 [6]. The SC technology for the storage rings was solidly established there.

In the 1990s, high intensity storage ring colliders, so-called "factories," were proposed in which ampere class beams of electrons and positrons collide with each other to attain the high peak luminosity. To accelerate these beams, an RF system with a smaller number of cavities with high accelerating gradient was necessary to reduce the total ring impedance for suppressing the beam instabilities. New, single cell cavities with sufficiently damped higher order modes (HOMs) and power couplers of hundreds of kilowatts were developed at both Cornell University and KEK for their B meson factories, i.e. CESR and KEKB [7, 8]. These cavities were commissioned in 1997 and 1998, respectively. Eight cavities of KEKB have provided an RF voltage of 13 MV to accelerate an electron beam of 1.4 A delivering an RF power of 2.4 MW.

Because of these successful results, the SC damped cavity technology was immediately applied to the mid size storage rings, i.e. the synchrotron light sources. The use of a few SC cavities increases the stored beam current and upgrades the generation of the existing machines. Today, many storage rings with a synchrotron light source have installed or plan to install SC cavities, such as CLS, NSRRC, DIAMOND, SOLEIL, IHEP, SSRF and NSLS-II [9–14].

A short pulse neutron scattering facility, the SNS accelerator complex, consists of a 2.5 MeV front injector system, a 186 MeV NC linac, a 1 GeV SC linac and an accumulation ring. The SC linac accelerates a proton beam in two stages, $\beta = 0.61$ and $\beta = 0.76$. Because of their high beta values, the SC technology developed for the lepton acceleration can be easily extended to these structures. Two kinds of 805 MHz six-cell cavities — 27 cavities of 0.61 and 80 cavities of 0.76 — provide an accelerating gradient of 10–15 MV/m in a pulse mode of 1 ms with a

repetition rate of 60 Hz. Another SC high intensity proton accelerator is LHC at CERN, where two independent sets of SC cavities provide an accelerating voltage of 16 MV to each beam of 7 TeV×0.6 A, corresponding to a gradient of 5.5 MV/m. The single cell cavities of 400 MHz are made from copper on which a thin film of a few microns of Nb is sputtered onto the inner surface. Commissioning has been scheduled for 2008 [15].

Recently the construction of an electron linac, based on SC technology having a high accelerating gradient of 25 MV/m, has started at DESY, namely EURO-XFEL [16]. Commissioning is scheduled for 2013. The purpose of the facility is to generate extremely brilliant and short pulses of spatially coherent x-rays of 0.1 nm. A pulsed beam of 0.7 ms with a repetition rate of 10 Hz is accelerated to 10–20 GeV by the 1.3 GHz SC linac of 1 km. The fundamental technology of this project is based on the one of the international TESLA collaboration (since 1992).

In 2004, the ITRP (International Technology Recommendation Panel) recommended that the main linacs of ILC (International Linear Collider) be designed based on the 1.3 GHz SC technology instead of the x-band NC system. The SC linacs of electrons and positrons include 560 RF units in total, each of which contains 26 nine-cell SC cavities. To achieve the goal of a 200–500 GeV center-of-mass system (cms), the linacs, with a combined length of 23 km, are designed with an average gradient of 31.5 MV/m. A pulse operation of 1.4 ms with a repetition rate of 5 Hz gives an average current of 10 mA. Although the Baseline Configuration Document (BCD) and the Reference Design Report (RDR) of ILC were published in 2005 and 2007 respectively, the accelerating gradient of the main linac is still under discussion.

On the other hand, the development of a CW SC linac with a rather low gradient of 15–20 MV/m is desired for an ERL (energy recovery linac). By returning the accelerated beam into the same accelerating linac through the reverse RF phase of 180°, the kinetic energy of the decelerating beam is transferred to the linac, exciting the electromagnetic field, and is reused to accelerate the fresh beam from the injector. Since the beam itself is not multiturned in the system, a CW beam with sufficiently low emittance can be expected. The original concept was offered by M. Tigner in 1965 [17]. The key technology of the

ERL is a DC photocathode gun as well as the high gradient SC cavity with sufficiently damped HOMs.

In this paper, a brief history of the technology development and some fundamentals on SC cavities are described and the details of the recent applications are presented.

2. Historical Overview

2.1. *High gradient CW linac*

Aiming for a high accelerating gradient in CW mode, fundamental research on the SC cavity started in the 1960s. RF loss in the cavity surface is in proportion to the square of the accelerating gradient and is a few MW per meter at a gradient of 10 MV/m in an NC copper cavity. Therefore the operation of NC cavities is restricted in a very short pulse or in CW mode with a very low gradient of < 2 MV/m so as to keep the loss within a few tens of kW per meter. On the other hand, the surface resistance of an SC cavity is typically five orders of magnitude less than that of an one NC, so that the RF loss drastically reduces to 10 W per meter. Therefore the study and development of the SC cavity technology has mainly been carried out in the field of particle accelerators, aiming to obtain a CW linac with a high accelerating gradient.

The first beam acceleration using an SC cavity was carried out by HEPL at Stanford University in 1964 [18]. W. M. Fairbank, H. A. Schwettman and P. B. Wilson accelerated a 1 μA electron beam using a 2856 MHz SC three-cell cavity with a CW field gradient of 2–3 MV/m. This cavity was made of lead-plated copper. Based on this result, the first SC linac, SCA, was planned at HEPL, in which a beam of 100 μA was to be accelerated to 2 GeV by 24 6-m-long structures of 1.3 GHz with a gradient of 14 MV/m. In 1971, the injector part went into commission. The system, including cryogenics, worked well except for the accelerating gradient, which was limited to 3 MV/m. Extensive measurements showed that the limitation was caused by electron-related phenomena, such as field emission and multipacting. This pioneering work, however, showed much fundamental knowledge and technology of SC cavities and specified the direction of the subsequent research and development. In 1968, J. P. Turneaure and N. T. Viet achieved a peak magnetic field of 1080 Oe using a single cell 8.6 GHz Nb cavity (shown in Fig. 1), which corresponded to an accelerating gradient of

Fig. 1.   8.6 GHz Nb cavity at HEPL [19].

20–30 MV/m [19]. Since then, Nb has been the main material used for SC cavities.

In the 1970s, a lot of research was concentrated on establishing a fabrication procedure to obtain a defect-free Nb surface. However, the maximum field gradient of multicell structures was still 2–3 MV/m. In material research, Nb_3Sn cavities were studied because of their superior SC potential ($T_c = 18.2$ K; $H_c = 5400$ Oe); the Nb_3Sn layer was deposited by vapor diffusion of tin into the Nb surface at 1050°C. However, neither the surface resistance nor the field strength of Nb_3Sn reached the level of pure Nb cavities [20]. On the other hand, U. Klein and D. Proch introduced a spherical cavity shape in 1978 to avoid electron multipacting [21]. The field limitation of 2–3 MV/m had been caused by multipacting discharge in the cavity. They simulated the electron trajectory in the cavity and optimized the cavity shape to avoid one-point multipacting. In a spherical cavity, the electron emitted at the high electric field region easily moves toward the equator within a few RF periods where no electric field exists, and stops without any resonant condition. Since then, the spherical shape has been the standard shape of SC cavities in high gradient applications. This shape has a further advantage of round and smooth curvature, making it easy to wash the inner surface after chemical treatment. The investigations during that period also made it clear that the clean work was important not only for the cavity surface but also for the whole cavity setup. These developments pushed the gradient up to ∼5 MV/m.

2.2. *Heavy ion accelerator*

The most successful application of the SC cavity in the 1970s was a heavy ion linac, where CW operation was essential for obtaining a precise beam,

Fig. 2. Split ring resonators for ATLAS with various beta [22].

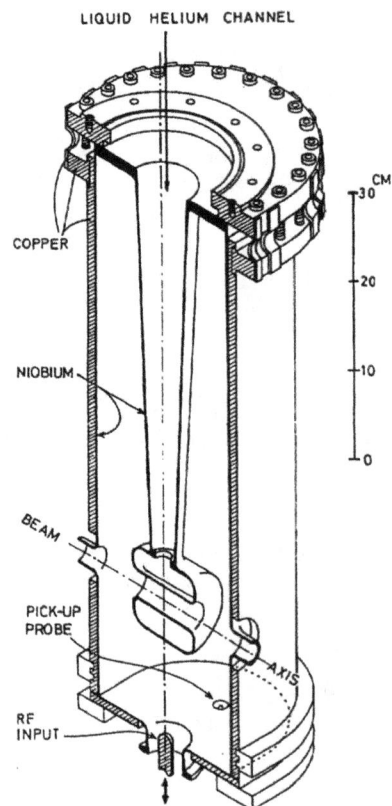

Fig. 3. Quarter-wave resonator at JAEA [24].

even though the achievable field was rather low at 2–3 MV/m. Furthermore, excellent power efficiency was possible, because the cavity loss became comparable to the beam power of several watts. On the other hand, the narrow accelerating gap for lowbeta particles at a high frequency becomes a cause of electric discharge easily at a very high field gradient. Therefore, the structure of low beta at a rather low frequency of 50–350 MHz had to be developed. Figure 2 shows the split ring structures used in ATLAS at ANL [22], which have drift tubes in the cavity to synchronize the RF phase with a beta of 0.06–0.16. Further developments on low beta cavities resulted in the quarter-wave resonator (QWR), shown in Fig. 3. The inner conductor was directly cooled by a liquid He but the outer cylinder, made of Cu/Nb, was cooled by conduction. Because particle velocity changes with beam energy, a variety of structures with a frequency of 99–145 MHz and a beta of 0.01–0.3 were combined in ATLAS. Since the commissioning in 1978, ATLAS has been operated as the first largescale application, and holds the world

record for longest running time. Other SC applications of lowbeta structures in operation are shown in Table 1.

2.3. High energy storage ring

In the 1980s and 1990s, in spite of the fact that the achievable gradient of multicell SC cavities was not higher than 5 MV/m, new application was proposed in a high energy accelerator field. High energy physics needed large lepton colliders using storage rings in which the circulating electrons and positrons lose the energy due to synchrotron

Table 1. Low beta SC systems in operation.

Place	Frequency (MHz)	Cavity	
ANL (ATLAS)	49–855	Nb, split-ring, QWR	$\beta = 0.009$–0.3 since 1978
Stony Brook (SUNY)	150	Pb/Sn, QWR split loop	$\beta = 0.068$–0.1 since 1983
Washington U.	150	Pb/Cu, QWR	$\beta = 0.1$–0.2 since 1987
Florida U	97	Nb, split ring	$\beta = 0.105$ since 1987
Kansas U.	97	Nb, split ring	Since 1990
JAERI	130	Nb, QWR × 46	$\beta = 0.08$ at 3–5 MV/m since 1994
INFN-LNL [23] (ALPI)	80–240	Pb/Cu, Nb, Nb/Cu, QWR × 70	$\beta = 0.056$–0.17 since 1994

radiation. Radiation loss increases as the fourth power of the particle energy as

$$\text{Synchrotron loss} = 0.0885 \frac{W^4}{\rho} (\text{MeV/turn}), \quad (1)$$

for a bending radius of $\rho(\text{m})$, where W is the particle energy in GeV. A powerful RF system in CW mode is required to compensate for this heavy beam loading due to synchrotron radiation. Hence, the development of SC multicell structures for $\beta = 1$ was strongly pushed at many laboratories, such as KEK (TRISTAN), DESY (HERA), CERN (LEP) and Cornell University (CESR). In general, storage rings use rather low frequency to obtain a large beam aperture size. On the other hand, high frequency is advantageous in suppressing multipacting. Therefore, Cornell University developed a noncylindrical cavity of S band, muffin-tin, which had a large opening horizontally to prevent synchrotron light striking the cavity walls, achieved 4 MV/m [25]. At CERN a Nb-sputtered Cu cavity was developed; a few-micron-thick Nb film was deposited by magnetron sputtering on the inner surface of Cu cavities, aiming to increase the thermal stability against local heating due to defects, and to save the cost of Nb materials for a large number of 352 MHz cavities. The improved thermal stability upgraded the gradient to 10 MV/m.

One of the most effective fundamental researches in the 1980s was the improvement of Nb purity, which increased the thermal conductivity of the cavity wall. The Nb ingots are purified using electron beam melting in the vacuum furnace. By repeating the melting process, the gas components of Nb can be reduced substantially. High purity Nb has good thermal conductivity, which increases thermal stability against local heating caused by surface defects, and consequently pushes up the achievable field gradient. The residual resistance ratio (RRR), which is defined as the ratio of electric resistivity between the temperatures 300 K and 4.2 K,

$$\text{RRR} = \frac{R(300\,\text{K})}{R(\text{normal state at } 4.2\,\text{K})}, \quad (2)$$

is usually used as an indicator of Nb purity and gives an approximate thermal conductivity λ at 4.2 K by $\lambda \sim 0.25 \times \text{RRR}$ [26]. Multimelting under precisely controlled vacuum pressure of the furnace improved the RRR of Nb from 40 to 300 in industrial production.

Another purification of Nb is the so-called yttrification or titanification, where Nb sheets or cavities are annealed with Y or Ti in a vacuum furnace at 1200–1400°C to deposit a layer of Y or Ti on the Nb surface, so that the gas components of bulk Nb move and diffuse into this layer. This layer is chemically removed after annealing [28, 29]. Figure 4 shows the thermal conductivity of Nb samples with various RRRs. As a result, the maximum field gradient of multicell structures was improved to 10 MV/m or higher in all laboratories.

In 1980, KEK launched a new development team regarding SC RF for storage rings, headed by Y. Kojima [30]. After fundamental research lasting three years on a 500 MHz single cell cavity, the team designed a new multicell structure with end cavities which had a slightly shorter cell length and a larger aperture size than for other cells, to obtain a flat field profile of accelerating mode. Some asymmetry was also given between the end cells to couple out the HOMs by couplers on the beam pipe. An antenna type 152D coaxial power coupler was developed, which had a ceramic disk and a water-cooled inner conductor so as to deliver a CW power of several tens of kW. For cavity fabrication, a lot of process controls were done in close cooperation with the industry, such as upgrading of the RRR of Nb material

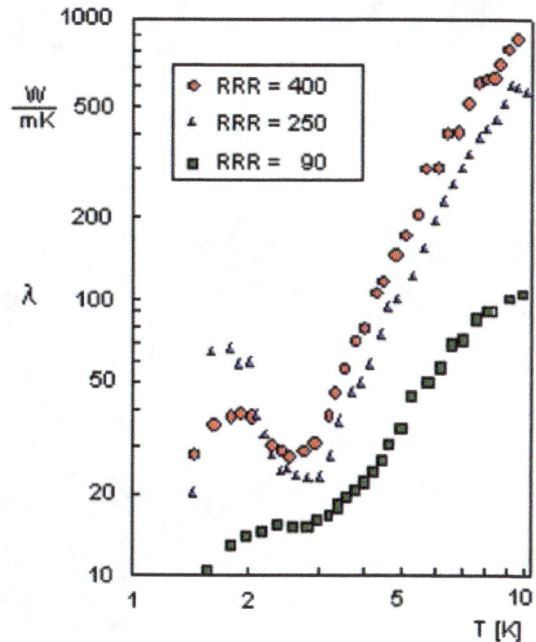

Fig. 4. Thermal conductivity of Nb with various RRRs [27].

and quality control of the forming and welding process. A horizontal rotating electropolishing system was established for the treatment of multicell cavities [31]. Heat treatment of 700°C was carried out in a vacuum furnace to remove the hydrogen that entered the Nb surface during the electropolishing. Thorough rinsing was done after the final electropolishing, using 7000 L of pure and ultrapure water for each cavity. Assembly of all the components around the cavity was done in a clean room of class 100.

In 1988, the SC system of 509 MHz five-cell cavities was commissioned at KEK-TRISTAN. The 32 cavities with a total effective length of 47 m supplied an RF voltage of 200 MV with a gradient of 4–6 MV/m and stored a 14 mA beam. Figures 5 and 6 show the five-cell cavity and the cryomodules in the TRISTAN tunnel. High beta SC systems at HERA

Table 2.   High beta SC systems.

	Tristan (KEK)	HERA (DESY)	LEP-II (CERN)	CEBAF (JLab)
Beam energy	32 GeV	32 GeV	100 GeV	1–6 GeV
Current	14 mA	40 mA	6 mA	100 μA
Frequency	509 MHz	500 MHz	352 MHz	1.5 GHz
Cavity type	Nb	Nb	Cu/Nb	Nb
Number of cavities	Five-cell 32	Four-cell 16	Four-cell 288	Five-cell 330
Effective length (m)	48	20	600	165
Total voltage operation	200 MV 1988–1995	50 MW 1990–2007	3.4 GV 1996–2000	800 MV 1996–

(DESY), LEP (CERN) and CEBAF (Jlab) were also completed and commissioned, as listed in Table 2.

CEBAF (Continuous Electron Beam Accelerator Facility) is an SC recirculating system for nuclear physics research, consisting of two SC linacs with a total length of 165 m. The continuous electron beam is accelerated to 4 GeV after five times of passage through the linacs. Four pairs of five-cell Nb cavities are grouped in one vacuum vessel. The power coupler and HOM coupler of the 1.5 GHz cavity are based on the waveguide type. The original beam energy of 4 GeV supplied by an average gradient of 5 MV/m was improved to 6 GeV and will be up to 12 GeV by adding the high accelerating gradient units and 2 K refrigerator power. Figure 7 shows the RF unit of the CEBAF module.

The 352 MHz four-cell SC system of CERN-LEP was commissioned in 1996. Most cavities were not made of Nb but of copper, sputtered with a thin film of Nb onto the inner surface, aiming to improve the thermal instability against the local heating due to surface defects and to save the material cost of Nb. The SC system of 600 m including 272 cavities upgraded the LEP energy from 45 to 100 GeV,

Fig. 5.   Five-cell cavity for TRISTAN.

Fig. 6.   TRISTAN cryomodules in the tunnel.

Fig. 7.   RF unit of a 1500 MHz CEBAF cavity [4].

Fig. 8. Assembly of the CERN-LEP cavity. The 352 MHz cavity was made of copper with a thin film of Nb. The vacuum vessel was composed of a set of sheets fixed on the frame [5].

Fig. 9. HOM-damped SC cavity for CESR. A 500 MHz single cell cavity has a fluted beam pipe and ferrite dampers. A waveguide power coupler is adopted [34].

providing an accelerating voltage of 3.5 GV with an average gradient of 6 MV/m. Four cavities were grouped in one cryostat. The vacuum vessel of the cryostat was made up of a set of wall sheets fixed on the frame, allowing easy access to the cavity during assembly. This cryostat structure is shown in Fig. 8.

2.4. *High current acceleration*

Besides the energy frontier such as LEP, the brilliance frontier is also the application field of an SC RF system. The superior characteristic of an SC cavity under a high intensity beam is based on the high CW field, high stored energy, and low R/Q of the cavity shape. In the 1990s, precise experiments supported by a "factory machine" were proposed, i.e. the B-meson factory and the τ-charm factory, in which ampere class beams are stored and collide with each other, obtaining a target luminosity of 1–10×10^{33} cm^{-2}s^{-1}. In such rings HOM impedance should be sufficiently damped to avoid the beam instabilities that are mainly excited in RF cavities. Hence, an RF system using a smaller number of SC cavities is advantageous in reducing the total HOM impedance of the ring. Cornell University and KEK developed a single cell HOM-free SC cavity separately for their B-factory storage rings, CESR and KEKB. The two cavities were commissioned in 1997 at CESR and in 1998 at KEKB respectively. In CESR, the current limitation of 0.35 A, which

occurred due to the instability caused by HOMs of NC cavities, was improved to 0.8 A by replacing the NC cavities with four SC cavities [32]. The Cornell cryomodule is shown in Fig. 9. All HOMs in the single cavity can propagate out through a "fluted" beam pipe and be damped by the absorbers located at the room temperature end. An RF power of 200 kW is fed through the waveguide coupler under the cryomodule, in which a ceramic plate isolates the cavity vacuum. In KEKB, eight SC damped cavities have supplied 13 MV, delivering 2.4 MW of RF power to an electron beam of 1.4 A. The maximum beam power of 380 kW has been achieved by each cavity so far [33].

Because of the successful operation of both the Cornell and KEK cavities, the technology of HOM-damped SC cavities has become an attractive way to upgrade mid size storage rings that have a limited RF space in the ring. The synchrotron radiation (SR) rings of CLS (Canada), TLS (1.5 GeV; Taiwan) and DIAMOND (UK) introduced the Cornell type 500 MHz SC damped cavities into their light sources. Cornell University has transferred the cavity technology to a company, so that the Cornell type has become commercially available, with a guaranteed performance of 2 MV with Q of $> 5 \times 10^8$. On the other hand, IHEP in China redesigned the 500 MHz SC cavity based on the KEKB 509 MHz cavity by making a modification — increasing the equator width by 23.7 mm [35]. The newly designed cavities were commissioned in 2006, and achieved the design current of 250 mA immediately, in the SR mode. The final goal of these cavities is a 950 mA beam of each electron and positron in the collision mode as a tau charm factory. The SOLEIL SC cavity is based on the Nb–Cu technology, which was developed for LEP cavities. A couple of rather large cavities of 352 MHz are housed in one cryostat. Parameters of these high intensity colliders are listed in Table 3, and SR rings are in Tables 4 and 5. Besides these rings in operation, new proposals of NSLS-II of BNL and TPS (3 GeV; Taiwan) of NSRRC have been approved, where two or three SC damped cavities are to be installed.

Another recent application of SC cavities is a high intensity proton linac for neutron physics and a nuclear waste transmutation system. The use of SC cavities is attractive for reducing the total length of linacs and the cost, because a high duty or CW beam

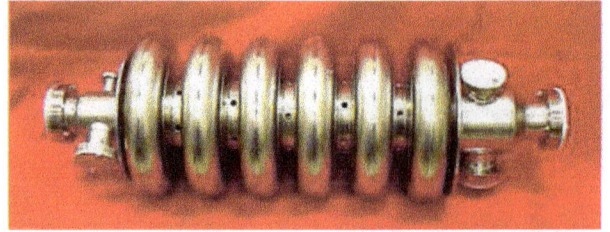

Fig. 10.   SNS six-cell cavity of beta = 0.61 [36].

Table 6.   Low beta cavities of SNS proton linac.

	Low beta	High beta
Beta	0.61	0.78
Energy (MeV)	186–387	387–1000
Operation temperature	2.1–4.5 K	2.1–4.5 K
Number of cavities	33	48
Frequency	805	805
Cavity voltage (MV/m)	10.1	15.9
E_{sp}/E_{acc}	2.71	2.19
R/Q (W)	279	483
Pulse	1.3 ms × 60 Hz	1.3 ms × 60 Hz

of 1–2 GeV is essential for obtaining the highest overall efficiency of the system. The spallation neutron source (SNS) has been completed and commissioned at Oak Ridge, providing the most intense pulsed neutron beams for materials science [36]. The proton accelerator complex consists of a 2.5 MeV front injector system, a 186 MeV NC linac, a 1 GeV SC linac and an accumulation ring. The SC linac accelerates a proton beam of 1.4 mA in two stages, $\beta = 0.61$ (Fig. 10) and $\beta = 0.76$. Two kinds of 805 MHz six-cell cavities — 27 cavities of 0.61 and 80 cavities of 0.76 — an provide accelerating gradient of 10–15 MV/m in a pulse mode of 1 ms with a repetition rate of 60 Hz (Table 6).

Figure 11 shows the operating gradient of all SNS cavities [37]. Not only the individual cavity performance but also the collective effect between the neighboring cavities determined the operation gradients. Then the low beta cavities exceeded the design gradient, but the ones with high beta could not provide the full performance. Further understanding of pulse operation is needed.

One of the most recent topics is a crab cavity at KEK. KEKB collides the electron and positron beams with a finite crossing angle of 22 mrad, as shown in Fig. 12, so as to achieve the minimum bunch spacing of 0.6 m, avoiding the first parasitic collision which is an unnecessary collision with the

Table 3.   SC cavities for high intensity storage rings.

	CESR	KEKB	BEPC-II
Physics	B	B	τ c
Ring	Single	Double	Double
Particles	e^+, e^-	e^+, e^-	e^+, e^-
Energy (GeV)	6	3.5/8	1.5/1.5
RF frequency	500 MHz	509 MHz	500 MHz
Number of cavities	4	8	1+1
Circumference	768 m	3016 m	238 m
RF voltage	5 MV	10–15 MV	1.5
Current (A)	0.8	1.4	0.95

Table 4.   SC-based synchrotron light sources (1).

	Chess	TLS	CLS	Diamond
Number of cavities	4	1	1	2
Energy (GeV)	5.3	1.5	2.5/2.9	3
Current (mA)	500	500	250	300
Frequency (MHz)	500	500	500	500
Cavity voltage (MV)	1.3	1.6	2.4	2.0
Power/coupler (kW)	160	82	245	270

Table 5.   SC-based synchrotron light sources (2).

	SSRF	BEPC-II	SOLEIL
Number of cavities	3	2	4
Energy (GeV)	3.5	2.5	2.75
Current (mA)	300	250	500
Frequency (MHz)	500	500	352
Cavity voltage (MV)	2.0	1.5	1.5
Power/coupler (kW)	250	96	150

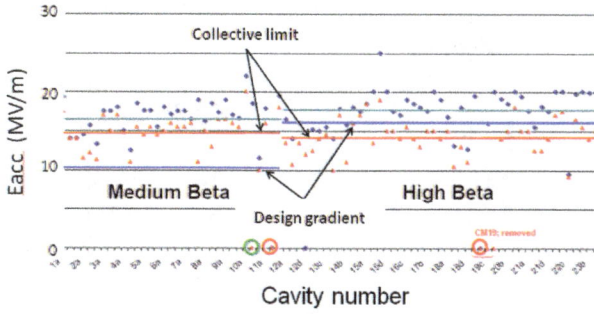

Fig. 11. Accelerating gradients of SNS cavities. The operation voltage of high beta cavities is lower than the design value because of a collective limit [37].

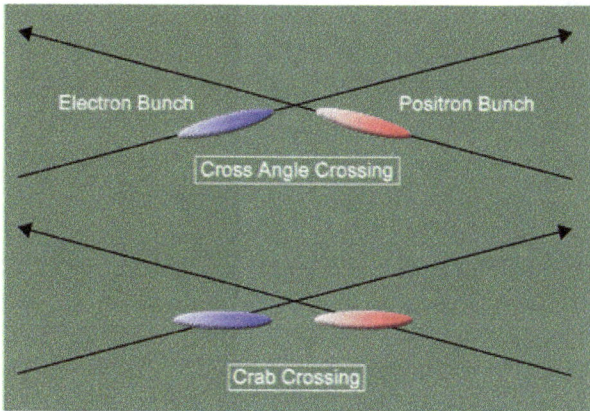

Fig. 12. Crab crossing scheme.

neighboring bunch. However, recent simulation study of crab crossing showed the possibility of luminosity enhancement not only by the geometrical effect but also by a beam–beam effect on the beam size. In the crab crossing scheme, the head and tail of each bunch are horizontally kicked in opposite directions by the crab cavity. Then bunches arrive at the IP, tilting its direction, and make a head-on collision, keeping the crossing angle of 22 mrad. To confirm the effect, two crab cavities were installed in both rings of LER and HER, one cavity for each ring. The kick voltage of each cavity is determined by the beta function and the phase advance of betatron oscillation at the cavity locations, typically 0.9–1.4 MV [38]. This crab kick scheme is useful for obtaining a short bunch length or a short laser pulse in storage rings tilting bunches in the vertical direction using a vertical crab kick.

Although the application of SC RF technology is expanding to various fields, high gradient application has been slow, since the production yield of

Table 7. Parameters of SC-based FELs.

	JAEA	JLab	ELBE	SDALINAC
Frequency (MHz)	500	1500	1300	3000
Energy (MeV)	17	120	40	130
Current (mA)	5 (pulse)	10	1 (pulse)	0.06
Cavity	5-cell	7-cell	9-cell	20-cell
Gradient (MV/m)	5	12	10	6
Commissioning	2000	1998	2004	1990

high gradient cavities is still low. Typical applications of SC linacs are FELs or ERLs, which are listed in Table 7.

Recently, the construction of EURO-XFEL was started in Hamburg, and it will be completed in 2013. The 1.3 GHz 20 GeV SC linac consists of 116 RF units, each of which contains eight TESLA nine-cell cavities. The specified average gradient is 23.6 MV/m, which is based on the long development history and technology experiences at the TESLA test facility (TTF) and an FEL facility, FLASH (Free Electron Laser in Hamburg). Development and information on industrial production of the modules will be a reference for future ILCs and ERLs [39–42].

3. Superconducting Cavity

The whole of the superior property of the SC cavity originates in the small surface resistance of the superconductor. This small resistance allows one to operate SC cavities in the CW mode, and to choose the cavity shape, which is impossible for NC cavities because of a large heating on the surface of the cavity.

3.1. *Shunt impedance*

The shunt impedance R_0, given below, relates the accelerating voltage V_c to the power dissipation P_c in the cavity wall as a Joule loss. To achieve a high accelerating field and good power efficiency, cavities with high shunt impedance are essential:

$$R_0 = \frac{V_c^2}{P_c} = \frac{V_c^2}{\frac{1}{2}\int_s R_s H^2 ds}.$$ (3)

Note that the other expression, $R_0 = V_c^2/2P_c$, is used in lumped circuit theory.

On the other hand, the quality factor Q_0, which is the ratio of stored energy and power loss per RF cycle, is written as

$$Q_0 = \omega \frac{U}{P_c} = \omega \frac{\frac{1}{2}\mu \int_v H^2 dv}{\frac{1}{2}\int_s R_s H^2 ds}.$$ (4)

This can be altered for a cavity, assuming uniform R_s, to

$$Q_0 = \frac{\omega\mu}{R_s}\frac{\int_v H^2 dv}{\int_s H^2 ds}$$

$$= \frac{\Gamma}{R_s}\left(\Gamma = \omega\mu\frac{\int_v H^2 dv}{\int_s H^2 ds}\right), \qquad (5)$$

where the factor Γ, which depends only on the cavity shape and not on the material properties, is a geometrical factor and is useful for obtaining the surface resistance R_s from the measured Q_0.

From these two parameters, the ratio of R_0 and Q_0 is given as

$$\frac{R}{Q} = \frac{V_c^2}{\omega U}. \qquad (6)$$

This R/Q is a figure of merit that relates the accelerating voltage to the stored energy of the cavity and depends on neither the cavity material nor the frequency. From these, the shunt impedance R_0 is written as

$$R_0 = \frac{R}{Q}Q_0 = \frac{R}{Q}\cdot\frac{\Gamma}{R_s}. \qquad (7)$$

The difference between SC and NC cavities is due to the surface resistance R_s. For NC cavities, R_s is given using the DC conductivity σ as

$$R_s = \sqrt{\frac{\omega\mu}{2\sigma}}. \qquad (8)$$

On the other hand, the theoretical surface resistance of SC cavities as given by the BCS theory is

$$R_{s,\mathrm{BCS}} = A\frac{\omega^2}{T}\exp\left(-\frac{\Delta(0)}{k_B T_c}\cdot\frac{T_c}{T}\right), \qquad (9)$$

where k_B is a Boltzmann constant, and Δ and T_c are the SC gap energy and the critical temperature of the cavity material respectively [43]. Coefficient A is a material parameter that includes the penetration depth (λ_{L0}), the coherence length (ξ_0), the Fermi velocity (v_F) and the mean free path. Fundamental properties of typical SC materials are shown in Table 8.

Table 8.   Fundamental properties of typical SC materials.

Material	$T_c(\mathrm{K})$	$\Delta/k_B T_c$	$H_c(\mathrm{Oe})$
Pb	7.2	2.2	800
Nb	9.2	1.9	2000
Nb$_3$Sn	18	2.2	5400
YBaCuO	93	~ 2	$> 10{,}000$

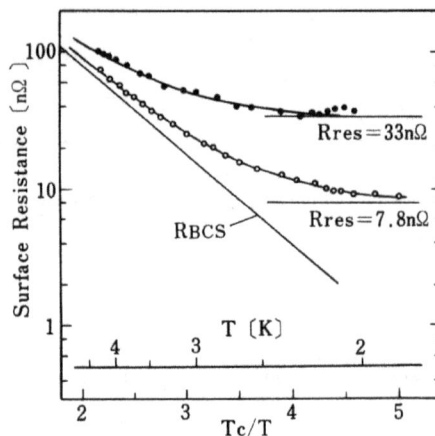

Fig. 13.   Surface resistance of 509 MHz Nb cavities. T_c is 9.25 K.

As a convenient expression, Eq. (9) is approximately written for Nb and $T < T_c/2$ as

$$R_{\mathrm{BCS}} = 10^{-4}\frac{f^2}{T}\exp\left(-\frac{18}{T}\right)\,[\Omega], \qquad (10)$$

where f is the RF frequency in GHz and T is in K. For instance, a 500 MHz copper cavity ($\sigma = 0.58 \times 10^8\,\Omega^{-1}\mathrm{m}^{-1}$) has a surface resistance of 5.8 mΩ, whereas the surface resistance of SC cavities is extremely small and is typically 10–100 nΩ for Nb.

3.2. Surface resistance

From Fig. 13, which shows the surface resistance of a 508 MHz Nb cavity at a low field gradient, it is evident that the real surface resistance can be written as the sum of the theoretical resistance R_{BCS} and an additional term R_{res} which gives the lower limit of the surface resistance. That is,

$$R_s = R_{\mathrm{BCS}} + R_{\mathrm{res}}, \qquad (11)$$

where R_{BCS} is given by Eq. (9). R_{res} is called the residual resistance and becomes dominant in the low temperature region.

3.2.1. Effect of external magnetic field

One of the loss mechanisms is the contribution of flux trapping of the residual magnetic field due to the geomagnetic field or magnetic impurities. A flux trapped on the surface during cooling forms a normal conducting zone of $\pi\xi^2$, where ξ is the coherence length and has the NC resistance R_n given in Eq. (8), instead of the SC resistance R_s. The number of flux-oids N trapped in the area A is proportional to the residual field H_{ext} as

$$N = \frac{\mu_0 H_{ext}}{\Phi_0} \cdot A, \qquad (12)$$

where Φ_0 is a quantum of flux. The critical magnetic field of type II superconductor H_{c2} is given as

$$H_{c2} = \frac{1}{\mu_0 2\pi\xi^2} \cdot \Phi_0. \qquad (13)$$

Thus the additional loss is

$$R_{mag} = \frac{N\pi\xi^2 R_n}{A} = \frac{H_{ext}}{2H_{c2}} \cdot R_n. \qquad (14)$$

The ratio R_n/R_s is roughly $\sim 10^5$ and the additional resistance R_{mag} reaches R_s at $H_{ext} = 2 \times 10^{-5} H_{c2}$. Because R_n has a frequency dependence of $f^{1/2}$, this trapping effect is more serious for higher frequency cavities.

3.2.2. *Q-disease*

Another mechanism is the loss due to hydride, which is called "Q-disease." In 1991, Q-degradation related to the cavity cooling procedure was found at Saclay [44]. Figure 14 shows the cooling procedure and the Q–E plot at each cooling cycle. The results show that the Q degrades more seriously with a longer time of keeping the cavity at 120–170 K. This phenomenon is now understood as the nucleation of Nb-hydride in the surface, which results in a weak superconductor and increases the surface resistance. Hydrogen moves in Nb at temperatures above 60 K and accumulates at nuclear sites under the appropriate concentration of hydrogen. Therefore, rapid cooling in the dangerous temperature region 120–170 K can avoid this degradation. Alternatively, degassing of hydrogen in a vacuum furnace at $> 700°$C is also effective.

Other surface conditions, such as dislocations, impurities, roughness, stress and chemical residue, others, may contribute to R_{res} but their effects are not clear. Although a well-treated surface has an R_{res} of 1–10 nΩ, the resistance increases, typically to 10–100 nΩ at a high field level, because of the additional loss due to heating or the effects of emitted electrons. From the frequency dependence in Eq. (9), R_{BCS} becomes dominant at 4.2 K for >1 GHz and is decreased to a level comparable with R_{res} by cooling the cavity. This is the reason that Nb cavities of high RF frequency are cooled below 2 K, whereas cavities

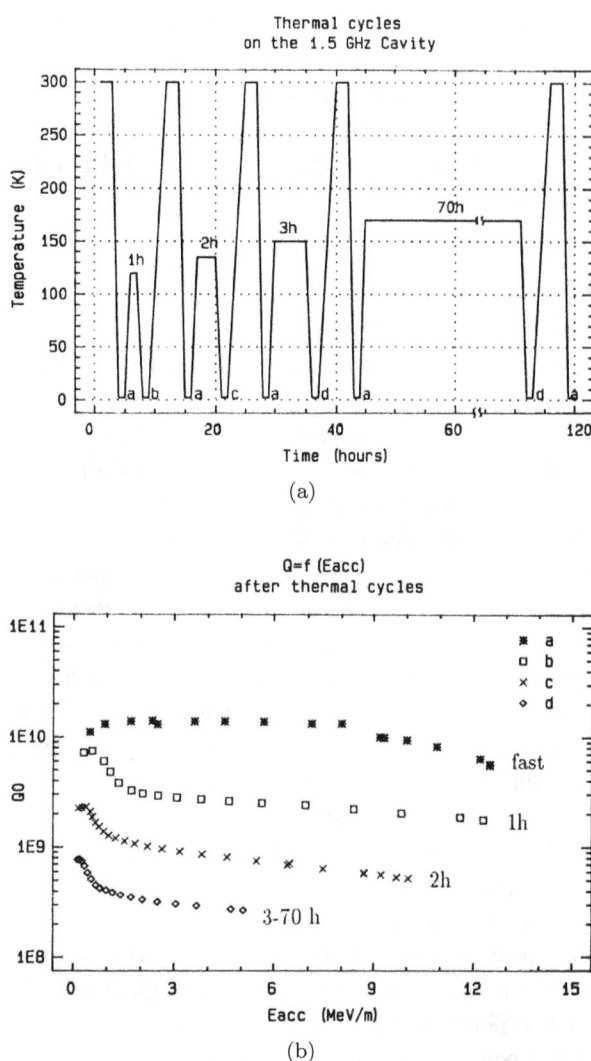

(a)

(b)

Fig. 14. *Q*-disease observed at Saclay. A variety of cooling patterns (a) gave different results for Q–E plots (b) [44].

with a frequency below 1 GHz are usually operated at 4.2 K.

3.3. *Field limitation*

The theoretical limit of the RF field is determined by the superheating critical magnetic field, H_{sh}, which is a little higher than the DC magnetic field, H_c, namely $H_{sh} = 1.2 \, H_c$ for Nb, where $H_c(T) = H_c(0)[1 - (T/T_c)^2]$. Since the typical shape of SC cavities has an H_{sp}/E_{acc} ratio of ~ 40 Oe/(MV/m), a gradient > 50 MV/m is expected for the H_{sh} of Nb cavities of 2400 Oe, and greater for high T_c materials, where H_{sp} is the maximum magnetic field on the cavity surface. However, heating and quenching due to various phenomena on the cavity surface prevent the gradient from reaching this theoretical limit. Up to now, the maximum gradient has been achieved by pure Nb cavities and not by alloys or high T_c materials. Even in Nb cavities, the gradient of 50 MV/m is obtained only by single cell cavities and not by multicell cavities.

3.3.1. *Multipacting*

An electron emitted from the cavity surface is accelerated by the RF field and impacts on the surface again, producing a second generation of electrons. These secondary electrons are also accelerated and produce the third generation at the next impact. If this cycle is continued resonantly and the secondary emission coefficient is larger than unity, the number of electrons increases exponentially. This electron avalanche dissipates the RF power, i.e. degrades the Q suddenly at the resonance level, heats up the impact spot, and eventually leads to SC breakdown. Since this process contains the many parameters of impact energy, the secondary emission coefficient, RF phase and amplitude, the direction of trajectories, and others, prediction of the resonant field levels by a trajectory calculation is difficult. However, a simulation on a simple model shows that resonance happens at narrow discrete field levels, and the onset field level is proportional to the RF frequency for one-point multipacting and quadratically for two-point multipacting. From this point of view, high frequency is advantageous in achieving a high accelerating field.

By using the spherical shape as described previously, one-point multipacting is no longer a serious problem for cavities, but it is still serious for other parts, such as input couplers and HOM couplers.

3.3.2. *Thermal breakdown*

Another field limitation is caused by local heating and quenching due to surface defects. Consider a half-sphere of NC metal ($2a$ in diameter) embedded in an SC surface of thickness d (see Fig. 15). Power dissipation at the defect is

$$\dot{Q} = \frac{1}{2} R_n H^2 \pi a^2, \qquad (15)$$

where R_n is the surface resistance of the NC metal given in Eq. (8). Under the assumption of $a \ll d$, the

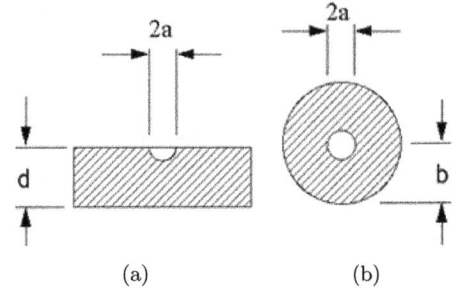

Fig. 15.   Half-sphere defect embedded in the SC surface (a), and a spherical defect surrounded by a sphere of SC metal (b).

model is equivalent to a spherical defect surrounded by a sphere of SC metal of diameter $2b$, as shown in Fig. 15. Here, we can assume that $a \ll b$. Then the spherical defect dissipates power of $2\dot{Q}$, and heat flow to the SC metal of thermal conductivity κ is

$$-4\pi r^2 \kappa \frac{dT}{dr} = 2\dot{Q}. \qquad (16)$$

By integrating from a to b,

$$\int_a^b \frac{dr}{r^2} = -\frac{2\pi\kappa}{\dot{Q}} \int_{T_a}^{T_b} dT, \qquad (17)$$

where T_a and T_b are the temperature at the defect surface and at the outer surface of the SC metal cooled by liquid He, respectively. Since $a \ll b$,

$$\frac{1}{a} = \frac{2\pi\kappa (T_b - T_a)}{\dot{Q}} \qquad (18)$$

Using Eq. (15),

$$H = \sqrt{\frac{4\kappa(T_a - T_b)}{a R_n}} \qquad (19)$$

As the temperature of the defect reaches T_c, i.e. quenching, H reaches its maximum,

$$H_{\max} = \sqrt{\frac{4\kappa(T_c - T_b)}{a R_n}} \qquad (20)$$

As an example, an impurity of $a = 100\,\mu$m and $R_n = 10\,\text{m}\Omega$ on the Nb surface with $\kappa = 50\,\text{W/m} \cdot \text{K}$ (this K corresponds to Nb of RRR $= 200$) and $T_c = 9.2\,\text{K}$ gives $H_{\max} = 3.2 \times 10^4$ A/m at $T_b = 4.2$ K. In other words, a copper particle on the Nb surface of RRR $= 200$ limits the maximum gradient to 10 MV/m at 4.2 K if the particle is located at the peak magnetic field $H_{\text{sp}} = H_{\max}$. The power dissipation of only 0.16 W is enough to limit the cavity field.

Figure 16 shows the Q–E of a 509 MHz Nb cavity at 4.2 K. At the first cooling, the cavity quenched

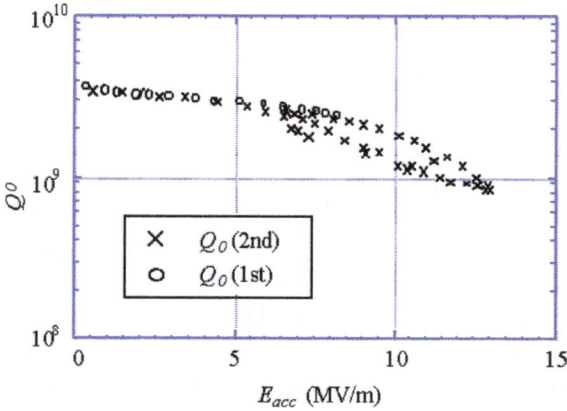

Fig. 16. *Q–E* plot of a 509 MHz single cell cavity.

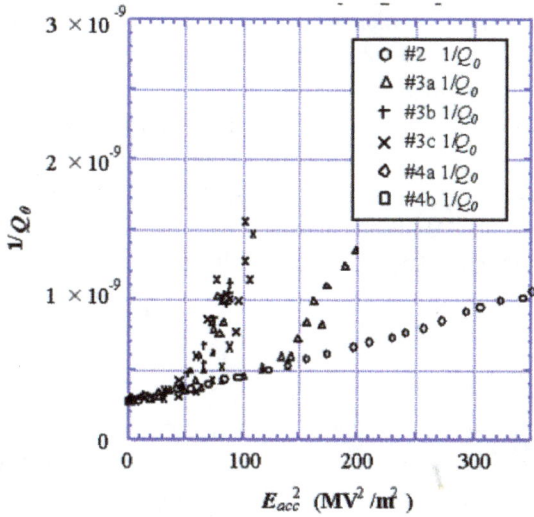

Fig. 17. $1/Q$ vs. E^2 of 509 MHz single cell cavities.

at 8 MV/m, showing local heating at the equator. After grinding off a 2-mm-long, 0.2-mm-deep defect and slight electropolishing, the maximum field was improved to 13 MV/m.

3.3.3. *Field emission*

Field emission is also a limitation due to emitted electrons in the cavity. Electron multipacting increases the number of electrons in a resonance process, but the increase follows the Fowler–Nordheim law [45] in this limitation. The field emission current from metal surfaces is described as

$$I \propto (\beta E)^{2.5} \exp\left(\frac{-\beta \phi^{1.5}}{\beta E}\right) \quad (21)$$

in the RF field, where E is the surface electric field, ϕ is the work function, and β is an enhancement factor

that may be related to the emitter shape, although this is not clear. The electron current increases exponentially as the field strength and degrades Q_0 by dissipating the RF power (electron loading) and consequently heating the cavity wall. This phenomenon becomes serious at a higher field level, and is dominant in the limitation of the recent high field cavities. Figure 17 is a modified Q–E plot of $1/Q$ vs. E^2, where the vertical axis is the power loss. The loss proportional to E^2 is understood to be the loss related to the resistance, and the exponential increase is due to electron loading.

4. Cavity Fabrication

In the previous sections, we showed the important factors for realizing high performance SC cavities, namely magnetic shielding, high thermal conductivity of the cavity material, and a defect-free surface. An impurity of 100 μm causes a heating spot and becomes a source of emitted electrons, which degrade the achievable field gradient seriously. However, in most cases the defects cannot be detected by the naked eye. Thus we have to concentrate our efforts on keeping the surface clean.

4.1. *Cavity shape*

In the design of NC cavities, the cell shape is optimized to obtain the highest R/Q or shunt impedance of R_0 to minimize the wall loss, so that an aperture with a small diameter is chosen together with a nose cone. On the other hand, in an SC cavity for a high beta structure, a simple spherical shape with a large aperture size is optimal at the operating frequency, because it avoids multipacting and makes it easy to polish and wash during the surface treatments. This shape decreases the R/Q by 1/2–1/3 compared to that of NC cavities, but the advantage of easy fabrication guarantees a high R_0 by realizing a high Q value. In the optimization process, the field ratios E_{sp}/E_{acc} and H_{sp}/E_{acc} must be kept as low as possible; E_{sp} and H_{sp} are the maximum electric and magnetic fields on the cavity surface.

4.2. *Fabrication*

It is difficult to define a "standard" procedure for fabrication processes; each laboratory has established its own. Typical fabrication steps based on electropolishing are as follows:

- Surface inspection and cutting Nb sheets of RRR 150–200.
- Forming half-cells by spinning or hydroforming.
- Mechanical polishing and dipping in a HCl bath to eliminate metal impurities; rust checking by dipping in a water bath.
- Electron beam welding at the equator and the irises.
- Grinding off the inner welding seams, on which impurities or microcracks might be left.
- Electropolishing or chemical polishing which removes $80\,\mu$m of the surface.
- Heat treatment at 700°C for 1.5 h to reduce the surface stress and hydrogen contamination during electropolishing.
- Frequency tuning by giving an inelastic deformation to the cavity length.
- Final, slight electropolishing to remove $10\,\mu$m.
- Rinsing with pure and ultrapure water, including high pressure rinsing of 4–11 MPa.
- Assemby in a CLASS 10–100 clean room.

Improvement and development of these processes are being continued to achieve a higher field gradient, to improve the reliability, and to lower fabrication costs.

4.2.1. Nb material

High purity of Nb is desired, but it lowers the mechanical strength. Typically, the yield strength of annealed Nb sheets decreases to ~ 40 MPa. Thus careful structural analysis of the cavity shape and choice of material thickness is needed. Rather deep scratches on the material surface are mechanically ground off one by one, because impurities might be embedded during the sheet rolling process and hide behind the scratches. An eddy current scanning system was developed to check the materials for TESLA cavities, by which iron inclusions in the surface were successfully detected [46]. Recently, single cavities made up of Nb sheets of single crystal or large grain size have been tested, aiming to reduce the grain boundary zone, which contains a lot of impurities. Furthermore, recent technology progress of the Nb purification increases the size of pure Nb ingots which have a sufficiently large diameter for forming L-band half-cells. This is advantageous in

eliminating the sheet rolling process and reducing the material cost.

4.2.2. Welding

The most dangerous process in cavity fabrication is welding, because many impurities may melt into the welding seams. Therefore only electron beam (EB) welding is used to weld the Nb in order to get defect-free welding seams. The optimum condition for making a smooth and nonsputtered seam is sensitive to the welding parameters of voltage, electron beam current, moving speed of the beam, and the tolerance of the welding materials. In general the tolerance of mismatching between the materials should be within 10% of thickness. From this point of view, the forming process of the half-cells is the most important for realizing the high gradient cavities. Usually the welding surface is smoother on the beam side than on the opposite side. Hence welding from the inside using a small electron gun is desired. R&D of a seamless multicell cavity which is formed from a seamless Nb tube by spinning is being carried out at various laboratories.

4.2.3. Surface treatment

In chemical polishing, a 1:1 mixture of concentrated hydrofluoric acid and nitric acid is used. For a large cavity, phosphoric acid is added to slow down the reaction speed. For a small cavity, cooled acid is also useful for reducing the reaction speed. In electropolishing, the Nb surface is polished in a mixture of concentrated sulfuric acid and hydrofluoric acid in a volume ratio of 85:10 at a DC voltage of 25–30 V between the cavity and the cathode inserted in the cavity, keeping a current density of ~ 50 mA/cm^2. In contrast to the orange-peel-like surface due to chemical polishing, a mirror-like surface is obtained by well-controlled electropolishing, which seems favorable for a high field gradient.

4.2.4. HPR

High pressure rinsing (HPR) is indispensable in the final process of high gradient cavity production [47]. Ultrapure water of 18 MΩ·cm (at 25°C) is pressurized to 4–11 MPa and sprayed through nozzles 0.4–0.6 mm in diameter to remove chemical residuals and dust particles from the cavity surface. Although this process is effective, especially in suppresing the field emission, the standard parameters of the HPR system, such as water pressure, flow of water, size and number of nozzles, and nozzle shape, have not been fixed yet [48].

5. SC Cavity for High Intensity Storage Ring

In the 1990s, high intensity beams were required for various fields. One of the most recent problems was a luminosity frontier in high energy physics, such as the study of CP violation and topics relevant to B-meson decays. To support this research program, storage ring colliders with ampere class beams were the solution for obtaining the high luminosity of $> 10^{34}$ cm^{-2}s^{-1}, where a powerful CW RF system with sufficiently lowered HOM impedance was needed to avoid the beam instabilities caused by high intensity beams. For this purpose, SC damped cavities have been developed at Cornell University and KEK. The wonderful results of both cavities demonstrated that this technology was also useful in medium-sized rings with a synchrotron light source. Today many laboratories have already applied or plan to apply this SC technology.

5.1. *SC cavities*

In general, RF cavities are the main source of the ring impedance which causes the beam instabilities and additional RF losses. Because of a high accelerating gradient, the use of a single cell SC cavity is advantageous in reducing the number of RF cavities and HOM modes, and consequently the total ring impedance and a required RF power. Another advantage of SC cavities is the cavity shape with a large beam aperture, where the HOMs can propagate easily out of the cavity. This shape has a rather low R/Q but is not serious for an SC cavity because of its high Q value.

From the viewpoint of beam–cavity interaction, the resonance frequency of the cavity has to be lowered as beam intensity increases in order to minimize the generator power. The amount of this frequency shift is proportional to the beam intensity as

$$\Delta f_0 = -\frac{I_b f_0}{2V_c} \left(\frac{R}{Q}\right) \sin \phi_s, \qquad (22)$$

where f_0, I_b, V_c and ϕ_s are the resonant frequency, the total beam current, the accelerating voltage and the synchronous phase. This frequency detuning has to be kept within the revolution frequency f_{rev}, to avoid the instability excited by the accelerating mode. In a large storage ring, the revolution frequency becomes small and gives the limitation of the frequency detuning. From this point of view, high voltage with low R/Q of an SC cavity is very advantageous.

Empty buckets between the bunch trains cause the phase oscillation of the accelerating voltage, and consequently modulate the longitudinal bunch position. This phase oscillation, $\Delta\phi$ is

$$\Delta\phi \propto \frac{\pi f_0}{V_c} \left(\frac{R}{Q}\right) I_b T_{\text{gap}}, \qquad (23)$$

where T_{gap} is the duration of the bunch gap. It is evident that the SC cavity with a high voltage and low R/Q is advantageous also in this phase oscillation.

5.2. *Cavity module*

Figure 18 is a drawing of the 509 MHz SC cavity for KEKB. In order to reduce the input coupler power and the number of HOM modes, a single cell structure was adopted. The large diameter of the beam aperture was optimized to obtain sufficient coupling of the beam pipe for the monopole modes of TM$_{011}$ and TM$_{020}$. A 300-mm-diameter cylindrical beam pipe is attached on one side to extract the lowest dipole modes, TE$_{111}$ and TM$_{110}$. On the other hand, the Cornell cavity adopted a fluted beam pipe, as shown in Fig. 9. In both schemes, all HOMs propagate out of the cavity along the beam pipes and are damped by HOM absorbers located outside the cryostat. This simple damping system is possible only for SC cavities, because in NC cavities the rather low R/Q of this shape causes serious power dissipation on the cavity wall. In SC cavities, the shunt impedance is kept sufficiently high by the high Q value. Figure 19 shows the cavity performance measured in the acceptance tests of the KEKB cavities.

5.2.1. *HOM impedance*

To avoid multibunch instabilities, the growth time (= 1/growth rate) of the fastest-growing multibunch beam mode due to each HOM must be longer than the energy damping time τ of the

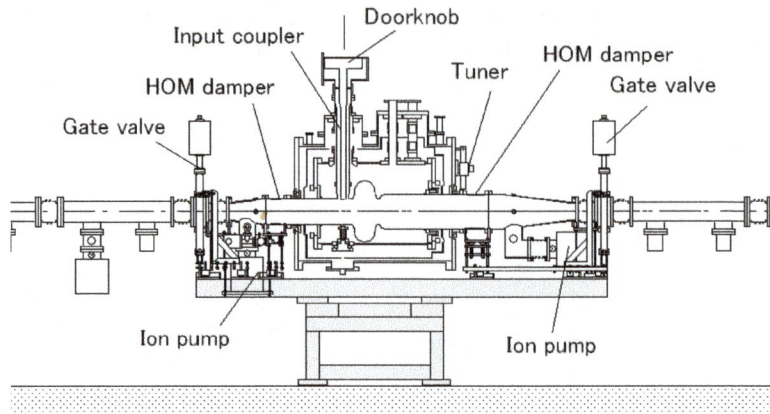

Fig. 18. Schematic drawing of the KEKB SC damped cavity. A large bema pipe 300 mm in diameter is designed to take out the lowest HoM. A coaxial antenna type power coupler can deliver a power of 400 kW to the beam. The inner conductor is cooled by water. To reduce the broad-band HOM, long taper sections are on both sides.

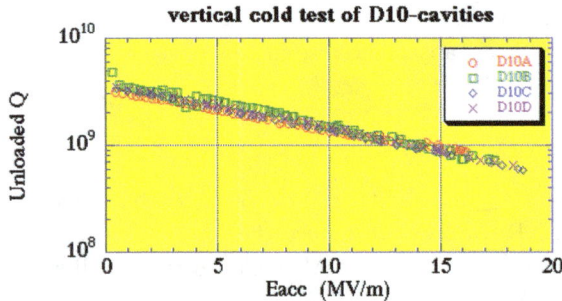

Fig. 19. Vertical test results of KEKB cavities of the D10 site.

ring. For longitudinal modes, the threshold of the impedance is

$$N_{\text{cav}} \cdot R_{\text{long}} < \frac{4 \cdot \tau_{\text{long}}^{-1} \cdot \nu_s \cdot E/e}{I_b \cdot f_{\text{HOM}} \cdot \alpha}. \quad (24)$$

In this equation, N_{cav} is the number of cavities, E/e the beam energy in V, ν_s is the syncrotron tune, α the momentum compaction factor, and I_b the beam current in A. To achieve this limitation, all HOMs have to be reduced its R/Q or the Q value.

5.2.2. Ferrite damper

The damping characteristics of absorbers depend strongly on the geometrical parameters, such as the distance from the cavity, the length and the thickness of the ferrite, and the tapering between the dampers and the beam duct. Figure 20 shows that the Q of TM_{011} depends strongly on the ferrite thickness and the distance from the cavity. These parameters can be simulated by the computer codes MAFIA,

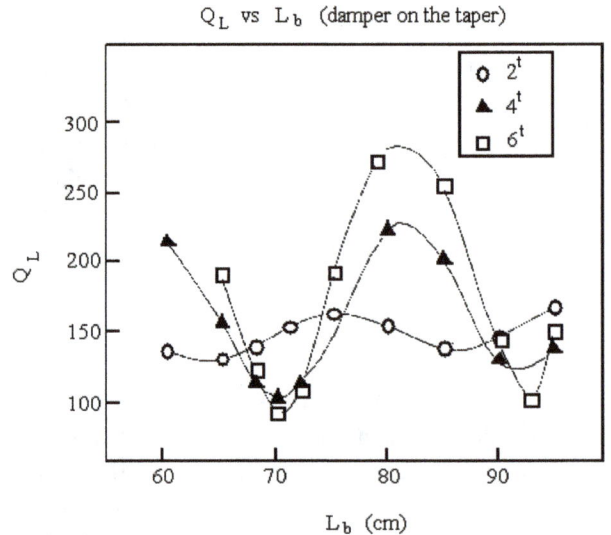

Fig. 20. The Q of TM011 for the ferrites 2, 4 and 6 mm in thickness. The horizontal axis is the distance between the ferrite damper and the cavity.

SEAFISH, HFSS and so on [49–51]. Figure 21 shows the ferrite damper the using the HIP (hot isostatic pressing) process. The powder of ferrite is packed into an iron vessel with a copper cylinder. After being evacuated, the iron vessel is heated to 900°C under a pressure of 1500 bar in a furnace. In the vessel the ferrite is sintered and bonded on to the copper cylindrical wall simultaneously under vacuum [52]. After polishing the ferrite to 4 mm in thickness, vacuum flanges are electron-beam-welded. Every damper is power-tested to 15 kW, monitoring the absorbed power and surface temperature. Figure 21 shows the damper of a Cornell cavity also.

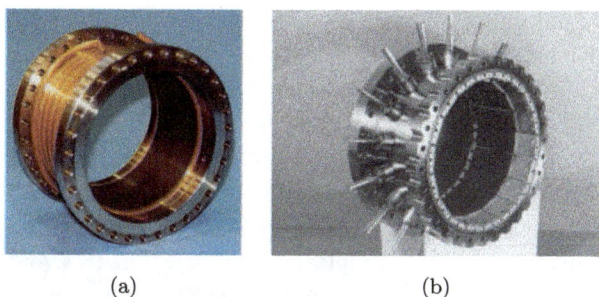

(a)   (b)

Fig. 21.   Ferrite HOM damper; KEKB (a) and CESR-B (b).

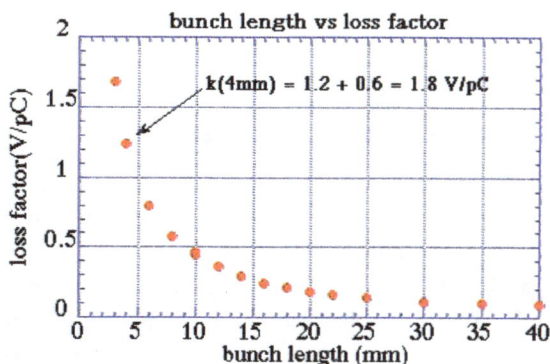

Fig. 22.   Loss factor k vs. bunch length of KEKB cavity.

Small ferrite tiles are brazed on the 19 copper plates and are assembled in a vacuum chamber. Each plate is power-tested to 600 W.

5.2.3. *Broad-band HOM*

To estimate the heat loading of a ferrite damper, one has to consider broad-band HOMs, which become dominant for a beam of short bunch length.

A charge q leaves a power of $k_m q^2$ for the mth HOM, where k_m is called the loss factor and has a unit of volt/coulomb. Thus k_m has to be summed up over the Fourier components of the bunch profile to obtain the total loss factor k. The total power loss is given by using k and the number of injecting bunches per second, i.e. I_b/q:

$$P_{\text{total}} = k \cdot q \cdot I_b \cdot 10^{12} \qquad (25)$$

where k is in V/pC, q is the bunch charge in C and I_b is the beam current in A.

Figure 22 shows the simulation results of the KEKB cavity using the ABCI code [53]. Although the loss factor of a cavity part is only 0.66 V/pC, it becomes 1.8 V/pC in total for a bunch length of 4 mm, since the loss factors of the tapers on both sides and the ferrite dampers themselves must be added. A beam of 1.1 A with a bunch charge of 8 nC creates an HOM load of 16 kW for each cavity module. An additional loss due to the trapped HOM of ~ 1 kW should also be of concern. At Cornell, the cavities are connected to each other by 240 mm beam ducts to reduce the number of taper sections. It should be mentioned that k increases rapidly for a bunch length below 1 cm (Fig. 22).

5.2.4. *Power coupler*

Acceleration of a high intensity beam with a smaller number of cavities imposes heavy loading on the input couplers. Actually, several hundred kW of power has to be supplied to the cavity in B-factory machines.

Two types of input couplers are available for high current SC cavities: coaxial antenna type couplers of KEKB and SOLEIL, and waveguide couplers of Cornell cavities. Coaxial type couplers have the following characteristics, which differ from those of waveguide couplers:

- Coupler size is determined by the desired RF power level and not by the frequency.
- Variable coupling strength is possible by changing the penetration of the inner conductor into the beam pipe.
- DC voltage is available between the inner and the outer conductor to suppress multipacting.
- A complex cooling system of inner and outer conductors.

The power coupler of KEKB has a 1-cm-thick alumina ceramic disk which is located $\sim 1.75\,\lambda$ way from the cavity side, so that the ceramic is not positioned at the maximum field both on and off resonance. Three signals near the window — vacuum pressure, emitted electrons and arcing — are used to protect the ceramic disk. The outer conductor, which is made of stainless steel with 30 μm copper plating, is cooled by a He gas flow. On the other hand, the inner conductor is cooled by water. A CW power of 800 kW was given to the coupler at a room temperature test stand, and no problem was observed [54]. DC voltage between inner and outer conductors is useful not only for suppressing multipacting but also for processing the coupler at room temperature under full reflection by providing the various resonance conditions of multipacting purposefully.

5.3. *Operation of high intensity SC cavities*

KEKB is an asymmetric electron–positron collider using 8 GeV (HER) and 3.5 GeV (LER) storage rings. The SC cavities were installed in HER with 12 NC cavities sharing the accelerating voltage and beam loading by giving an offset to the RF phase between the SC and NC cavities. The maximum stored beam of HER reached 1.4 A, with a bunch charge of 8 nC providing the peak luminosity of 1.7 $\times 10^{34}$ cm^{-2}s^{-1} [55]. Each cavity delivers a power of 350 kW to the beam with a gradient of 5 MV/m, and an HOM of 16 kW is absorbed by a pair of ferrite dampers. The RF trip rate is about 0.5/day for eight cavities.

Today, six SC-based storage rings for the light source are in operation: CESR-CHESS, CLS, TLS, SOLEIL, DIAOND and BEPC-II. Fourteen SC cavities of three different types provide a gradient of 4–8 MV/m, delivering an RF power of 80–270 kW to beams of 250–500 mA. The CLS cavity delivered a power of 270 kW to a 300 mA beam. The recent trip rate of the TLS cavity is 0.5 per week. The power couplers of SOLEIL cavities were conditioned to 200 kW under full reflection and the gradient reached 2.5 MV in both cavities. The great success of the SC cavity application to the synchrotron light source is largely based on the pioneering work of Cornell and KEK demonstrating the following advantages of the SC cavity: a high gradient of CW to reduce the number of cavities, large beam pipes with beam-line dampers for a simple and sufficient HOM-damping scheme, and power couplers of hundreds of kW CW.

6. Application to RF Deflection, Crab Cavity

The first beam deflection using the SC cavity appeared in 1977, which was the RF separator of 2856 MHz at CERN [56]. In 1988, new beam deflection of crab crossing was proposed by R. Palmer to obtain a substantial head-on collision of the electron and positron beams, keeping a finite crossing angle at the colliding point in linear colliders [57]. K. Oide and K. Yokoya proposed the use of this scheme in storage ring colliders [58]. Furthermore, A. Zohlents showed the possibility of obtaining sub-picosecond x-ray pulses by tilting the long bunches using a vertical crab kick [59]. Therefore, a crab cavity is useful not only in high energy accelerators but also in light sources, and technical establishment is strongly expected.

KEKB has the finite crossing angle of 22 mrad at the collision point. This crossing scheme has the advantage of the simple beam separation to avoid the parasitic collision, keeping the shortest bunch spacing. Besides the geometrical effect, however, the simulation of the head-on collision with the crossing angle, the so-called crab crossing, predicted a luminosity enhancement by a factor of 2 [60]. Therefore new crab crossing of the single cavity scheme was proposed, in which one crab cavity is located at any place in each ring and provides the kick voltage determined by the beta function and the phase of betatron oscillation at the cavity location. In this scheme, the tilt of bunches is not localized at the colliding point but propagates around the whole ring. The crab cavity provides a kick voltage of 1.4 MV in CW but basically delivers no RF power to the beam.

6.1. *Crab cavity*

Conceptual design of the KEKB crab cavity followed the idea that was given by K. Akai *et al.*, in 1992 [61]. The beam bunches are kicked in the horizontal by the deflecting mode of TM$_{110}$, which has a resonant frequency of 509 MHz. By choosing a nonaxially symmetric shape, the frequency of the degenerate mode is pushed up to 700 MHz so that the mode can propagate out through the large beam pipe toward the HOM absorber located on the room temperature side. Another trapped mode of TM$_{010}$, which has the lowest frequency of 410 MHz, is coupled out by a long coaxial coupler inserted into the beam pipe. This coupler also works as a frequency tuner by changing the position on the beam axis. A schematic view is shown in Fig. 23.

Because of the large cavity size of about 1m × 0.5 m, the cell shape was carefully determined from the viewpoint of both RF performance and the mechanical property. The half-cells made up of Nb sheets with an RRR of 200 were electron-beam-welded and barrel-polished, removing the inner welding seam by 100 μm. After electropolishing of 100 μm followed by annealing of 750°C, the final electropolishing and high pressure rinsing was performed.

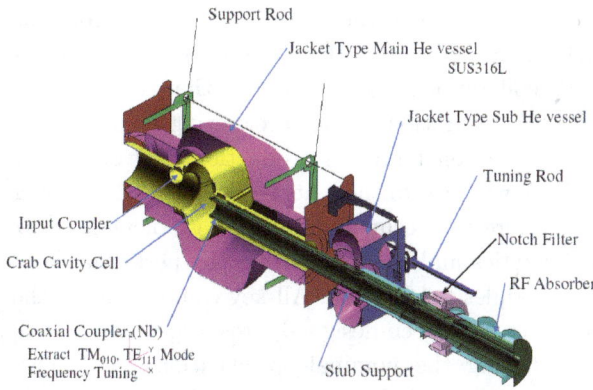

Fig. 23. Cross-sectional view of the KEKB CRAB cavity. The lowest mode can be extracted through a coaxial coupler.

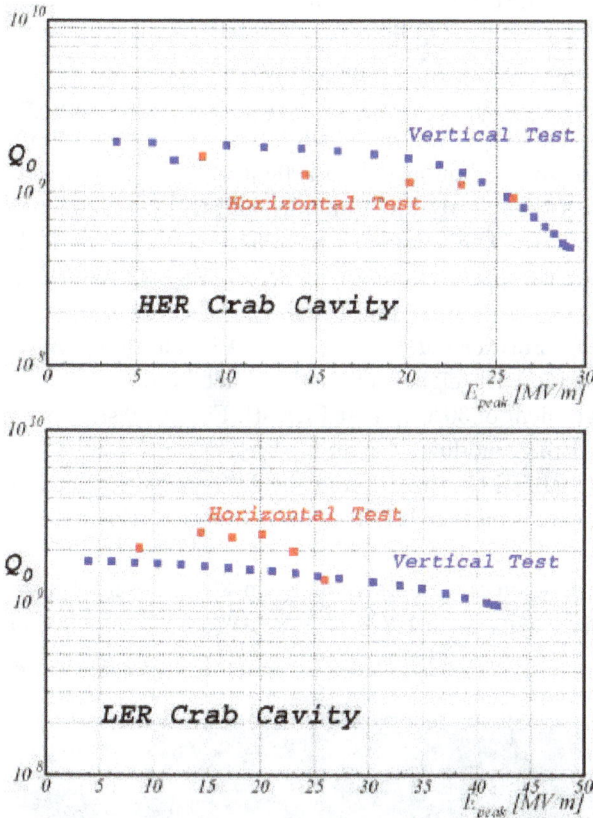

Fig. 24. Cold test results of KEKB-CRAB cavities; cavity tests in a vertical cryostat (blue) and full-assembl tests (red).

Figure 24 shows the cavity performance in the vertical cold tests [62].

6.2. *Optimization of the coaxial coupler*

The structure of the coaxial coupler is somewhat complicated. It includes four parts: a superconducting coaxial tip, a stub support with the He cooling channel at the middle, and a notch filter and a ferrite damper on the room temperature side. All monopole modes in the cavity propagate out through the coupler and are damped by the ferrite absorber. The diameter of the coaxial tip has a dipole cutoff frequency of 600 MHz, which has an attenuation of 60 dB/m for the crab mode. Therefore, a long coupler of 1 m is required for the crab mode not to be damped by the HOM absorber. To support such a long conductor against mechanical vibration, a stub support is located at the middle of the coupler. Although the frequency of the crab mode is lower than the dipole cutoff, this mode can easily pass through the coupler and reach the ferrite absorber, if the conductor is misaligned from the central axis. Therefore, an RF filer is attached to reflect the mode back to the cavity.

6.3. *Commissioning*

Two crab cavities were installed in 2007 (Fig. 25) [37]. The performance of the cavities and the effect of the crab kick were carefully confirmed with beams of a small current. The tilt of bunches observed by streak cameras as shown in Fig. 26 agreed well with the calculated results.

After the fundamental research on the crab crossing, high luminosity operation has been continued with crab crossing. In collision, high specific luminosity at low beam current agreed well with the simulation. Figure 27 shows the specific luminosity per bunch as a function of the product of bunch currents. However, the reason for the rapid fall with the increase in the current is still not known.

Fig. 25. Crab cavity in the KEKB tunnel.

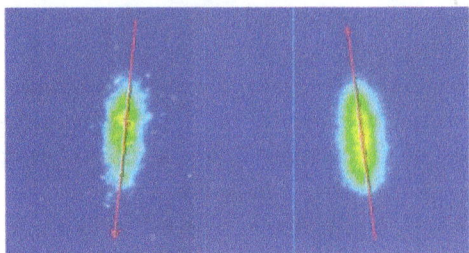

Fig. 26. The crab kick was confirmed by a streak camera. The kick angle agreed well with the calculated one.

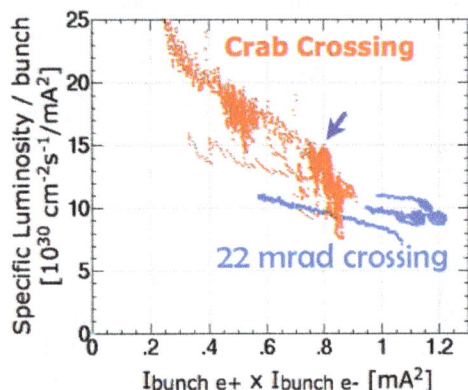

Fig. 27. Normalized specific luminosity at low current agrees with the simulation, i.e. double the infinite crossing. The reason for the rapid drop as the beam intensity increases is not known yet.

Table 9. Operation parameters of the KEKB crab cavities the second round.

	LER	HER	Unit
Beam current	1600	950	mA
Number of bunches	1389	1389	
Crab voltage (max.)	1.1	1.8	MV
Crab voltage (operation)	0.9	1.45	MV
Phase stability	0.1	0.1	deg
HOM power	11.5	12.0	kW
R/Q	46.7	46.7	W
E_{sp}/V_{kick}	14.4	14.4	MV/m/MV
Q_L	2.0	1.6	$\times 10^5$

In the commissioning of one year, the crab cavities have worked well, providing a stable kick voltage. The main parameters achieved by both cavities are shown in Table 9.

7. Application to High Gradient Linacs, ILC and ERL

7.1. International Linear Collider (ILC)

The TeV Energy Superconducting Linear Accelerator (TESLA) collaboration was created in 1989. In the following year the first TESLA workshop was held at Cornell University. The purpose of this worldwide collaboration was to discuss the parameters of TESLA, limitation of the accelerating gradient, and cost reduction for mass production. A test facility (TTF) was established at DESY in 1992, aiming to explore the technology of high gradient 1.3 GHz SC cavities and to demonstrate the performance of SC modules for TESLA. All key components of the modules have been designed, prototyped and tested, establishing the nominal specifications. The optimized nine-cell TESLA cavity is shown in Fig. 28 and the cavity parameters are listed in Table 10. The results of these researches were summarized as the TESLA Design Report (TDR) in 2001, concluding with the SC linacs of 500 GeV cms with a combined length of 33 km and an XFEL facility as an option [63]. This option developed into the project of EURO-XFEL, on which a technical design report (TDR) was published in 2006 [64]. The construction has been started in Hamburg and will be completed in 2013. The accelerating gradient measured at TTF for the past ten years, shown in Fig. 29, indicates that the desired gradient of 23.6 MV/m is promising. Furthermore, recent TESLA modules equipped inside with eight cavities have achieved an average gradient of 30 MV/m at FLASH. Figure 30 shows the TESLA module of eight cavities.

In ILC, the International Technology Recommendation Panel (ITRP) evaluated the TESLA TDR and concluded that the ILC linacs were feasible of based on the SC technology in 2004. Even with an SC cavity with a Q_0 of 1×10^{10}, a CW operation

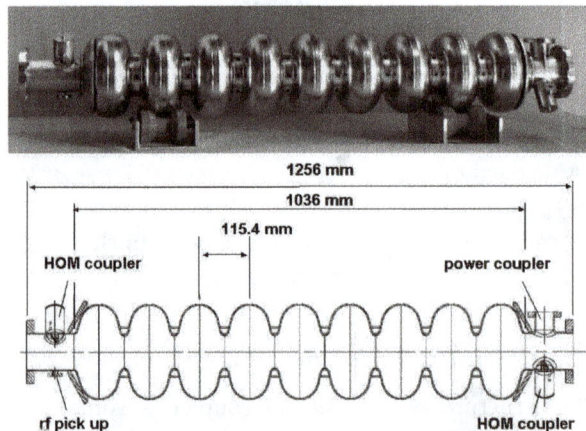

Fig. 28. TESLA nine-cell cavity.

Table 10. Geometrical parameters of the TESLA nine-cell cavity.

Frequency	1300	MHz
Active length	1.038	M
Gradient	23.6	MV/m
Quality factor Q_0	10^{10}	
Iris diameter	70	Mm
R/Q	1036	Ohm
E_{sp}/E_{acc}	2.0	
H_{sp}/E_{acc}	42.6	Oe/MV/m
Tuning range	± 300	kHz
Q_{ext} of power coupler	4.6×10^6	

Table 11. Parameters of ILC and EURO-XFEL.

	500 GeV	XFEL
Beam energy (GeV)	250×2	20
Total length (km)	11×2	1.2
Gradient (MV/m)	31.5	23.6
Number of bunches per RF pulse	2625	3250
Bunch spacing (ns)	369	200
Repetition rate (Hz)	5	10
Number of e per bunch (10^{10})	2	1.82
$\varepsilon_x{}^*/\varepsilon_y{}^*$ (10^{-6}m rad)	10/0.04	12/0.025
$\beta_x{}^*/\beta_y{}^*$ (nm)	20/0.4	25/0.5
$\sigma_x{}^*/\sigma_y{}^*$ (nm)	639/5.7	618/4.0
σ_z (mm)	0.3	0.5
Number of accelerator modules	1680	116
Number of cavities	14560	928
Number of klystrons	560	29
Klystron peak power (MW)	10	5.2
RF pulse (ms)	1.56	1.38
Luminosity (10^{34} m^{-2}s^{-1})	2	—

at 25 MV/m or more is impossible due to a huge refrigerator power of > 50 W/m, so that a duty factor must be reduced to $\sim 1\%$. Nevertheless, high luminosity operation is expected in ILC, because even at 1% the duty cycle is much higher than that of NC linacs, 0.001%. The long pulse operation of SC linacs can reduce the peak power of RF sources. Furthermore, a rather large aperture size of 1.3 GHz cavities is advantageous for beam transportation in long distance linacs. On the other hand, pulsed operation at a high gradient causes a serious frequency shift called Lorentz force detuning. The radiation pressure of the electromagnetic field deforms the cavity and shifts the resonant frequency, which is proportional to the square of the accelerating gradient and is approximately 600 Hz for operation at 31.5 MV/m [65]. This frequency shift has to be compensated for by a feedforward system using a fast tuner to stabilize the accelerating field.

The ILC Global Design Effort (GDE) was established and a worldwide collaboration was organized, including three regions — the Americas, Asia and Europe. The Baseline Configuration Document (BCD) [66] and the ILC Reference Design Report (RDR) [67] were published in 2005 and in 2007. Figure 31 shows a schematic layout of the ILC complex for 500 GeV, which consists of a polarized electron source, an undulator-based positron source, two damping rings with a circumference of 6.7 km,

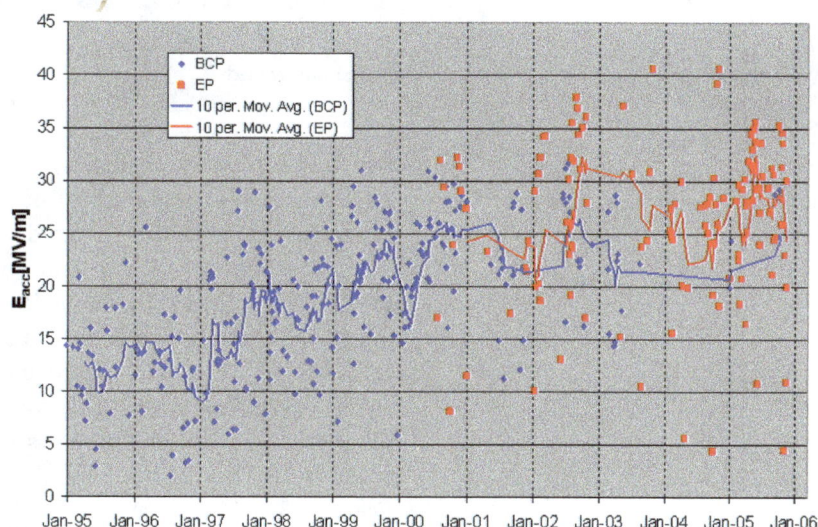

Fig. 29. Maximum field gradient measured at TTF for the past ten years using 120 cavities. In the first five years, the cavities were mainly chemically polished. On the other hand, in the latter five years, the cavities were electropolished.

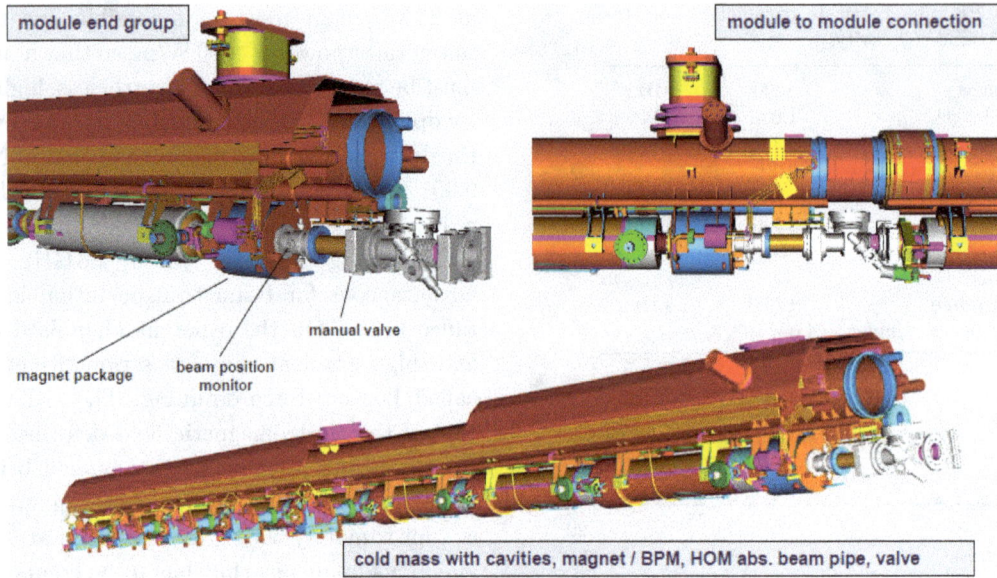

Fig. 30.　Cryomodule of EURO-XFEL. Eight nine-cell cavities are grouped in one cryomodule [64].

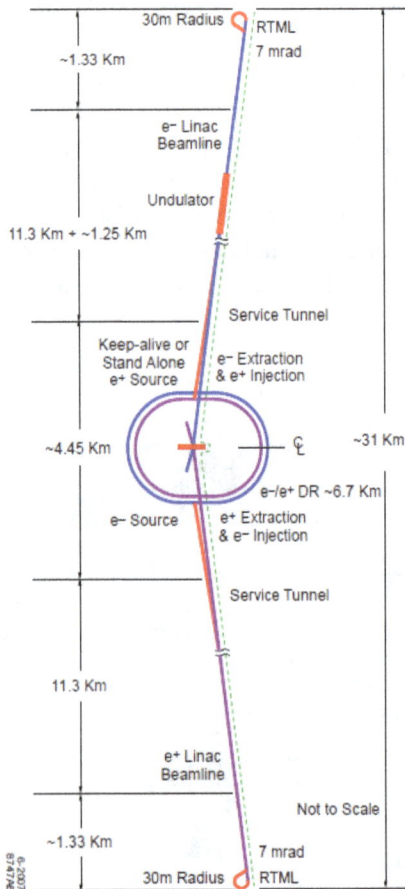

Fig. 31.　Layout of ILC [67].

Fig. 32.　STF at KEK. The first prototype has been completed and cooled.

two 1.3 GHz SC main linacs 11 km in length with an average operating gradient of 31.5 MV/m and an interaction point with a 14 mrad crossing angle.

Following the ITRP recommendation of the SC-based ILC main linacs, large R&D infrastructures are under construction at both FNAL (SMTF) and KEK (STF) (Fig. 32). These include facilities for cavity processing, vertical performance tests, module assembling and module tests with beam, aiming to optimize the TESLA module for the ILC specification, 31.5 MV/m, and to improve the reliability and cost of all other components, such as frequency tuners and power couplers. Finding breakthroughs to upgrade the production yield of nine-cell cavities of 35 MV/m to 80% is the most important issue for these new facilities. They are to be the core in each region and support the worldwide collaboration on the SC cavity technology for a global linac [68].

7.2. *Energy Recovery Linac (ERL)*

Fundamental studies of CW linacs based on TESLA 1.3 GHz technology are in progress at various laboratories, aiming for the ERL-FEL. Because of the CW operation (100% of the duty factor), a moderate gradient of 15–20 MV/m is chosen, considering the cryogenic capacity. The operation at a high accelerating gradient makes short the total linac length and reduces the static loss, which is inversely proportion to the total length of the linac; however, the RF dynamic loss per unit length becomes large in proportion to the square of the accelerating gradient. Hence an optimum operation gradient with the minimum cryogenic load exists. An ERL contains two SC linacs: an injection linac without energy recovery and a main linac with energy recovery. From beam dynamics at the merger section, the injector energy is chosen at 5–10 MeV. Therefore, power delivery of 1 MW in CW and its cooling are the key issue in obtaining an injection beam of 100 mA. On the other hand, the input power of 10 kW is enough for each cavity of the main linac, even if the operation gradient is 20 MV/m with a beam of 100 mA, because the RF power is supplied by the decelerating beam. Therefore, the key issue of the main linac is to obtain the extremely stable high gradient CW field with neither dark current nor discharging, which causes the cavity trips. High Q of the cavity is absolutely necessary in ERL, as well as the

Fig. 33. Assembly of the two-cell injector cavity of Cornell University. A double coupler scheme can reduce the power per coupler and eliminate the asymmetry caused by a coupler [69].

Table 12. Large scale ERL.

	Cornel	APS	KEK	RHIC
Energy (GeV)	5	7	5	54
Bunch charge	77 pC	77 pC	77 pC	5 nC
Average current (mA)	100	100	100	50×2
Bunch length (fs)			2 ps	2 ps	7.8 mm
Bunch rep. rate (MHz)	1300	1300	1300	9.4
Cavity type	7-cell	7-cell	9-cell	5-cell
Injection energy (MeV)	5–15	10	10	4.7
Frequency (MHz)	1300	1300	1300	704
R/Q (Ω)			897	403
Cell-to-cell coupling	1.89	1.89	3.8	3

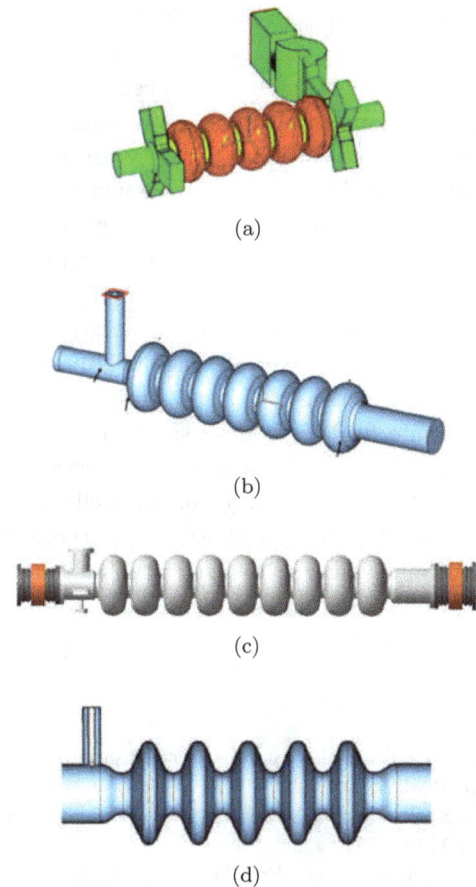

(a)

(b)

(c)

(d)

Fig. 34. Various ERL SC cavities: (a) JLab five-cellcavity with a waveguide coupler, (b) Cornell seven-cellcavity using a beam-line damper, (c) KEK nine-cellcavity with beam-line dampers and (d) RHIC 704 MHz five-cell cavity.

sufficiently damped HOMs induced by both accelerating and decelerating beams. Furthermore, a short bunch length of ~ 1 mm causes additional heating of broad-band HOMs. Besides the SC cavity issue, a DC photocathode gun and its drive laser are also key components. The precise and intense beam of

100 mA with sufficiently low emittance determines the performance of the total ERL system.

For the technology demonstration of ERL, Cornell University has constructed a test injector of 5–15 MeV. Figure 33 shows the assembly of the 1.3 GHz two-cell cavity with HOM dampers on both sides [69]. Recent proposals for ERL are summarized in Table 12, with candidates for the cavity shapes in Fig. 34 [70–74].

8. Conclusion

The challenge to obtain a defect-free surface has been contining for more than four decades as the main issue of SC cavity technology. An enormous effort regarding SC application has advanced the new fields at every developmental stage. As a result, various accelerators based on the SC cavity are in operation, such as low beta linacs for heavy ions and protons, high beta RF for high energy storage rings, high intensity storage rings and beam deflection cavities. Today the most challenging issue is a high gradient linac aiming for ILC or ERL. A collaborative organization of a worldwide scale is being built among three regions: the Americas, Asia and Europe. Although the maximum field gradient of single cell cavities is very close to the intrinsic limitation of a critical magnetic field of Nb, the limitation of multicell cavities is still far below the theoretical one. The purpose of the international collaboration is to know the reason for this gap and to find breakthroughs so as to improve the production yield of high quality cavities.

References

[1] R. C. Pardo et al., Proc. PAC93 (Washington, 1993), pp. 1694–1699.
[2] Y. Kojima et al., Proc. PAC89 (Chicago, 1989), pp. 1789–1791.
[3] B. Dwersteg et al., Proc. EPAC94 (London, 1994), pp. 2039–2041.
[4] C. Reece et al., Proc. PAC95 (Dallas, 1995), p. 1512.
[5] G. Cavallari et al., Proc. EPAC94 (London, 1994), pp. 2042–2044
[6] P. Brown et al., Proc. PAC01 (Chicago, 2001).
[7] S. Belomestnykh et al., Proc. EPAC96 (Sitges, 1996), p. 2100.
[8] T. Furuya et al., Proc. 9th SCRF Workshop (Santa Fe, 1999).
[9] M. S. de Jong et al., Proc. APAC04 (Korea, 2004).
[10] J. R. Chen et al., Proc. APAC04 (Korea, 2004).
[11] R. P. Walker, Proc. EPAC06 (Edinburgh, 2006).
[12] C. Zhang and Q. Qin, Proc. APAC07 (Indore, 2007).
[13] Z. T. Zhao and H. J. Xu, Proc. APAC04 (Gyeongju, 2004).
[14] CDR of NSLAII (2006).
[15] L. Evans, Proc. APAC07 (Indore, 2007).
[16] M. Altarelli et al. (eds.), TDR of EURO-XFEL (2007).
[17] M. Tigner, Nuovo Cimento 37, 1228 (1965).
[18] H. A. Schwettman et al., Proc. 5th Int. Conf. High Energy Accelerators (Frascati, 1965), p. 690.
[19] P. Turner and N. T. Viet, Appl. Phys. Lett. 16, 333 (1970).
[20] B. Hillenbrand et al., IEEE Trans. Magn. Mag. 13, 491 (1977).
[21] U. Klein and D. Proch (Wuppertal, Nov. 1978), WU B 78–31.
[22] K. W. Shepard, Proc. 2nd SCRF Workshop (CERN, 1984), p. 9.
[23] A. Pisent et al., Proc. LINAC06 (Knoxville, 2006).
[24] S. Takeuchi et al., Proc. 8th SCRF Workshop (LNL-INFN, 1997) LNL-INFN, (Rep) 133/98, p. 237.
[25] H. Padamsee and A. Joshi, J. Appl. Phys. 50(2), 1112 (1979).
[26] H. Padamsee et al., RF Superconductivity for Accelerators (John Wiley & Sons USA, 1998).
[27] G. Muller, Proc. 3rd SCRF Workshop, ANL, 1987, ANL-PHY-88-1, p. 331.
[28] H. Padamsee, IEEE Trans. Magn. 21, 1007 (1977).
[29] P. Kneisel, J. Less-Common Met. 139, 179 (1988).
[30] Y. Kojima et al., Proc. 4th SCRF Workshop (KEK, 1989), KEK Rep. 89–21 (Jan. 1990), A, p. 89.
[31] K. Saito et al., Proc. 4th SCRF Workshop (KEK, 1989), KEK Rep. 89–21 (Jan. 1990), p. 635.
[32] S. Belomestnykh et al., Proc. EPAC00 (Vienna, 2000).
[33] T. Furuya et al., Proc. 9th SCRF Workshop (LANL, 1999).
[34] H. Padamsee et al., Proc. 5th SCRF Workshop (DESY, 1991), p. 138
[35] Z. Q. Li et al., Proc. 13th SCRF Workshop (Beijing, 2007).
[36] G. Ciovati et al., Proc. PAC01 (Chicago, 2001).
[37] I. E. Campisi et al., Proc. PAC07 (Ailbuquerque, 2007).
[38] K. Akai et al., Proc. 13th SCRF Workshop (Beijing, 2007).
[39] T. Shizuma et al., Proc. LINAC00 (Monterrey, 2000).
[40] G. R. Neil et al., Phys. Rev. Lett. 84, 662 (2000).
[41] P. Michel et al., Proc. FEL06 (Berlin, 2006).
[42] A. Richiter, Proc. EPAC96 (Sitges, 1996).
[43] A. A. Abrikosov, L. P. Gor'kov and I. M. Khalatnikov, Sov. Phys. JETP 8, 182 (1959).
[44] B. Bonin and W. Roth, Proc. 5th SCRF Workshop (DESY, 1991), DESY-M-92-01, p. 210.
[45] R. H. Fowler and L. Nordheim, Proc. R. Soc. London. A, Math. Phys. Sci. 119, 173 (1928).
[46] W. Singer et al., Proc. 8th SCRF Workshop (LNL-INFN, 1997), LNL-INFN (Rep.) 133/98, p. 850.

[47] P. Kneisel and B. Lewis, *Part. Accel.* **53**, 97 (1996).

[48] D. Sertore *et al.*, *Proc. PAC07* (Albuquerque, 2006).

[49] www.cts.de.

[50] www.ansoft.com.

[51] M. de Jong *et al.*, *J. Microwave Power & Electromagnetic Energy* **27**(3), 136 (1992).

[52] T. Tajima, thesis, KEK Rep. 2000-10 (Sep. 2000).

[53] Y. H. Chin, *User's Guide for ABCI Version 8.8* (LBL-35258, UC-414, Feb. 1994).

[54] S. Mitsunobu *et al.*, *Proc. 7th SCRF Workshop* (CEA-Saclay, France, 1995), p. 735.

[55] KEK B-Factory Design Report, KEK Rep. 95-7 (Aug. 1995).

[56] A. Citron, *et al.*, *Nucl. Instrum. Methods* **164**, 31 (1979).

[57] R. B. Palmer, SLAC-PUB-4707, 1988.

[58] K. Oide and K. Yokoya, *Phys. Rev. A* **40**, 315 (1989).

[59] A. Zholents *et al.*, *Nucl. Instrum. Methods A* **425**, 385 (1999).

[60] K. Ohmi and K. Oide, *Proc. EPAC06* (Edinburgh, 2006), pp. 616–618.

[61] K. Akai *et al.*, *Proc. PAC93* (Washington, 1993), p. 769.

[62] K. Hosoyama *et al.*, *Proc. 13th SCRF Workshop* (Beijing, 2007).

[63] TESLA Technical Design Report (TDR), Mar. 2001.

[64] The European X-ray Free-Electron Laser. Technical Design Report (TDR), Jul. 2007.

[65] M. Liepe *et al.*, *Proc. PAC01* (Chicago, 2001).

[66] ILC Baseline Configuration Document (BCD) (Dec. 2005).

[67] ILC Reference Design Report (RDR) Aug. 2007.

[68] Pagani, *Proc. APAC07* (Indore, 2007).

[69] M. Liepe *et al.*, *Proc. 13th SCRF Workshop* (Beijing, 2007).

[70] M. Liepe *et al.*, *Proc. 11th SCRF Workshop* (Travemunde, 2003).

[71] C. Wang, *Proc. 13th SCRF Workshop* (Beijing, 2007).

[72] T. Furuya *et al.*, *Proc. 13th SCRF Workshop* (Beijing, 2007).

[73] R. A. Rimmer *et al.*, *Proc. PAC07* (Albuquerque, 2007).

[74] I. Ben-Zvi *et al.*, *Proc. EPAC06* (Edinburgh, 2006).

Takaaki Furuya is a professor of accelerator science at the High Energy Accelerator Research Organization (KEK). He is an expert on superconducting RF technology and its applications in TRISTAN and KEK-B. He is interested in the energy recovery linac (ERL) and is conducting research on the development of a superconducting cavity for the main linac. He enjoys playing tennis with his colleagues on weekends.

Reviews of Accelerator Science and Technology
Vol. 1 (2008) 237–257
© World Scientific Publishing Company

Cooling Methods for Charged Particle Beams

V. V. Parkhomchuk and A. N. Skrinsky*

*Budker Institute of Nuclear Physics,
Novosibirsk, Russia
skrinsky@inp.nsk.su

We review the basic methods for cooling charged particle beams in storage rings and accelerators. Applications of cooled beams are discussed.

Keywords: Beam cooling; beam instability; collider.

1. Introduction

Experimental studies in nuclear physics and elementary particle physics use accelerated beams of charged particles. The success of the experiments is strongly determined by the quality of the beams: small energy spread, small angular distributions and small size at high intensities. These attributes enable one to place detectors near the interaction point and to observe in detail vertices from short-lived particle decays. The most impressive recent example of this is the experiments on CP parity violation using colliding beams at the B-factories at KEK and SLAC. Those facilities have high intensity and small beam sizes. Such beams can be obtained only by cooling and storing the beam, during which time the phase space density is increased by many orders of magnitude. This beam cooling and increase in phase space density is especially important for antiparticles, which do not exist naturally — antiprotons, positrons, or some rare nuclei that can be produced by the irradiation of targets by nuclei. As a rule, when produced, these particles have a low initial density in phase space, and several cooling methods have been invented to increase that density and store the beams.

An increase in the beam phase space density cannot be achieved by the use of any given external electromagnetic fields, i.e. fields independent of the motion of the individual beam particles. The Liouville theorem is the key concept. The particle distribution function in phase space is constant along the system trajectory — or, equivalently, the particle density near a given point in the system traveling through phase space is constant in time. In accelerator physics this assertion is formulated as follows: phase space density along the beam trajectory is constant. It is easy to show that if the particle distribution function shrinks, for example along the coordinate p_i, it stretches in the conjugate coordinate q_i in such a way that the product $\Delta p_i \Delta q_i$ is constant. Aberrations can strongly distort the shape of the phase space volume and thus decrease the effective density by an effective increase in the volume from vacant places in phase space. For an increase in density it is necessary to introduce dissipative forces that act on the individual particles that are dependent on their motion, and directed against the deviation of the particle from the beam equilibrium velocity.

Various diffusion processes — those related to the particular cooling method, and those of an "external" character — provide beam-heating power. Among them are residual gas scattering, fluctuations in the cooling system itself, intrabeam scattering (IBS), and particle scattering in an internal target or from a counterrotating colliding beam. The balance of heating and cooling power determines the equilibrium beam parameters.

There are five known methods of charged particle cooling in accelerators, which differ by how the dissipative portion of the friction force is produced.

Radiation cooling is based on electromagnetic radiation of light particles (electrons and positrons) traveling in a magnetic field. If the average energy loss of the equilibrium particles is compensated for, the particle oscillations with respect to the equilibrium orbit are gradually damped in a properly ring magnetic structure. Radiation cooling is widely used in the storage of electrons and positrons and in experiments with colliding beams. The energy of the proton beam at the 7 on 7 TeV LHC collider is high enough so that radiation cooling becomes significant.

Ionization cooling is related to the energy transfer to atomic electrons during the motion of charged particles through matter. Interaction of particles with nuclei — scattering, bremsstrahlung or nuclear absorption — limits the field of application for ionization cooling to the cooling of muon beams. Muon beams can be used for the generation of intense neutrino beams and for muon colliders. Ionization cooling is quite topical and awaits practical development and implementation.

Electron cooling of heavy particles is based on beam interactions with an electron beam traveling in a common orbit together with the same average velocity in a straight part of the storage ring. These systems enable the storage, cooling and confinement for a long time of beams of antiprotons and unique nuclei produced in targets. Electron cooling is especially effective for highly charged ions. Studies of rare ions and nuclei at GSI with the use of electron cooling have shown the great potential of this technique. In 2005, at FNAL, an electron cooler with an electron beam energy of 4.3 MeV was commissioned for the storage and formation of antiproton bunches and the luminosity of the Tevatron collider was substantially increased.

Stochastic cooling is based on the use of wideband feedback systems. Wideband amplifiers amplify signals induced by particles in the pickup electrodes and particle motion is corrected using feedback kickers. A wide bandwidth of a few GHz enables attenuation of the relative deviations of neighboring beam particles and provides rather rapid cooling. This method of cooling has been used at present to implement antiproton beam storage for the powerful proton–antiproton colliders. Intensive R&D has been carried out on using stochastic cooling at the RHIC ion–ion collider to maintain its high initial luminosity.

Stochastic cooling is promising for the primary cooling of rare exotic secondary ion beams in Project FAIR at GSI.

In laser cooling of ions, atomic shell electrons interact resonantly with the laser beam. This method is mainly used in atomic physics experiments. The uniquely low equilibrium temperatures of those beams open up new prospects for fundamental experiments in atomic physics.

This review provides a brief summary of each of these methods and a comparison of their capabilities and fields of application.

2. Radiation Cooling

Synchrotron radiation occurs during relativistic particle motion along the curved trajectory in a bending magnetic field [1]. The radiation field forms near each particle a friction force responsible for the average energy loss. Generally, within the range of the decelerating force (radiation reaction) from an individual particle, there are many other neighboring particles in the beam. So, each particle is affected not only by its radiation reaction field but also by the reaction fields produced by the neighboring particles [2]. However, if there is no correlation between the particle positions, the time-averaged decelerating force comprises only the intrinsic field of the particle. If the particles are concentrated in short bunches, radiation with a wavelength λ greater than the bunch (or microbunch) length, $\lambda > L_b$, becomes coherent and the radiation power increases quadratically with the number of particles in these microbunches, N_b^2. Consequently, the electric field related to this portion of radiation acting on each particle of a microbunch increases as N_b. However, as a first approximation, the field does not influence the relative particle motion and, gives zero contribution to cooling. At present, such regimes are realized in free electron laser systems. In these devices, a high quality electron beam with a small energy spread is bunched by interaction with the optical field and generates coherent radiation during the time of flight through the interaction region. But, in this case, the energy spread of such a beam increases strongly and, after radiation of a small fraction of its energy, the beam (in the case of energy recovery accelerators) is directed to the energy recovery system, where it returns the remaining energy to the cavities accelerating the new beam with a small energy spread.

Proposals have been made for superfast cooling based on the coherent radiation from a particle bunch [3, 4]. In these concepts, it is assumed that one can obtain cooling at high coherent radiation power increasing as N^2, and with a large number of particles one can achieve a much higher rate of cooling. However, the problem with these proposals is that the radiation reaction should be counterdirected with respect to the individual particle and only in this case will cooling occur [5]. Denote the radiation reaction as $-\vec{F}(\vec{V}_0 + \Delta \vec{V}_i)$. With coherent radiation, the force is determined by the average motion of the particle inside the microbunches $\vec{V}_0$ and independent of the individual particle motion $\Delta \vec{V}_i$. As a result of the development of oscillations in such a coherent system, the particle velocity spread increases even as the decrease in the total beam energy is transformed to coherent radiation.

Cooling of the betatron and synchrotron oscillations is based on compensation of the average loss of each particle passing through an RF cavity. The longitudinal RF field returns only longitudinal energy losses for equilibrium particles on the equilibrium orbit. Losses by the transverse and longitudinal oscillations are not compensated for and this produces damping of these deviations from the equilibrium orbit.

Energy losses are discrete rather than continuous processes and occur by emitting relatively high energy quanta. These jumps occur at random times and the energy value of these individual quanta increases rapidly with magnetic field and particle energy. The particle, which moves along the equilibrium orbit, suddenly loses a portion of its energy and the equilibrium orbit changes accordingly. As a result, these jumps excite longitudinal and transverse oscillations about the new orbit. The betatron oscillation amplitude is proportional to the difference between the new and old orbits — due to the dispersion function at the point of photon emission. To obtain a low value for the beam emittance, the frequency of the transverse betatron oscillations should be as high as possible. As a result, most of the circumference of the ring is occupied by quadrupoles and not by bending magnets. Designs with many alternating polarities of the magnetic field along the beam orbit have become popular for control of the cooling [6]. These (sometimes superconducting), insertions (Fig. 1) are installed in low value dispersion function regions of the lattice and help decrease the growth in transverse emittance.

While the vertical emittance in short magnets is determined, as a rule, by coupling effects between

Fig. 1. Superconducting multipole wiggler with a 4 T magnetic field installed in the Canadian Light Source as a powerful x-ray source (designed, produced and commissioned by BINP, Novosibirsk).

the radial and vertical oscillations, for an ideally flat orbit the vertical amplitudes excited during radiation of quanta because of the spread of the quantum deflection angle are noticeably smaller than the radial ones, and other effects such as intrabeam scattering and coupling of oscillations dominate.

Radiation cooling has become quite common and is an important method in high-energy physics research. The main application is the storage of large electron and positron currents and conducting colliding beam experiments. It cannot be overstated that the presence of this "natural" cooling was decisive in the rapid development of this field, which is presently providing a significant amount of knowledge in elementary particle physics, The first experiments were performed on electron–electron interactions at Stanford and Novosibirsk (1965) and electron–positron interactions at Novosibirsk (1967).

A necessary evolution to linear electron positron colliding beams at very high energies — hundreds of GeV — is caused by the dramatic increase of energy losses in synchrotron radiation in circular machines. This evolution will make the role of radiation cooling even more important. To obtain the required luminosity, one has to compress the intense bunches of electrons and positrons at the interaction point down to a fraction of a micron squared, which requires the development of dedicated damping rings at energies of a few GeV. These damping rings should be equipped with strong superconducting wigglers for intensifying radiation cooling and suppressing intrabeam scattering of intense short bunches.

The direct use of synchrotron radiation over a wide frequency range from THz up to high energy γ quanta has become more and more popular because of the high brightness of these light sources based on low emittance electron beams [2].

Radiation cooling has not yet played a noticeable role for proton beams but the imminent commissioning of the 7 on 7 TeV proton–proton LHC collider will change the current situation, and even more so if the beam energy is increased further beyond 7 TeV.

3. Ionization Cooling

High energy muon storage rings and the possibility of muon colliders are of primary importance for the future of elementary particle physics [7–14]. These are the drivers for the development of ionization cooling. Using existing technologies, it seems easier to build a hadron (proton) collider for the same energy and for the same luminosity of the fundamental parton–parton interactions. However, to realize this goal, the hadron–hadron energy should be ten times as high as the lepton–lepton one. Hadrons introduce in the collision their complex structure of quarks and gluons, whereas at lepton colliders (electrons/positrons or muons) we deal with the fundamental (unstructured to our present knowledge) incident particles. Additionally, in hadron–hadron collisions the fundamental interactions are not monochromatic (100% energy spread!), and each fundamental interaction is accompanied by many interactions of the remnant hadrons.

Even in the framework of the Standard Model, muon colliders and e^+e^- (linear) colliders are not just technically different versions of *lepton colliders* of the same energy (hundreds or thousands of GeV). The difference is related to the much higher "parasitic" radiation in the electron case.

(1) The radiation of individual initial colliding particles is much higher for electrons (and increases with the energy). Tagging of the photons (and, moreover, measuring their energies) is almost

impossible for the options with high luminosity per bunch, currently under development.

(2) Coherent fields in the collision region are so high that the synchrotron radiation of the colliding electrons/positrons consumes a substantial fraction of their energies.

Hence, instead of pure electron–positron collisions with a narrow energy spectrum we will get a wide initial spectrum, many parasitic photons and photon–photon and photon–electron collisions. Eliminating this background will be a major headache.

If we go to higher energies, the $e^{+/-}$ collider becomes more and more difficult technically. To prevent the growth of synchrotron radiation in the coherent field of the colliding bunches and to keep increasing the luminosity we need to make the horizontal beam size big enough, and thus need to make the vertical size of the interaction spot one nanometer or smaller! There is no such limitation for a muon collider since the rest mass of the projectiles is much greater.

In addition, perhaps, at energies of around 1 TeV and higher some *new physics* will appear, and the muon will be discovered to be more than just a heavy electron, but instead a particle with completely new interactions. In this case, studying muon collisions would bring fundamentally complementary information to the study of electron collisions, and the muon collider becomes a *must*.

Muon storage rings required for muon colliders would give birth to excellent muon and electron neutrino/antineutrino beam sources. These are called neutrino factories.

Why would neutrino factories be very useful?

(1) Narrow (cooled) intense muon beams in storage rings of sufficiently high energy produce narrow (with a transverse momentum of around 30 MeV/c) ν_e, ν_μ, anti-ν_e and anti-ν_μ beams, and enable complete studies of neutrino interactions (behind perfect muon shielding!).

(2) Such beams are perfect for long distance neutrino studies (neutrino oscillations and related topics).

Accelerator technologies for muon colliders and neutrino-factories are similar in many respects. They both use ionization cooling, but the cooling requirements are weaker for neutrino factories. So this may make for an easier "stage 1" option on the road to an eventual muon collider. Both projects require fast muon acceleration. But the magnetic field requirements for the muon decay ring are easier to obtain for the neutrino factory.

We at Novosibirsk started — already in the 1960s [7–10] — to consider options for lepton colliders reaching energies of hundreds of GeV (and even higher) — linear electron–positron colliders and muon colliders. We worked on these concepts in parallel.

At first glance, the main disadvantage of muons is their very short lifetime (2.2 μs at rest). Of course, the muon lifetime grows proportionally to its energy, $E_\mu/E_{\mu 0}$, but still remains short when considered in the context of a complete accelerator facility. Hence, the cooling of muon beams and their acceleration to the energy required must be fast. It is easy to calculate that the required acceleration rate is well within technical limits. The main problem is fast and sufficient cooling of the muon beams. The only method is ionization cooling. Several authors have considered ionization cooling for stable particles (say, protons) [15, 16]. But it is easy to demonstrate that the protons (as well as any other strongly interacting particles) would be lost from nuclear interactions faster than their cooling rate, so that under practical conditions ionization cooling is not useful. This statement is even more valid for electrons (positrons) because of the high level of bremsstrahlung.

But for muons — and for muons only! — ionization cooling is what we need: muons have normal ionization losses, but no strong interactions and negligible bremsstrahlung (below 1 GeV, where muon ionization cooling is of interest). In our early reports we always presented ionization cooling as an essential part of the muon collider [7–14].

After we presented a reasonably complete theoretical consideration of ionization cooling as an inherent part of the muon collider (1981) [7–21], the approach attracted interest, and now many groups in different laboratories throughout the world are actively developing different options for various stages of the muon collider. International collaborations for the muon collider were established and are working toward developing and demonstrating ionization cooling.

The principal idea of ionization cooling is quite simple and — in some aspects — similar to the very familiar radiation cooling (at least, for the transverse part of the six-dimensional emittance). The friction force due to ionization loss is directed opposite to the full velocity of individual muons, but only lost momentum parallel to the equilibrium orbit is restored by an external electric field. In practice this is done with an accelerating RF field.

But, as is often the case, *life* is more complicated than the *idea*:

(1) For high luminosity we need to cool in all of the six phase space dimensions, including longitudinally; but in the longitudinal direction "natural" ionization cooling is relatively very slow, or even negative at low energies (heating instead of cooling).

(2) In addition to have useful energy losses, muons will also experience multiple scattering on nuclei and electrons in the stopping media, as well as fluctuations in the ionization losses.

Let us evaluate (as a first approximation) the ionization decrements and the equilibrium 6-emittance of a muon beam under ionization cooling.

For the case of cooling due to "full energy losses" of any origin, the increment of six-dimensional density (or the sum of decrements for all the three emittances) is equal to

$$\delta_{\Sigma 0} = \frac{2P_{\text{fr}}}{p_\mu v_\mu}\left(1 - \frac{p_{\mu\text{long}}}{2v_\mu}\cdot\frac{dv_\mu}{dp_{\mu\text{long}}}\right) + \frac{1}{v_\mu}\cdot\frac{dP_{\text{fr}}}{dp_{\mu\text{long}}}.$$

(Energy losses of the "equilibrium particle" are assumed to be compensated for by an external source.) Here P_{fr} is the ionization loss power, p_μ, v_μ are the muon momentum and velocity, and $\delta_{\Sigma 0}$ is the decrement of the total six-dimensional emittance.

The power of ionization energy losses by a charged particle, and consequently the cooling decrement, is proportional to the electron density — hence, we need to use high density materials as a stopping material.

Consequently, the sum of decrements, expressed now in cm^{-1} of the cooling material (for small angles in the beam) in lithium is as shown in Fig. 2.

As is seen in the figure, to cool the six-dimensional emittance by a factor of *a million times* — say, at 200 MeV (kinetic energy) — we need to travel through about 15 m of lithium with

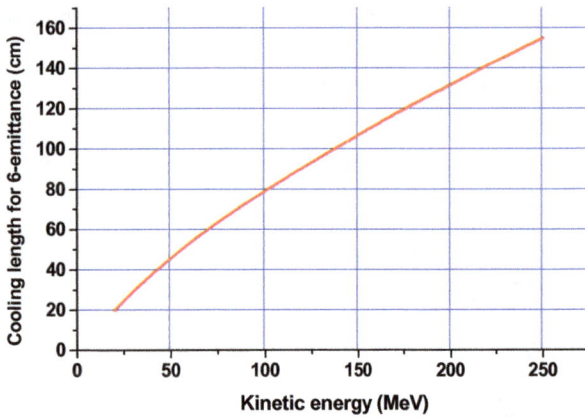

Fig. 2.   Cooling length for the six-dimensional emittance (i.e. the product of all three emittances) in lithium (Li).

"continuous" energy recovery. In this example, 10% of the muons will decay; the losses due to single scattering in this case are negligible.

But there are longitudinal problems! Figure 3 shows the "natural" longitudinal decrement as a function of the cooling energy.

It can be seen that "natural" longitudinal cooling is really too slow — many tens of meters are required to cool. And at lower energies the longitudinal *decrement* even converts into an *increment* (the situation is different when the muon velocity is lower than this of the atomic electrons, but this is not an interesting practical case).

Hence, it is obligatory to effectively *redistribute* the sum of decrements in favor of the longitudinal degree of freedom — this is a real challenge for inventors!

One option looks very simple — in principle:

- Cool transverse emittances of the whole single bunch beam (maybe not up to the desired limit at this point),
- Divide up the longitudinal emittance into 10–20 independent beamlets with different energies and as short as possible,
- Equalize energies and time position of all these beamlets,
- Combine all the beamlets back into one bunch,
- Repeat transverse cooling on the newly combined bunch.

Transverse cooling asymptotically squeezes the muon angular spread down to the equilibrium one, i.e. to the multiple scattering angle acquired at a distance of one transverse emittance cooling length (if we neglect the velocity dependence of energy losses, and other complications). It is easy to see that θ_{eq}^2 depends on Z, but not on the average electron density or the focusing (if uniform); its dependence on the cooling energy is presented in Fig. 4.

The corresponding relative energy spread at equilibrium (due to the balance of fluctuations of the ionization losses and the effective longitudinal cooling) is shown in Fig. 5 (the length of the final transverse cooling is assumed equal to $3 \cdot \delta_{tran1}$).

As one can see, the equilibrium angles at energies of interest — upon full cooling — are not small. Hence, the usual paraxial beam optics calculations will work poorly — an additional headache!

To reach small-enough emittances at the final cooling stage — and, hence to reach acceptable

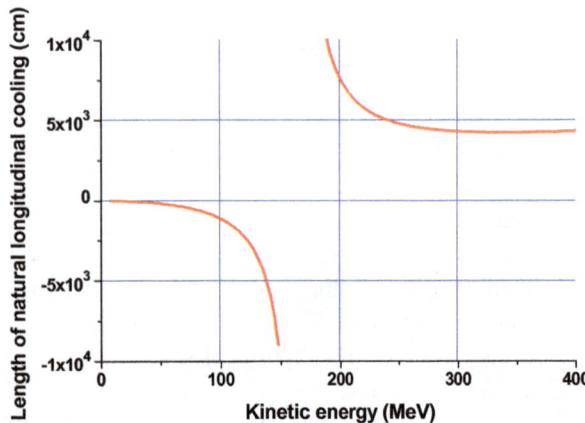

Fig. 3.   Inverse longitudinal decrement (length of cooling).

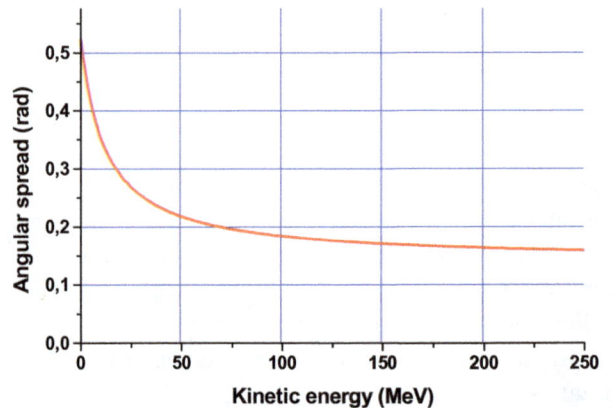

Fig. 4.   Equilibrium angular spread resulting from ionization cooling.

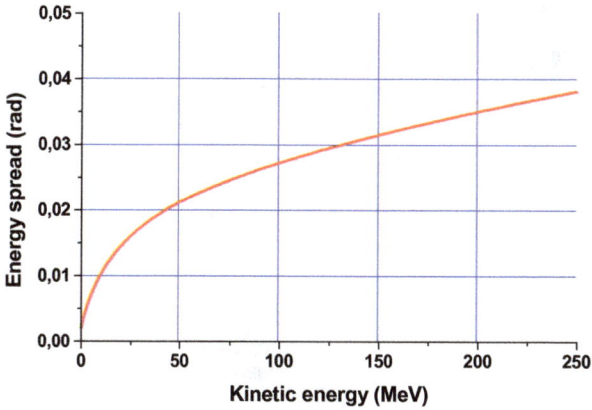

Fig. 5. Resultign relative energy spread for the cooling scheme discussed in the text.

collider luminosity — we need to use very strong focusing in all three directions — as strong as practically possible.

For these final stages the best option for ionization cooling up to now seems to be the use of lithium rods with a strong current along the rod to provide as strong transverse focusing as possible. This scheme is presented in Fig. 6.

Since high repetition rates are necessary, we really need to use liquid lithium to remove the ohmic heat. The muon beams are focused by the azimuthal magnetic field gradient, The limit is the magnetic field on the surface (10 T, or somewhat higher in the pulsed operation mode). For the parameters under discussion, $\beta_{Li} \to 1\,cm$, which is quite acceptable.

The whole device is a very long lithium lens (in total), developed at Novosibirsk for positron and antiproton collection decades ago, and still in use (INP, CERN, FNAL), plus appropriate RF insertions for keeping the muon beam energy constant on average. The improved — and the first liquid — version of a lithium lens was successfully tested at Novosibirsk.

The acceptance of the lithium rod should be large enough to accept the injected beam

without losses and be several times larger than the equilibrium emittance.

For the single transverse direction, the ratio of the lithium rod normalized acceptance and equilibrium emittance of the cooled beam in a typical case is presented in Fig. 7.

At the final cooling stage it is possible to reach a very small muon beam emittance (one-dimensional emittance, defined here for a Gaussian beam as the product of spatial dispersion and angular dispersion). See Fig. 8.

In order to reach the highest possible luminosity for any (affordable) number of muons (as was shown in Ref. 8, for example) we need the equilibrium normalized six-dimensional emittance to be as small as possible.

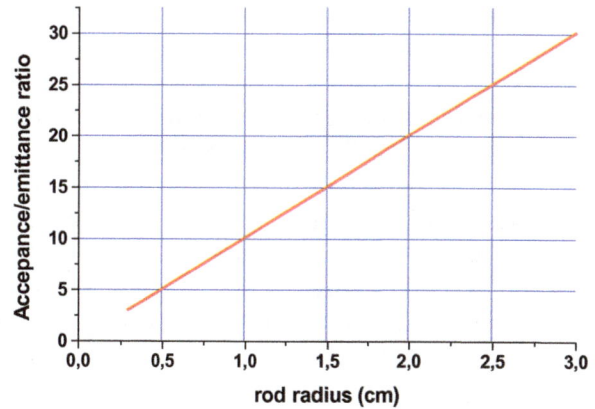

Fig. 7. Example of the intermediate stage ratio of normalized acceptance ε_{nA} to the equilibrium emittance ε_{neq} as a function of the rod radius (10 T field, 200 MeV kinetic energy).

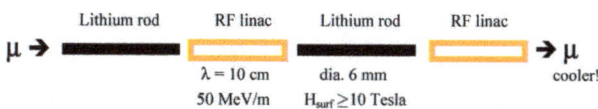

Fig. 6. Scheme for the final stages of muon transverse cooling.

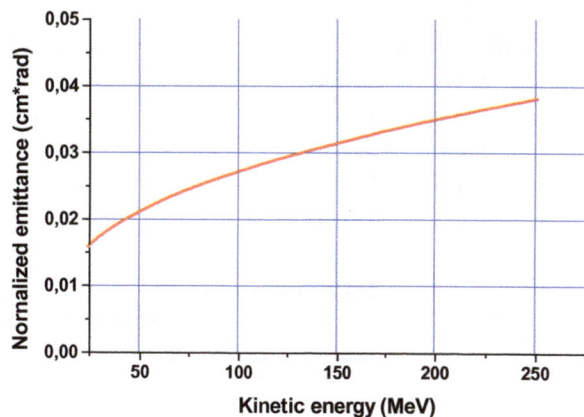

Fig. 8. Normalized transverse emittance after final cooling.

We see here again that β^2_{tran} and effective β_{long} (which can be considered as local focal lengths) in the cooling medium should be as small as possible. It is not clear at the moment which option of effective redistribution of decrement in the six-dimensional emittance will insure optimal longitudinal cooling in a real project.

One of the most difficult problems is proper matching of focusing in the sequential moderation/acceleration sections from the exit of one moderator section to the entrance of the next. High angular spreads (and nonmonochromaticity) can lead to unacceptably high nonlinear and chromatic aberrations, requiring modest beta functions in matching sections. The focal lengths of most familiar individual lenses — short solenoids or quadruple doublets — are linearly proportional to the square of the momentum of the particles they are focusing. But the focal lengths of lithium and plasma lenses are just proportional to the momentum; hence, their use should make it easier to reach low-enough chromatic aberrations in matching sections. One option that seems promising is shown in Fig. 9.

The short and strong lithium lenses at the exit and entrance of the lithum rods are necessary to make the beta function a few times larger than in the rods and to ease the matching into the plasma lenses with longer focal lengths. However, using lithium lenses instead of plasma lenses at much higher beta values inside accelerating structures is impossible. Multiple scattering results in unacceptable emittance growth.

The resulting spherical and chromatic aberrations are acceptably small — quite comfortable for the acceleration and emittance gymnastics needed later.

At many stages of the muon beam cooling it will be necessary to transform the shape of the muon beam phase space distribution and to transform its transverse emittance into a longitudinal one and vice versa — keeping the six-dimensional emittance as unperturbed as possible. Such gymnastics may be

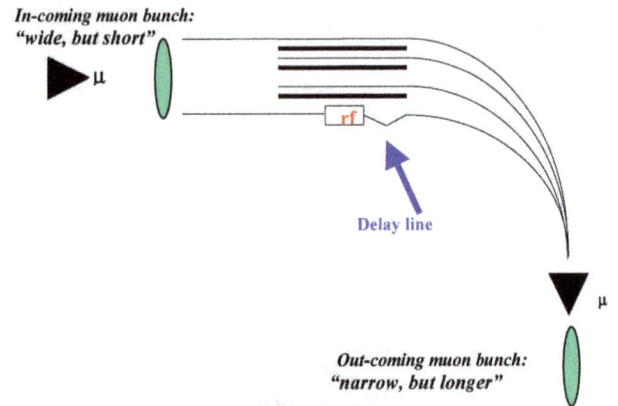

Fig. 10.   Example of "bunch gymnastics" needed to maximize the luminosity of a very high energy muon collider.

needed to reach the luminosity and monochromaticity goals of a real muon collider.

For the purpose of emittance gymnastics, we can use a combination of:

- Dispersive elements,
- Septum elements,
- RF accelerating/decelerating structures,
- Delay lines

(but not ionization components, which degrade the six-dimensional phase space density by scattering).

Such a transition should be arranged at some convenient energy of the muon beam. An option is presented in Fig. 10.

One can find many considerations related to muon colliders and neutrino factories in reviews [22–29] and other publications of the International MICE and MuNu collaborations. Using ionization cooling, it seems possible to construct very intense ν_e, ν_μ fluxes and even a multi-TeV muon collider with luminosity in the 10^{35} cm^{-2}s^{-1} range assuming the number of muons in the counterrotating bunches 10^{12} [22].

4. Electron Cooling

4.1. *NAP-M, the first storage ring with electron cooling*

The history of the development of electron cooling began at the Institute of Nuclear Physics (Novosibirsk) just after the first successful experiments there with electron–electron and electron–positron colliding beams. Radiation cooling plays a

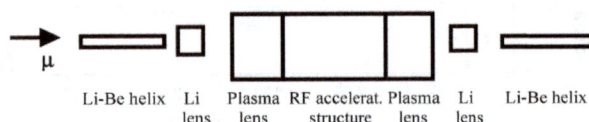

Fig. 9.   Schematic of a matching section between two consecutive cooling sections (geometrically not to scale).

decisive role in the achievement of high luminosity in electron–electron and electron–positron colliders. Its importance is clearly seen for positrons, which are generated in a target irradiated by high energy electrons. The positrons are produced with large energy and angular spreads so that at injection the entire storage ring aperture is filled by a hot positron beam. After cooling the positrons so that they occupy a small phase space volume, one can multiply inject to reach the required number of antiparticles.

In the case of protons, the beam injectors available in the 1960s could not provide the required brightness and it was necessary to find an analog of radiation cooling for more massive particles. Cooling based on ionization losses in matter was suggested but interaction with the target nuclei did not allow the application of this method because it would make the beam lifetime too short. The idea of using electron cooling, proposed by G. I. Budker in 1965 [30], was to shift from cooling with a stationary target to the use of a pure beam of electrons (without nuclei). The electrons would travel with the same average velocity V_0 as the proton beam. Of course, the electron beam density n_b is much smaller than the electron density in condensed matter n_t, but in this case, electrons are traveling together with the proton beam and the interaction efficiency between the two beams depends only on the spread of relative velocities of the protons and the electrons ΔV. The interaction efficiency is inversely proportional to the third power of relative velocities. The cooling rate decrease is compensated for by decreasing the electron beam density: $\lambda_b/\lambda_t = (n_b/n_t)/(V_0/\Delta V)^3$. For the relative spread $\Delta V/V_0 = 10^{-4}$ this factor is equal to 10^{12}.

Since the electron mass is only 1/1836 of the proton mass, the electron beam energy required to match the proton velocity is much less. This makes the whole scheme technically feasible. For example, to cool 100 MeV protons, electrons with an energy of only 54 keV are required. According to preliminary estimates, the cooling time is about 1 s for typical beam parameters [30].

Progress in storing proton beams of the required intensity and emittance continued, especially after the INP successes in the development of hydrogen negative ion sources [31, 32] and charge exchange injection [33]. These advances made the use of

electron cooling in proton colliders of lower priority. However, cooling remained essential for the realization of successful proton–antiproton experiments.

Electron cooling work began in 1967 with theoretical studies [34–37] and the development of an electron beam facility [38]. These were aimed at verification of the electron cooling concept. First, an experimental station was constructed for developing the methods for power generation and power recovery of intense electron beams. In this system, two power supply sources are used: the high voltage source connected to the cathode, which determines the electron velocity in the cooling section and at the collector, and a rather low voltage source (of a few kilovolts) providing the difference of potentials between the cathode and the collector. The load current to the high voltage source circuit is determined by electron beam losses due to scattering between the cathode and the collector and reflection from the collector. The power from the total electron beam current flowing in the collector rectifier circuit can amount to tens of kilowatts. During operation, it is necessary to transmit electric power to the high potential and remove the power from the collector under the total potential (Fig. 11). Successful experiments at the test stand on generation of a cold electron beam and power recuperation enabled the start of the development of the facility for checking the electron cooling idea — this prototype electron cooler is called EPOCHA (from the Russian "Electron Beam to Cool Antiprotons"). It became a test bench for testing the basic ideas proposed for an electron cooling device and was specially designed to work with the proton storage ring NAP-M (Storage Ring of Anti-Protons — Model).

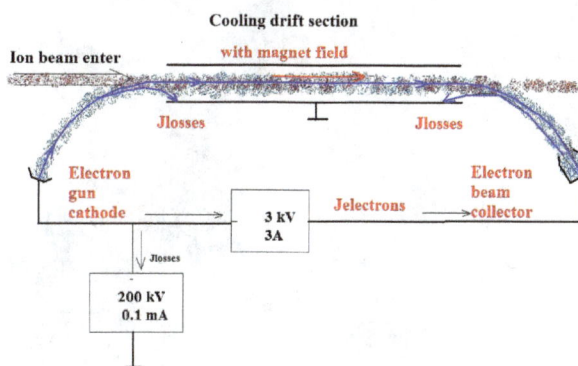

Fig. 11. Layout of the electron cooling facility with power supply sources.

An electron gun was put into a solenoid producing the longitudinal guiding magnetic field, which accompanied the beam until it reached the collector. The longitudinal magnetic field enabled us to pass an electron beam over the required, rather long (a few meters) distance without the inevitable beam dilution and without angular excitation using axially focusing lenses. Sections with a toroidal magnetic field were used for merging the proton and electron beams. For compensation of centrifugal drift, we used a horizontal magnetic field superposed in the toroidal sections. After the cooling section, the electron beam was decelerated and captured by the electron collector. In order to provide a high vacuum, ion pumps were installed in all accessible places. To correct the dipole kick from the toroidal section, which strongly distorted the proton orbit, we installed dipole correctors powered in series with the solenoids and toroids, thus lowering the ripple requirement in the power supply. Figure 12 is a photograph of the EPOCHA electron cooling facility.

These first experiments with electron cooling carried out in Novosibirsk in 1974 demonstrated the possibility of obtaining high cooling rates and low stable temperatures [39, 40].

Since that time, electron cooling has been important in advancing elementary particle and nuclear physics, and even for atomic physics. Many experiments have benefited from our groundbreaking innovative work on many methods of cooling, and their number is still growing.

4.2. Cooling diagnostics

Detection of cooling required the development of systems for measurement of the ion beam parameters. The most interesting of these diagnostic tools was the proton beam profile monitor using a thin jet of magnesium vapor. By mechanical scanning across the proton beam we measured the ionization current and obtained the beam profile both prior to and after cooling. See Fig. 13 [40].

Setting the magnesium vapor jet at the center of the beam (without mechanical scanning) allowed one to carry out fast measurements with millisecond resolution of the proton beam density during the cooling process. This was required to further study fast electron cooling. The magnesium vapor jet was set up at a distance from the cooled proton beam. Measuring the proton beam arrival time Δt after a jump of electron energy, opened up the possibility of a detailed study of the friction force $F = \Delta p / \Delta t$ and investigation of the effect of longitudinal flatness of the electron velocity distribution. (See Subsecs. 4.3.1 and 4.4.3). For high current ion beams and with the high ionization of residual gases (for heavy ions) in the vacuum chamber, a clear signal from the profile monitor was obtained without an additional jet. The multichannel profile monitor using residual gas was employed for commissioning electron cooling on the LEIR ring [42].

For measurements of beam dimension equilibrium values, we used a thin (2–3 microns) quartz wire crossing the beam. This enabled us to obtain much better spatial resolution [40]. Since the wire motion was at the high velocity of 5 m/s, most of the proton beam was retained and we could repeat several measurements at essentially the same proton beam intensity.

Fig. 12.   First electron cooling installation in the NAP-M storage ring.

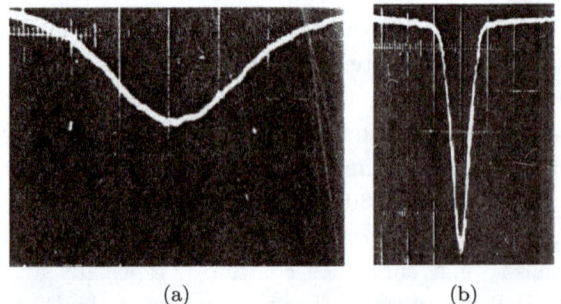

(a)   (b)

Fig. 13.   Historic first measurement of proton beam profiles before (a) and after (b) electron cooling at the NAP-M storage ring.

Monitoring Schottky noise is very important for measurement of the longitudinal momentum spread in the ion beams. The method is based on the measurement of the signal spectrum induced on pickup electrodes during motion of the beam particles. Modern spectrum analyzers with parallel processing of signals enable us to measure both the longitudinal and transverse emittances of beams.

4.3. *Basic technical parameters of electron cooling devices*

4.3.1. *Electron beam energy*

At rather low voltages, it seems most natural to use direct electrostatic acceleration of electrons [43]. Electron cooling devices of this type are operated in the range of 500 V to 4.3 MV (FNAL) [44]. With voltages up to 100 kV, one can use separately located power supply sources. But at voltages higher than 200–300 kV, the high voltage generators and accelerating tubes are put into special vessels filled with compressed SF_6 gas, which increases the maximum electric field before sparking. At still higher voltages > 5–10 MV, we have to move to structures using electron linear accelerators. Accelerator recuperators of this kind are used presently in the construction of free electron lasers. A chain of rather low frequency cavities (500–1300 MHz) can be used for generation of a 50 MeV electron beam, which is required, for example, in the electron cooling project for gold ions at RHIC [45]. Preliminary hopes for systems with coasting electron beams were noticeably diminished after the first experiments at the LEPTA machine [46]. The problem of obtaining a long and stable circulating intensive electron beam with longitudinal fields turned out to be rather serious.

4.3.2. *Electron current*

Operating values of electron current are determined by the cooling rate required. With increase in the electron current the influence of space charge of the electron beam grows. A drift rotation of an electron beam occurs with respect to the axis, which is proportional to the electron density and inversely proportional to the longitudinal magnetic field. For weak magnetic fields, the electron drift velocity reaches the velocity spread of the cooled particles with a relatively low electron current. As a rule, the cooling rate grows initially with the current and then drops when

the transverse velocities of the electrons become close to the ion transverse velocities.

Interaction of the electron and intense ion currents necessitated control of electron density in the beam cross-section. For this reason, special electron guns are developed, which allow noticeably decreasing the electron beam density at the center where the intense ion beam is stored [47, 48] See Fig. 14. Such a design enables the distribution of the cooling decrement values over the beam cross-section and optimizes the ion beam storage.

4.3.3. *Electron beam temperature*

The electron beam emitted from the cathode has a temperature close to the cathode temperature T_k — about $1000\ K \sim 0.1$ eV. With acceleration in the electrostatic field and a rather strong longitudinal field, the transverse momentum is retained and the longitudinal velocity spread becomes very small since the energy spread is in the laboratory system. Already, at an electron beam kinetic energy of $E_{kin} = 10$ keV, the cathode initial temperature contribution to the effective temperature of the electron's longitudinal motion becomes about 0.01 K and it is natural that the longitudinal temperature will be determined by other factors. The most essential factor seems to be the relaxation of the initially arbitrary mutual position of the electrons. Immediately after acceleration, relaxation of the energy due to the electrons mutual repulsion results in a longitudinal temperature approximately equal to $T_\parallel = 2e^2 n_e^{1/3}$,

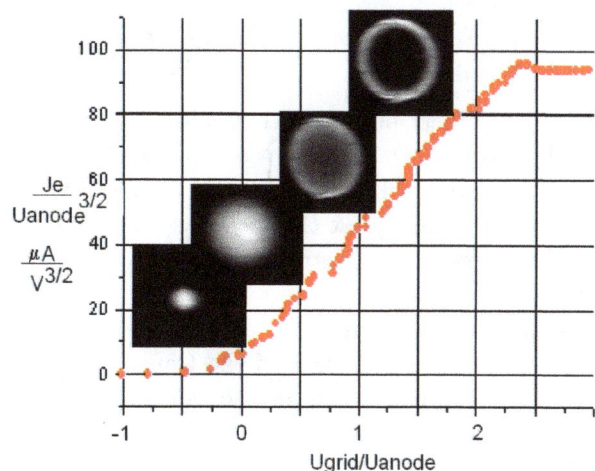

Fig. 14. Perveance of an electron gun $P = J/U^{3/2}$ with a variable electron beam profile.

where e is the electron charge and n_e is the electron beam density.

Some schemes are being developed for obtaining supercooled electron beams by slow (adiabatic) acceleration from the cooled surface of the photocathode [49]. These beams can be used for experiments in atomic physics but they are not yet used in electron cooling systems because of the problems in obtaining high intensity electron beams.

During the storage and cooling of highly charged ions, the main limiting factor is ion recombination and loss from the beam. Already, in the first experiments at NAP-M, after magnetized electron cooling, it was experimentally demonstrated that the increase in the electron beam transverse temperature caused a weak decrease in the cooling rate but noticeably reduced recombination between protons and electrons. For the electron cooling project in the RHIC collider, this effect turned out be rather important. Special experiments have been carried out to verify the effect of reducing recombination by high electron temperature for the highly charged ions at GSI in the ESR storage ring [50]. In the RHIC collider, the lifetime of ion beams should be many hours, with rather fast cooling. For suppression of recombination, it was suggested using a "transversely hot" electron beam in a strong magnetic field [45]. The temperature of transverse motion of an electron beam should be increased up to 100 eV but the cooling time should not be substantially longer. Some alternative proposals include the use of the forced transverse motion of electrons in an alternating magnetic field wiggler in the cooling section.

4.3.4. *Magnetic field*

A magnetic field provides electron beam formation in the electron gun and directs the beam to the ion beam orbit, while matching the dimensions in the electron gun and at the cooling section. In the toroidal sections, the electron beam travels with curvature radius R. For compensation of transverse drift a centripetal force is required. In the first machines with electron cooling, we used the transverse magnetic field to compensate for the centrifugal force. But the small electron flow reflected from the collector has a reverse velocity and instead of drift compensation, a double amplitude transverse shift results. In novel electron cooling devices, an electric field is used

for drift compensation independent of the direction of motion. That strongly decreases the electron current loss. Electrons reflected from the collector travel along the main beam and have the possibility of reaching the collector after a few oscillations along the main beam in the cooler.

At the cooling section, the magnetic field directs the electrons along the ion beam orbit and suppresses the effects of space charge from the intense electron beam. The high magnetic field allows a higher density of electron beam and an increase in the cooling rate. For effective electron cooling, linearity of the magnetic field lines of force is very important. An external transverse magnetic field leads to transverse motion of electrons with respect to the ion orbit. The angular deviation $\theta = \Delta B/B$ behaves as an effective temperature in the electron beam. As an example, an angular spread in the force line direction $\theta = 10^{-4}$ at an electron energy of 4 MeV produces the rather high effective temperature of 0.4 eV $= 4000\,\mathrm{K}$. In order to measure the magnetic field direction with an accuracy of $\sim 10^{-6}$, a special setup in the form of a magnetic probe with a mirror reflecting a laser beam is utilized. By moving the probe along the solenoid, we measure deviations of the reflected laser beam. The detector signal of the beam position is used for correction of the transverse fields along the solenoid [51–53]. We learned that deviations in our 4-m-long solenoid at the level of 10^{-5} must be corrected approximately once a fortnight because of the slow temperature deformation of the building basement. It is necessary to have such a measuring device inside the vacuum system and regularly correct the magnetic field to obtain high rates of electron cooling.

The collector magnetic field matches the size of the electron beam to be absorbed and distributes optimally the beam power over the collector surface. The presence of the quickly dropping magnetic field in the collector forms a magnetic mirror preventing electrons reflected from the surface from returning to the cooling section.

4.3.5. *Vacuum*

The vacuum system must meet the requirements for obtaining an acceptable ion beam lifetime. Ions colliding with residual gas atoms easily knock out electrons and subsequently the ions lie outside the acceptable aperture of the storage ring. The presence of an intense electron beam leads to the generation of a great amount of desorbed gas when the vacuum chamber and collectors are bombarded by electrons. With a desorption coefficient of 10^{-3}, an electron current of 1 A will produce 7×10^{15} atoms/s in the vacuum chamber. To obtain a gas density of 3.5×10^5 1/cm^3 (vacuum 10^{-11} Torr), the required pumping rate is 1.7×10^7 1/s, which is not realistic. Therefore, to obtain an acceptable vacuum, a thorough degassing of the collector is done by a long cleaning procedure using the electron beam itself. Inside the electron collector, we succeeded in achieving a desorption coefficient from the copper surface of less than 10^{-7}, but in the main part of the vacuum chamber at the cooling section, which is subjected to action of the collector-reflected loss current, it is difficult to achieve such good surface cleaning. In modern electron cooling devices, electrostatic bending plates are used, which suppress the electron beam losses down to the level of less than 10^{-6}, thus suppressing the desorption gas from the vacuum chamber walls down to the level of 7×10^9 atoms/s. In such conditions, after cleaning, the positive action on the vacuum by an electron beam was demonstrated. The pumping effect related to residual gas ionization and ion absorption exceeded the negative effect of desorption from the vacuum chamber wall. This resulted in improving the vacuum when the electron beam was switched on [54].

4.3.6. *General arrangement of the electron cooling system*

Figure 15 is a schematic diagram of the electron cooler of the LEIR facility at CERN, designed for the storage of the lead ions [42].

The diagram shows the series of magnetic structure components from the electron gun to the collector. In the toroidal sections, drift compensation plates and nonevaporable getter (NEG) ion pumps are installed. At the cooling section entrance and exit ends, there are pickup electrodes for determining the electron and ion beam positions.

4.4. *Kinetics of electron cooling*

4.4.1. *Friction force*

It is well known that in the Coulomb interaction, the exchange of energy between charged particles

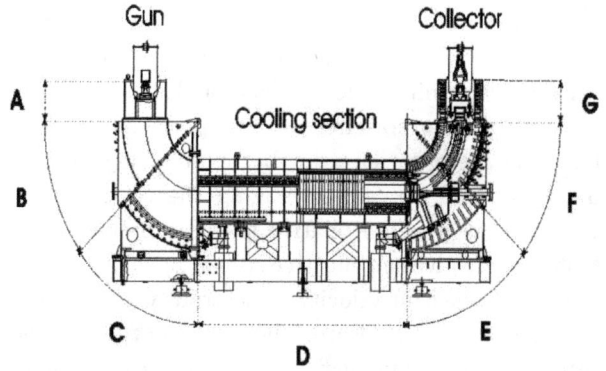

Fig. 15. Schematic diagram of the LEIR cooler. A is the gun solenoid with a longitudinally varying magnetic field for expanding the electron beam radius, B and F are small, 45° toroids with electrostatic deflecting plates, C and E are the main toroids for the joint ion and electron beams, D is the cooling straight section, and G is the collector solenoid.

logarithmically diverges in the limit of large impact parameter ρ. Therefore, in each case one has to find a $\rho_{\max}$, beyond which the interaction becomes effectively reduced compared to the pure Coulomb interaction [34]. The minimum impact parameter is a function of relative velocity $\rho_{\min} = e^2 Z/mV^2$. It is clear that in collisions of heavy particles with electrons in a magnetic field B under condition where the Larmor rotation radius satisfies the equation, $\rho_L = \frac{mV_{e\perp}c}{eB} \ll \rho_{\max}$, a substantial contribution to the collision integral could come from impact distance ρ, with $\rho_L < \rho < \rho_{\max}$. With these impact parameters, if the proton velocity with respect to the Larmor radius $\vec{V}_A = \vec{V} - \vec{V}_{e\parallel}$ is small, protons interact not with the free electron but with the Larmor radius since, for the time of interaction, the electron has time to execute many revolutions in the magnetic field:

$$\tau_s = \frac{\rho}{|\vec{V}_A|} \gg \frac{\rho_L}{|V_{e\perp}|} = \frac{1}{\omega_L}.$$

Because of the small spread of electron longitudinal velocities, cooling efficiency grows rapidly at low proton velocities. Stated more figuratively, in this case the effective electron temperature is not 1000 K but only a few K. A simple empirical expression for the friction force in the case of magnetized cooling suggests that [56]

$$\vec{F} = -\frac{4n_e e^4 Z^2 \vec{V}}{m_e (V^2 + V_{\text{eff}}^2)^{3/2}} \ln \left(\frac{\rho_{\max} + \rho_L + \rho_{\min}}{\rho_L + \rho_{\min}} \right),$$

where eZ is the charge of the ion traveling in the electron "gas" with density n_e, m_e is electron mass, and

$\vec{V}_{\text{eff}}$ is the effective velocity spread of the electron Larmor circle. As a rule, the effective spread in velocity determines the velocity of drift $V_d = c \times (2\pi e n_e x)/B$ Larmor circle at the space charge electric field intensive electron beam and longitudinal magnetic field B [56]. Averaging the cooling rate over the ion position relative to the center of electron beam x gives an effective spread in velocity $\vec{V}_{\text{eff}}$. Increasing the drift velocity of electrons with increasing intensity of the electron beam limits the cooling rate for K low magnetic field. Numerical calculations [55–57] show that the contribution to the friction force from incomplete collisions is noticeable. Special experiments show good agreement between this empirical equation for cooling force and experimental results [58].

4.4.2. Cooling time

The cooling oscillation time is characterized by the damping decrement. This is equal to the ratio of the mean dissipative power loss to the energy of oscillations. Consider the damping of the hot ion beam just after injection where the intrinsic electron motion can be neglected. For an ion beam having an emittance ε, the transverse velocities in an accompanying system of coordinates are $V_i = \gamma\beta c\sqrt{\varepsilon/\beta_\perp} = \gamma\beta c\theta_i$, where $\beta_\perp$ is the beta function value in the cooler, θ_i is the ion beam angular spread, and the cooling decrement in the laboratory system of coordinates is

$$n_e = n_{\text{elab}}/\gamma \cdot \delta = \frac{1}{\tau_{\text{cooling}}} = \frac{4(J_e/e)\eta_c r_e r_i}{\pi a_e^2 \gamma^5 \beta^4 \theta_i^3} L n_c,$$

where J_e/e is the ratio of the electron beam current to the electron charge, η_c is the fraction of the cooler at the ion orbit, $r_e = e^2/m_e c^2$, $r_i = (eZ)^2/M_i c^2$ are the classical radii of the electron and ion, and a_e is the radius of the electron beam. The cooling rate grows rapidly as the beam angular spread θ_i decreases. For example, if a beam 2 cm in size is cooled in 1 s, a beam 2 mm in size will reach its equilibrium in 1 ms. Such accelerated cooling leads to stringent requirements for the injected beam quality during storage. After the effective angular spread in an ion beam is decreased to the angular spread of the magnetic field force lines θ_e, the cooling decrement reaches its maximum value. Further cooling proceeds with a constant cooling rate to equilibrium, where diffusion becomes equal to the cooling rate. That

corresponds to the equality of the ion beam temperature and the "temperature" of the effective electron motion:

$$\theta_{\text{ieq}} = \theta_e \sqrt{\frac{m}{M}}.$$

Typical parameters for a low energy cooler (50 keV) are electron beam effective angular spread $\theta_e = 2 \times 10^{-5}$, and effective electron temperature in the accompanying system of coordinates that is only ~ 0.2 K. The ion beam equilibrium angular spread for $^{209}\text{Bi}^{+67}$ is $\sqrt{M/m} = 600$ times less than the effective angular spread in the electron beam that corresponds to the surprisingly small angles in the ion beam of $\theta_{\text{ieq}} = 4 \times 10^{-8}$. The low equilibrium temperature of the ion beam after magnetized electron cooling was (and will be) the basis for many interesting experiments with crystallization and ordering inside ion beams.

4.4.3. Recombination

The capture of electrons into atomic shell bound states is an important effect to be considered for electron cooling. In cooling low-charged ions, the ion orbit is changed strongly and the ion is lost. In the case of storage rings of large acceptance and high-charged ions $Z \gg 1$, situations are possible where the ion is not lost. Radiation recombination causes the most serious problems in the use of electron cooling for the long storage and confinement of high-charged ions. In cooling gold ions at RHIC with an electron beam peak current of 10 A, an energy of 50 MeV, and a beam diameter of 2 mm, the cooling time is 20 s but the lifetime is only 300 s. Magnetized cooling enables one to increase the transverse electron temperature from 0.1 to 1000 eV without significant degradation in the cooling rate but, in this case, the beam lifetime can be extended by two orders of magnitude (up to 10 h) [45]. For fast electron cooling the magnetic field in the cooling section should be more then 5 kG, so that the Larmor radius of 0.02 cm for hot electrons (1000 eV) does not exceed the ion path in the cooling section (in the beam reference system). This circumstance makes electron cooling feasible for use in large ion–ion colliders. This is called magnetized electron cooling with a high value of magnetic field in the cooling section. A variant is to use a low magnetic field but in this case the maximum useful cooling rate will be limited by electron drift. A special profile of the electron distribution with a minimum density at the center can also be used for optimization of the beam losses.

4.5. *Supercooled ion beams*

4.5.1. *Transverse emittance of the cooled beam*

Even in the first experiments at NAP-M at rather low DC currents, limitation of proton beam compression by space charge was observed. The mutual repulsion of ions inside the beam prevented further compression of the transverse emittance at a level determined by the value of the betatron tune shift due to space charge of the ion beam. As was learned from many experiments, with the increase in the beam intensity the tune shift is increased to $\Delta Q = 0.1$–0.15 and then the emittance grows proportionally to the ion beam current so that the beam density remains constant. It is of interest that in such a situation, the ion energy spread inside the beam grows very slowly with intensity.

4.5.2. *Longitudinal momentum spread*

The proton beam at NAP-M could actually be cooled down to 1 K and, in this case, some interesting effects were observed in relation to the fact that the electrostatic interaction began to compress the temperature fluctuations. This was displayed most vividly in the change in the spectral shape of the proton beam thermal noise. The voltage induced in an integer pickup electrode is proportional to the local density of the ion beam.

Interaction between particles of the cooled beam led to a mutual shielding of fields and to decrease of fluctuations in the ion beam [59]. Thermal fluctuations along the beam began to propagate not with the particle velocity but with the velocity of the longitudinal waves both along and counter to the beam motion. As a result, the spectrum around the nth harmonic was split into two peaks (Fig. 16), with their separation determined by the ion beam intensity.

4.5.3. *Electron "heating"*

Electron cooling is designed to decrease beam emittance and extend its lifetime. The decrease in the energy spread shifts the ion beams closer to the

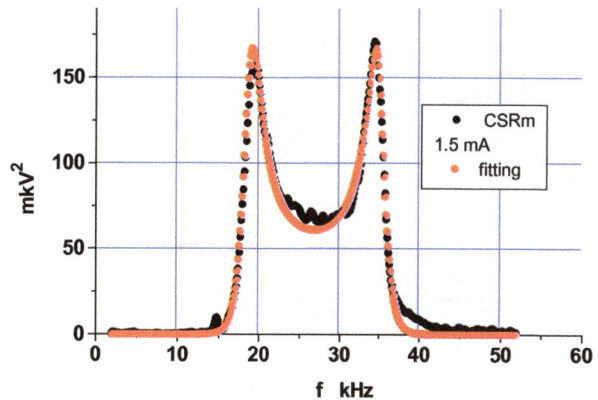

Fig. 16. Noise spectrum of a stored beam of carbon ions in CSRm with a current of 1.5 mA at the 100th harmonic of the revolution frequency (22 MHz) with electron cooling compared to a fit to the theoretical model.

threshold of coherent instabilities whose development is accompanied by beam loss [60]. These instabilities have been successfully suppressed by feedback systems. The most powerful feedback system was stochastic cooling and its action improved the situation dramatically. However, at CELSIUS, large losses of the proton beam were observed with electron cooling, which were not accompanied by clearly observed coherent signals. These events were called "electron heating" [61] and resembled colliding beam effects, or what are now popularly called "electron cloud" effects. As studies have shown, this effect was not directly related to decreasing momentum spread of the ion beam by cooling. The presence of an electron beam at the ion beam orbit with a strongly deflected energy results in the absence of cooling and losses in the intense proton beam.

Electron beam space charge can cause distortion of the storage ring optics and introduce nonlinear resonances. These effects can limit the useful aperture of the storage ring. But, as some experiments showed, strongly expanding the electron beam so that it occupied all of the useful aperture did not improve the situation very much. It occurred to us that the ion–electron wideband oscillations, which are developed at the injection of ions into the electron beam, are responsible for the beam loss. Outside the electron beam, ions can execute normal plasma oscillations with a relatively low frequency determined by the ion beam density and ion mass. Beyond the cooling section, electrons execute similar oscillations but of much higher frequency

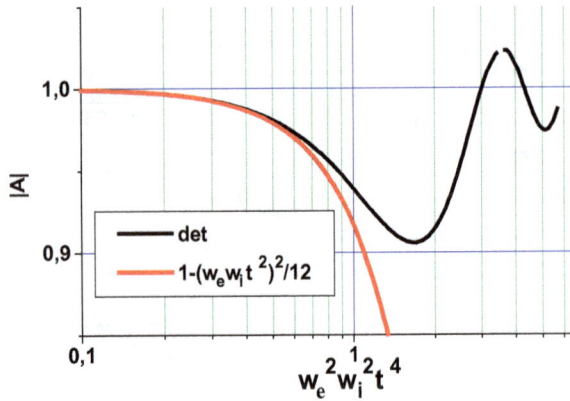

Fig. 17.　Changes of the ion plasma oscillation matrix determinant during the flight of the electron beam with an increase in the electron beam density.

because of the small electron mass. Outside the cooling section, where $n_e = 0$, single component plasma oscillations with a frequency of $\omega_i = c\sqrt{4\pi n_i r_i}$ are observed, and at the entrance into the electron beam, mutual two-component oscillations with a common frequency of $\omega = c\sqrt{4\pi n_i r_i + 4\pi n_e r_e}$ are generated. The matrix determinant value det is not equal to 1 since at each revolution we have fresh electrons and energy exchange occurs. With a low electron density, the damping of ion plasma oscillations will be fast (det < 1). But with high density beams, where det > 1, the plasma oscillations begin to increase their energy at each passage through the cooling section.

Figure 17 shows changes of the matrix determinant with an increase in the electron beam density. These parameters are typical for cooling a proton beam of density $3 \times 10^8 (1/\mathrm{cm}^3)$ at an energy of 100 MeV in a 3-m-long cooler. This is similar to the time limit of stochastic cooling due to the mutual influence of particles in the interaction region. Each ion, interacting with an electron beam, gets a friction force kick in the flight time τ of the electron beam. The parameter for this interaction is the product of the square of the plasma phase shifts for electron and ion beams during the time of passage through the cooling section τ: $\frac{4\pi e^2 n_e}{m} \times \frac{4\pi e^2 Z_i^2}{M} \times \tau^4 = \omega_e^2 \omega_i^2 \tau^4$ [56]. It is clear that a decrease in the electron beam density facilitates shifting this instability and increasing to an acceptable ion beam density in the central zone. This was the basic idea in producing electron coolers with a variable density profile of the electron beam for keeping a high cooling rate at large amplitudes.

At present, under construction are several large projects with electron cooling. The most ambitious project for cooling ions at an energy of 100 GeV is being developed for the RHIC ion–ion collider. A 50 MeV electron beam with a peak current of about 1 A is required. It will use a superconducting energy recovery linac, where the electron bunches after interaction with ion bunches will return their energy to the linear accelerator after passage through the cooling section.

Development of cooling methods is making rapid progress. There is also noticeable progress in the use of stochastic cooling for intense ion beams [62, 63]. Which of these cooling methods will become optimal will be known in the near future.

5. Stochastic Cooling

Stochastic cooling, suggested by Van der Meer [62], is based on the use of active feedback systems interacting with a beam. A transverse cooling system is illustrated in Fig. 18. The longitudinal cooling has been observed with systems similar in concept.

A particle path through the pickup 1 produces a short $\Delta t = 1/W$ pulse of voltage at the output of the preamplifier: $U = ZqWk_1\frac{x}{A}$, where Z is the impedance of the pickup electrodes, W the frequency band of the amplifiers, q the particle charge, k_1 the amplification factor of the preamplifier, and A the characteristic of the pickup electrode approximately equal to the aperture. Through the delay line (3) and the power amplifier (4), a signal is applied to the correction plates (5) (kicker) simultaneously with the arrival of the particles. The distance between the measurement and the kick along the orbit is selected to give a particle time to execute 1/4 of a betatron oscillation, so the maximum position deviation detected by the pickup electrode can be transferred into the maximum velocity kick

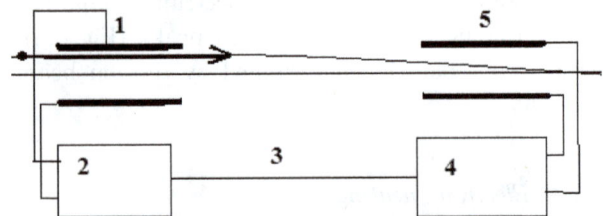

Fig. 18.　Layout of the transverse stochastic cooling. (1) pickup electrodes, (2) preamplifier, (3) delay line for synchronizing ions at kicks, (4) kicker power amplifier, (5) kicker.

in the kicker. As a result of the interaction with the field in the kicker, the particle oscillation amplitude $x = a \sin(Q_x \omega_0 t + \varphi)$, after averaging over the initial phases φ, will be changed as $\Delta a^2 = -\lambda a^2 + \frac{\lambda^2}{2} a^2$, where $\lambda \approx k_1 \frac{r_p \beta_x}{A^2 \beta \gamma}$ is the interaction parameter of the feedback per beam single passage. As is seen from this equation, oscillations of a single particle are damped most rapidly at $\lambda = 1$, but in order to attain fast cooling we need a technically unattainable amplification coefficient. But, since we have to damp not a single particle but a beam with many particles, this will limit the admissible values of the amplification coefficient and make it more reasonable and technically attainable. With the total number of particles in a beam given by N, within the pulse duration $\Delta t = 1/W$, only $N^* = N \frac{f_0}{W}$ particles can participate simultaneously in the interaction. In this case, the equation taking into account the mutual interaction is modified to become $\Delta a^2 = -\lambda a^2 + \frac{\lambda^2}{2} a^2 N^*$. Here we assume that there is no correlation between particles and, on average, the sum of signals from all the particles is equal to 0 and the sum of squared signals grows as the number of particles N^*. In this case, the optimal value for the feedback coefficient turns out to be equal to $\lambda = 1/N^*$, and the damping decrement of the stochastic cooling system will be limited by the value

$$\delta = \lambda f_0 = \frac{W}{N}.$$

Note that in this case the joint (coherent) oscillations of the beam as a whole will be damped in the time of a single turn, $\lambda N^* = 1$.

Despite the decrease in the required amplification coefficient, gain values for k_1 remain rather high. Assuming that at the amplifier input the power is determined by thermal fluctuations at the input impedance, the output power of such a system will be $P = 4kTWk_1^2$. So, if one aims to achieve a cooling time of 1 s at an energy of 8 GeV ($\gamma\beta \approx 8$) and frequency band $W = 5 \times 10^9$ Hz = 5GHz, with revolution frequency $f_0 = 10^6$ Hz, the required value will be $\lambda = 10^{-6}$, and the required amplification factor reaches $k_1 = 5 \times 10^7$. In this case, the system power will be 200 kW, which is still higher than reasonable. In order to solve the problem, a large number of input pickup electrodes are used. When their signals are summed up, the signal–thermal noise ratio becomes noticeably better. So, if we simply take $n = 100$ in parallel in the operating system, for achieving the

same cooling time the amplification factor and the total power of all the amplifiers will be decreased by $1/n$, which enables the development of realistic stochastic cooling systems.

Recently, progress was made in the realization of stochastic cooling for bunched ion beams [63]. In the cooling regime for large colliders, in order to increase luminosity, the beams are grouped in short bunches of a length not exceeding the beta function value at the interaction point. The powerful coherent signal of such bunches is a major obstacle to the development of stochastic cooling systems with high amplification factors for a bunched beam. Successful experiments in the RHIC collider have shown that, in principle, the use of new filters and high frequency amplifiers demonstrated a solution to these problems.

6. Laser Cooling

Experiments in atomic physics with partially ionized ion beams are amenable to laser cooling. Laser cooling is based on the interaction of such ion beams with laser beams resonant at the frequency of atomic transitions. Let us confine ourselves to a simple model of an ion as an oscillator with intrinsic frequency ω_r and absorption line width determined by dipole radiation $\Gamma = \frac{2r_e}{3c} \omega_r^2$ [1]. In this model, the differential cross section of the light quanta is equal to

$$\sigma(\omega) = \frac{8\pi}{3} \frac{r_e^2 \omega^4}{(\omega_r^2 - \omega^2)^2 + \omega^2 \Gamma^2}.$$

At the maximum, the cross section reaches $\sigma_{\max} = 6\pi\lambda^2$. By absorbing the laser beam radiation, the ion receives an impulse from each quantum in the direction of the laser beam propagation equal to $\hbar\omega/c$. Therefore, the ion is subjected to the force $F = \frac{jW}{c} \sigma(\omega)$, where jW is the laser radiation power density at the ion location. In the simplest scheme of laser cooling, in the reference frame accompanying the ion beam, one can produce two laser beams with radiation frequency shifted down by approximately the width of the absorption line $\omega_1 = \omega_r - \Gamma/2$, but oppositely directed (along and counter to the ion beam motion). Then the ion, by moving toward the photon with a frequency less than the resonance frequency (because of the Doppler effect), will see a frequency closer to the resonance and the ion will decelerate more strongly than it will be accelerated by absorption of the accompanying beam quanta far

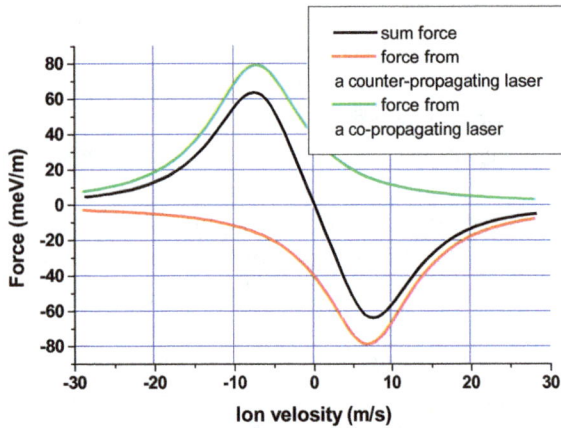

Fig. 19. The total and the two (opposite) laser beam forces acting on the ion depending on the velocity in an accompanying reference system.

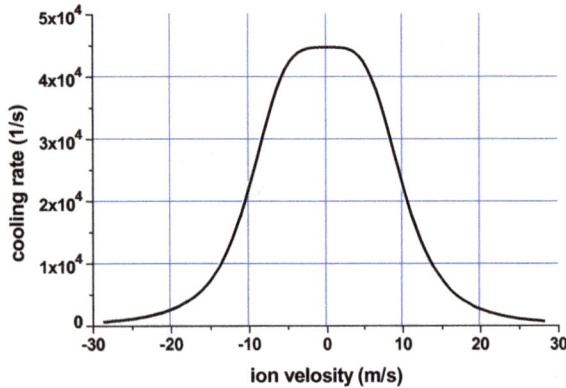

Fig. 20. The cooling rate versus velocity deviation from its equilibrium value.

from the resonance. For illustration, consider the friction force acting on the ion in two laser beams of 1 cm diameter and counterdirected with a beam power of 10 mW each (Fig. 19). The resonance wavelength of the laser light was taken to be 2500 Å and an ion mass of 24, close to the parameters of the Mg^{+1} ion experiments [45–47].

Figure 20 shows the ion cooling rate $\delta = \tau^{-1} = F(V)/M_i V$ as a function of the velocity deviation from its equilibrium value. Near the equilibrium $|V| \ll c\Gamma/\omega_r$, the cooling rate will be

$$\delta = \frac{jW\sigma_{max}}{M_i c^2}\frac{2\omega_r}{\Gamma}.$$

It is seen that, even if the ion moving along the storage ring orbit is subjected to the laser beam action 1% of the time, the cooling time will be rather

short — 2 ms. The processes of radiation and absorption of light quanta are also accompanied by ion heating because of fluctuations of the number of quanta radiated in a given direction. Let us assume that we continuously cool the ion in all three directions by six laser beams. In actual experiments, different systems are used for redistribution of cooling decrements. The ion energy equilibrium equation in one of three directions will have the form

$$\frac{dE}{dt} = -2\delta \times E + \frac{1}{3}\frac{dN}{dt}\frac{(\hbar\omega_r/c)^2}{2M_i},$$

where dN/dt is the number of absorbing and emitting quanta from the six laser beams per second. With heating/cooling equilibrium, the energy spread in one direction in the coordinate system accompanying the beam is $E = \frac{\hbar\Gamma}{2}$. Within the framework of this approximation, it follows that the equilibrium temperature reaches the very low value of 2.7 mK by a very narrow resonant line, $\Gamma/\omega_r \ll 1$. The feasibility of reaching such low temperatures was demonstrated in the laser cooling experiments at Heidelberg and Arhus [64–66]. Obtaining such low temperatures opens up the prospect for fundamental studies in the physics of atoms and in obtaining unique crystalline beams. Of course, handling ions having electrons on the atom shells requires a very high vacuum and not too high beam energy because of the losses and short lifetime of these ions in bending magnets.

It is worth mentioning that the above approximation does not take into account saturation effects, which limit the admissible laser power. If the number of quanta absorbed and irradiated in unit time dN/dt approaches the absorption line width Γ, one should switch to quantum-mechanical calculations, which account for the probability of finding the ion in an excited state. Quite recently, at the new storage ring S-LSR (Japan) [67], special laser cooling experiments have begun. The specific feature of such a storage ring is the use of electrostatic bends with compensation for the ion's longitudinal shift. Such optics will facilitate the study of crystalline beam phenomena under novel conditions.

7. Some Examples of Using Beam Cooling

At present, the most important application of cooling is the storage of positrons and antiprotons for colliding beam experiments. Colliders provide the major part of data in high energy physics. The most important recent developments in beam cooling are in muon beams for neutrino factories and even muon–muon colliders. But these projects require a budget of the order of a billion US dollars and many decades for their realization. This list of the highlights in beam cooling includes using electron cooling to prepare dense bunches of lead ions [42] for ion-on-ion collisions at LHC and very intensive cooling for the internal target experiment of the FAIR project at GSI.

One of the initial applications for cooling was to increase the luminosity in internal target facilities [68, 69]. The luminosity in experiments with a superfine target is determined by the storage rate of particles $\dot{N}$ and cross-section σ_{tot} of their consumption at the target because of scattering and nuclear absorption $L = \dot{N}/\sigma_{tot}$. By decreasing particle consumption, cooling enables the storage of more particles and hence increases luminosity. For example, a study of the sizes of nuclei requires measurements of interference of electromagnetic and nuclear scattering at small scattering angles. If small angle scattering with an angular resolution of $\Delta\theta$ is studied, in the absence of cooling, multiple scattering at the target would increase the angular spread in a beam rapidly so that the beam becomes unsuitable for these experiments in a short time. By decreasing the particle amplitude oscillations, cooling returns particles to the beam so that the only consumption is due to their nuclear interaction σ_n with the target. Therefore, the luminosity increases. For $Z = 100$, $\gamma\beta = 1$, $\Delta\theta = 10^{-4}$ the increase achieved is more than 400 times and this is the main argument for developing experimental storage rings of the COSY-type in Jülich [70], ESR and NESR at GSI [71], and CSR at IMP (China) [72]. The use of a combination of stochastic cooling of rare and hot ions escaping the target and electron cooling confining the beams during the interaction with the target opens up new experimental possibilities.

Cooling opens up interesting possibilities in studies of heavy ions by keeping the ions in the storage rings. In this case, some experiments have already demonstrated the occurrence of radioactive beta decays of nuclei where electrons exiting from a nucleus are captured by free atomic shells. In these processes, a monochromatic neutrino is radiated,

which can be used for precision experiments. Observation with Schottky signals from individual ions is possible since the total ion charge is not changed and a slight variation of its mass leads to a small shift of the revolution frequency between the initial nucleus and the newly generated ion. A new class of experiments with the use of cooling is very attractive for studies in a broad field of physics, ranging from elementary particles to the physics of atoms [73].

Acknowledgments

This article could not have been written without stimulation from Alex Chao and Weiren Chou. We are grateful to Ernie Malamud for assistance in transforming our Russian English into more readable English. We also thank our BINP colleagues and our friends from other laboratories in the cooling community for their many contributions to this field.

References

[1] L. D. Landau and E. M. Lifshits, *Theory of Fields* (1967).

[2] CERN Accelerator School 98-04. Synchrotron radiation and free electron lasers.

[3] H. Ikegami, Coherent microwave cooling (CMC) of electron and ion beams, in *Workshop on Beam Cooling and Related Topics* (4–8 Oct. 1993; CERN 94-03, 26 Apr. 1994), pp. 81–101.

[4] V. V. Berezovsky and L. I. Menshikov, *Lett. JETPH* **86**, 411 (2007).

[5] S. van der Meer, Discussion of Ikegami's paper, in *Workshop on Beam Cooling and Related Topics* (Montreux, 4–8 Oct. 1993); CERN 94-03, 26 Apr. 1994, Proton Synchrotron Division, p. 123.

[6] M. G. Fedurin, M. V. Kuzin, N. A. Mezentsev and V. A. Shkaruba, *Nucl. Instrum. Methods Phys. Res. A* **470**, 34 (2001).

[7] G. I. Budker, *Proc. 7th Int. Conf. High Energy Accelerators* (Yerevan, 1969), p. 33.

[8] G. I. Budker, extract in *Physics Potential and Development of $\mu^+\mu^-$ Colliders: Second Workshop*, ed. D. Cline, *AIP Conf. Proc.* **352**, 4 (1996).

[9] A. N. Skrinsky, Intersecting storage rings at Novosibirsk, talk given at *Int. Sem. Prospects of High-Energy Physics* (Morges, Switzerland), printed at CERN (unpublished) (Mar. 1971).

[10] A. N. Skrinsky, extract in *Physics Potential and Development of $\mu+\mu-$ Colliders: Second Workshop*, ed. D. Cline, *AIP Conf. Proc.* **352**, 6 (1996).

[11] A. N. Skrinsky, *XX Int. Conf. High Energy Physics*, invited talk (Madison, 1980).

[12] A. N. Skrinsky, *Proc. N.Y.* **2**, 1056 (1981).

[13] A. N. Skrinsky, *AIP Conf. Proc.* **68**, 1056 (1980).

[14] A. N. Skrinsky, *Usp. Fiz. Nauk* **138**(1), 3 (1982). (in Russian). *Sov. Phys. Usp.* **25**(9), 639 (1982). (transl.).

[15] A. A. Kolomensky, *At. Energ.* **19**, 534 (1965).

[16] Yu. M. Ado and V. I. Balbekov, Ionization cooling, *At. Energ.* **39**, 40 (1971).

[17] A. N. Skrinsky and V. V. Parkhomchuk, *Sov. J. Part. Nucl.* **12**, 223 (1981).

[18] V. V. Parkhomchuk and A. N. Skrinsky, *Proc. 12th Int. Conf. High Energy Accelerators* (Batavia, Illinois, 11–16 Aug. 1983), pp. 485–487; reprinted in *Sausalito 1994: Physics Potential and Development of $\mu^+\mu^-$ colliders*, pp. 7–9. Published in *AIP Conf. Proc.* **352**, 7 (1996). Also in *Proc. 12th Int. Conf. High Energy Accelerators*, pp. 485–487, and *Sausalito 1994: Physics Potential and Development of $\mu^+\mu^-$ Colliders*, pp. 7–9.

[19] A. N. Skrinsky, Ionization cooling and muon collider, in *Proc. 9th ICFA Beam Dynamics Workshop: Beam Dynamics and Technology; Issues for Muon-Muon Colliders* (Montauk, New York, 1995).

[20] A. N. Skrinsky, *AIP Conf. Proc.* **372**, 133 (1996).

[21] A. N. Skrinsky, *Nucl. Instrum. and Methods A* **391**, 188 (1997).

[22] A. N. Skrinsky, *Proc. Long Beach 2002: Physics and Technology of Linear Accelerator Systems* (Accelerator School, 2002), pp. 322–355.

[23] M. Charles Ankenbrandt *et al.*, Status of muon collider research and development and future plans, BNL-65623, FERMILAB-PUB-98-179, LBNL-41935, LBL-41935, Aug. 1999. 95.

[24] M. Charles Ankenbrandt *et al.*, *Phys. Rev. ST Accel. Beams* **2** (0810001), 1999. e-Print: physics/9901022.

[25] R. B. Palmer, D. Neuffer and J. Gallardo, *Advanced Accelerator Concepts: 6th Annual Conference*, ed. P. Schoessow, *AIP Conf. Proc.* **335**, 635 (1995).

[26] D. Neuffer and R. B. Palmer, *The Tamura Symposium*, ed. T. Tajima, *AIP Conf. Proc.* **356**, 344 (1996).

[27] A. N. Skrinsky, Muon collider at 10 TeV to 100 TeV, in *HEMC'99 Workshop* (Montauk, 1999).

[28] A. N. Skrinsky, *AIP Conf. Proc.* **530**, 311 (2000).

[29] C. Rubbia, A. Ferrari, Y. Kadi and V. Vlachoudis, *NIM Phys. Res. (A)* **568**, 475 (2006).

[30] G. I Budker, *At. Energ.* **22**, 346 (1967).

[31] G. I. Budker, G. I. Dimov, A. G. Popov, Yu. K. Sviridov, B. N. Sukhina and I. Ya Timoshin, *At. En.* **19**(6), 507 (1965).

[32] G. I. Budker, G. I. Dimov and V. G. Dudnikov, *Sov. At. En.* **22**, 384 (1967).

[33] G. I. Dimov and V. G. Dudnikov, *Phys. Plasmas* **4**(3), 1978.

[34] Ya. S. Derbenev and A. N. Skrinsky, Preprint INP 1968, n. 225, *Particle Acceleration* **8**, 1 (1977).

[35] Ya. S. Derbenev and A. N. Skrinsky, *Plasma Phys.* **4**(3), 492 (1978) (in Russian).

[36] Ya. S. Derbenev and A. N. Skrinsky, *Particle Accelerators* **8**(4), 235 (1978).

[37] Ya. S. Derbenev and A. N. Skrinsky, *Sov. Sci. Rev., Sec. A. Phys. Rev.* 65 (1981).

[38] V. I. Kudelainen, I. N. Meshkov, V. V. Parkhomchuk, R. A. Salimov, A. N. Skrinsky and V. G. Fainshtein, *J. Tech. Phys.* **46**(8), 1678 (1976).

[39] G. I. Budker, N. S. Dikansky, V. I. Kudelainen, I. N. Meshkov, V. V. Parkhomchuk, A. N. Skrinsky and B. N. Sukhina, *IEEE Trans. Nucl. Sci.* **22**, 2003 (1975).

[40] G. I. Budker, N. S. Dikansky, V. I. Kudelaineen, I. N. Meshkov, V. V. Parkhomchuk, D. V. Pestrikov, A. N. Skrinsky and B. N. Sukhina, *Particle Accelerators* **7**(4), 197 (1976).

[41] G. I. Budker and A. N. Skrinsky, *Usp. Fiz. Nauk* 561, 1978.

[42] G. Tranquille, *Proc. COOL07*, http://bel.gsi.de/cool07/PAPERS/TUM1I01.PDF

[43] T. Shirai, S. Fujimoto, M. Ikegami and A. Noda, *Proc. COOL05, AIP* **821**, 103 (2006).

[44] S. Nagaitsev, A. Bolshakov, D. Broemmelsiek and A. Burov, *Proc. COOL05, AIP* **821**, pp. 39–45, 2006. http://conferences.fnal.gov/cool05/Presentations/Monday/M07_Nagaitsev.pdf

[45] Electron cooling for RHIC, C-A\AP 47, Apr. 2001, www.bnl.gov/cad/ecooling/docs/PDF/AP_notes/ap_note_47.pdf, http://www.bnl.gov/cad/ecooling

[46] A. Kobets, Y. Korotaev, V. Malakhov and I Meshkov, *AIP Conf. Proc.* **821**, 95 (2005).

[47] A. Ivanov, A. Bubley, A. Goncharov, E. Konstantinov, S. Konstantinov, A. Kryuchkov, V. Panasyuk, V. Parkhomchuk, V. Reva, B. Skarbo, B. Smirnov, B. Sukhina, M. Tiunov, M. Zakhvatkin and X.-D Yang, *Proc. EPAC02*, 1356 (2002).

[48] A. Bubley, V. Panasyuk, V. Parkhomchuk and V. Reva, *Instrum. Exp. Tech.* (1), 91 2006.

[49] P. Logachev, PhD thesis (BINP, 1996).

[50] P. Beller, K. Beckert, B. Franzke, C. Kozhuharov, F. Nolden and M. Steck, *Nuclear Instru. Methods Phys. Res. Sec. A: Accelerators, Spectrometers, Detectors and Associated Equipment* **532**, 427, Issues 1–2, 11 October 2004.

[51] L. N. Arapov, N. S. Dikansky, V. I. Kokoulin, V. I. Kudelainen, V. I. Lebedev, V. V. Parhomchuk, B. M. Smirnov and B. N. Sukhina, *Proc. 13th Int. Conf. High Energy Accelerators.* (7–11 Aug. 1986), Vol. 1, pp. 341–343.

[52] V. Bocharov, A. Bubley, S. Konstantinov, V. Panasyuk and V. Parkhomchuk, *AIP Conf. Proc.* **821**, 360 (2005).

[53] V. Tupikov, G. Kazakevich, T. Kroc, S. Nagaitsev, L. Prost, A. Shemyakin, C. Schmidt, M. Sutherland and A. Warner, *AIP Conf. Proc.* **821**, 375 (2005).

[54] V. Reva, V. Parkhomchuk and B. Skarbo, *Instru. Exp. Tech.* **49**(3), 446 (2006).

[55] V. V. Parkhomchuk and A. N. Skrinsky, *Rep. Prog. Phys.* **54**(7), 919 (1991).

[56] V. V. Parkhomchuk, *Proc. COOL 99, NIM Phys. Res. A* **441**, 9 (2000).

[57] D. L. Bruhwiler and G. I. Bell, *Proc. COOL07*, http://bel.gsi.de/cool07/TALKS/WEM2C04_TALK.PDF

[58] A. V. Fedotov, B. Galnander and V. N. Litvinenko, *Proc. COOL05 AIP, Proc. Conf.* **821**, 265 (2006) http://conferences.fnal.gov/cool05/Presentations/Thursday/R05_Fedotov.pdf

[59] V. V. Parkhomchuk and D. V. Pestrikov, *JETP* **50**, 1411 (1980).

[60] J. Bosser and C. Carli, *Proc. COOL05, Nucl. Instrum. Methods Phys. Res. A* **441**, 1 (2000).

[61] D. Reistad *et al.*, *Proc. Workshop on Beam Cooling and Related Topics* (Montreux, Oct. 1993; CERN 94-03), p. 183.

[62] D. Mohl, G. Petrucii, L. Thorndahl and S. van der Meer, *Phys. Rep.* **58**, 73 (1980).

[63] J. M. Brennan and M. Blaskiewicz, *Proc. COOL07*, http://bel.gsi.de/cool07/PAPERS/MOA1I01.PDF

[64] S. Schroder *et al.*, *Phys. Rev. Lett.* **64**, 2901 (1990). J. S.Hangst *et al.*, *Phys. Rev. Lett.* **67**, 1238 (1991).

[65] J. S. Hangst, *NIM Phys. Res. A* **441**, 196 (2000) ECOOL'99.

[66] A. Noda *et al.*, *Proc. COOL07*, http://bel.gsi.de/cool07/PAPERS/FRM1I01.P DF

[67] G. I. Budker and A. N. Skrinsky, *Proc. Xth Int. Acceleration Conf.* (Protvino 1977), Vol. 2, p .141.

[68] G. I. Budker and A. N. Skrinsky, *Proc. of G. I. Budker Articles* 470 (1982).

[69] COSY ring, http://www.kfa-juelich.de/ikp/ikp-general/cosyh_e.html

[70] FAIR project, http://www.gsi.de/fair/index_e.html

[71] CSR project, http://www.impcas.ac.cn/zhuye/en/main.htm

[72] Fritz Bosch, *Proc. COOL07*, http://bel.gsi.de/cool07/TALKS/THA2I01_TALK.PDF

Vasily Parkhomchuk is a Senior Staff Scientist at the Budker Institute of Nuclear Physics. He has been concentrating on electron cooling research, beginning with the first electron cooling experiment at BINP, which was his Ph.D. thesis, and continuing to the present. He participated in developing the electron cooling systems for GSI (Germany), IMP (China) and CERN (LEIR), and is now involved in the development of new projects. Parkhomchuk is a Corresponding Member of the Russian Academy of Sciences.

Alexander Skrinsky is Director of the Budker Institute of Nuclear Physics. He is author or co-author of many accelerator ideas and developments: e^+e^- colliders, proton–anti-proton colliders, electron cooling, experiments using polarized beams, and measurements with record accuracy of elementary particle masses using resonant depolarization. He has contributed to linear e^+e^- colliders, μ^+ μ^+ colliders using ionization cooling, and is coinventor of the optical klystron and the high mean power free electron laser based on accelerator–recuperators. He is very fond of cross-country skiing in Siberia. Skrinsky is an Academician of the Russian Academy of Sciences, a Foreign Member of the Swedish Royal Academy of Sciences, and a Fellow of the American Physical Society.

Reviews of Accelerator Science and Technology
Vol. 1 (2008) 259–302
© World Scientific Publishing Company

The Supercollider: The Pre-Texas Days
A Personal Recollection of Its Birth and Berkeley Years

Stanley Wojcicki

Department of Physics, Stanford University,
382 Via Pueblo Mall, Stanford, CA 94305, USA
sgweg@slac.stanford.edu

This article describes the beginnings of the Superconducting Super Collider (SSC). The narrative starts in the early 1980s with the discussion of the process that led to the recommendation by the US high energy physics community to initiate work on a multi-TeV hadron collider. The article then describes the formation in 1984 of the Central Design Group (CDG) charged with directing and coordinating the SSC R&D and subsequent activities which led in early 1987 to the SSC endorsement by President Reagan. The last part of the article deals with the site selection process, steps leading to the initial Congressional appropriation of the SSC construction funds and the creation of the management structure for the SSC Laboratory.

Keywords: Accelerator; collider; superconductivity; SSC; CDG; superconducting magnets.

1. Introduction

This article is the first in a two-part account of the history of the Superconducting Super Collider (SSC), or the Supercollider for short. Its story starts in the early 1980's, when initial stage-setting events for the SSC took place, and runs to the end of 1988, when the activities of the Central Design Group (CDG), responsible for the oversight and coordination of the early SSC efforts, started being phased out and the SSC focus shifted from Berkeley to Texas. The account of the years from 1989 to 1993 when the SSC was terminated, is the subject of a companion article [1]. This combined account relies to a large extent on numerous relevant documents and contemporary articles, but is also very much shaped by my personal recollections.

The SSC was to be the biggest and costliest basic research instrument ever constructed. The proposing high energy physics (HEP) community argued persuasively that such a large scale instrument was essential for attacking the frontier problems in particle physics. Because of its unprecedented size and cost, the SSC generated a lot of controversy both in scientific circles and among the public at large. Thus the history of the SSC has to involve not only science and technology, but also politics and public relations.

I do not attempt to provide a detailed technical description of the SSC and of the technical challenges that were faced, and to a large extent met, during its brief history. Instead, I try to present to a rather general scientific audience my recollections of the scientific, technological and political background which led to the SSC proposal, of the efforts to make the SSC a reality, and finally (in the second article) of the events that led to its demise. I had strong involvement with and in the SSC over most of that time; in such circumstances development of some biases is probably unavoidable. I have tried to maintain objectivity in this article, but some prejudices may flavor this account and the reader might keep that in mind [2]. In writing this article I have profited from discussions with many colleagues involved with the CDG.

My association with the SSC had two distinct phases. During the period discussed here, I was intimately involved in SSC, and my perspective is one of an insider. On the other hand, during the Texas days, most of the time my involvement was from outside, in my role as a High Energy Physics Advisory Panel (HEPAP) member and later as its chair. I did spend eight months at the SSC Laboratory (SSCL) on my sabbatical in 1993, just before the SSC cancelation, and obtained additional insight from that vantage point.

The story and the lessons of the SSC might be especially timely today, since the international HEP community is proposing to build a new facility, the International Linear Collider (ILC). There are many parallels with the SSC — in the cost, the technological challenge, and the difficulty of its realization. Study of what went wrong, but also what went right, with the SSC might be instructive as we grapple with how one might best achieve the HEP goals.

2. The Early 1980's

It is quite common today in federally sponsored research for the funding agencies, in deciding on allocation of funds, to rely heavily on advice from one or more peer review groups specially constituted for this purpose. In HEP in the US, the principal body providing this function is HEPAP. It was initiated in 1967 — with Victor Weisskopf, a professor of physics at MIT and former director-general (DG) of CERN, as its chair — as an advisory group to the Atomic Energy Commission (AEC), the federal agency responsible for the stewardship of the major part of the US HEP program. This advisory structure carried over to the Energy Research and Development Administration (ERDA) as AEC was phased out, and more recently to the Department of Energy (DOE) as that department absorbed ERDA. The other major US HEP funding agency, the National Science Foundation (NSF), has always had access to HEPAP deliberations and recommendations, but it was only in the last decade that both agencies assumed formally the responsibility for HEPAP membership and activities.

HEPAP has typically 15–20 members, each serving for 3 years, and meets 3 or 4 times a year. Its membership is composed mainly of relatively senior US HEP physicists, with a few members from abroad and from related fields or industry. HEPAP, as a chartered federal panel, has to meet in public. That requirement makes the debates open to all interested parties and thus restrains the discussion of controversial issues. This feature and its infrequent meeting schedule means that HEPAP is not ideal for debating and deciding on recommendations for the more complex and controversial issues.

The mechanism for dealing with such topics is the appointment of HEPAP subpanels, with a very specific charge and a more limited term of appointment for its members. Each subpanel, after its deliberations, which can be held in closed sessions, writes a report that is presented to HEPAP, which discusses it and can accept it or reject it (so far always the former, if I remember correctly).[a] Most often the main issue presented to the subpanels has been long range planning, frequently focusing on relative priorities of new accelerator facilities being proposed. The first of these subpanels was formed in 1974 and was chaired by Weisskopf. Such a subpanel, named the Subpanel on Long Range Planning for the US High Energy Physics Program, was formed in 1981 with George Trilling, a professor of physics at UC Berkeley, as its chair, to address the anticipated key issues in the field in the 1980's.

Before discussing the charge of this subpanel and its recommendations, it is useful to recall the overall situation in HEP in the early 1980's. The mood among the US HEP community at that time could be characterized as one of malaise. The successes of the 1970's — the concurrent J/ψ discovery at Brookhaven National Laboratory (BNL) and Stanford Linear Accelerator Center (SLAC), the τ lepton discovery and elucidation of the charm sector at SLAC, the startup of Fermilab several years ahead of SPS (400 GeV proton synchrotron) at CERN — were things of the past. The community became used to being pre-eminent in the field, but the future could not promise that this would continue. All indications were that the torch of leadership was passing to Europe, where it might well stay for the forthcoming decade, with the US lagging a poor second behind [3].

The European e^+e^- high energy collider, PETRA, was completed a year ahead of its US competitor, PEP, and was the first one to provide conclusive evidence for gluons. CERN SPS physics was eclipsing that from the Fermilab accelerator, both in neutrino physics and in high energy muon scattering studies. CERN was about to succeed in achieving high energy proton–antiproton collisions in its SPS and thus very likely to observe both W and Z

---

[a]This practice has changed in recent years. HEPAP now receives the almost final version of the reports some time before its formal meeting and its members can provide initial feedback to the Subpanel.

bosons (predicted by the Standard Model of particle physics, which had passed all the other experimental tests to date). In contrast, Fermilab was involved in an ambitious program to convert its accelerator to a superconducting fixed target accelerator [4] and eventually to a 2 TeV center-of-mass energy superconducting collider. This was, however, a long range and risky endeavor which might or might not succeed. The construction of the CERN LEP collider, which would be a Z factory and a means of studying e^+e^- interactions at energies beyond the Z, was to be initiated in 1984. The ambitious SLAC Linear Collider (SLC) project, however, was still in an R&D phase and its complexities led many to doubt whether its design goals would even be approached.

Simultaneously, serious strains and cracks were developing in the US HEP structure. US HEP at that time was effectively based on three principal DOE-supported accelerator laboratories: BNL, Fermilab and SLAC (with an important smaller effort at Cornell, supported by NSF). This reflected the fact that the accelerator facilities, to be on the frontier, had to be so big and hence so expensive that they had to be few in number and built on a regional (or even national) basis. By early 1980's, with HEP budgets fixed (or decreasing in terms of purchasing power), many in the community questioned whether the three-lab model was viable for the long term future.

A significant factor responsible for the difficult situation in US HEP was a mismatch between the expected and the actual funding. In 1978 an agreement was reached between the Office of Management and Budget (OMB) and DOE as to the levels of funding for HEP in the future. Those levels, plus ones a few percent above and below, were given to the HEPAP subpanels as the yardsticks for planning future programs. Since the actual appropriations invariably turned out to be lower than in even the lowest scenarios, it is not surprising that the recommended and generally adopted programs could not be fitted into the appropriated budgets.

This dilemma was brought to the fore by the issue of ISABELLE, a p–p collider with 800 GeV total energy in the center of mass [5]. This collider, using superconducting magnets, was proposed by BNL in 1976 and endorsed by a HEPAP subpanel [6] in 1977. The project was included by DOE in its budgetary submission and approved by Congress in

1978. By 1981 the conventional construction, including the tunnel, was essentially finished but on the technical front serious problems developed [7]. When the project was first proposed, the superconducting technology for accelerator magnets was still in its infancy. BNL gambled on the so-called "braid" superconducting cable for the magnets. Even though successful in few early "handmade" prototype magnets, the cable did not appear to be adequate for mass production and thus the technical side of the project was delayed. The braid was originaly developed at BNL, so there was probably some sentimental attachment to it. An alternative, called Rutherford cable, was developed in that laboratory and had great promise, but no significant data were available as yet at that time. Fermilab started an effort, using this cable, to upgrade its 400 GeV synchrotron to a superconducting accelerator, with initially a fixed target capability and later evolving toward a proton–antiproton collider. Besides the technical concerns, there was an additional question whether these two facilities were sufficiently complementary to warrant construction of both of them within the limitations of the projected HEP budgets.

In trying to understand the sociological background in US HEP in the early 1980's, one must also look at the overall national scene. The late 1970's were years of malaise also on the overall national scale. The oil embargo, high inflation and high interest rates, the Iran hostage situation, all contributed to a great deal of self-doubt in the US population at large. Undoubtedly it was this atmosphere and Reagan's promise to reverse it that played a major role in Carter's defeat in 1980.

George Keyworth was appointed as Reagan's science adviser and Director of Office of Science and Technology Policy (OSTP). He is best known as an advocate of the Strategic Defense Initiative (SDI), known as Star Wars. He believed in bold steps in scientific initiatives and was concerned that the US was squandering resources by adhering to "entitlement" programs for various labs rather than trying to optimize the program for the best science by setting priorities [8, 9]. It is not surprising that he believed that ISABELLE was a mistake and one should strive for a more ambitious project.

It was in this climate that DOE convened the Trilling Subpanel. As expected, its charge dealt with the optimization of the US HEP program [10]. But it also emphasized the request for a recommendation regarding ISABELLE; three different budgetary levels were put forth as the framework for specific recommendations. The most relevant parts of the charge are quoted below:

> "Develop a strategy and long-range plan for the U.S. Program over the next decade under the following funding constraints:
>
> 1. The DOE/OMB agreement of 1978, which is equivalent to a DOE level of $423M in FY 1982 dollars.
> 2. A level of 10 percent lower.
> 3. A level of 10 percent higher.
>
> ...submit an interim report which evaluates the priority of ISABELLE in each of the three constraint cases, taking into consideration the physics potential of the facility, the funding and schedule required to complete construction, and competing needs and opportunities...."

The deliberations of the Trilling Subpanel took up a large part of 1981. Even though the Subpanel members do not represent their regions or their home laboratories, it is difficult in such deliberations to completely divest oneself of one's more parochial interests, whether they be connected with one's scientific area of interest or with the laboratories hosting one's experimental program. Thus it is not surprising that there would be a strong difference of opinion on the most controversial issue, i.e. ISABELLE. The final report represented a carefully worded compromise; it was not finalized until January 1982.

In retrospect, three Subpanel recommendations and suggestions [11] are especially important in their relevance to the topic of this article:

(1) Regarding ISABELLE, the Subpanel concluded that in the middle level scenario "construction of this scientifically valuable facility would have to be abandoned in its present scope."
(2) In the case of termination of the ISABELLE project, the report gave a recommendation: "Start by the mid-1980's on a new high energy construction project of more modest financial scope." Some possible projects were mentioned in the report.
(3) The Subpanel also addressed the issue of long range planning within HEP, pointing out that

some felt disfranchised in this currently HEPAP-dominated process. The Subpanel suggested that it would be appropriate if "DPF (Division of Particles and Fields) on its own made a larger effort in the planning area. This effort might involve such activities as sponsorship of appropriate workshops...."

This last recommendation, made without much discussion or controversy, had the most immediate impact. Charles Baltay, a Physics professor at Yale and an incoming (January 1982) chair of the Division of Particles and Fields (DPF) of the American Physicsal Society (APS), took this message to heart and got the DPF to organize (and funding agencies to sponsor enthusiastically) a planning workshop in the summer of 1982 in Snowmass, Colorado. The previous HEP workshops tended to address physics at a specific facility and thus were sponsored by single laboratories. This one was organized quite differently: different physics topics, deemed as worthy of investigation, were examined from the point of view of how effective different facilities would be in addressing them. This extremely useful exercise provided a yardstick by which different potential facilities could be measured. A possible downside was that some would accept the results unequivocally, forgetting that a lot of these conclusions were highly model-dependent.

A multi-TeV machine was a topic of two previous workshops in the late 1970's, sponsored by the International Committee on Future Accelerators (ICFA) [12, 13]. ICFA was chartered by the International Union for Pure and Applied Physics (IUPAP) with a charge to organize workshops, exchange information and hold joint studies on a Very Big Accelerator (VBA). The VBA was conceived in the 1950's not only as an ambitious future facility requiring global participation in construction but also as a means of bringing together scientists from different nations to work jointly on a scientific project and thus further the cause of peace [14]. But little concrete progress was made, and even at the time of the ICFA workshops the theoretical motivation for the needed energy was not well established and the required technology was still far from proven. The Snowmass '82 workshop was thus the first one where a collider in the 20 TeV energy/beam range was discussed in some detail [15] and with the hope that such a facility might be feasible. From the theoretical point of view, this was the machine called for. The Standard Model of particle physics was getting established as an excellent paradigm for explaining phenomena in the sub-100 GeV energy region. But the model was incomplete and there were strong theoretical arguments that new phenomena must occur in the TeV or below constituent energy range.

From the experimental point of view, a machine probing this energy would represent a huge leap forward, and addressing those challenges was the main focus of the accelerator working group at the workshop. Several ideas were put forth as to how this energy regime could be attained. They spanned semi-conventional approaches based on high field magnets ($\sim$10 T) to (hopefully) lower cost approaches based on lower field ferromagnets with superconducting coils. Maury Tigner, who was the coordinator of this group, summarized the outcome of the discussions [16]: "...we were able to give some hope to the idea that a well designed and concentrated R/D program, elaborating much further on technologies we now possess, might bring a 20 TeV facility within our national reach." He highlighted three central challenges for this R&D program:

- Achievement of virtual automation of superconducting magnets, accelerator housing and other accelerator component manufacture and installation;
- Achievement of a thorough understanding of the field–cost relation for superconducting magnets;
- Achievement of a thorough understanding of the luminosity–aperture–energy relation.

The outcome of the Snowmass '82 workshop was a strong desire and enthusiasm for exploring the multi-TeV region in the next generation of hadron colliders. Many individuals became strong proponents for such an action; Leon Lederman was one of the strongest and most vocal advocates of building a multi-TeV hadron collider and at the same time terminating the ISABELLE construction.

3. Formation of a New Subpanel

Several disparate factors generated the driving force for another subpanel, not even a year after the previous subpanel's report was finalized. There was

an awakening enthusiasm for a bold new initiative, lingering unease about what to do about ISABELLE (renamed now the Colliding Beam Accelerator, CBA) and fear that US HEP was losing its pre-eminence. Keyworth shared all of these sentiments and was probably quite influential in creating this subpanel, hoping that it would suggest killing CBA and starting on a multi-TeV collider. The charge of the Subpanel [17] verbatim, was:

> "Consider and make recommendations relative to the scientific requirements and opportunities for a forefront United States High Energy Physics Program in the next five to ten years. Make specific recommendations with respect to possible new construction items for FY1985. In its deliberations the subpanel should consider opportunities that may be available later as it weighs its specific recommendations. The report of the Subpanel should include a definite recommendation concerning the proposed Colliding Beam Accelerator at Brookhaven National Laboratory. The Subpanel should estimate the program funding required to realize the scientific opportunities associated with the recommendations, and should give the relative priorities of the actions it recommends."

This charge differed dramatically from those given to the previous subpanels. There was no funding constraint within which the recommendations had to fit; rather, the Subpanel was being invited to specify the funds needed to restore a forefront US program. The reference to "possible construction items" was also rather open-ended; traditionally subpanels reacted to proposals from the laboratories rather than initiating something on their own. Finally, one might contrast the word "definite" in front of CBA recommendation with "...evaluates the priority of ISABELLE...." in the 1981 charge.

I was recruited to be the chair of this Subpanel in the fall of 1982, via a phone call from Bill Wallenmeyer, Director of the Office of High Energy Physics in DOE. I was a somewhat curious choice for this role. I had never been a member of HEPAP or of any of its subpanels. But I had been involved quite extensively in various committees dealing with science policy decisions and specific US HEP programs: I had served as a chair of both the Fermilab and SLAC Program Advisory Committees, chaired one of the subgroups of the recently concluded DOE comprehensive review of the US HEP university program

(TACUP), and was a member of a recent committee to review the *raison d'être* of both the LBL and Argonne HEP programs. There was probably a desire to have someone in that chair position young enough to feel as if he or she had a stake in the outcome of the process and in the future of the US HEP program.

The Subpanel had 17 members, and its membership represented an effort to include all segments of the US HEP community [18]. There was one non-HEP member, Arthur Kerman, a respected nuclear physicist from MIT. There were two Europeans on the Subpanel. One was John Adams, a well-known British accelerator physicist and a former CERN (DG). The second was Carlo Rubbia, the leader of the UA1 experiment at CERN, probing for the first time high energy p-pbar collisions with the goal of discovering the predicted carriers of the weak force, charged W and neutral Z bosons. At that time Rubbia had a joint appointment at CERN and on the Harvard faculty. Jack Sandweiss, as the HEPAP chair, attended all the Subpanel meetings. An important and intentional difference in the composition of this Subpanel was that there were no members from the DOE proton accelerator laboratories. The purpose was to exclude people with potential conflicts; the drawback was exclusion of some expertise on accelerator matters.

The schedule outlined for the Subpanel's work was very tight. After an organizational meeting in Washington on February 25/26, 1983, there were to be three meetings at Fermilab, BNL and SLAC, with a final meeting at Woods Hole, Massachusetts, on June 5–11, 1983, to write the report. There was a full HEPAP meeting scheduled for June 29/30 to discuss this report. As usual, the timeline was driven by the US budgetary schedule.

From the very beginning it was clear what the main issues were going to be [19]:

(1) Does one press for a bold new initiative?
(2) What should be the recommendation on the CBA?
(3) How does an intermediate construction project (*à la* the CBA) interact with the potential new initiative? Is it needed for the health and vitality of the US HEP program?

Thus the deliberations of the Subpanel focused to a large extent on these three issues.

4. New Subpanel Activities and Related Events

The first, organizational, meeting in Washington created the structure for the Subpanel's work and included presentations from Alvin Trivelpiece (Director of DOE Office of Energy Research), Keyworth and Marcel Bardon (Head of the Physics Division at NSF). The tenor of those presentations can be summarized in a few lines taken from my summary of that meeting written for the record and for the benefit of the members unable to attend [20]: "...the message conveyed was quite clear and unequivocal: the Panel should ignore any political, geographical or institutional issues and concentrate on making a recommendation based on objective scientific issues. In addition, the budgetary questions should not drive our recommendations; we should concentrate on what is needed to reestablish American leadership and preeminence in the field during the next decade.... Dr. Keyworth emphasized that he will support good, sound scientific judgement that is presented in a well-reasoned way... on the specific issue of CBA, it was stressed that what is needed is a firm unequivocal decision, arrived at on purely scientific grounds... we should not shy away from 'thinking big' when writing a recommendation."

I divided the Subpanel membership into three task forces, each charged with acquiring and evaluating information on a specific topic relevant to the overall charge of the committee. However, putting all this information together and arriving at recommendations would be the responsibility of the whole Subpanel. The three task forces and their chairs were: Accelerator Technical Task Force (Maury Tigner, professor at Cornell, chair), CBA Physics Task Force (H. H. Williams, professor at University of Pennsylvania, chair), Overall Program Task Force (C. Baltay, chair). "CBA Physics" was somewhat of a misnomer, since that group was not only to look at CBA physics but also to compare its potential with that of other possible intermediate time scale machines. The Program Task Force was to estimate future trends in the field and evaluate risks, benefits and costs of possible scenarios. It was also charged with appraising HEP plans abroad and evaluating US HEP manpower data.

It was important to get input from all members of the community. I later wrote a letter [21] to all DPF members describing the charge and our plans and inviting them to express their opinions by writing to the Subpanel. The four lab directors (Leon Lederman, Fermilab; W. K. H. Panofsky, SLAC; Nick Samios, BNL; and Boyce McDaniel, Cornell) were invited to make presentations to the Subpanel describing programs at their laboratories and giving their vision of the future. They were evenly split on the CBA, with Panofsky and Samios advocating its continuation. Former and current HEPAP chairs (Weisskopf; Sid Drell, Deputy Director at SLAC; and Sandweiss) were also invited to present their views at subsequent meetings. I was charged also with writing to John Adams (absent from this first meeting) and asking him to prepare a document outlining the European plans for the next 5–15 years.

The general format of the laboratory meetings — Fermilab, Brookhaven and SLAC — was similar but with somewhat different emphasis at each lab. Naturally, at Brookhaven the focus was on the CBA — its physics reach, relevant technical issues, and costs. The meeting at SLAC was the last laboratory meeting, so considerable time was spent on discussing the issues stated in the charge. At the evening town meeting there was an opportunity for Q&A between the audience and the Subpanel members.

Several new ideas and concepts were put forward during the Subpanel's deliberations. Fermilab proposed construction of a dedicated collider (DC) — a 2 TeV per beam p-pbar collider based on Tevatron-style magnets and using proven technology [22]. It would have better reach than the CBA in a number of different areas. BNL put forward a new cost estimate for completion of the CBA, somewhat lower than the previous ones: US$218 million with the first colliding beams in October 1987. They also suggested that the CBA machine could be a stepping stone to a high energy collider in the 10–30 TeV range to be built at or contiguous to BNL — a "Sandatron" — using the CBA as an injector [23]. SLAC presented the possibility of an e^+e^- linear collider in the 1 TeV range [24], but it was clear that such a machine required considerably more R&D than a high energy pp machine.

A number of relevant and important events took place during the first half of 1983, in parallel with the Subpanel's work. The most important ones were:

(1) DPF workshop at Berkeley (February 28–4 March) on detector issues in a high radiation environment, which would be expected either at the CBA or at a multi-TeV hadron collider.

(2) DOE review of the CBA technology and costs. On the whole it was quite positive; the magnet problems appeared to have been solved, mainly through adoption of Fermilab Rutherford cable, with Bob Palmer playing a key role here; the cost and schedule projections were reasonable. The new management appeared to have the situation well in hand.

(3) Workshop on R&D for a 20 TeV Collider, held at Cornell form March 28 to April 2. Organized by Maury Tigner, it had as its goal the identification of the most important R&D issues relevant to the construction of a 20 TeV collider [25]. A rough construction cost estimate was also made for such a machine — US$1.72 billion. This cost, however, included no R&D, no preoperation, no detectors, no contingency and no escalation.

(4) Discovery of the W and Z bosons [26–29] at CERN in 1983.

The Subpanel received over 200 letters from the community addressing the issues spelled out in the Subpanel's charge. Almost without exception, they supported the idea of building a high energy collider; many of them felt that was essential. A flavor of these opinions is well represented by the following quote from one of the letters: "We should place as the highest priority of the American program the design and construction of a super high energy (>20 TeV) hadron collider."

The comments regarding the CBA varied, however. The majority of the letters were against it; some, mainly from the traditional BNL users, argued for its construction. The two quotes below, one pro and one con, are representative of these two sets of opinions:

"I believe the CBA should be built and be ready for experiments before the end of 1987. The physics potential justifies it...."

"The decision to build an intermediate energy machine, with the concomitant delay in acquiring a truly high energy collider, is a decision to remain second to European physics for all or most of the rest of our careers."

The possibility of international collaboration on the high energy collider did not play a prominent role in the Subpanel's deliberations. Up to that time collaboration on an international scale in the exploitation of facilities was the accepted norm, with the accelerator construction done on a national or regional basis. Even LEP, by far the largest accelerator facility planned up to that time and potentially providing a tunnel for a VBA-like machine, was always seen as an exclusively European construction project.

The Subpanel was well aware of the situation abroad and the plans for accelerator facilities in other regions. It was partly their ambitious plans (like LEP) that made many believe that the US must have its own frontier energy facility. Sir John Adams wrote a very thoughtful memo describing his analysis of the pros and cons of going ahead with the CBA and/or a large hadron collider from the point of view of its impact on the international science [30]. He pointed out that the proposed SSC would be a significant perturbation on the currently planned program on the worldwide scene. I want to quote two excerpts from this memo on issues that turned out to be quite prophetic and are also very relevant today for the ILC. The first one dealt with the cost estimates: "I note with dismay the tendency which has grown up in recent years to err on the side of optimism. This daring approach can have unfortunate consequences which put at risk the future of large physics communities if the optimum turns out to be unrealistic... governments lose confidence in the competence of the planners and their laboratories." And on the long term planning: "... predicting government reactions several years ahead is an uncertain art particularly since it may be another government in power by then which is not usually committed to the promises of its predecessor."

The perceived need to restore US eminence in HEP through the construction of a frontier collider facility dominated the Subpanel's attitudes toward all the issues in its charge. It also dominated some influential public opinion, as indicated by an editorial at that time in *The New York Times* [31], titled "Europe 3, US not even Z-Zero." I quote several characteristic sentences from that editorial: "The bad news is that Europeans have taken the lead in the race to discover the ultimate building blocks of matter.... European accelerators have established a better record of success than any of the three American laboratories... American accelerators should be designed to win or not be built at all.... The 3–0 loss

in the boson race cries out for earnest revenge. The physics team needs to try harder, and coach Keyworth should reward any sensible new strategy with management's full support."

This sentiment came through quite clearly in a meeting I had with Keyworth in May, arranged at the suggestion of Jim Leiss, Associate Director of the DOE Division of High Energy and Nuclear Physics. The meeting was also attended by Doug Pewitt and Al Trivelpiece. Quoting from the notes I wrote right after the meeting: "GK stressed the political difficulties in getting a new project approved... he said that the key word is *competition*, as the thing that the Congress will understand. Hence a project that will put us ahead (e.g. "desertron") will stand a better chance of being approved than CBA, which will only make us equal.... He is sure that a lot of people in Congress would rather spend that money on things like biomedical technology." Keyworth was clearly enamored with the possibility of a new multi-TeV collider. He offered to fly to Woods Hole for the Subpanel's last meeting but I managed to dissuade him from it, arguing that it would prejudice the integrity of Subpanel deliberations.

Before that meeting I spent about half an hour alone with Pewitt discussing the funding prospects. Quoting again from my notes: "DP stressed the political realities today. He said I should be aware that nobody in Washington can really 'deliver' a 3-billion dollar package." I mention this because some members of the HEP community (as well as some outside it) rather naively thought that all one had to do was to convince Keyworth about the value of the project. It is my opinion, however, that the Subpanel members were not influenced (at least not significantly) by such considerations and evaluated the issues on the grounds of physics potential and technical feasibility.

The Subpanel met at Woods Hole on June 5–11 for what was supposed to be the final meeting. However, it became clear, rather early, that the process still had not converged sufficiently. The Subpanel had already agreed in its earlier meetings to recommend initiation of a vigorous R&D effort on the high energy collider (in the 10–20 TeV per beam energy range), with a goal of being able to proceed to construction as soon as possible. It also agreed that in spite of its significant scientific potential, the DC proposed by Fermilab would be too much of a distraction both to the Tevatron effort and to the recommended R&D program on the multi-TeV collider. But there was a pronounced division on the CBA issue, mainly regarding the impact of the CBA on the multi-TeV collider program. The CBA straw votes oscillated between 10 to 7 and 9 to 8 against continuation. There were only two or three members who still appeared somewhat flexible on this issue.

Several of us were very concerned about a potential serious rift in the community, regardless of which recommendation the Subpanel adopted. To forestall this, it was suggested (I recall by Dave Jackson) to invite the four lab directors to Woods Hole, brief them on the current status of deliberations, and make a plea for their active help in keeping the community together after the report was made public. We did this, meeting with each director individually, and it probably did help at some level. One very important point (not discussed to any significant extent previously) was made by Panofsky, who expressed concern that the rest of the scientific community might react very negatively to an expensive new HEP initiative and that one had to do whatever possible to forestall that.

We had to schedule another meeting to resolve the CBA situation and finish the Subpanel's work. It turned out to be possible to delay the HEPAP meeting until early July and find several contiguous days beforehand that would suit the great majority of the Subpanel members. The final meeting of the 1983 Subpanel was scheduled for June 30–July 2 at Nevis Laboratories of Columbia University.

5. The Subpanel's Recommendation

An initial straw vote on the CBA at Nevis showed no significant change of opinion among the Subpanel members. The decision was made to report a negative recommendation and state that it was decided by a small majority (10 to 7 against on a final vote). It was also decided to state that in spite of an agreement on most of the facts regarding the CBA, different Subpanel members drew different conclusions from these facts. These conclusions, essentially arguments pro and con the CBA, were to be included in the report.

The rest of the meeting was devoted to writing the report, focusing especially on potentially politically sensitive comments about the CBA. There was some discussion of the name for the proposed new collider. The adopted name, Superconducting Super Collider (SSC), was proposed by David Jackson.

There were five recommendations [32] in the report: on the SSC, on the CBA, on the DC, on the ongoing program, and on the need for support of technology research and development. The main ones, on the SSC and the CBA, stated:

"The Subpanel unanimously recommends the immediate initiation of a multi-TeV high luminosity proton–proton collider project with the goal of physics experiments at this facility at the earliest possible date"; and

"By a majority vote, the Subpanel recommends that the Colliding Beam Accelerator (CBA) project at Brookhaven not be approved."

Immediately after the CBA decision was made, I had to call Nick Samios and inform him of it. It was not a pleasant task. Nick appeared stunned by the decision. Since the report was still being drafted I did not elaborate to any great extent on that or any other recommendation but did say that the decision on the CBA was final.

The discussion regarding international collaboration on the SSC in the Subpanel's deliberations was rather limited. The need to restore energy frontier research in the US was viewed as one of the strong arguments for the SSC. The parameters of the SSC were chosen so that it would complement the existing and proposed facilities around the world. It was felt that the energy chosen had to be significantly higher than what could be achieved in a potential hadron collider in the LEP tunnel at CERN. The report strongly endorsed the customary international collaboration in HEP experiments, ending with the following statement: "...the U.S. should welcome foreign involvement in exploitation of new U.S. facilities, in either a general way or for a specific capability."

Toward the end of the meeting, the two European members of the Subpanel, John Adams and Carlo Rubbia, asked that the report state that they dissociate themselves from the statements about restoration of American pre-eminence in the field. A statement to that effect was included in the report [33].

We were able to achieve a high level of confidentiality about the result of the deliberations, at least for a few days afterward [34]. *The New York Times* published an article on July 2, 1983, under the byline of Walter Sullivan, titled "Panel of experts dead-locked on next U.S. atom smashers," which stated that the panel "could not agree on what to recommend" and that its "noncommittal report" would be discussed by full HEPAP on July 11.

One important event occurred before that forth-coming HEPAP meeting. It was announced by Leon Lederman in a cable [35] to the physics community. The cable stated: "At 3:27 AM today, July 3, we successfully accelerated protons to an energy of 512 billion electron volts in the Fermilab supercon-ducting energy saver." This was a new record for man-made acceleration of elementary particles and provided an important boost to the credibility of the Subpanel report regarding the technical feasibility of the SSC.

At the HEPAP meeting the discussion of the Subpanel report was started with my presentation as the Subpanel chair. As expected, there was some concern about the risk of going "for broke" with the SSC. Bob Palmer of Brookhaven, a HEPAP member, referring to the projected SSC costs, argued: "That is a very optimistic number. And it's not promising that Congress will appropriate it. I am appalled at the kind of risk you're recommending." During the following floor comment period, several prominent physicists — Nick Samios, Sam Ting, T. D. Lee and Bob Marshak — expressed their dismay at the rec-ommended cancelation of the CBA, which they saw as a facility with very promising physics prospects. In the end, however, the argument for a united front and aggressive pursuit of the energy frontier prevailed. All HEPAP members voted in favor of acceptance of the Subpanel report with all of its recommenda-tions [36].

With the benefit of hindsight one can speculate about the scientific impact that the CBA would have had if it had been completed. My opinion is that its exploitation would have been much more diffi-cult than anticipated. The experience with detec-tors from both the Tevatron and the LHC sug-gests that the technology available in the 1980's and early 1990's would not have been able to han-dle the luminosities required for frontier investiga-tions. On the other hand, cancelation of the CBA, a half-completed project with no apparent techni-cal difficulties, established a very bad precedent and raised the bar for the level of readiness that would be required in the future for project construction approval.

On July 12, 1983, Jack Sandweiss, as HEPAP chair, formally submitted the Subpanel report to Dr. Alvin Trivelpiece, stating that HEPAP unani-mously endorsed all of the Subpanel's recommen-dations. The concern that construction of the SSC might negatively impact federal support of other scientific research was addressed in the transmit-tal letter [37] as follows: "HEPAP realizes that this project demands effort and resources unprecedented for a single tool in basic research. It would consti-tute a new national commitment that should signal the intention of the U.S. to strengthen its support of basic science in all its aspects. As a community of high energy physicists, we would not wish this project to have a negative impact on support of other branches of science." In light of future developments, one cannot overemphasize this last point. The SSC would be a project an order of magnitude greater in scope and cost than the others put up for consid-eration (CBA, DC). The construction price tag was about ten times the annual support level of HEP and hence would require doubling the HEP budget during SSC construction. Thus it was likely that scientists in other fields might be concerned that it would have a negative impact on funding levels in their fields.

The letter also endorsed the steps that the Sub-panel identified as being required in the next stage: "It is clear that a first phase of project planning, cost definition and R&D, followed by a second phase of construction, must be a broad-based national effort, centrally managed from the outset. HEPAP concurs with the Subpanel belief that a successful approach will call for an innovative collaboration between the universities, the national laboratories, and indus-try with the Government. HEPAP recommends that steps to define this process be initiated immediately."

6. Pre–Central Design Group Events

The US HEP community had unequivocally spoken in favor of the SSC. What remained was to make it a reality. One view of the enormity of this chal-lenge was summarized in this paragraph of a July 1983 article in *Science* [38]: "The community is gam-bling its future on a program for which there is no explicit proposal, no design, no site, no research and development plan, no management plan, no manage-ment team, no director, no budget, and no guarantee of long-term federal support. Moreover, the super-collider is an enormous extrapolation from current experience." From an interview for that article I was quoted: "The course is bold, risky, and perhaps foolhardy."

Doug Pewitt, speaking at BNL, cautioned the HEP community not to underestimate the challenge [39]: "A machine of the order of 10–20 TeV challenges us on multiple fronts — technically, politically, and financially... any commitment to proceed with a next-generation accelerator will require that all those factions — the Executive Branch, Congress, the physics community, the science community in general — agree on its importance... such an initiative is possible only with greater national commitment of resources to basic research... the path through Congress can hardly be smooth."

In spite of these great challenges (or maybe because of them) things moved surprisingly quickly after the HEPAP recommendations. Already on August 11, 1983, Al Trivelpiece, in a letter to Jack Sandweiss [40], requested additional input from HEPAP, specifically "advice and recommendations on the content and implementation of the FY1984 R&D effort preliminary to the Superconducting Super Collider." The mechanism was to be a new "HEPAP Subpanel to consider and provide advice on these matters."

The Subpanel was formed within days, with Panofsky as its chair. There was no formal charge but the guidelines presented to the Subpanel by the Chair included recommending allocation of funds (to become available for the SSC as a result of CBA termination) to institutions interested in SSC R&D and providing a consultative process among the participants. The Subpanel held its only meeting on September 7–9, 1983, where it heard presentations from various groups proposing a total program of US$28.3 million. The Subpanel made an allocation recommendation [41] to DOE and strongly recommended that "an interim line manager be selected on as rapid a time scale as possible."

The preliminary allocation for FY84 to BNL for the CBA work was US$23 million. To initiate SSC work on a fast track, DOE wanted to reprogram US$18 million of these funds to the SSC R&D work. This required agreement from the chairs of the four congressional committees (appropriations and authorization, House and Senate) dealing with science research. At the request of Rep. William Carney, in whose district BNL was located and who not

surprisingly was very unhappy with the suggested CBA termination, the House Science and Technology Committee held a hearing on October 19 on the HEPAP recommendations and DOE's plans for their implementation. DOE plans were announced in letters from DOE Secretary Donald Hodel, sent to the four committee chairs on October 18. Al Trivelpiece, Jack Sandweiss and I were asked to testify at that hearing, via both written testimony prepared beforehand and discussion during the hearing [42].

The hearing started with Rep. Carney showing the CBA construction already completed, emphasizing the large scale of the project. He stated that the appropriations for the CBA by Congress to date were largely based on previous HEPAP recommendations, in 1974, 1975 and 1977: "We have gone along with it because you asked us to." But now that the high energy community has reversed itself in an unprecedented way, "How can I get a consensus from American taxpayers that this is a prudent step?" The questions afterward, however, were not hostile and were focused more on the feasibility of the SSC within a reasonable cost envelope, rather than on the recommended termination of the CBA. We tried to make the point that one cannot extrapolate the SSC cost from the Tevatron or CBA magnets since an optimized design and additional R&D would reduce the cost significantly. Following the HEPAP recommendation a reasonably strong consensus had developed that the CBA could not be revived, at least not as a proton–proton collider, and hence that issue was pretty much moot; the important question was whether one should embark on SSC R&D. After the hearing, I was told by a staff member that it had gone well; this was confirmed in a letter to Jack and me from Trivelpiece a few days later, thanking us for our efforts. Finally, the few written "softball" questions I received from the committee afterward confirmed that general impression. The initial congressional hurdle appeared to have been passed over successfully [43].

To obtain a more reliable estimate of the SSC cost, the laboratory directors chartered in January 1984 a Reference Design Study, sponsored jointly with URA and DOE. This study, led by Maury Tigner and held at Lawrence Berkeley Laboratory (LBL), brought together some 160 experts in the field of accelerator physics. The goal was a site-independent review of the technical and economic feasibility of the SSC. It used three different super-conducting magnet designs, with different magnetic field strengths. The study, completed in May 1984, concluded that no major obstacles existed; cost estimates for a 20 on 20 TeV collider ranged from US$2.7 to 3.05 billion, depending on the type of magnet chosen. This estimate was in 1983 dollars and was for the collider complex alone [44].

The community's interest in the SSC was demonstrated by a number of workshops relevant to the SSC design, construction and utilization. The major ones were at the University of Michigan on accelerator physics issues [45], at BNL on cryogenic systems for the SSC, at Texas Accelerator Center on the fixed target option, on the p-bar p option [46] for the SSC (jointly organized by Argonne and the University of Chicago), and at LBL on electroweak symmetry breaking. In parallel there was a series of meetings examining principal issues connected with doing experiments at the SSC — called Physics at the SSC (PSSC) [47]. The culminating effort of all these activities was a DPF-sponsored workshop on the design and utilization of the Superconducting Super Collider, held from June 23 to July 13, 1984 in Snowmass, Colorado [48].

There was a very important theoretical paper [49] by E. Eichten, I. Hinchliffe, K. Lane and C. Quigg, "Supercollider Physics," published in *Reviews of Modern Physics* in 1984. This work used parton distribution functions to calculate cross sections at different collider energies for various processes of interest. It provided firm phenomenological footing for the next generation of calculations and rationalized the argument for a high luminosity, multi-TeV collider.

A top priority task was establishment of the management structure for the SSC activities. A real concern was that if one started this process from scratch with a new legal entity, much time would be lost. The optimum seemed to be for the Universities Research Association (URA)[b] to perform this management task as an addendum to its contract. The Executive Committee of the Council of Presidents urged Guyford Stever, President of URA, to submit a proposal to DOE for organizing the SSC

R&D effort. A letter to that effect was sent [50] by Stever to DOE Secretary Donald Hodel on July 14, 1983. Subsequently a Memorandum of Understanding (MOU) was drafted under URA auspices and signed by seven entities with heavy involvement in HEP and with planned involvement in the SSC.[c] It stated: "The Signatories of this Memorandum, all having serious commitments to the future of high energy physics, agree to provide a climate for development of an effective organization designed to operate and manage the R&D and conceptual design of the SSC project by the DOE."

A Board of Overseers (BOO) for SSC R&D and conceptual design would be created which would have "principal responsibility and authority for management oversight of the R&D and conceptual design phases as described by the DOE." This arrangement was designed to avoid any potential conflict of interest between Fermilab and the SSC effort. The Board was to be composed of "eleven members chosen from the community with significant interests in the SSC program." It was to "reflect the legitimate interests in this program on the part of universities, national laboratories, industries, and government." The initial membership[d] included seven members affiliated with the seven signatories of the MOU and four additional members from outside HEP, mainly industry with potential SSC interest. At the Board's first meeting on March 23, 1984, Boyce McDaniel was elected as the chair. Also at that meeting a subcommittee was formed (Bjorken (chair), Cronin and Trilling), charged with addressing the issue of the selection of the Director of the R&D effort [51]. Based on this arrangement, in March 1984, DOE appointed the URA to manage the R&D and design studies for the SSC (Phase 1 of the SSC).

The Director's search committee reported on its progress at the next three meetings. A number of

---

[b]URA was started in 1965 as a consortium of 25 US research universities to provide a nationwide management organization for Fermilab. By 1983 it had grown to 54 universities, including some from abroad.

[c]The signatory institutions were: Associated Universities, Inc (operator of BNL), Cornell University, Stanford University (operator of SLAC), Texas University Consortium (operator of Texas Accelerator Center, TAC), URA (operator of Fermilab), University of California (operator of LBL) and University of Chicago (operator of Argonne National Laboratory).
[d]The initial members were: J. D. Bjorken (Fermilab), J. Cronin (University of Chicago), J. Deutsch (MIT), H. Furth (Princeton), J. Hulm (Westinghouse), B. McDaniel (Cornell), G. Pake (Xerox), S. Treiman (Princeton), G. Trilling (LBL), S. Weinberg (University of Texas, Austin) and S. Wojcicki (Stanford).

outside individuals were consulted and DOE input was obtained on the top candidates. Based on these reports and extensive discussions within the Board, the Board decided, at its fourth meeting on May 12, 1984, to offer the position of the Director of the Central Design Group to Maury Tigner, and he accepted [52]. The CDG was to be the national organization charged with coordinating the SSC R&D work and the related activities. I was asked by the Board and Maury to assume the position of the Deputy Director, and I agreed to do so.

Before the CDG could get to work, it needed a home. Proposals were solicited for offers to host the group. Nine different proposals were received, including ones from the major accelerator laboratories. The choice was up to Tigner, who recommended LBL as the one which offered the most advantages and the fewest negatives [53]. LBL had a strong accelerator physics group and several of its members could be counted on to contribute to the CDG work. It had a strong technical group with involvement in superconducting magnets focused mainly on the task of cable development. It had an intellectually stimulating atmosphere and was an attractive place, thus facilitating recruitment. The experience of carrying out the Reference Design work there was quite positive. The laboratory and its Director, Dave Shirley, were eager to host the CDG and were willing to provide the necessary space (including 20 parking spots). Even though CDG home would be at LBL, it would be an independent entity, reporting to the Board of Overseers rather than the LBL Director. Both the Board and DOE approved this choice.

During May 8–11, 1984, DOE performed an in depth review of the Reference Design Study [54]. This review supported the cost estimates given in that document. Based on it, DOE Secretary Don Hodel authorized the release of US$20 million for the SSC R&D [55]. The work of the Central Design Group was officially launched.

7. Central Design Group (CDG) Overview

The CDG was to oversee and coordinate all the SSC activities and control the SSC R&D budget until SSC Laboratory's management structure was established. It would perform some of the tasks inhouse and coordinate and direct the R&D efforts at national laboratories and universities. Its work could be grouped into several categories:

- R&D and design in technical areas. The main focus was the magnet development but equally important were efforts in other technical systems and in accelerator physics.
- Conventional systems work. The principal activity was definition of required facilities for the future accelerator laboratory and preparation of the site requirement document for the anticipated site competition.
- Interactions with the government agencies, both the executive and the legislative branch. This involved budgetary issues, reviews, preparation of required documents, congressional testimonies and personal interactions.
- Interactions with the general public and the high energy community (both in the US and abroad).

The starting point for CDG's work was definition of the principal design parameters for the SSC. As discussed earlier, there was a strong conviction in the US HEP community, which crystallized at the Snowmass '82 workshop, that the main motivation for this facility had to be exploration of the physics beyond the Standard Model. The scale of electroweak symmetry breaking indicated the need to probe the 1 TeV energy scale at the constituent level. This argument motivated the 1983 HEPAP recommendation of a multi-TeV hadron–hadron collider. The paper by Eichten *et al.* [49] provided tools for quantitative estimates of different processes as a function of energy. The consensus opinion in the community was coalescing around 20 TeV as the energy in each beam and a luminosity of 10^{33} cm^{-2} s^{-1}. There was, however, no clear threshold that had to be attained. The high luminosity required pretty much excluded an antiproton–proton option.

Tigner and I drew up a formal proposal for primary physics parameters of the SSC, which was presented to the BOO for their approval [56]. The suggested parameters were:

- Proton–proton collider
- 20 TeV in each beam
- Luminosity of 10^{33} cm^{-2} s^{-1}
- six experimental areas

The Board approved these goals but also added the following statement: "The CDG will continue to perform the studies... which examine tradeoffs between primary physics design goals (energy, luminosity, and number of experimental areas) and total project costs...." This statement is worth noting in light of the subsequent debate in 1989 about possible energy reduction to limit the cost increase.

The CDG was going to be led by the Director, Maury Tigner, who had two Deputy Directors — Dave Jackson, responsible for operations, and myself, responsible for external relations. Bob Matyas from Cornell served as Management Advisor to the Director. There was an External Advisory Group, headed by Paul Reardon from BNL. The CDG activities were divided into several departments, headed respectively by:

- Accelerator Physics — Alex Chao, SLAC
- Superconducting Magnets — Victor N. Karpenko, formerly from Livermore Lab
- Accelerator Systems — Peter Limon, Fermilab
- Injector, Steve Holmes — Fermilab (somewhat later)
- Conventional Systems — Jim Sanford, BNL
- Management and Administration — Tom Elioff, LBL
- Detector R&D unfilled initially

The Directorate members and the Department Heads formed the SSC Steering Committee, which was to have (as charged by Tigner) the responsibility to "sell" the SSC and make it successful. The first meeting was on October 10, 1984. About a year later, the group was enlarged and renamed the CDG Coordinating Committee, providing a forum for discussion of and decisions on various CDG-overviewed activities.

An early task undertaken by the CDG was the preparation of the Phase I (R&D and design) schedule and milestones. This was based on rather optimistic budgets, doubling each successive year. The most important early milestones were preparation of Site Parameters Document (April 1985), Magnet Design Type Selection (October 1985), Conceptual Design Report, CDR (March 1986) and Initiation of the Magnet Systems Test (October 1986) [57]. The hoped for construction start was to be in October 1987.

The CDG started its activities at LBL in October 1984 with only a few members — its telephone list dated October 22, 1984 had seven scientists and two supporting staff members. The CDG existed as a separate entity for over four years, until the end of 1988, when SSC laboratory was founded and the CDG became for about a year one of SSCL's Divisions. The amount of work that was accomplished during the CDG's existence is truly impressive. There is no doubt that Tigner deserves the lion's share of the credit for its achievements. He set the tone for the organization, provided inspired leadership, and set an example by working long days, frequently seven days a week. CDG goals were achieved with a minimum amount of bureaucracy and in a stimulating intellectual atmosphere, fueled and supported by the weekly "Hadrons and Cheese" Friday seminars. In this brief summary, one cannot possibly do justice to all the work that was done, and in the following sections I will only be able to briefly describe the CDG's most important activities and accomplishments.

8. Magnet Design and Other Technical Activities

The superconducting magnets were going to be the heart of the SSC; they were the largest and most expensive subsystem of the collider and also the most challenging. Their design determined the other parameters of the machine, in both the conventional and the technical areas. Thus the main priority among the CDG activities was the R&D on the magnets and selection of the design to be used. This was a very intense effort, involving a large number of institutions. Remarkable progress was made in that area during the lifetime of the CDG; I limit my discussion to a few highlights.

When CDG activities started, superconducting magnet technology could be characterized as becoming mature [58, 59]. It had been over a year since protons were first accelerated to 512 GeV in the Energy Saver and the Tevatron was slated to run soon at 800 GeV in the fixed target mode and 900 GeV in the collider mode. BNL solved its magnet problems and had a successful systems test. DESY Laboratory in Hamburg, Germany, was about to start construction of a new electron–proton collider, HERA, with the proton ring based on superconducting magnet technology, with magnets built by industry. It was clear that the next generation high energy hadron colliders would be based on superconducting magnets but their optimum design was uncertain. The goal of CDG magnet work was to improve the technology as much as possible and define the design.

In 1984 this technology was based on the Rutherford cable utilizing niobium–titanium (NbTi) filaments embedded in a copper matrix. The shape of the field was determined by placement of the coils made with this cable; the structure of the positioned coils gave the name "cos θ" to these magnets. Such magnets were the basis of both the Tevatron / Energy Saver and HERA, and they were designed for a B field around 4.5 T. Two other general concepts were also being put forward. One was a superferric magnet; most of the field would be generated by the iron magnetized by a superconducting coil [60]. The field would be only 2–3 T but less conductor would be required. The other concept was based on a higher field magnet using a niobium–tin superconductor. But this design was still in the development stage and was viewed as unrealistic for the SSC.

By 1984 the only viable options appeared to be superferric and cos θ NbTi magnets. But even here there were distinct options. In the Tevatron magnets, the iron surrounding the coil was at room temperature; in the HERA design it was held at liquid helium temperature. Both HERA and Tevatron magnets required only one superconducting channel — for protons in HERA and a common channel for both p-bars and protons in the Tevatron. A proton–proton collider would need two separate channels with opposite magnetic fields. There were two options for such a machine: housing these two channels in the same cryostat (2-in-1 design), as proposed by Bob Palmer of BNL, or separately (1-in-1 design).

The main players in the superconductiing magnet effort were Fermilab, BNL, LBL and Texas Accelerator Center (TAC). The last was a new organization, created near Houston with support from George Mitchell (a Texas oilman), by Russ Huson and Peter McIntyre, both former Fermilab employees, who wanted to build a research center focusing on accelerator issues. In addition, there was very important work at the University of Wisconsin on superconducting cable under the leadership of P. Larbalestier.

The most difficult part of the magnets was the ends; a high fraction of the cost was in the connections there. Thus, in the early stages of the R&D, relatively short (∼1 m) magnets were built so as to understand and optimize the most difficult elements. Cost optimization of the machine, however, would eventually require magnets significantly longer than those used in the Tevatron. Initially, three different magnet types were worked on [61]:

- Design A — 2-in-1, cold iron, high field (6.5 T); pursued at BNL and LBL;
- Design B — 1-in-1, warm iron, high field (5 T); pursued at Fermilab;
- Design C — 2-in-1, low field (3 T, superferric); cold iron; pursued at TAC.

Subsequently, work on designs A and B indicated that a better option would be a 1-in-1, cold iron, high field (6.5 T) magnet, called design D. The main focus of the magnet activities at BNL, LBL and Fermilab shifted toward the work on that design. In parallel, the TAC group started working on design C*, a 1-in-1 version of design C. Early in 1985, Tigner appointed a Technical Magnet Review Panel (TMRP) to review and help guide the magnet R&D program. This panel, chaired by Alvin Tollestrup of Fermilab, was composed of senior members from each institution working on the magnets as well as additional scientists without direct institutional involvement in this work.

Naively, one might view the magnet development work as having purely technical scope. However, there were also strong political overtones. The state of Texas was very much interested in hosting the eventual SSC Laboratory and even offered, to that end, to consider providing the site and conventional facilities at no cost to the federal government. Adoption of a low field magnet, requiring a larger diameter machine and hence a larger overall site, would help Texas in the future site competition by reducing the total number of proposals satisfying the site requirements.

The work on superconducting cable was being carried out through a collaborative effort between DOE, universities and industrial concerns, with the University of Wisconsin playing a leading role. The manufacturing process was quite complex and was based on many years of development work [62]. Since later on new, as-yet-undeveloped, higher-temperature, ceramic superconductors were being suggested for the SSC, it is useful to review briefly the complex manufacturing process. Several thousand NbTi 1/8″ diameter rods are stacked into billets within a copper matrix. The billets are about 75 cm in height and 30 cm in diameter. They are then drawn into wire, about 0.8 mm in diameter, with original rods becoming fine filaments about 2–3 microns in diameter. The resulting wire is then wound in a helical pattern into a cable about 9 mm wide.

One of the most important parameters of the wire is its current-carrying capacity at the operating field and temperature. The cable used for the Tevatron had a capacity of 1800–2000 A/mm^2 at 4.2 K and 5 T. The Reference Design had assumed that with additional R&D one could raise that to 2400 A/mm^2. The R&D during the initial years of the CDG resulted in filaments with a capacity up to 2750 A/mm^2, allowing 6.5 T magnets in the CDR design. Besides the work on the wire, new cable winding machines had also been developed in the laboratories and the acquired know-how transferred to industry.

There were a number of other technical R&D activities that were essential for producing the collider design. Among the most important ones were the accelerator physics calculations which provided important guidelines for the design not only of the magnets but also of the accelerator as a whole. Accelerator physics studies were a key input determining the aperture of the magnets. In the SSC the projected lifetime of the beam was of the order of 1–2 days, limited mainly by proton–proton inelastic interactions at the collision points; for the projected circumference of the SSC of about 80 km, the protons would execute typically about 10^9 revolutions after injection. Ray-tracing of individual particles over such a long time would clearly not be possible. Thus other calculational methods had to be developed for this problem [63, 64].

The method used in the SSC design relied on the concept of linear aperture, i.e. the size of the physical region in which the motion of the particles around a closed orbit is approximately linear. Because of nonlinear effects, the actual aperture required would be larger than this linear aperture. To obtain the relationship between the two, one has to take into account such effects as irregularities in the magnetic field, the size of beams, variations in the injection position, alignment errors and correction schemes.

The quality of the magnetic field around the orbit executed by the protons is determined mainly by the dipoles, which are the most numerous

components in the lattice. This quality is characterized by the size of the multipole components, i.e. deviations from the perfect dipole field. Their effect can be reduced or eliminated by correction coils; the optimum design of the machine will then be a compromise between the effort and cost required to reduce the magnetic field inhomogeneities and their variation from magnet to magnet and the complexities of the required correction coils.

The SSC was to be the first proton machine in which synchrotron radiation from the circulating protons would be significant [65]. At the design energy and luminosity each beam would radiate about 9 kW in synchrotron power. This would have a positive effect of increasing the luminosity during the first day after injection due to synchrotron damping. On the negative side, the heat from the absorbed radiation would have to be removed, placing additional load on the cryogenic system. In addition, the energetic photons upon striking the beam pipe walls would desorb gas molecules "frozen" on those surfaces. If they are not removed, pressure would build up in the beam pipe, shortening the beam lifetime. To investigate these effects, several experiments were performed at the National Synchrotron Light Source at BNL to study the desorption due to synchrotron radiation.

The design of the cryogenic system was another major challenge addressed in the early years of the CDG. The SSC would be by far the largest cryogenic system held at the liquid helium temperature of 4.3 K. To keep operating costs low it was essential to reduce the heat leakage into the magnets from its surroundings. This was achieved by having an intermediate shield at liquid nitrogen temperature and optimizing placement of the insulation and support mechanism. Measurements of heat leakage in a 12 m test cryostat confirmed preliminary estimates that the heat leakage could be kept at 0.25 W/m, roughly a factor of 5 below the Tevatron value.

All of the above activities, plus measured performance of magnet prototypes, formed the input for the selection of the magnet design type. To evaluate all of this information, Tigner appointed several different task forces. The Aperture Task Force, composed of worldwide experts in this area, evaluated all the relevant experimental and theoretical results with a view to understanding whether the apertures proposed were adequate [66]. The Cost Comparison

Task Force had as its charge obtaining detailed cost estimates of all magnet-dependent SSC subsystems for all of the candidate magnet types. It received input on the cost issue both from the proponents of different magnet styles and from industrial firms contracted by the CDG for cost evaluation and, based on this input, made its own independent cost evaluation. The Operations and Commissioning Task Force was to look at operational issues connected with each candidate magnet type.

A Magnet Selection Advisory Panel (MSAP) was then appointed to review the reports of all the task forces as well as the final report [67] of the previously mentioned TMRP, and to put forward a recommendation on the magnet type to be selected. The Panel was chaired by Frank Sciulli, a professor of physics at Columbia University, and included experts on large technical systems, superconductivity technology, accelerator design and operation, use of accelerators for particle physics and/or underground construction or who had experience with complex science and technology policy issues. In addition, three industrial consultants, with expertise in the technology of superconductivity and magnets based on it, were appointed to advise the Panel.

The Panel met at Berkeley on August 25–28, 1985, to hear presentations from proponents and reports from various CDG analyses. The meeting was also attended by representatives of all four R&D centers as well as representatives from DOE and URA. The final report, dated September 9, 1985, was unanimous (including the three consultants) in recommending a high field, $\cos \theta$, cold iron, single-channel magnet, i.e. design D, as the basis for the SSC. The basic arguments for the $\cos \theta$ design were its well-understood behavior and predictable cost. The high field was favored for operational and cost reasons. Maury Tigner accepted this recommendation.

The possibility for future upgrades of the machine was not one of the criteria given to the Panel. Some people argued for the largest possible tunnel so as to have the option in the future to go to higher energies by using next generation high field magnets. Tigner did not find this argument persuasive, arguing that the scientific and technical uncertainties in predicting the future some 15 years in advance were so large that it would not be wise to depart from the optimum design for those reasons [68].

Even though the magnet selection process was finished and recommendation made and accepted, the issue was not put to rest. The decision did not sit well with the TAC or with the Texas Congressional delegation. A rather far-ranging House Energy and Water Subcommittee hearing was held on October 29, 1985, part of which was devoted to the magnet selection process [69]. Joe Barton, Representative from the Texas 6th Congressional district, tried through his questions to raise doubt about the quality of the selection process, focusing mainly on the tunneling costs, but did not appear to make any significant impact on the subcommittee members.

At the time of the MSAP meeting, TAC had completed and measured only relatively short superferric magnets. A 28 m magnet was being produced by General Dynamics Corp but was not finished. A few months later, a finished magnet was delivered to TAC and cooled down; a number of measurements of the quality of its magnetic field were then performed. The TAC leaders, F. R. Huson and P. M. McIntyre, found the results so promising that on April 7, 1986, they wrote a letter [70] to William Wallenmeyer, stating: "This new information leads us to the conclusion that the superferric design is the best and least expensive choice for the SSC... project cost can be reduced by US\$1 billlion compared to the high field design."

This letter triggered a request from Al Trivelpiece to Jack Sandweiss, as HEPAP chair, to convene a subpanel with a charge to review the new information from TAC to determine whether the evidence was sufficient to warrant reopening the magnet selection process. This new subpanel "on Review of Recent Information on Superferric Magnets," chaired by Burt Richter from SLAC, met twice, on April 24 and May 8–10, 1986, when it heard presentations from TAC and had discussions with members of TAC and CDG. Its deliberations on the status of the TAC magnet R&D program and on conventional cost issues led to its report [71], dated May 20, 1986, drawing three conclusions:

(1) The TAC magnet R&D program is still about two years behind the high field magnet program. The TAC magnet is not ready for industrialization.
(2) Cost uncertainties in TAC technical components are large, and superferric magnet costs cannot be

estimated with confidence at the present state of R&D.

(3) We do not believe that the new information presented warrants reexamination of the CDG magnet-selection decision.

The magnet selection issue was finally settled; design D magnet was the focus of subsequent R&D work [72] and the basis for the CDR.

9. Conventional Facilities Plans

It was anticipated from the time of the 1983 HEPAP Subpanel that there would be a nationwide competition for the SSC site. This process would be organized and directed by the federal government but the CDG was to play an important supporting role. The first CDG task in this area was the preparation of the "Siting Parameters" document, scheduled for completion in April 1985. Preparing this document and directing the development of conceptual design for the conventional SSC facilities were the main tasks of the Conventional Facilities Department of the CDG during the first two years.

The conventional facilities of the SSC could be divided into five components: the main SSC tunnel for the 20 TeV storage rings, the injector complex consisting of several lower energy accelerators, the interaction regions, the laboratory campus and the laboratory infrastructure (e.g. roads, power distribution). Because of its length, some 80–90 km, and the requirement that the tunnel be at least 20 below the surface over its whole circumference, the first component was most expensive and was the one that imposed the most restrictive conditions on the suitability of a given site.

Based on the above components of the SSC, a "Siting Parameters" document was prepared for DOE [73]. The function of the report was summarized in its abstract: "The principal objective of this technical advisory is to delineate for the DOE those factors governing the choice of a site for the SSC essential to the creation of a laboratory." The report addressed only the technical issues; the definition and execution of the site selection process, including the preparation of the formal "Site Selection Criteria" document, was going to be the responsibility of DOE.

The "Siting Parameters" document used the accelerator parameters suggested in the Reference Design Study and considered a facility based both on

3T and 6T magnets. It summarized the technical criteria that DOE might use in evaluating the candidate sites for the SSC and addressed the requirements on the topology and geography of the site and the auxiliary conditions, like availability of power and water, proximity to large population centers, ease of access via public transportation, etc. Thus, for example, the required overall surface area under lab control was estimated at 11,000 acres; the main ring tunnel should be level within 1°; up to 2000 gal/min of water should be available and up to 250 MW of electrical power, with separate feeds. The geology of the site had to have been extensively characterized; man-made excessive noise and vibration should be avoided. The environmental issues had to be addressed so that compliance with National Environmental Protection Agency (NEPA) standards could be achieved.

The next step in the work of the Conventional Facilities Department was selection of an architectural and engineering (A&E) group of companies to assist in developing the conceptual designs for the laboratory campus and interaction region buildings, injector complex, utility distribution, required roads, etc. This group was also charged with evaluating different tunneling methods in common use and estimating the costs of tunneling in various soil conditions. The RTK joint venture (Raymond Kaiser Engineers, Tudor Engineering Company, Keller & Gannon-Knight) was selected for this task.

The underground component of the SSC was going to be a major part of the project in terms of effort, cost and time. Thus expert advice and oversight during the design and construction phase appeared beneficial. To that end, Tigner appointed an Underground Tunneling Advisory Panel (UTAP) with the advice and under the leadership of Bob Matyas. A related effort was getting the US National Committee on Tunneling Technology of National Research Council to initiate a study of the contractual relationships between different parties in large underground engineering efforts. At the request of DOE, who sponsored the study, the SSC was used as an example of such an effort. The subcommittee appointed for this task was composed of 12 individuals from all over the US, with expertise covering all aspects of such projects. Their report [74], "Contracting Practices for the Underground Construction of the Superconducting Super Collider," became a bible of the tunneling industry and is still in wide use today. Several members of the CDG Conventional Facilities Department made major contributions to this work.

10. The Conceptual Design Report, CDR (1986)

The work described above formed the basis for the CDR, scheduled to be completed in April 1986. This would provide an up-to-date vision of what the SSC would look like, what its performance would be, how long it would take to build it, and what it would cost. It would be the basis of the proposal to DOE to build the SSC. The initial budgetary assumptions were much more optimistic than what was eventually realized. Instead of doubling every year, the annual appropriation for the SSC R&D for the first three years remained constant, around US$20 million. In spite of that, the realized progress was impressive across the board. The full length magnets were still a few months away at the time of the CDR but the experience obtained from the shorter magnets gave one confidence in the successful conclusion of that R&D program.

The actual preparation of the CDR was the focus of the CDG members and its visitors during the last few months of 1985. Six US national laboratories and 13 universities were involved. In addition CERN, DESY and KEK (Japanese HEP laboratory) contributed. Dave Jackson accepted the monumental task of serving as the editor of the report. Consisting of a 712-page main volume and 4 attachments, the report [75], described the physics at the SSC, the relevant experience with previous facilities, the R&D programs over the previous three years, the technical design, the relevant accelerator physics issues, the conventional facilities required, the technically driven construction schedule and a cost estimate down to WBS level 4. The basic parameters were: proton–proton collider, 20 TeV per beam, and design luminosity of $10^{33}\,\mathrm{cm}^{-2}\,\mathrm{s}^{-1}$. The CDG also made a study of the pbar-p option (advocated probably must strongly by Leon Lederman) and found that the resulting relatively small cost savings did not warrant a factor of 10 loss in luminosity and additional operational complexity wich would decrease reliability [76]. Since the site was yet to be chosen, a variety of different realistic sites were considered for which efficient construction methods would exist. The design

was based on 16.54-m-long magnets, a peak field of 6.6 T and a 4-cm-diameter aperture. The last was the result of extensive accelerator physics studies and overall cost optimization.

The collider would have an injector complex consisting of four accelerators. A linac would accelerate H^- ions to 600 MeV (kinetic energy), which would be stripped at injection into the low energy (8 GeV/c) booster (LEB). Subsequently, the medium energy booster (MEB) and high energy booster (HEB) would bring the protons to energies of 100 and 1 TeV, respectively. The 1 TeV protons would then be injected into the collider rings. The HEB would be a superconducting machine to minimize its power usage and its circumference; the other preaccelerators would operate at room temperature.

The main collider tunnel would be in the shape of an oval with a circumference of 82.944 km. Two semistraight sections of the oval would house injection and utility regions as well as six interaction halls. The main campus buildings would be near the injector complex. The parameters of main interest to the experimentalists were: 4.8 m separation between bunches, 1.4 average number of interactions per bunch crossing at peak luminosity, luminosity lifetime of about 1 day and injection time of about 1 hour. Four test beams (up to 1 TeV energy) would be available from the HEB.

The cost estimate for the total accelerator complex was US$3.01 billion (in 1986 dollars). The average contingency was 21.4% of the base cost, rather low by the standards used today, especially considering the significant increase in scale over previous machines. These costs did not include detectors, R&D, preoperations or escalation. This was the standard costing practice for DOE accelerators in 1986. Furthermore, no site acquisition costs were included (on the assumption that the site would be provided free by the proponents). The construction time was estimated to be six-and-a-half years.

The CDR was delivered to DOE on April 1, 1986, and underwent an extensive review on April 26–30 at LBL involving 83 reviewers and consultants [77]. The conclusion was: "The DRC [DOE Review Committee] concludes that the design set forth in the CDR is technically feasible and properly scoped to meet the requirements of the US high energy program in the period from the mid-1990's to well into the next century.... The DRC finds that the SSC CDR cost estimate is credible and consistent with the scope of the project." A parallel review of the costs was performed by the Independent Cost Estimators (ICE) group. They agreed with the general cost estimate but expressed concern that some of the estimates and the level of contingency might be on the optimistic side.

To obtain an estimate of the potential cost of the initial detector complement and associated computing, three advisory panels were appointed. The first one, called the Detector Cost Model Advisory Panel, chaired by George Trilling of Berkeley, had as its charge definition of a likely complement of detectors that would be able to appropriately respond to the physics needs a decade hence. The second one, chaired by Roy Schwitters of Harvard, was asked to estimate the cost of such a set of detectors. The third one, chaired by Stu Loken of LBL, was to estimate the associated computing costs [78]. This work was completed by April 20, 1986. The total cost estimate obtained for detectors from these efforts was in the US$558–865 million range, and for computing it was US$70 million. The shortcoming of this process was that it did not allow for optimization of the physics output per dollar in either the choice or the design of the detectors [79].

11. The CDG and Government Agencies

Interaction with government entities and individuals was a very important part of the CDG activities. One key element of these interactions was the organization of, preparation for, and participation in reviews. DOE held both quarterly and annual reviews of the CDG, lasting typically two or three days. They covered the full spectrum of the CDG activities, focusing on technical work for the collider itself, but also covering budgets, detector R&D and various outreach activities. In addition, there were other reviews of major pieces of work, such as the Conceptual Design Report.

Another important activity was providing information, in either written or oral form, to various government agencies. There was a constant flow of information from the CDG to DOE about the work being done and the results. There were regular monthly and quarterly reports detailing work done and budgetary status. Trip reports from international travel kept DOE informed about our activities in the international collaboration area. There was

the regular annual testimony at the Congressional authorization and appropriation hearings (about 2–4 per year), generally given by Maury Tigner. Briefing papers and SSC fact sheets were prepared for distribution to the Members of Congress and Congressional staff. There were other hearings when specific issues came up, such as after the magnet selection decision, where CDG members were asked to testify. There were periodic interactions with DOE officials outside of reviews. In addition, there were many interactions with members of Congress or members of their staff, generally at their request but also frequently initiated by the CDG.

The CDG was often asked to prepare briefing documents. Early in 1985 at the request of the OSTP, the National Academy was asked to provide a briefing on "Scientific Frontiers and the Superconducting Super Collider" and I was asked to chair the panel to prepare that report [80]. Subsequent to that I gave oral briefings on the substance of the report to G. K. Keyworth, NSF Director E. Bloch, DOE officials and staff of the relevant Congressional committees. There were also CDG briefings about the SSC to the National Science Board of NSF. In later years, as the SSC was receiving more attention at higher levels of the federal government and beginning to be treated more seriously, we were frequently called on by DOE to provide information for briefing books for various officials about different aspects of the project: physics goals, technical details, technological spinoffs, international situation, budgets and schedules, etc.

12. The CDG and the US HEP Community

From the very beginning of the SSC, the US HEP community exhibited great interest in it and great enthusiasm about its physics potential. The CDG has actively worked with the community, with three main thrusts in mind:

- Involving members of the community in the SSC design and R&D activities;
- Enlisting the community to generate grassroots and Congressional support for the project;
- Working with the community to ensure the maximum possible state of readiness to do physics with the SSC upon its completion.

The appropriations for the first four years (FY84–FY87) of SSC R&D stayed constant at about US$20 million/year; that level was certainly inadequate for the effort that was anticipated at the beginning of the CDG. Thus the CDG activities had to rely heavily on the community participation. There were a number of long term visitors at the CDG who came for a few months or a year, both from the US and from abroad, and made important contributions to various technical or planning issues. Experts in various areas contributed generously through their membership in various task forces that were being organized all the time.

One of the mechanisms for communicating with the community and other interested parties was *The Lattice*. This was a bimonthly, 8–12-page newsletter published by the CDG under URA auspices.[e] It described the main happenings at the CDG, provided relevant news about the SSC R&D results and a calendar of upcoming events, and listed the recent SSC reports published.

A formal way to get the community involved in planning for physics at the SSC and at the same time form an advocacy group for the SSC was the creation of the Users Organization of the SSC (UOSSC), open to all scientists with a professional interest in the SSC [81]. This organization was formed in the spring of 1987 by the Division of Particles and Fields of the American Physical Society at the request of the CDG and the community as a whole. The first meeting of the UOSSC was held on May 20–21, 1987, at LBL.[f] Lee Pondrom of the University of Wisconsin was elected the first chair of the UOSSC. On July 16, 1987, as part of the Workshop on Experiments, Detectors, and Experimental Areas, held in Berkeley, a town meeting was organized at which Pondrom outlined the planned activities of the organization and Brig Williams (Penn) and Satoshi Ozaki (KEK) discussed the ongoing detector R&D activities.

Numerous periodic workshops were an efficient means to both keep the community informed and solicit their input. They were generally organized by a combination of DPF, national laboratories, universities and the CDG. To give a flavor to these activities, I will mention here some of the larger and/or more important ones:

- Workshop on Triggering, Data Acquisition and Computing for High Energy/High Luminosity Hadron–Hadron Colliders, held November 11–14, 1985, at Fermilab.
- 1986 workshop in Snowmass, Colorado on the "Physics of the Superconducting Supercollider" [82], attended by 264 people. This was the first major workshop after publication of the CDR, so it gave participants a chance to critique it and also use the parameters to start thinking seriously about generic detector designs. At this meeting a request was formulated for a separate program to fund generic detector R&D.
- Workshop on Experiments, Detectors, and Experimental Areas, held July 7–17, 1987, in Berkeley [83]. Over 260 people attended, with 24 of them from outside the US.
- Summer Study on High Energy Physics in the 1990's, held June 27–July 15, 1988, in Snowmass, Colorado [84]. This study, with 536 participants, had a somewhat broader scope, examining anticipated opportunities in high energy physics in the 1990's, both at the SSC and at other facilities.
- Workshop on Triggering and Data Acquisition for Experiments at the SSC at the University of Toronto, Canada, held January 16–19, 1989.
- Workshop on Calorimetry for the SSC, held March 13–19, 1989, at the University of Alabama, Tuscaloosa, Alabama. The choice of this location was part of a broad effort to generate and/or increase interest in the SSC in areas without traditionally large involvement in HEP.

Early in 1987, the CDG (in consultation with DOE, NSF and the community) established a Generic Detector R&D Coordinating Office [85]. Its mission was to coordinate and focus the national effort on generic (rather than detector-specific) detector R&D relevant to the SSC and maintain appropriate contact with similar research being done elsewhere in the world. Full utilization of luminosities in the 10^{33} range required sophisticated detectors with hitherto-unavailable capabilities [86]. Thus generic R&D was needed to identify the problems and find solutions. M. G. D. Gilchriese, a professor at Cornell, was

---

[e]The initial name, *SSC Newsletter*, was found objectionable by the DOE on the grounds that it might imply a stronger commitment to the SSC than DOE was prepared to make at that time. The more innocuous name *The Lattice* was accepted as a compromise.

[f]There was an earlier, informal meeting of potential SSC users in May 1985, at LBL, organized by the CDG, but no formal organization was formed at that time.

appointed the Detector R&D Coordinator. In addition, an International Advisory Committee, under the chairmanship of H. H. Williams from the University of Pennsylvania, was appointed to provide advice and guidance to the Coordinator regarding this R&D program.

The program was to be proposal-driven, with the proponents making submissions to funding agencies, which generally would refer them to the CDG for recommendation. The Coordinator, assisted by the Advisory Committee, would examine them and advise the funding agencies regarding their potential support and its level. The Coordinator would also make periodic recommendations to the funding agencies regarding the required funding level for these activities, periodically publicize the R&D required in hitherto-neglected areas, and organize workshops as deemed necessary.

At its first meeting in April 1987, the Committee considered 31 tasks requesting US$3.7 million and recommended funding 11 of them with a total sum of US$545,000. DOE followed this recommendation. The program grew in the following years: in FY88, US$2.25 million were disbursed (US$8.9 million were requested and a US$3.4 million funding level was recommended); in FY89, US$2.9 million were allocated for new efforts and US$3.4 million for continuation of ongoing ones. During the existence of the CDG, 7 national labs, 29 universities and 8 industrial concerns were involved in the program. R&D in many diverse areas was supported; some of the topics studied were radiation-hardened electronics, pixel detectors, triggering and data acquisition, computing, large mechanical systems (e.g. for calorimeters) and the test beams required [87]. Once SSC Lab was established, the program would be absorbed by the Lab and would gradually transform itself into specific detector design and R&D.

13. The CDG and the General Public

To make the SSC a reality, the project had to enjoy significant popular support so as to get the required Congressional approvals. Accordingly, from the very beginning of the CDG's existence there was a concerted effort to inform the general public about the SSC and to create enthusiasm for it. The sending-man-to-the-moon project in the 1960's, which fired up the general population's imagination, was a model to emulate.

One of the means toward that end was publication of various brochures describing SSC in general or focusing on some of its specific aspects. Some of the brochures were:

- "To the Heart of Matter," a general description of the SSC and its physics motivation, first published in 1984. Later, two updated versions were printed, when the CDR was finished and when SSC Lab was established in Texas.
- "Supercollider R&D, The First Two Years," published in December 1985, describing the initial SSC R&D, its goals and achievements. This brochure was geared toward industrial concerns which potentially might be involved in that work.
- "The Superconducting Super Collider," a leaflet distributed to the states interested in hosting the SSC, describing SSC physics motivation, the accelerator complex and its potential impact on the community.
- "SSC Detectors — Looking to the Future," describing the challenges presented by the future SSC detectors, also aimed at potentially interested industrial enterprises.

These booklets were put together by members of the CDG and LBL staff, with Rene Donaldson, CDG editor/publisher, playing a major part in producing the publications. We were helped in these and other public outreach efforts by the prizewinning science writer K. C. Cole. In 1988, Chris Quigg, theory group leader at Fermilab, who replaced J. D. Jackson as Deputy Director for Operations in May 1987, and Leon Lederman compiled a book titled *Appraising the Ring*, which put together short statements of support for the SSC by various individuals prominent in science or industry [88].

In parallel, there was a concerted effort on the part of the whole HEP community to give presentations about the SSC in various forums: colloquia, schools, conferences, etc. Fermilab Director Leon Lederman was one of the most active scientists in promoting the SSC through numerous talks and articles. Following up on the request from AAAS, we organized a symposium on the SSC at their annual meeting in the spring of 1985, with Lederman, Jackson and Tigner as the speakers. The CDG made a strong effort to contact physics teachers about the SSC by sending "To the Heart of Matter" to all science teachers in the country. We also organized an exhibit about HEP and the SSC at the Annual Science Teachers Convention on April 7–10, 1988, in St. Louis. A traveling exhibit about the SSC was prepared and shown at various science centers through collaboration with the Association of Science and Technology Centers.

There was also a broad effort to get articles published in various popular journals and newspapers about the SSC either by members of the HEP community or by journalists, based on interviews. Some of the popular articles written by the HEP community were:

- "Elementary Particles and Forces," by Chris Quigg, in *Scientific American*, April 1985.
- "To Understand the Universe," by Leon Lederman, in *Issues in Science and Technology*.
- "The Superconducting Supercollider," by J. D. Jackson, Maury Tigner and S. Wojcicki, in *Scientific American*, March 1986, describing the SSC and its physics goals.
- "The Case for the Supercollider," by J. Cronin, in *Bulletin of the Atomic Scientists*, May 1986, pp. 8–11.
- *Science* article "Elementary Particle Physics and the SSC," by Chris Quigg and Roy Schwitters, March 1986.
- *American Politics* article on the SSC in July 1986 by Roy Schwitters; it was reported to have made a significant positive impression on Secretary Herrington.

These outreach activities intensified significantly during the site selection time, 1987–1988. Officials in interested states wanted to become better-informed about the nature and scientific goals of the SSC, its site requirements, and what the SSC would mean to a given state if the SSC was located there. As a national group, the CDG could not express favoritism toward any single proponent, but if requested we did go to various events connected with the site selection all over the country and tried to answer all the questions. During that time the CDG also produced a short videotape discussing the SSC and geared to the parties interested in hosting the SSC [89].

A very important issue was interaction with the non-HEP scientific community. There was an understandable concern there about potential negative impact of the SSC on other sciences. Very simplistically, the reaction of other scientists fell into two broad categories: some, like Phil Anderson, a Nobel laureate and a professor of theoretical physics at Princeton, Jim Krumhansl, a condensed matter theorist at Cornell, and Rustum Roy, a materials scientist at Penn State, argued strongly against the SSC, being convinced that it was bound to hurt their fields; others supported the SSC's scientific goals and hence its construction, provided that this would not impact adversely the support for their own fields. I was on the APS Council for three years during the CDG days; the Council shied away from any positive statements about the SSC, because a number of its members expressed these fears. I discuss this issue in more detail and provide some specific examples in the following sections. This potentially negative reaction was a serious concern of the CDG and the BOO; an effort was made to alleviate the concern and emphasize the SSC's scientific value through colloquia, popular articles, rebuttals of unjustified anti-SSC articles, etc. John Deutsch, one of the BOO members and a chemist by training, emphasized frequently the need for such intensified efforts.

As the SSC project became better known [90] and began to look like it might materialize, the industrial interest intensified. We tried to mobilize potential industrial participants in SSC construction to help with getting the SSC approved. The example of NASA and the International Space Station, where industrial support in getting Congressional approvals was very helpful, was a good model. An important meeting along those lines was the National Symposium on the SSC held on December 3–4, 1987, in Denver, Colorado [91]. This was during the site competition, so interest from industry and state officials was very high. There were about 600 participants.

It was always planned to fabricate the SSC superconducting magnets in industry. To accomplish this, the technology in this area, developed in the national labs, had to be transferred to industrial concerns. The CDG and DOE formulated a three-phase Magnet Industrialization Program to accomplish this, the three phases being Technology Orientation, Tooling Design and Magnet Preproduction, and Magnet Production. Phase I was initiated by the publication of an announcement in the *Commerce Business Daily* on July 21, 1988. During Phase I, representatives from participating companies had the opportunity to visit national laboratories and interact there with the magnet experts. The initial Technology Orientation meeting was held at the CDG on January 9–10, 1989.

The next big industrial meeting, organized by a consortium of industrial companies, national laboratories, universities and government agencies, was held in New Orleans on February 8–10, 1989 [92, 93]. It focused on physics, technology and politics relevant to the SSC. DOE Secretary Herrington addressed the audience at that meeting.

We tried to get a professional assessment of the impact of HEP research on innovation, new products, and new industrial activity. This was a topic that frequently came up in Congressional debates and various articles. Such study, focused on benefits from CERN contracts, was performed at CERN in 1984, and found that for every franc spent by CERN on high technology contracts, three francs of secondary economic benefits were generated [94]. The methodology used was frequently questioned, so we initiated interaction with two US economists, David Mowery of Carnegie Mellon University and W. Edward Steinmueller of Stanford, to investigate the general question of economic benefits coming from basic research. They submitted a formal proposal [95] to DOE for such a study. The proposal was funded and the resulting report received moderate attention in Washington but had no major impact on the SSC. The main results of the research were published in a journal and that article [96] has been cited frequently.

14. International Collaboration Issues

The issue of potential international collaboration on SSC construction is quite complex. When one looks superficially at the global situation in the area of science and technology in the early and mid-1980's, one might surmise that the situation was ripe for significant collaboration. At the 1982 Versailles summit meeting of the heads of state of seven principal industrial nations, Mitterand proposed exploration of a possibility of closer cooperation between the major industrial powers in the area of science and technology. The first step was to set up working groups in 18 project areas, with lead countries identified in each. The US would serve as the lead country in 5 areas, including HEP, this to be chaired by Alvin Trivelpiece [97]. The activities of these groups were

endorsed by the heads of state at their meeting two years later [98]. However, the lofty goals, as expressed in the relevant resolutions, were submerged, at least as far as the SSC was concerned, by practical and parochial considerations having to do with perceived needs for regionally based programs and preservation of the existing labs.

Early in the SSC days there was resentment in some circles that the SSC decision was made unilaterally by the US and that the process completely bypassed ICFA.[g] However, the facts were that while ICFA did organize workshops and did discuss issues, it was never involved in planning any construction project and, in my opinion, was not structurally designed to do that. At the May 1984 ICFA meeting at KEK in Japan, it was formally recognized that ICFA should not be a body for planning joint accelerator undertakings but rather should focus its efforts on sponsoring workshops, facilitating coordinated research and development efforts in key HEP technologies and organizing worldwide inclusive meetings on future plans for regional facilities. The key term of reference was modified to read: "... to promote international collaboration in all phases of the construction and exploitation of very high energy accelerators" [99].

The SSC started and evolved as a US domestic project and one of its expressed goals was restoration of US pre-eminence in HEP. This made the subsequent efforts to obtain contributions from abroad to its construction difficult, but not impossible. To understand the complexities involved, one has to look at the two most important potential contributors, Western Europe and Japan, where situations were quite different.

When the SSC was proposed, the two large HEP laboratories in Western Europe were initiating major accelerator projects. CERN in Geneva was starting construction of a large electron–positron collider, LEP, in a tunnel 27 km in circumference. It was scheduled for completion in spring 1989 and CERN

---

[g]There is some controversy as to how damaging this was, especially with the Japanese HEP community. It is my opinion, based on numerous trips to Japan and conversations on this topic with a wide spectrum of Japanese HEP physicists, that it was not very relevant. The only person I ever heard complaining about this issue was Yamaguchi, chair of ICFA at the time, who voiced his unhappiness about this at the 1987 ICFA seminar held in Serpukhov. But, for a somewhat different point of view, see Refs. 2 and 14.

construction funds were committed through 1992. The other laboratory, DESY in Hamburg, was in the process of constructing the world's first electron–proton collider, HERA, scheduled to be finished around 1990.

The long term plans for the future of CERN provided additional complication. Already, when the LEP collider was being designed the possibility was kept in mind that the tunnel could eventually be used for a hadron–hadron collider. Very little work had been done on this possibility prior to 1984, CERN being very much occupied with LEP. The idea was, however, revived very shortly after the HEPAP recommendation for the SSC, spurred by the substance of this recommendation. A March 1984 workshop at Lausanne studied the feasibility of hadron colliders in the LEP tunnel but it was almost entirely devoted to physics issues [100]. At the May 1984 meeting of the AAAS, CERN DG Herwig Schopper described plans to build a 5-on-5 TeV proton–proton collider by adding current technology superconducting magnets on top of the low field LEP magnets. He argued that the price tag would only be US$500 million and invited the US to participate [101].

As subsequent developments showed, that cost figure was completely unrealistic and the idea of building and running the collider contemporaneously with LEP operation was problematic on operational grounds. But it was true that a Large Hadron Collider (LHC) in the LEP tunnel was an obvious next step for CERN to take once the physics at the LEP was fully exploited. Thus the SSC, or any other high energy hadron collider elsewhere, was viewed by many as a threat to CERN's future and it was only natural that the DG of CERN would be concerned. A few quotes from Schopper's prepared testimony [102] before Congress on April 7, 1987, illustrate this: "The Committee of Council is of the opinion that a hadron collider in the LEP tunnel should be seriously considered as the next step in the exploration of the microcosmos... hadron collider in the LEP tunnel would cover the interesting energy range at a fraction of the projected cost of the SSC." And, later on, during the SSC discussion: "... if we would propose such a project of that size, it would not go through because the emphasis is on cost efficiency, cost effectiveness."

On the whole, however, Europeans did express support for the SSC, indicating at the same time that the likelihood for significant European contribution to the SSC construction was slim. Volker Soergel, DG of DESY, in his prepared testimony [103] for the same hearing, wrote: "I think SSC is a great project and I hope very much that a machine with those characteristics can be realized.... For financial collaboration, I am not as optimistic that a sizable amount of money for these big projects can come from abroad." The most likely possibility for contributions from Western Europe was Italy. Italy had a strong tradition of supporting physics in general and HEP in particular; it also had a heavy involvement in the Collider Detector at Fermilab (CDF). It had already contributed to foreign accelerator construction by providing superconducting magnets for HERA.

The situation in Japan was somewhat different. The Japanese were in the process of constructing an e^+e^- collider called TRISTAN (with an energy up to about 35 GeV in each beam), due to be finished in 1986. Thus, from the funding point of view, they were a much more likely partner in any new large particle physics facility. They had also established a tradition of working at US accelerators, the most significant being their large involvement in the CDF at the Tevatron. Keyworth realized this and as early as the summer of 1984, on his trip to Japan, broached the subject of Japanese participation in the SSC [104].

Keyworth's intervention had both positive and negative sides to it. On the plus side, the Japanese government circles were happy to be approached at early stages of the project, before full design was settled on. On the other hand, the Japanese HEP community felt bypassed. They wanted the initial contacts to be made on the physicist-to-physicist level, so that they would have had the opportunity to decide whether this was a project they want to be involved in. When I made my initial trip to Japan in the summer of 1985 and discussed the SSC with the Japanese HEP community members, that sentiment was emphasized to me quite strongly, especially by M. Koshiba.

The Japanese community was split regarding their future. The accelerator physicists wanted to pursue an e^+e^- linear collider as the next initiative; that presented a greater challenge, since the SSC was considered a relatively uninteresting machine from the accelerator physics point of view. The HEP experimental community, on the other hand, wanted to take advantage of nearer term opportunities that would be provided by the SSC. The latter argument appeared to prevail, to a large extent thanks to Panofsky, who made strong arguments about very different time scales for these two efforts.

There were several other possibilities. Canadian physicists expressed strong interest in experiments at the SSC and were suggesting a cross-border site for the SSC, spanning northern New York State and southern Quebec. Should that materialize some contribution could be expected, financial or in-kind, as well as provision of cheap electric power during operations. The potential difficulty was that TRIUMF, the Canadian HEP laboratory in Vancouver, was proposing construction of a high intensity, medium energy proton synchrotron called KAON with a price tag of around US\$400 million. Small contributions might have been forthcoming from China, Russia and India. Russia, however, was trying then to build its own facility, a 3 TeV superconducting proton synchrotron, UNK, at Serpukhov, to be completed no earlier than the mid-1990's. Initially a fixed target machine, there were vague plans to eventually convert it to a collider. Subsequently, a very ambitious proposal was put forward to build a 1 TeV electron–positron linear collider in Protvino [105].

The issue of international collaboration was discussed extensively by the SSC BOO, starting already in 1984. In his memo to the Board, Panofsky outlined four possible ways in which one could proceed with the international construction of a high energy hadron collider:

- Financial participation by other nations in SSC construction authorized by the US and located in the US.
- Contributions in kind by other nations to an SSC authorized by the US and located there (essentially a HERA model).
- International collaboration on exploitation of the SSC, i.e. in construction and operation of the detectors (the standard mode of collaboration to date).
- Abandon the idea of the SSC in the US and instead participate in construction and exploitation of a hadron collider in the LEP tunnel.

In the first draft Panofsky's personal preference was for the third option. The subsequent draft stated that construction of the SSC should be carried out within the national US program but one should seek participation from foreign countries at all levels of R&D, design, engineering and construction. It also advocated internationalization of SSC Laboratory during its operational phase. It was unlikely that the second option, if successful, would reduce significantly the US part of the SSC cost. According to Soergel's Congressional testimony, 20% (by cost) of the total components for HERA were contributed by countries other than Germany, mainly Italy, France and Canada. The actual cost offset was less than that, however, because of additional management complications, such a scenario required and because of loss of competitive bidding. On the other hand, success in that alternative might be important politically, providing an outside endorsement of the value of the SSC.

During the mid-1980's there was a considerable effort both to "spread the word" about the SSC outside of the US and to involve physicists from abroad in various SSC activities. Numerous visits to foreign universities and laboratories were undertaken on a relatively informal basis by the members of the CDG, of the BOO, and of the HEP community in general. One of my principal responsibilities as Deputy Director for External Relations was to oversee, nurture and participate in these activities. During the CDG days I probably made about ten trips to Japan, a large fraction of them as part of official US delegations once the SSC received official endorsement from the executive branch. Contacts were also being made with industrial concerns abroad; thus, for example, Tigner made a trip to Japan to visit Japanese companies with a potential interest in the SSC.

In parallel, we made efforts to include scientists from abroad in US conferences and workshops on the SSC, and in different committees and task forces dealing with SSC matters. Thus, for example, there were European and Japanese physicists on the Detector Cost Model Advisory Panel as well as on the Advisory Committee for the Generic Detector R&D. The latter had fewer political barriers, since the generic nature of the efforts discussed was such that these efforts would find application in many different projects. On the SSC Users' Executive Committee, 4 of the 11 members were scientists from Europe, Japan or Canada.

15. Presidential Approval — "Throw Deep"

The excellent review of the Conceptual Design Report convinced DOE at the level of the Office of Energy Research (i.e. Trivelpiece) that the SSC was very likely a viable and well-understood project. It was also clear that it would be difficult to obtain an increase in the level of SSC R&D funding without the go-ahead for construction. Thus, after completion of the CDR, the efforts were intensified to convince the DOE Secretary to endorse SSC construction. Many members of Congress were strongly in favor of the SSC, but the cost of the project was so high that to be built it needed a strong push from the Administration. Thus, for example, on April 11, 1986, 91 Members of Congress signed a letter to the President urging him to support the SSC [106]. Congress was unwilling to keep doling out the R&D money without some positive action from the Administration that they would go ahead with it [107]. The House Appropriations Committee in its FY87 appropriation language stated: "The Committee is concerned by the lack of commitment by the Administration with regard to the Superconducting Super Collider." It reduced the HEP appropriation by US$27 million from the President's budget request, as a way of putting pressure on the Administration to make a decision.

There was a concerted effort by the SSC proponents to convince Secretary Herrington to make the SSC a high priority within DOE. Herrington was very close to Reagan and his opinion would carry significant weight with the President. Vigorous activism was needed since the opposition to the project from scientists in other fields was intensifying. The Association of American Universities (AAU), an organization composed of some 50 presidents of the nation's leading universities, approved a letter to Herrington urging him to back the project [108]. A number of prominent scientists and industrialists[h] sent letters to the key people in government arguing for the SSC approval. On the whole, the mood in the country at that time was still relatively favorable toward the SSC. A *Los Angeles Times* editorial stated: "The Super Collider would provide the next breakthrough in our understanding of matter. That alone is a very good reason to build it... What the scientists are proposing should rank high on our list of national objectives [109]." An important meeting was a lunch Herrington had with several well-known scientists[i] to solicit their views about the SSC.

The first review of the SSC by the Domestic Policy Council took place in December 1986; another one in the same forum was held shortly before Christmas. Herrington and Trivelpiece presented the SSC case. The meetings were inconclusive and the decision was postponed till January 29, when Reagan presided over the meeting [110]. Trivelpiece gave the main presentation, about the history and goals of HEP and the connection between that field and technology. As expected, there were concerns about budget implications for other sciences. This was at least partly defused by proposing the first major expenditure of US$348 million only in FY89. Subsequently US$600–700 million would need to be spent annually for five years but that would be during the next Administration.

After the presentation and a question period, Reagan recalled an anecdote. The Oakland Raiders' well-known quarterback, Ken Stabler, was reputedly asked the meaning of a poem by Jack London that London considered as his credo:

> I would rather be ashes than dust
> I would rather that my spark
> Should burn out in a brilliant blaze
> Than it should be stifled in dry rot
>
> I would rather be a superb meteor
> Every atom of me in magnificent glow
> Than a sleepy and permanent planet

The credo meant "Throw deep," according to Stabler, and Reagan directed Herrington to follow that motto by going ahead with the SSC [111, 112]. The next morning, the White House formally announced its approval of the SSC.

16. Site Selection

Shortly after Reagan's decision to "throw deep," Secretary Herrington announced on February 10, 1987, the mechanics of site selection process. The initial

[h]Some of the industrialists who wrote were John Akers of IBM, Douglas Danforth of Westinghouse Electric, Edward Jefferson of DuPont and Roger Smith of General Motors [111].

[i]Some of the scientists attending were Solomon Bucksbaum, Roy Schwitters, H. Guyford Stever, Charles Townes and Steven Weinberg (Ref. 111 and Schwitters' notes in *Fermilab Archives*).

step was establishment of an SSC Site Task Force, reporting to the Director of DOE's Office of Energy Research, and chaired by Wilmot N. Hess, Associate Director of the Office of High Energy and Nuclear Physics, who replaced Jim Leiss in that position.[j] The Task Force issued on April 1, 1987, an "Invitation for Site Proposals for the Superconducting Super Collider (SSC)." This 76-page document [113] described the SSC, spelled out the requirements for site proposals, outlined the site selection process with the timelines for different steps, and described all the information that had to be provided. The technical aspects of the invitation were based very closely on the work done in the conventional facilities area by the CDG, and several CDG physicists participated in its preparation, but as individuals rather than members of the CDG. Tigner wanted to make sure that the CDG would not be involved in what might be a politically charged process. The underlying principle of the selection process, as stated in the invitation, was: "The goal in evaluating sites is to select a site that will permit the highest level of research productivity and overall effectiveness of the SSC facility at a reasonable cost of construction and operation and with minimum adverse impact on the environment."

The schedule for the process, outlined in the invitation, was very tight. The deadline for submission of proposals was August 3, 1987, 2:00 p.m. (later extended to September 2, 1987). The first filter of the proposals was to be conducted by DOE with the goal of eliminating proposals that did not meet the requirements in the invitation. The remaining proposals would then be handed over in September 1987, to a broadly based committee, selected by the National Academy of Science (NAS) and the National Academy of Engineering (NAE), which would evaluate them based solely on provided written documentation (i.e. no site visits). That committee was to produce a Best Qualified List (BQL) by December of that year; the list was to be unranked and it was up to the committee to decide on the number of sites making that list. Subsequently, a DOE-appointed group would conduct visits of the sites on the BQL and perform additional detailed evaluation of the BQL proposals. These evaluations would be

based on both technical and cost information, the latter being the life cycle cost (LCC), i.e. construction cost plus operating cost for a 25-year period. Based on this information, the Secretary of Energy would identify the preferred site. When all the environmental requirements were satisfied, the site would become the final selected site.

The invitation was rather specific as to the format of the proposals and the information that needed to be provided. The six general criteria, in order of importance, were: Geology and Tunnelling, Regional Resources, Environment, Setting, Regional Conditions, and Utilities. It was estimated that the proposal length would be about 200 pages.

The proponents were required to donate the proposed site to the federal government and were encouraged to provide financial and other incentives. The invitation stated: "The site proposed must be entirely located in the United States of America...." The transfer of all the required land (about 16,000 acres) was to be completed by April 1, 1990. Preference was expressed for locating the ring in a horizontal plane but a tilt up to 0.5° would be permissible. The water and utilities requirements were similar to those in the original "Site Requirement" document produced by the CDG in 1985.

A great deal of interest was generated in response to this invitation all over the country [114]. The funds spent by different states on proposals were reported as ranging from US$300,000 to about US$10 million. Several states published glossy brochures extolling the virtues of their sites.[k] However, there were concerns and complaints, expressed by a number of states, that the schedule was too tight, the four months allocated for preparation of the proposals being insufficient. In addition, objections were raised to the provision that the financial incentives in the offer would influence the decision. It was argued that this provision favored larger states. Sen. Pete Domenici from New Mexico introduced an amendment, enacted on July 11, 1987, which required DOE to use only qualities of the sites as the selection criteria. DOE amended the invitation to conform with this act and extended the proposal deadline by one month [115].

---

[j]Jim Leiss retired toward the end of 1985 and for about a year Bill Wallenmeyer held that position on an acting basis while continuing as Director of the Division of High Energy Physics.

[k]I have seen glossy brochures produced by Arizona, Colorado, Illinois, Ohio and Washington.

DOE received 43 proposals. Seven of them were judged as not meeting the basic qualification criteria set forth in the invitation. One of those was the transborder proposal submitted by New York State jointly with the province of Quebec, it not being located entirely in the US. In addition, the Wallkill Valley, New York, proposal was withdrawn in October 1987. Thus 35 proposals were transmitted on September 27, 1987, to the NAS/NAE committee for evaluation.

The NAS/NAE committee had 21 members [116, 117] and was chaired by Edward A. Frieman from the Scripps Institution of Oceanography and the University of California, San Diego. The membership was very diverse, including high energy physicists (both experimental and theoretical), accelerator experts, high level university and national laboratory officials and representatives from industry, the Army Corps of Engineers and public institutions. The Academies provided staff support under the leadership of Raphael G. Kasper, who served as Project Director. In addition, the committee had the benefit of comments from a number of other colleagues at different subgroup meetings.

The committee established seven working groups, each one focusing on one of the six technical criteria and the costs. Each working group was composed of committee members with specific expertise in the area of focus of that working group. Besides the meetings of the working groups, the whole committee met as a whole on three occasions, June 30–July 1, October 8–10 and November 13–14, 1987. At the committee's final meeting there was a full discussion on the strengths and weaknesses of each of the 35 proposals. No ranking of chosen sites was made and no *a priori* decision was made as to the appropriate number of sites to be placed on the BQL. The final BQL included all the sites on which the committee consensus for inclusion was reached.

The report summarizing the committee's deliberations and conclusions was forwarded to DOE on December 24, 1987, and published later by the National Academy Press [118]. It listed the eight sites being put on the BQL and for each gave a short description of its strengths and weaknesses, as seen by the committee. These sites were, in alphabetical order: Arizona/Maricopa, Colorado, Illinois, Michigan/Stockbridge, New York/Rochester, North Carolina, Tennessee and Texas/Dallas–Fort Worth

[119]. The DOE Energy System Acquisition Advisory Board accepted this list without any modifications on January 15, 1988. On the same day, the New York site was withdrawn by the proposers because of significant local opposition.

The Illinois proposal presented a special situation. The potential usage of the Fermilab accelerator complex and other laboratory infrastructure could offset partially the cost of the SSC construction. DOE had performed a study of the technical feasibility and potential cost savings of using the Tevatron complex as the injector for the SSC. The report concluded that it was reasonable to expect that the Tevatron complex could meet the SSC requirements. It was not clear, however, whether that complex could meet the SSC reliability criteria and what fraction of the existing components would have to be replaced.

The next step was a detailed evaluation of all the BQL sites by the DOE Task Force. The evaluation started with week-long visits to all the sites by the staff of a DOE contractor, RTK, with a DOE representative, to gather additional data required for the eventual EIS. Subsequently, between April and June, the Task Force visited each BQL site and then assigned a ranking on each technical subcriterion for each proposal. At this time the Task Force also updated the life cycle cost analyses based on additional data submitted by the proponents, the information obtained during the site visits, and the geotechnical investigations of the sites. The costs ranged from US$10.7 billion to US$11.5 billion for the six non-Illinois sites. The LCC for the Illinois site was estimated to be US$11.4 billion if no credit was given for the Fermilab facilities, and between US$10.4 billion and US$10.9 billion if credit was given, with the range being due to uncertainties in estimating the suitability and lifetime of the Tevatron complex.[1] A very different cost estimate was obtained by Illinois site advocates, who argued that siting the SSC at Fermilab would save US$3.28 billion over the lifetime of the project [120]. Finally, the Task Force determined that all the BQL sites met the

---

[1]In an April 27, 1987, letter to Wallenmeyer, responding to his inquiry, Tigner stated that he could not see any reasons why Tevatron would not be suitable as the SSC injector. Any additional modification costs were said to be small compared to the estimated cost of the injector systems of US$341.4 million (FY88 $).

equal opportunity requirements in the original invitation [121].

Because of the many complaints received subsequently about procedures for the process, the General Accounting Office (GAO) was asked by Congress to assess the fairness of the process by examining a number of different issues. GAO interviewed most of the key participants in the process and recommended in its report [122] dated January 1989 that "the Secretary of Energy ensure for any future site selection process similar to the SSC that potential site proposers be given the maximum information possible in the invitation about the relative importance of the selection criteria." DOE responded that the relative importance of different criteria was clearly defined in the invitation.

The report of the Task Force, providing the evaluations of the BQL sites and their LCC estimates, was forwarded by Hess to Robert Hunter, Director of DOE Office of Research, on November 7, 1988. Shortly afterward, Secretary Herrington announced the choice of the Texas/Dallas–Fort Worth site as the preferred site [123–125]. In December of that year, DOE issued the final EIS for the SSC, and on January 18, 1989, finalized the decision to site the SSC in Texas [126].

The fairness of the site selection process has been intensively debated in the subsequent years, with accusations that Texas was selected because of its political clout and its planned contribution to offset the cost of the SSC [127]. Claims were made that some negative aspects of the Texas site were ignored, such as fire ants [128]. The financial offer (US$1 billion) was not part of the official proposal but it was well known that Texas, if selected, would make a contribution of this magnitude. I was a member of the NAS/NAE panel and saw no evidence for the above allegations. The Texas site clearly belonged on the BQL and it ranked very high in the minds of many panel members. But it was clear that states tried to exercise as much political pressure as possible [129]. I have no inside knowledge as to how much impact this had on the final decision.

But there were flaws in the site selection procedure. The most important one was ignoring the sociological and practical implications of starting a new laboratory on a green site, much more important than the issue of costs. Such a site meant starting from scratch in all respects, the most important

being recruitment of staff and creation of technical support facilities. Furthermore, a green site would provide no fiscal flexibility in case of construction stretch-out. All of the costs at the laboratory had to be charged against the construction project, since there were no other laboratory activities to which one could shift the idled personnel.

Another deficiency was the use of the life cycle costs. Even though cost differences in downstream years were discounted, this procedure underestimated the importance (if only political and emotional) of the initial construction cost. It was the initial construction cost that would determine how fast the SSC could be built and how vulnerable it would be to potential cancelation during its construction. An estimate of construction and operating costs separately, possibly with a guideline to the Task Force regarding the relative weight to be allotted to each, would have been much more appropriate.

Finally, the decision not to consider the transborder site in New York State had long range repercussions. Rep. Sherwood Boehlert (R-NY) stated formally his objection to this decision in a comment [130] appended to a House authorization bill: "Our pleas for international cooperation are going to sound awfully hollow if we turn down this opportunity to reduce the cost of this hugely expensive project." Boehlert later became one of the strongest and most influential opponents of the SSC in Congress. Undoubtedly the site decision influenced his stance but it has also been widely reported that Krumhansl (who was in Boehlert's district) played a large role in turning him against the SSC.

17. Post-CDR Technical Activities

The technical work at the CDG continued at an intensive pace after the CDR was finished. The main focus was on extending the studies made for the CDR and on the R&D for long lead time critical path hardware systems such as superconductor, magnets and cryogenic systems. Here I only mention some highlights.

After magnet type selection was made, further development of the baseline dipole design was a shared responsibility between BNL, Fermilab, LBL and CDG. LBL focused on cable development and short magnets, BNL on cold mass production for long magnets, and Fermilab on cryostats and magnet testing. The CDG provided direction and coordination of this effort. This was quite challenging in light of these geographically highly dispersed efforts. Nominally Vic Karpenko, as Magnet Division Head, had the principal responsibility in this area, but his experience in a somewhat different culture at Livermore generated some difficulties in his interactions with the labs. A scheme adopted by Tigner was to have a "Gang of Four" — Tigner, Goldwasser, Limon and Karpenko — assume joint management responsibility and visit periodically all the labs. Ned Goldwasser was the former Deputy Director at Fermilab who was spending two years at the CDG as Associate Director, starting in September 1986. This scheme was a temporary solution lasting about a year, until late in 1987, when John Peoples from Fermilab was recruited to take over as Magnet Division Head for a year (he was then replaced by Tom Kirk). Peoples took an experimental approach and continued and enlarged a program of producing a number of magnets, each with a somewhat different design. That strategy, coupled with extensive instrumentation on the magnets, gave much better understanding of potential problems. By the end of 1988 a dipole coil configuration was developed which was believed to provide optimum magnetic characteristics. It was anticipated that the R&D baseline magnet design would be obtained by the end of 1989 [131].

There was also extensive work in other areas, such as bore tube corrector development, design and fabrication of the required tooling, cryostat design and testing, and accelerated life tests of the magnets. There was also work on collider quadrupole design and on conceptual design of HEB and injector magnets. Another important activity pursued in parallel was the development and initiation of the magnet industrialization program. The cable R&D focused on optimization of the superconducting material, improving the critical current operating margin, and on technology transfer to the cable industry. A cabling machine was designed, produced in industry, and after successful tests shipped to a potential cable manufacturer — New England Electric Wire. In October 1988, the machine made cable successfully at its design speed. Cable sufficient for six dipoles was produced during December 1988 and January 1989.

The studies of desorption continued both at the Synchrotron Light Source at BNL and in a collaborative experiment at KEK. There was also work on neutron backgrounds in the tunnel and experimental halls, both via simulations and experimental measurements at the Tevatron. Understanding the aperture requirement continued, with theoretical work and with focused experiments at the Tevatron. [132] There was additional work on various lattice issues that were site-dependent. Similarly, more detailed work was commenced in the area of conventional facilities, specifically on the design of the experimental halls, on determination of the tunnel cross section, injector concepts and footprint studies. It was clear, based on the initial detector designs, that the experimental halls would need to be larger than was suggested in the CDR.

FY89 appeared to be the first year with a significant increase in SSC R&D funding — US\$100 million in the President's request (US\$95 million were eventually appropriated). About half of that sum would be allocated to the magnet program, with US\$16 million of that slated for the industrialization activity. There were also plans to increase significantly generic detector R&D and initiate preliminary engineering and design (at a US\$16 million level).

18. Starting the SSC — Getting Congressional Approval

Obtaining a site for the potential location of the SSC still left two major issues to be resolved before SSC Laboratory could become a reality. The first one was obtaining Congressional appropriation of the initial construction funds. The second one was creating a management organization for SSC Laboratory.

Reagan's endorsement of the SSC had an important symbolic value but, from the practical point of view, did not represent a major step toward making the SSC a reality. He was approaching the end of his term and his influence on subsequent appropriations was waning. As a matter of fact, Congress did not appropriate any construction funds for the SSC for either FY88 or FY89. Thus our challenge was to persuade Congress of the value of the SSC and to convince the future Administration (as well as the subsequent ones) to pursue the SSC [133].

The overall funding climate in 1987 was already significantly different from that in 1983. There was a deeper concern about the budget deficits and growing realization that the "credit card" financing of federal expenditures could not go on forever. Many scientists in other disciplines began to worry seriously that construction of the SSC would drain funds away from their own fields.[m] Thus substantial opposition began to develop in scientific circles against the SSC even though there was very little evidence that at this time the SSC had a negative impact on other basic research; the SSC might even have been pushing funding up in other areas [134] (for example, NSF was embarking on a budget-doubling five year program [135]). The expected foreign commitments to SSC construction were not materializing and there was a strong sentiment in the House to postpone construction approval until definite agreements were reached on cost sharing [136]. It was realized that some of the support for the SSC was rather soft and could disappear once the site was selected [137]. When the NAS committee reduced the number of possible SSC sites to seven, many previous backers (Rep. Boehlert is the best-known example) became either neutral or even anti-SSC. Furthermore, two new developments fueled the opposition.

The first of these was increased publicity for potential construction of the LHC in the LEP tunnel at CERN. In his April 1987 testimony before the US House Science, Space and Technology Committee, Schopper still advocated a 5–6 TeV collider in the LEP tunnel, claiming that it could be built for 1/3 to 1/4 of the SSC cost [138]. The situation in Europe, however, was far from rosy at the time and there were many pressures to reduce HEP funds. Great Britain's Kendrew report was recommending scaling back its high energy budget [139] by 25%; other countries were rumored to want to follow suit. The CERN Council set up a management committee, with Anatole Abragam as chair, to evaluate the CERN future under various funding scenarios. At the same time, a Long Range Planning Committee was

appointed, with Carlo Rubbia as chair, to look at different options for a new post-LEP accelerator facility at CERN. Preliminary reports [140–142], presented to the Council by these two committees in June 1987, discussed the possibility of a hadron collider in the LEP tunnel, but it was far from a certainty and one certainly did not know at that time its parameters, its cost, or its potential time scale. It was very likely that if the US would proceed expeditiously with the SSC, CERN might focus on an electron–positron collider as its future facility. Nevertheless, in the US, the LHC was being put forward by some as an obvious alternative to the SSC at one-quarter of the cost.

The other event was the exciting new discovery in the field of superconductivity [143]. Late in 1986, Bednorz and Muller, working at IBM Laboratory in Zurich, discovered superconductivity in previously unexplored materials. Lanthanum–barium–copper oxide was shown to become superconducting up to temperatures over 30 K. Shortly afterward materials were found with even higher critical temperatures [144–146]. Chu and his collaborators at Houston, working with a group at the University of Alabama, observed stable superconductivity at temperatures as high as 95 K, significantly above the boiling temperature of nitrogen at 77 K. The potential significance of these new discoveries was not lost on either the proponents or the opponents of the SSC [147, 148].

In response to these developments Tigner convened a CDG Task Force to explore potential applications of this new technology to the SSC and to understand the challenges that had to be overcome before it could be utilized [150]. The Task Force identified a large number of problems that had to be solved before practical applications could be realized, and estimated that even under optimistic assumptions the use of these materials could reduce the overall SSC cost by only 3%.

There were significantly different claims made, however, by some of the SSC opponents. James Krumhansl, serving at that time as the President-elect of the APS, in his testimony [151] to Congress on August 7, 1987, said: "In regard to the SSC these superconducting materials offer the possibility of obtaining much higher magnetic fields than those now planned... it might be possible to achieve the projected design energy with a ring perhaps 10 miles rather than 50 miles in circumference... a pause in

---

[m]A good example is Krumhansl's letter to Herrington [152]. From the very beginning, however, the HEP community argued that the SSC would have to be built with "new" money and hopefully stimulate increased support for all basic research; a good example is an editorial by J. D. Bjorken, *Phys. Today* **41**, 136 (1987). This was also the DOE position; its Energy Research Advisory Board (ERAB) stated, speaking of the SSC: "It cannot be undertaken without a multibillion dollar incremental commitment to basic science over the next decade." On the whole, basic research fared reasonably well during the Reagan administration; see C. Norman, *Science* **231**, 785 (1986).

the commitment to SSC and its siting should be given serious consideration." In a letter of February 19 to Secretary Herrington he was even more optimistic [152]: "They unquestionably have the potential to save billions of dollars in construction and operation of particle accelerators like SSC. Because they are easily fabricated, I have little hesitation that they will be brought to technological usability in three to five years."

Philip Anderson included in his testimony [153] at the same hearing: "It is possible to anticipate that one can either omit Helium cooling or reduce the size of the ring or both within a matter of 2–3 years." But he was not completely self-consistent, stating also: "Any good engineer will tell you that the last thing in the world you want for your large engineering project to depend on is truly innovative technology.... The delays and mistakes inevitable in new technology are just too common."

To provide an objective assessment of the potential benefits of high temperature superconductors and the possible time scale, the NAS convened a panel with a charge to provide a Research Briefing on High-Temperature Superconductivity. This panel, chaired by John Hulm of Westinghouse, concluded in its 1987 report [154]: "The complexity of the materials technology and of many of these applications makes a long term view of research and development essential for success in commercialization. The infectious enthusiasm in the press and elsewhere may have contributed to premature public expectations of revolutionary technology on a very short time scale."

In parallel, on April 23, 1987, Alvin Trivelpiece requested the Basic Energy Sciences Advisory Committee (analog of HEPAP in materials science) to study the possible use of these materials in particle accelerators. The panel convened for this purpose [155] was chaired by Albert Narath of AT&T Bell Laboratories and reached conclusions similar to those of the NAS Committee: "On the basis of rather optimistic assumptions, it appears possible to develop the necessary technology and demonstrate acceptable filamentary conductors in about 4 years. However, the panel believes that under realistic conditions an additional 8 years would be needed before full-scale production of magnets equivalent to the SSC benchmark could commence." It became clear already in early 1989 that even these "realistic" panel

reports were much too optimistic [156–158]. And today, some 22 years and many man-years of intensive scientific and engineering effort later, we are still far away from any significant large scale application of those materials.

Even though objective analyses showed that the new superconducting materials were not ready for applications to accelerator magnets, the seeds planted by the SSC opponents found a resonance in the popular press. Numerous articles and editorials appeared, with titles like "New Findings Could Make Supercollider Obsolete." *The New York Times*, in an editorial [159] titled "Super Hasty on the Supercollider," cited Philip Anderson's claim that "new materials could make magnets 10 times more powerful... permitting reduction of circumference from 50 miles to 5 miles." It is not surprising that such reports made an impact on the US Congress. Undoubtedly influenced by these developments and the associated publicity, the Senate Budget Committee requested the Congressional Budget Office (CBO) to prepare a report evaluating the risks and benefits of building the SSC. Their report [160], published in October 1988, formally expressed no recommendations but only options for Congressional actions. But it treated highly futuristic possibilities like an e^+e^- collider in the TeV range and a hadron collider in the LEP tunnel on the same footing as the SSC. Thus it gave credibility to the notion that there were real alternatives to the SSC in the same time frame. DOE made an effort [161] to rebut some of the CBO assessments but they did not appear to be 100% effective.

The advocacy for the SSC was weakened by the departure from the Administration of two key SSC supporters, Keyworth and Trivelpiece. Keyworth [162] left at the end of 1985 and his position was not filled for a year, until William Graham, former NASA Administrator, was confirmed by the Senate. Trivelpiece, so influential in getting the SSC to this point, left DOE in April 1987, to assume the position of Executive Director of AAAS [163]. He had established a good relationship with Congress and thus his departure made subsequent interactions with Congress more difficult. During the next five years the position of DOE OER Director was held a large fraction of the time, on an acting basis, by James Decker, a career DOE official, since invariably there would be a long time interval between the resignation of one director and the confirmation of his or her successor.

These concerns about the SSC design and a proposed steep rise in the future annual SSC budgets (US$363 million request for FY89) contributed to Congressional hesitancy in starting construction of the SSC. The House Committee on Science, Space, and Technology authorized [164] funds for SSC R&D and construction for FY88, FY89 and FY90 (US$878,000 total) on October 15, 1987, but only R&D funds were appropriated for FY88 and FY89 — US$33 million and US$95 million, respectively. The significant increase for FY89, the words associated with it, and the House authorization gave hope, however, that Congress might approve SSC construction in the following year. The Report of the Senate Energy and Water Development Appropriation Bill, 1988, stated [165]: "The Committee provides $35,000,000 for the SSC in FY88. This includes funds to continue the SSC R&D program, support site selection studies, and undertake other necessary activities in support of the project. This will keep the design team together and support other non-site activities for one more year while the administration explores means of funding this project."

19. Starting the SSC — Management Issue

The other very important issue was management of SSC Laboratory. Examining the precedents, one could see three distinct modes as possibilities:

- Management by a university or a university consortium. The prime examples at that time of this mode were SLAC, LBL and Fermilab.
- Management by an industrial concern, examples being Sandia and Oak Ridge.
- Formation of a new government entity like NASA or NIST.

There was a general concern in the government circles that the SSC construction was simply too big a task to entrust to academics. As a result there was some sentiment to decouple the construction and operation of the SSC: the SSC would be built by an industrial venture or a government entity like the Army Corps of Engineers and then its operation turned over to another organization charged with subsequent management of the SSC. Panofsky argued forcefully and persuasively against this choice

on the grounds that an accelerator laboratory is a "living" entity, constantly being upgraded and modified as new needs arise or new technologies become available. Thus the experience gained in the construction phase is essential for subsequent operation.

The HEP community wanted the management contract to be given to URA, for both construction and operation. URA submitted an unsolicited bid for such a contract on March 2, 1987 (and another, somewhat modified version on February 22, 1988), prepared under the direction of Ned Goldwasser. That proposal was never acted on by the DOE [166]. To avoid concerns about a potential conflict of interest between Fermilab and SSCL, early in 1988 URA had reorganized: a Board of Trustees was to be responsible for corporate URA affairs and two separate Boards of Overseers, one for Fermilab and one for SSCL, would be responsible for laboratory oversight. On June 1, 1988, McDaniel, in his capacity as chair of the SSC BOO, wrote to Joseph Salgado, Undersecretary of DOE, urging that "DOE take immediate action to set up an SSC organization where management responsibility is centralized in some entity drawn from the scientific community." Salgado's response of July 22, 1988, stated: "We are now developing a plan of action for establishing a permanent management structure for the SSC" [167].

On August 3, 1988, DOE announced an open competition for management of SSC design, construction and operation [168, 169]. The actual Request for Proposals (RFP) was not issued till August 22, with a preproposal conference held on September 8. November 4, 1988, was the deadline for proposal submission; the selected contractor was to take over SSC management in FY89. There were two qualification criteria the proponents had to satisfy: they needed to have had recent significant involvement/association and experience with the US particle physics community and be willing to accept the contract regardless of the location chosen for the SSC. Ed Knapp, URA President at that time, hired Doug Pewitt, Keyworth's deputy in the early 1980's, and Francis Allhoff from EG&G, as the principal consultants to lead the URA effort of proposal preparation.

This process was organized and directed by URA and the BOO, with significant technical support from scientists in the CDG and the labs but as individuals. Formally, the responsibility rested with Ed Knapp as the URA President. URA had made a decision to team up with industrial concerns and submit a partnership proposal. It would be the prime contractor and one or more industrial concerns would be chosen as subcontractors to perform a substantial part of the work. This concept was pushed by Pewitt, with the argument that this would increase the political clout of the URA/SSC team, both in getting the contract and in subsequently getting the Congressional go-ahead.

In September 1988, URA invited several interested companies to make presentations in Washington.[n] EG&G and Sverdrup were eventually chosen as proposed subcontractors to be included in the proposal. EG&G had extensive experience as a DOE contractor based on their management role at the Nevada Test Site and Idaho Engineering Laboratory and their support operations for NASA's Kennedy Space Center. Their main asset for the SSC appeared to be the ability to provide professional support personnel on short notice. Sverdrup was a company with previous experience in construction of large, advanced technology facilities, such as the TRIDENT Support Facilities at King's Bay, Georgia, and the Space Shuttle Launch Complex #6 at Vanderburg AFB, California.

Time was short, so the proposal had to be written very hurriedly. The BOO established the SSC Oversight Committee, chaired by Martin Blume from BNL, to oversee writing of the proposal. The close-to-final draft was circulated to the full BOO in October 20, with a recommendation that the BOO approve the proposal. The time scale was so abbreviated that two of the members did not even have a chance to read the proposal when it was forwarded [170] in October 26 by McDaniel to John Marburger, President of the State University of New York, Stony Brook, and chair of the URA Board of Trustees, with a recommendation that the Board accept the proposal and submit it to DOE.

The proposal submitted to DOE [171] identified four individuals for the key personnel positions which were defined in the DOE-issued RFP:

- Roy Schwitters, Professor of Physics at Harvard, was designated as SSC Director. He held a leadership role in the very successful SPEAR experiment

at SLAC and subsequently was one of the co-leaders of the CDF group at Fermilab.
- Helen Edwards was selected as the head of the Accelerator Systems Division. She was serving in that role at Fermilab and was one of the leaders in the Tevatron project.
- Bruce Chrisman was designated as the head of Laboratory Administrative Services. Chrisman, a high energy physicist by training, was Associate Director for Administration at Fermilab and previously Vice President for Administration at Yale University.
- Robert Robbins was designated as the head of the Conventional Construction Division of SSC Laboratory. He was then Vice President of the Sverdrup Corporation and Manager of Operations for Sverdrup's 850-member Central Group.

The RFP actually defined four other key positions: Deputy Director, Project Manager, Head of Magnet Systems, and Head of Physics Research. Maury Tigner, Director of the SSC CDG, was designated in the proposal as Deputy Laboratory Director but no detailed information (e.g. a CV) was included for him like for the other four positions. Tigner's specific duties were not defined in the proposal except to say that he would play a key role in the SSC/CDG-to-SSCL transition. There were no designees for the other three key positions.

The proposal was submitted to DOE on November 4, 1988. No one else made a bid. Long and painful negotiations followed regarding the division of responsibilities between URA and DOE. DOE wanted to have much more control over all the decisions than was customary [172]. The contract was finally signed on January 16, 1989, two days before the site was officially finalized. The Superconducting Super Collider Laboratory could now be officially launched once Congress appropriated construction funds.

These decisions were crucial to the future of the lab and, not surprisingly, were controversial. One question is whether teaming up with industrial partners was a wise or necessary decision. Marilyn Lloyd, a Representative from Tennessee and the chair of the House Science, Space and Technology Committee, wrote to Joe Salgado, Undersecretary of DOE, complaining that the terms of the RFP were biased in favor of URA [173]. So industrial partnership might

---

[n]Nine different companies made presentations at a meeting held in URA offices in Washington on Sep. 8 and 9, 1988.

have been needed to satisfy Congress. The fact that one of the consultants, Allhoff, was an employee of EG&G raised concerns about a potential conflict of interest. Both Sverdrup and EG&G could have been hired as contractors later on (if their services were necessary). This decision to go into partnership was the beginning of a drift away from the historical HEP culture in laboratory management. Maybe such a change was unavoidable given the scope of the SSC. But the ensuing clash between this new industrial/military culture and the traditional HEP culture contributed to many subsequent SSC difficulties. Later on, the SSCL was most faulted in the area of documentation and lack of adequate cost and schedule systems; it was unfortunate that the need for expertise in this area was not anticipated in choosing industrial partners.

The other obvious question had to do with the selection process and the choice of the top SSCL personnel. The search process for the Director is described in the URA M&O proposal submitted to DOE [174]. It states that the planning process began in May 1988, when a list of attributes desired in a Director was drawn up and a charge was formulated for the Search Committee. At the August 20, 1988, BOO meeting, each member of the Committee agreed to interview eight or more persons by phone and report on their views at the subsequent meeting. 115 persons comprising "prominent physicists, statesmen of science, past government administrators, laboratory directors, university presidents, accelerator builders, and industrialists" were interviewed and the results reported at the August 30 BOO meeting. 58 candidates were suggested; the ensuring discussions reduced the list to 11 and then to 5. A secret ballot was then taken, with the candidates being ranked. There was significant preference for three of them and they were acceptable to all present. Another ballot was taken, with these three ranked in preferred order; Roy Schwitters emerged as the top-ranked person, and his name and those of the other two top-ranking candidates were submitted to the Executive Committee of the Board of Trustees. There were no interviews with any of the top candidates. The Executive Committee approved this ranked list by acclamation at its September 1 teleconference meeting [175]. The full Board approved the nomination on September 8, 1988, and shortly afterward Schwitters accepted the position.

I would like to expand on, comment on and clarify the above description of the selection process. My comments are based on the BOO and Search Committee minutes, Boyce McDaniel's personal notes, correspondence and other relevant records, and discussions with several former BOO or Search Committee members. At the May 7–8, 1988, BOO meeting referred to above, Panofsky stressed the importance of having a search committee in place as soon as possible and Trilling presented a suggested procedure for selection of the SSC Director. It was similar to typical search processes, with advertisements, invitations for applications and nominations, and broad consultations. In spite of the expressed urgency of proceeding with the establishment of the search committee, the only relevant BOO action at that meeting was the appointment of the Search Procedures Committee, with George Trilling as chair [176]. The next meeting was scheduled for three-and-a-half months later, on August 19–20.

At that meeting Trilling distributed a sheet containing "Desirable Attributes of SSC Director" and "Proposed Charge to Director of Search Committee" and suggested several individuals for the Search Committee. This list of attributes and the charge were adopted by the BOO pretty much verbatim at that meeting (rather than at the May one). On the grounds that there was very little time available to prepare the proposal, the BOO established itself as the Search Committee and for that purpose excused Trilling and Schwitters (as potential candidates) and invited Jerome Friedman, Frank Sciulli and Martin Walt (from Lockheed Missiles and Space Co.) to serve on the Committee [177].

The only in-person meeting of the Search Committee took place eight days later, on August 28, 1988, at O'Hare Hilton. The minutes [178] are rather terse and just state that after reports from all members of the Committee on their outside consultations and a following discussion, "a ranked list of three candidates was developed and agreed to by a vote of the Committee." The other two top candidates, in that order, were Nick Samios and Leon Lederman.° The only other meeting of the Search Committee was via teleconference two days later, lasting only

---

°Neither the minutes nor the M&O proposal mentions these names or the ranking order. My information comes from McDaniel's notes and conversations with BOO members.

one hour and with 7 of the 12 Search Committee members participating [179]. At the beginning Jerry Friedman reported on his phone conversations with 11 individuals who had worked with Schwitters on the CDF. There was apparently no effort to contact other former Schwitters collaborators or make similar inquiries about the other two top candidates. Afterward, the motion was offered to recommend to the URA Board of Trustees the appointment of Roy Schwitters as the director designate. The motion was agreed to by acclamation.

On the whole, I would classify the search as rather hurried and narrow, and suffering from a lack of openness (the former justified as being due to the very short time frame available.)ᵖ I have not seen a full list of the 115 individuals reportedly consulted, but from what I have learned, I would conclude that it was not very representative of the HEP community. Such a high number appears somewhat inconsistent with what I was able to learn unless a large majority of the interviews were by Panofsky and/or Treiman. Furthermore, not all interviews occurred early enough to be useful for the selection discussion. One BOO member made his interview phone calls on September 7. Strangely, no current member of the CDG was contacted for suggestions or advice during the process.

Panofsky informed Tigner of the Board's decision on September 4, 1988. The members of the CDG staff were not formally informed until October 30, when the BOO held an Overseers' meeting at LBL. Marburger, McDaniel and Knapp met at that time with the senior staff of the CDG to inform them of the decision. It was not a pleasant event and a lot of acrimony was exchanged, mainly having to do with the search process.

There is no doubt that Schwitters was an outstanding physicist with a lot of experience in detector construction and operation. He was also politically astute and was viewed as having a personality that would probably resonate with Congress

and senior government officials. But his contributions were always within the framework of an existing laboratory and on projects considerably smaller; starting from scratch a new, large-scale, multi-billion-dollar accelerator laboratory on a green site might well call for quite different talents and expertise.

One of the seven points in the adopted list of attributes desired in a Director [180] was: "Strong management abilities, including understanding of engineering and construction, excellent judgement and the ability to attract outstanding individuals to the SSC Laboratory." A large number of people felt that Tigner was a more appropriate choice based on that criterion, especially in light of his excellent performance as leader of the CDG. Another criterion in the attributes list was "Ability to promote and defend the SSC in all the appropriate forums, and to work effectively with the DOE." There was a feeling in some circles (BOO, URA and DOE) that Tigner would not meet this criterion well because his primary interest was in accelerators rather than in experimental particle physics and he did not have sufficient breadth of support in the HEP community and in parts of DOE.

Several members of the BOO and of the URA management felt that Tigner's personality was such that he would not be an appropriate Director. Everyone recognized Tigner's technical excellence but there was a feeling that his rather dry general manner would not go over well with members of Congress and high level officials in the executive branch. I do not share that viewpoint. Tigner and I participated in a number of joint meetings with Government officials and reporters and members of editorial boards of newspapers and magazines. Even though Tigner did not come across as a casual, easygoing person, he conveyed a very authoritative, but not threatening or condescending, manner and left no doubt about his mastery of facts and issues. His testimony before Congressional committees, which I heard on several occasions, tended to be on the dry side but was very logical and clear and generally well received by the audience. q

I always believed that the duties of the SSC Laboratory Director (especially initially) could not be

---

ᵖThe most recent Director's search processes in HEP, first at Fermilab and later at SLAC, were quite formal, lasted several months, and had as committee members individuals not directly associated with the nominating body. I can speak from personal experience as a member of the most recent Fermilab Director's Search Committee. We held about 6–8 two-day meetings, interviewed or heard from a large number of individuals and had half-day-long in-person interviews with the top six candidates.

---

qRep. W. Carney was quoted in Ref. 52: "Tigner is one of the most informed and imperturbable witnesses I have seen on Capitol Hill."

handled by only one person. One might have investigated splitting the responsibilities for building the accelerator and creating the lab (with associated political interactions). The two-equal-co-directors model was tried successfully by CERN with Adams and van Hove when the SPS was being constructed.[r] A Schwitters/Tigner team might have been the optimum solution for the SSC Laboratory construction. It would have made the CDG-to-SSCL transition significantly smoother and put a truly dedicated hands-on person in charge of the accelerator design and construction. I do not believe that the BOO ever considered such a coequal directorship. Their ideal was a Schwitters/Tigner team but with Schwitters as the Director and Tigner as the Deputy and/or Technical Director. The personalities and the views of the two people involved might not have made a codirectorship possible and it would have been an unorthodox arrangement; it is not clear whether DOE would accept it at that time.[s] Nevertheless, it was worth a try.

Acknowledgments

I would like to express my thanks to a number of individuals who have helped me in the preparation of this article. I am very grateful to Adrienne Kolb, Fermilab archivist, for facilitating my access to the Fermilab Archives, the current repository for a lot of original SSC materials. The SLAC Library personnel, especially Abraham Wheeler, have been extremely helpful in retrieving articles and government documents not readily available on the shelves. I thank Maury Tigner and David Corson for facilitating access to the materials in the Cornell Library Archives. Ezra Heitowit and URA staff were very helpful in allowings me to see several relevant URA documents from the SSC era.

I have profited greatly from a number of interviews, in person or by phone, with many of my current or former colleagues involved in this stage of the SSC project. I would like to acknowledge very informative conversations with Marty Blume, Bruce Chrisman, Jerry Friedman, "Gil" Gilchriese, Ezra Heitowit, J. David Jackson, Tom Kirk, Neal Lane, Chris Laughton, Bob Matyas, John Peoples, Chris Quigg, Jim Sanford, Frank Sciulli, Roy Schwitters, Jim Siegrist, Maury Tigner and George Trilling. I thank them all for their generosity with their time and for their willingness to share with me their recollections.

Finally, I would like to thank Gilchriese, Jackson, Siegrist, Tigner and Trilling for reading and commenting on an early draft of this article. I also thank Lillian Hoddesen, Adrienne Kolb and Michael Riordan, for doing the same on an almost-final draft. The responsibility for any factual errors or erroneous conclusions is, however, entirely mine.

This work was partially supported by the NSF grant PHY-0354945.

Appendix A. Glossary of Abbreviations

AAAS: American Association for Advancement of Science
A&E: architectural and engineering
AEC: Atomic Energy Commission
APS: American Physical Society
BNL: Brookhaven National Laboratory
BOO: Board of Overseers (for the SSC)
BQL: Best Qualified List
CBA: Colliding Beam Accelerator (originally ISABELLE)
CBO: Congressional Budget Office
CDF: Collider Detector at Fermilab
CDG: Central Design Group
CDR: Conceptual Design Report
CERN: European Organization for Nuclear Research (the acronym is for a former name in French: Conseil Europeen pour la Recherche Nucleaire)

---

[r]The situation at CERN in 1971 was not quite identical to that at the SSC even though there were many parallels. CERN-Meyrin and the "new" CERN were defined as two separate laboratories by the CERN Council in 1969 (*CERN Courier*, Jun. 1969, pp. 162–165) when other sites than the Geneva area were being contemplated for the new machine. Each laboratory was to have its own director-general. When the decision was made to build the SPS across the border in France, this structure was preserved (*CERN Courier*, Feb. 1981, pp. 36 and 37), even though now there was a very strong coupling between the two laboratories: one responsible for construction of a new facility (eventually to be linked with the old facilities), the other for operating a physics program at an ongoing laboratory.

[s]O'Leary's solution for "SSC management problems" from late 1993 — separate contractors, one for design and operations and another for construction and installation — had some features in common with this scheme. Another relevant observation is that the standard practice today for large experiments is to have two cospokespersons with equal overall authority and responsibility. In most cases, this system seems to work well.

DESY : Deutches Electronen-Synchrotron (high energy physics laboratory in Hamburg, Germany)

DG : Director-General

DOE : Department of Energy

DPF : Division of Particles and Fields

DRC : DOE Review Committee

EIS : Environmental Impact Statement

ERDA : Energy Research and Development Administration

FY : fiscal year

GAO : General Accounting Office

HEB : High Energy Booster

HEP : high energy physics

HEPAP : High Energy Physics Advisory Panel

HERA : Hadron Electron Ring Accelerator

IBM : International Business Machines

ICE : Independent Cost Estimators

ICFA : International Committee for Future Accelerators

IG : Inspector-General

ILC : International Linear Collider

ISABELLE : proposed pp collider at BNL; later renamed CBA

IUPAP : International Union of Pure and Applied Physics

KAON : Kaon–Antiproton–Other hadron–Neutrino Factory

KEK : National Laboratory for High Energy Physics, in Tsukuba, Japan (Japanese acronym)

LBL : Lawrence Berkeley Laboratory

LCC : life cycle cost

LEB : Low Energy Booster

LEP : Large Electron–Positron collider

LHC : Large Hadron Collider

MEB : Medium Energy Booster

MIT : Massachusetts Institute of Technology

M&O : management and operations

MOU : memorandum of understanding

MSAP : Magnet Selection Advisory Panel

NAE : National Academy of Engineering

NAS : National Academy of Science

NASA : National Aeronautics and Space Administration

NEPA : National Environmental Protection Agency

NIST : National Institute of Standards and Technology

NSF : National Science Foundation

OER : Office of Energy Research (in DOE)

OMB : Office of Management and Budget

OSTP : Office of Science and Technology Policy

PAC : Program Advisory Committee

PEP : Positron–Electron Project

PETRA : German acronym for a 30 GeV positron–electron collider at DESY

R&D : research and development

RFP : Request for Proposals

RTK : Raymond Kaiser Engineers, Tudor Engineering Company, Keller & Gannon-Knight joint venture

SDI : Strategic Defense Initiative

SLAC : Stanford Linear Accelerator Center

SLC : SLAC Linear Collider

SPC : Scientific Policy Committee

SPS : Super Proton Synchrotron

SSC : Superconducting Super Collider

SSCL : Superconducting Super Collider Laboratory

TAC : Texas Accelerator Center

TMRP : Technical Magnet Review Panel

TRISTAN : 60 GeV positron–electron collider in KEK Lab, Tsukuba, Japan

TRIUMF : TRIUniversity Meson Facility (Canada's National Laboratory for Particle and Nuclear Physics)

UOSSC : Users' Organization of the SSC

URA : Universities Research Association

UTAP : Underground Tunneling Advisory Panel

VBA : Very Big Accelerator

References

[1] The account of the SSC's Texas days will appear in the *Reviews of Accelerator Science and Technology*, Vol. 2 (2009).

[2] For an alternative description of these events, written from a very different vantage point, see: L. Hoddeson and A. Kolb, The Superconducting Super Collider's frontier outpost, 1983–1988, *Minerva* **38**, 271 (2000).

[3] J. Irvine *et al.*, *Phys. Today* **39**, 27 (1986).

[4] L. Hoddeson, The first large-scale application of superconductivity: the Fermilab Energy Doubler, 1972–1983, *Historical Studies in the Physical and Biological Sciences* **8/1**, 25 (1987).

[5] For a detailed history of ISABELLE see: R. P. Crease, Quenched! The ISABELLE saga, in

Physics in Perspective **7/3** (2005), part I, pp. 330–376; and part II, pp. 404–452.

[6] Report of the 1977 Subpanel on New Facilities of the High Energy Physics Advisory Panel, ERDA 77-71, Jun. 1977. The initial BNL proposal, considered and approved by the 1975 Subpanel, was for 200 GeV in each beam. It was subsequently modified in light of scientific and technological developments.

[7] *Phys. Today* **34**, 17 (1981).

[8] P. David, *Nature* **303**, 465 (1993).

[9] G. Keyworth, editorial in *Science* **219**, 801 (1983).

[10] Report of the Subpanel on Long Range Planning for the US High Energy Physics Program of the High Energy Physics Advisory Panel (DOE-RE-0128, Jan. 1982), App. A.

[11] Report of the Subpanel on Long Range Planning for the US High Energy Physics Program of the High Energy Physics Advisory Panel (DOE-RE-0128, Jan. 1982), App. A., pp. 48–55; *Phys. Today* **35**, 51 (1982).

[12] *Proc. Workshop on Possibilities and Limitations of Accelerators and Detectors* (Batavia, Illinois; Oct. 15–21, 1978), ed. D. Edwards.

[13] *Proc. 2nd ICFA Workshop on Possibilities and Limitations of Accelerators and Detectors* (Les Diablerets, Switzerland), CERN Report RD/450-1 (1979), ed. U. Amaldi.

[14] For a comprehensive discussion on the history of the VBA, see A. Kolb and L. Hoddeson, The mirage of the world accelerator for world peace and the origins of the SSC, *Historical Studies in the Physical and Biological Sciences* **24**, 101 (1993).

[15] *Proc. 1982 DPF Summer Study on Elementary Particle Physics and Future Facilities* (June 23–July 16, 1982, Snowmass, Colorado), eds. R. Donaldson, R. Gustafson and F. Paige; *Phys. Today* **36**, 19 (1983).

[16] M. Tigner, in *Proc. 1982 DPF Summer Study on Elementary Particle Physics and Future Facilities* (June 28–July 16, 1982, Snowmass, Colorado), eds. R. Donaldson, R. Gustafson and F. Paige, pp. 50–53.

[17] Report of the 1983 HEPAP Subpanel on New Facilities for the US High Energy Physics Program (DOE/ER-0169, July. 1983), App. A.

[18] Report of the 1983 HEPAP Subpanel on New Facilities for the US High Energy Physics Program (DOE/ER-0169, July. 1983), App. B

[19] M. M. Waldrop, *Science* **220**, 809 (1983).

[20] Memo from S. Wojcicki to members of 1983 HEPAP Subpanel on New Facilities after the Organizational Subpanel meeting in Feb. (undated)

[21] "Dear Colleague" letter from S. Wojcicki to DPF membership (Mar. 11, 1983).

[22] Proposal for a Dedicated Collider at Fermi National Accelerator Laboratory (May 1983).

[23] P. Reardon's letter to S. Wojcicki, May 17, 1983, on the occasion of the Subpanel's meeting at BNL, and the enclosure "Sandatron possibilities at BNL." The estimated additional cost for a 15 on 15 TeV collider was US\$883 million ± 250 million.

[24] B. Richter, presentation to the 1983 Subpanel at its meeting at SLAC, May 20, 1983. The statement was made based on a very preliminary study ($\sim$10 people for a few weeks), that a 2 TeV e^+e^- collider could be built using existing technology (20 MeV/m gradient) for US\$3.29 billion, with power consumption of 400 MW. This did not include R&D, detectors, contingency or escalation.

[25] Report of the 20 TeV Hadron Collider Technical Workshop (Cornell, Mar. 28–Apr. 2, 1983).

[26] *Phys. Today* **36**, 17 (1983).

[27] UA1 Collab. (G. Arnison *et al.*), *Phys. Lett.* **B 122** 103 (1983).

[28] UA2 Collab. (G. Banner *et al.*), *Phys. Lett.* **B 122**, 476 (1983).

[29] UA1 Collab. (G. Arnison *et al.*), *Phys. Lett.* **B 126**, 398 (1983).

[30] Memo from J. Adams to Subpanel members: Some remarks about new facilities for high-energy particle physics research in the USA (Apr. 22, 1983).

[31] Europe 3, US not even z-zero, *New York Times* editorial (Jun. 6, 1983).

[32] Report of the 1983 HEPAP Subpanel on New Facilities for the US High Energy Physics Program (DOE/ER-0169, July. 1983), pp. 5–14.

[33] Report of the 1983 HEPAP Subpanel on New Facilities for the US High Energy Physics Program (DOE/ER-0169, July. 1983), p. 67.

[34] W. Sullivan, Panel of experts deadlocked on next U.S. atom smashers, *New York Times* (Jul. 2, 1983).

[35] Cable from L. Lederman to chairs of physics departments in the US (Jul. 3, 1983).

[36] *Phys. Today* **36**, 17 (1983).

[37] Cover letter from J. Sandweiss to A. Trivelpiece, Report of the 1983 HEPAP Subpanel on New Facilities for the US High Energy Physics Program (DOE/ER-0169, July. 1983), App. A., pp. i–iii.

[38] M. M. Waldrop, *Science* **221**, 1038 (1983).

[39] Remarks by D. Pewitt to the Third US Summer School on High Energy Particle Accelerators at Brookhaven National Laboratory (Jul. 15, 1983).

[40] Letter from A. Trivelpiece to J. Sandweiss (Aug. 11, 1983).

[41] Letter from W. K. H. Panofsky to J. Sandweiss (Sep. 9, 1983).

[42] Transcript of hearing before the Subcommittee on Energy Development and Applications of the Committee on Science and Technology, US House of Representatives (Oct. 19, 1983).

[43] *Phys. Today* **36**, 41 (1983).

[44] Report of the Reference Designs Study Group on the Superconducting Super Collider, DOE/ER-0213, May 8, 1984; the DOE Independent Cost Estimating (ICE) staff obtained estimates that were higher by US$531 million, US$403 million and US$1060 million for the three designs; *Phys. Today* **37**, 17 (1984).

[45] *Proc. Ann Arbor Workshop on Accelerator Physics Issues for a Superconducting Super Collider*, UM HE84-1 (Ann Arbor, Michigan, Dec. 12–17, 1983), ed. M. Tigner.

[46] J. E. Pilcher and A. R. White (eds.), Pbar-p Options for the Supercollider (University of Chicago, 1984).

[47] *Summary Report of the PSSC Discussion Group Meeting*, eds. P. Hale and B. Winstein (FNAL, 1984).

[48] *Proc. 1984 Summer Study on the Design and Utilization of the Superconducting Super Collider* (June 23–July 13, 1984, Snowmass, Colorado), eds. R. Donaldson and J. Morfin.

[49] E. Eichten *et al. Rev. Mod. Phys.* **56**, 579 (1984); also an addendum in *Proc. 1984 Summer Study on the Design and Utilization of the Superconducting Super Collider* (June 23–July 13, 1984, Snowmass, Colorado), eds. R. Donaldson and J. Morfin, pp. 99–103.

[50] Letter from H. G. Stever, President of URA, to D. F. Hodel, Secretary of the US Department of Energy (Jul. 14, 1983).

[51] Minutes of the SSC R&D and Conceptual Design Board Meeting (Mar. 23, 1984; Washington, DC).

[52] I. Goodwin, *Phys. Today* **37**, 69 (1984).

[53] Memorandum of Understanding between LBL, the host, and URA with regard to location and support of the SSC Design Center (Sep. 1984).

[54] Report of the DOE Review Committee on the Reference Design Study for the SSC (DOE/ER Report; May 18, 1984).

[55] *Phys. Today* **37**, 21 (1984).

[56] Minutes of the Board of Overseers Meeting, at Brookhaven National Laboratory (Sep. 10–11, 1984).

[57] A more detailed listing of Phase 1 Milestones is given in the first issue of the *SSC Newslett.* (Nov. 12, 1984).

[58] R. Palmer and A. V. Tollestrup, *Annu. Rev. Nucl. Part. Sci.* **34**, 247 (1984).

[59] *Phys. Today* **37**, 17 (1984).

[60] R. Wilson, in *Proc. 1982 DPF Summer Study on Elementary Particle Physics and Future Facilities*, (June 28–July 16, 1982, Snowmass, Colorado), eds. R. Donaldson, R. Gustafson and F. Paige, pp. 330–334.

[61] *Phys. Today* **38**, 63 (1985).

[62] D. Larbalestier, G. Fisk, B. Montgomery and D. Hawksworth, *Phys. Today* **39**, 24 (1986).

[63] R. Talman, in *Proc. 1984 Summer Study on the Design and Utilization of the Superconducting Super Collider* (June 28–July 13, 1984, Snowmass, Colorado), eds. R. Donaldson and J. Morfin, pp. 349–352.

[64] SSC Aperture Estimates for Cost Comparisons (SSC-SR-1012, Aug. 1985).

[65] L. Jones, in *Proc. 1982 DPF Summer Study on Elementary Particle Physics and Future Facilities*, (June 28–July 16, 1982, Snowmass, Colorado, 1982), eds. R. Donaldson, R. Gustafson and F. Paige, pp. 345–346.

[66] Aperture Estimates. Presentation to the SSC MSAP by A. W. Chao on Aug. 26, 1985.

[67] Magnet Technical Review Panel Report (SSC-SR-1010, Jul. 1985).

[68] The selection process and the recommendation are discussed in detail in M. Tigner's memo to file, dated Sep. 13, 1986, App. C, in Report of the HEPAP Subpanel on Review of Recent Information on Superferric Magnets (DOE/ER-0272, May 1986).

[69] B. Schwarzschild, *Phys. Today* **38**, 58 (1985).

[70] Letter from F. R. Huson and P. M. McIntyre to W. Wallenmeyer dated Apr. 7, 1986, and attached as App. D in Report of the HEPAP Subpanel on Review of Recent Information on Superferric Magnets (DOE/ER-0272, May 1986).

[71] Report of the HEPAP Subpanel on Review of Recent Information on Superferric Magnets (DOE/ER-0272, May 1986).

[72] B. Goss-Levi and B. Schwarzschild, *Phys. Today* **41**, 17 (1988).

[73] *Phys. Today* **38**, 53 (1985).

[74] US National Committee on Tunneling Technology, *Contracting Practices for the Underground Construction of the Superconducting Super Collider* (National Academy Press, Washington, DC 1989).

[75] Conceptual Design of the Superconducting Super Collider by the SSC Central Design Group (SSC-SR-2020, Mar. 1986).

[76] An Assessment of the Antiproton–Proton Option for the SSC (SSC-SR-1022, May 1986).

[77] Report of the DOE Review Committee on the Conceptual Design of the Superconducting Super Collider (DOE/ER-0267, May 1986).

[78] Detector cost and off-line computing advisory panels created, *The Lattice* (Mar./Apr. 1986), p. 5.

[79] Cost Estimate of the Initial SSC Experimental Equipment (SSC-SR-1023, June 1986).

[80] Committee on Science, Engineering, and Public Policy, *Report of the Research Briefing Panel on Scientific Opportunities and the Super Collider* (National Academy Press, 1985).

[81] SSC users organization established, *The Lattice* (Aug./Sep. 1987), pp. 5–6.

[82] *Proc. 1986 Summer Study on the Physics of the Superconducting Supercollider* (Snowmass, Colorado, Jun. 23–Jul. 11, 1986), eds. R. Donaldson and J. Marx.

[83] *Proc. Workshop on Experiments, Detectors, and Experimental Areas for the Supercollider* (Berkeley, California, July 7–17, 1987), eds. R. Donaldson and M. G. D. Gilchriese.

[84] *Proc. Summer Study on High Energy Physics in the 1990's* (Jun./Jul. 1988), ed. Sharon Jensen (World Scientific).

[85] SSC establishes detector R&D coordinating office, *The Lattice* (Mar./Apr. 1987), p. 4.

[86] R. Huson, L. M. Lederman and R. Schwitters, in *Proc. 1982 DPF Summer Study on Elementary Particle Physics and Future Facilities*, (June 28–July 16, 1982, Snowmass, Colorado, 1982), eds. R. Donaldson, R. Gustafson and F. Paige, pp. 361–368.

[87] Detector research and development and computing. Presentation by M. G. D. Gilchriese at the DOE/SSC Annual Review (Jan. 30–Feb. 1, 1989).

[88] *Appraising the Ring: Statements in Support of the Superconducting Super Collider.* compiled and introduced by L. M. Lederman and C. Quigg (Universities Research Association, 1988).

[89] SSC videotape documentary available, *The Lattice* (Jun./Aug. 1988), p. 7.

[90] An article on the SSC even appeared on Jul. 23, 1985, in *American Way*, the in-flight magazine of the American Airlines.

[91] Colorado hosts SSC symposium, *The Lattice* (Jan./Feb. 1988), p. 4.

[92] "First SSC Industrial Symposium, Feb. 8–10, *The Lattice* (Sept.–Dec. 1988), p. 2.

[93] *Proc. Int. Industrial Symposium on the Supercollider* (Feb. 8–10, 1989), ed. M. McAshan (Plenum, 1989).

[94] M. Bianchi-Streit *et al.*, *CERN* **84-14** (Dec. 11, 1984).

[95] D. C. Mowery and W. E. Steinmueller, Economic payoffs from basic research: an examination of high energy physics (Feb. 5, 1986).

[96] P. A. David, D. C. Mowery and W. E. Steinmueller (1992), *Economics of Innovation and New Technology* **2**(1), 73 (1991).

[97] D. Dickson, *Science* **220**, 1252 (1983).

[98] D. Dickson, *Science* **224**, 1317 (1984).

[99] G. Flugge, *Particle Physics Facilities from the User's Point of View: The Role of ECFA and ICFA, Large Facilities in Physics, eds., M. Jacob and H. Schopper (World Scientific, 1995), pp. 79–98.*

[100] G. Brianti *et al.*, Summary Report of Workshop on the Feasibility of Hadron Colliders in the LEP Tunnel (CERN/LEP, Mar. 21–27, 1984).

[101] D. Dickson, *Science* **224**, 1216 (1984).

[102] H. Schopper's testimony at the hearing of the House Science, Space and Technology Committee on Apr. 7, 1987.

[103] V. Soergel's prepared testimony at the hearing of the House Science, Space and Technology Committee on Apr. 7, 1987.

[104] M. M. Waldrop, *Science* **225**, 490 (1984).

[105] M. Crawford, *Science* **238**, 16 (1987).

[106] M. Waldrop, *Nature* **232**, 571 (1986); this letter was organized by V. Fazio and R. Packard, representatives from California.

[107] I. Goodwin, *Phys. Today* **40**, 55 (1986).

[108] C. Norman, *Science* **232**, 705 (1986).

[109] *Los Angeles Times* editorial (Jan. 21, 1987).

[110] R. Pear, Plan offered for vast atom smasher, *The New York Times* (Jan. 19, 1987).

[111] I. Goodwin, *Phys. Today* **40**, 47 (1987).

[112] George Will's syndicated column "The Super Collider" (Feb. 15, 1988).

[113] Invitation for Site Proposal for the Superconducting Super Collider (SSC) (DOE/ER-0315, Apr. 1987).

[114] C. D. May, States politicking in race for atom smasher, *The New York Times* (Apr. 27, 1987).

[115] I. Goodwin, *Phys. Today* **41**, 47 (1987).

[116] The list of members is given in the Academy's report *Siting the Superconducting Super Collider* (National Academy Press, Washington, DC, 1988).

[117] I. Goodwin, *Phys. Today* **41**, 52 (1987).

[118] *Siting the Superconducting Super Collider* (National Academy Press, Washington, DC, 1988).

[119] I. Goodwin, *Phys. Today* **42**, 69 (1988).

[120] SSC at Fermilab means $3.28 billion saving. *SSC for Fermilab Newsletter*, Spring 1988.

[121] SSC Site Evaluations: A Report by the SSC Site Task Force (DOE/ER-0392, Nov. 1988).

[122] Determination of the Best Qualified Sites for DOE's Super Collider (GAO/RCED-89-18, Jan. 1989).

[123] Secretary Herrington's Preferred Site Selection statement dated Nov. 10, 1988.

[124] M. Crawford, *Science* **242**, 1004 (1988).

[125] IT'S HERE, *Waxahachie Daily Light* (Nov. 10, 1988).

[126] I. Goodwin, *Phys. Today* **43**, 95 (1989).

[127] M. D. Lemonick, A controversial prize for Texas, *Time* (Nov. 28, 1988).

[128] P. Weingarten, Collider, fire ants head for collision, *Chicago Tribune* (Dec. 18, 1988).

[129] The Texas efforts to land the SSC are thoroughly described in: P. T. Flawn, *The Story of the Texas National Research Laboratory Commission and the Superconducting Super Collider* (University of Texas at Austin, 2003).

[130] Additional Views of Hon. Sherwood Boehlert, appended to the SSC Authorization Act of 1987, HR3228.

[131] The superconducting magnet program during this time frame is discussed extensively in: B. Goss-Levi and B. Schwarzschild, *Phys. Today* **41**, 17 (1988).

[132] A. Chao *et al.*, *Phys. Rev. Lett.* **61**, 2752 (1988).

[133] D. Dickson, *Science* **236**, 246 (1987).

[134] M. Crawford, *Science* **237**, 22 (1987).

[135] J. Walsh, *Science* **235**, 1458 (1987).

[136] M. Crawford, *Science* **244**, 24 (1989).

[137] D. Thomsen, *Science News* **132**, 374 (1987).

[138] Schopper's testimony, presented at the hearing of the House Committee on Science, Space, and Technology House (Apr. 7, 1987).

[139] *Phys. Today* **38**, 67 (1985).

[140] D. Dickson, *Science* **235**, 1567 (1987).

[141] W. Sweet, *Phys. Today* **40**, 71 (1987).

[142] C. Sutton, *New Scientist* 30 (1987).

[143] J. G. Bednorz and K. A. Muller, *Z. Phys. B* **64**, 189 (1986).

[144] M. K. Wu *et al.*, *Phys. Rev. Lett.* **58**, 908 (1987).

[145] A. Khurana, *Phys. Today* **41**, 17 (1987).

[146] W. Sullivan, Team reports breakthrough in conductivity of electricity, *The New York Times* (Febr. 16, 1987).

[147] *Science* **236**, 247 (1987).

[148] I. Goodwin, *Phys. Today* **41**, 50 (1987).

[149] J. Gleick, Advances pose obstacles to atom smasher plan, *The New York Times* (Apr. 14, 1987).

[150] New superconductors examined for SSC, *The Lattice* (Jun./Jul. 1987), pp. 3–4.

[151] J. Krumhansl, prepared testimony for the hearing before the House Science, Space and Technology Committee on Apr. 7, 1987.

[152] J. Krumhansl, letter to Secretary John S. Herrington, dated Feb. 19, 1987; quoted in *Phys. Today* **41**, 50 (1987).

[153] P. Anderson, prepared testimony for the hearing before the House Science, Space and Technology Committee on Apr. 7, 1987.

[154] *Report of the Research Briefing Panel on High-Temperature Superconductivity*, National Academy Press, Washington, D. C. 1987

[155] Report of the Basic Energy Sciences Advisory Committee, Panel on High-T_c Superconducting Magnet Applications in Particle Physics (DOE/ER-0358, Dec. 1987).

[156] R. Pool, *Science* **243**, 162 (1989).

[157] R. Pool, *Science* **244**, 914 (1989).

[158] R. Pool, *Science* **245**, 1331 (1989).

[159] Super hasty on the supercollider, *The New York Times editorial*, Apr. 28, 1988.

[160] *Risks and Benefits of Building the Super Collider; A Special Study by the Congressional Budget Office* (Washington, DC, Oct. 1988).

[161] Department of Energy Response to the Congressional Budget Office Report on the Superconducting Super Collider, Febr. 1989; letter from Robert Hunter, Director, Office of Research, DOE, to James Blum, Acting Director, CBO (Nov. 9, 1988).

[162] E. Marshall, *Science* **230**, 1249 (1985).

[163] B. Culliton, *Science* **235**, 840, (1987).

[164] Superconducting Super Collider Authorization Act of 1987, HR3228.

[165] Energy and Water Development Appropriation Bill, 1988, (Sep. 16, 1987).

[166] Proposal to Serve as Contractor for the Construction and Operation of the Superconducting Super Collider Laboratory. Submitted by URA (Mar. 2, 1987).

[167] Jun. 1, 167, 1988, letter from Boyce McDaniel, for the SSC Board of Overseers, to Joseph Salgado, Undersecretary of Energy, DOE, and Salgado's response of Jul. 22, 1988.

[168] Request for Proposals DE-RP02-88ER40486 for the Selection of a Management and Operating Contractor for the Establishment, Management, and Initial Operation of the Superconducting Super Collider Laboratory, US DOE, Chicago Operations Office.

[169] I. Goodwin, *Phys. Today* **42**, 49 (1988).

[170] Letter from Boyce McDaniel, chair of BOO, to John Marburger, chair of the URA Board of Trustees, dated Oct. 26, 1988, and the enclosed attachments.

[171] Proposal for the Selection of a Management and Operating Contractor for the Establishment, Management, and Initial Operation of the Superconducting Super Collider Laboratory. Submitted by URA, Washington, DC (Nov. 4, 1988).

[172] For more detailed discussion of these negotiations, see : L. Hoddeson and A. Kolb, The Superconducting Super Colliders frontier outpost, 1983–1988, *Minerva* **38**, 271 (2000). and : L. Hoddeson and A. Kolb, The Superconducting Super Colliders frontier outpost, 1983–1988, *Minerva* **38**, 271 (2000).

[173] Aug. 11, 1988, letter from Marilyn Lloyd, chair of the House Subcommittee on Energy, Research and Development, to Joseph Salgado, Undersecretary, DOE; an article in the Aug. 15, 1988, issue of Inside Energy, titled "DOE announcement on M&O contract for collider seen favoring URA," discussed the concerns along those lines both in Congress and among some of the potential industrial bidders.

[174] Proposal for the Selection of a Management and Operating Contractor for the Establishment, Management, and Initial Operation of the Superconducting Super Collider Laboratory. Submitted by URA, Washington, DC (Nov. 4, 1988), Vol. 1, pp. 4.1–3.

[175] Minutes of the Executive Committee meeting of the URA Board of Trustees, Sep. 1, 1998.

[176] Minutes of the Meeting of the SSC Board of Overseers, May 7–8, 1988, O'Hare Hilton, Chicago, Illinois.

[177] Minutes of the Meeting of the SSC Board of Overseers, Aug. 19–20, 1988, O'Hare Hilton, Chicago, Illinois.

[178] Minutes of the Search Committee Meeting, Aug. 28, 1988.

[179] Minutes of the Search Committee Meeting, Aug. 30, 1988.

[180] Proposal for the Selection of a Management and Operating Contractor for the Establishment, Management, and Initial Operation of the Superconducting Super Collider Laboratory. Submitted by URA, Washington, DC (Nov. 4, 1988), Vol. 1, 4.1–3.

Stanley Wojcicki is a professor of physics at Stanford University. During the early SSC days he took a 4-year leave of absence to work at Berkeley in the Central Design Group. His current professional interests focus on study of neutrino oscillations. He served as chairman of the DOE High Energy Physics Advisory Panel in 1990–1996. His hobbies include hiking and travel to exotic places.

Reviews of Accelerator Science and Technology
Vol. 1 (2008) 303–317
© World Scientific Publishing Company

Accelerators and the Accelerator Community

Andrew Sessler

*Lawrence Berkeley National Laboratory,
Berkeley, CA 94720, USA
AMSessler@lbl.gov*

Ernest Malamud

*University of Nevada, Reno, NV 89557, USA
ernestmalamud@comcast.net*

In this paper, standing back — looking from afar — and adopting a historical perspective, the field of accelerator science is examined. The subjects explored are: how it grew, what were the forces that made it what it is, where it is now, and what it is likely to be in the future. Clearly, many personal opinions, are offered in this process.

Keywords: Accelerators; accelerator community; accelerator history; accelerator applications.

1. Introduction

It is informative, interesting, fun, and even wise on occasion, to step back and examine our field of accelerator science. This first issue of a review journal is an appropriate time for just such an examination. What are the roots of our science? How did it start? What were the driving forces that made it what it is? How did it evolve? What was and is the nature of the field? Where is it now? Where is it going?

In this short essay we will attempt such an examination. Clearly, a comprehensive job cannot be done in just a few pages, but perhaps, if we look from afar, we can, in broad strokes, paint a picture of our accelerator science. We hope this essay stimulates the reader to think about many different aspects — those raised as well as those ignored. Perhaps these thoughts will lead to a deeper examination of the many technical aspects touched upon, as well as stimulating the reader to reflect on the sociology and communications in the field, and the many diverse subfields that promise to become ever more important in the future. This list of subfields might include novel methods of acceleration (lasers and plasmas), sophisticated developments in external beam cancer therapy, and applications of accelerators to the energy problem, where accelerators drive power reactors and burn up the long-lived components of nuclear waste.

In Sec. 2, we examine the forces that drove this field: nuclear physics, high energy physics, condensed matter physics, chemistry and biology, national defense, medicine, and industry. In Sec. 3, we examine the sociology of the field: the people, the laboratories, and communication and instruction. Finally, in Sec. 4, we attempt to look into the future.

References have not been added, for the sheer number would fill many pages. Rather, we present concepts, laboratories, universities, and the names of some individuals, all well known to the reader. References, and a different view on the subject of this essay, may be found in the recent book *Engines of Discovery: A Century of Particle Accelerators* by A. Sessler and E. Wilson (World Scientific, 2007).

2. Accelerators

In this section are discussed the fields of science that drove the development of accelerators.

2.1. *Nuclear physics*

The development of accelerators began when Ernest Rutherford asked if it was possible to produce nuclear reactions artificially, i.e. not to have to use the energetic products of natural radioactivity. Thus started a race to build machines that would produce particles of sufficient energy to create radioactive isotopes. John Cockcroft and Ernest Walton

Fig. 1.   Ernest Orlando Lawrence (1901–1958). (*Courtesy of Lawrence Berkeley National Laboratory*)

developed, they lost the race to first produce artificial radioactivity.

Rutherford also inspired R. J. Van de Graaff, an engineer by training, who developed a machine that converted mechanical energy into electrical energy. During the 1930's Lawrence built everlarger cyclotrons. Just prior to WorldWar II, Donald Kerst was able to make an electron accelerator, the betatron, and in a subsequent paper with Robert Serber brought sophisticated quantitative mathematical analysis to these machines. With these devices nuclear physics flourished in the years prior to WWII.

After WWII, Edwin McMillan and Vladimir Veksler invented phase focusing. With this invention ever-more-powerful machines could be built for the study of nuclear physics. The 184-inch cyclotron (Fig. 2) produced pions, and the field of particle physics in the laboratory, as contrasted with cosmic ray studies, was born. At first, particle physicists studied particles interacting with nuclei, and nuclear properties when the bombarding particles were very energetic. Those projectiles were initially protons and later other nuclei, and then various mesons.

went the route of electrostatic accelerators, while Ernest O. Lawrence (Fig. 1), stimulated by a paper by Rolf Wideröe and working with his student M. Stanley Livingston, developed the cyclotron. In

During this same period nuclear physics drove the development of linear accelerators, first for

Fig. 2.   The magnet of the 184-inch cyclotron, the largest that Lawrence built. (*Courtesy of Lawrence Berkeley National Laboratory*)

protons (Luis Alvarez, and then many others) and then for electrons (William Hansen, Edward Ginzton, Pief Panofsky, and then many others). The culminations of these developments were the proton accelerator LAMPF at Los Alamos and the 2-mile-long electron accelerator at SLAC. The interesting physics of electron-nuclear collisions led to the desire for a very long electron pulse, which was accomplished in the mid-1990s at Jlab's SRF-based CEBAF. Nuclear physics also drove the development of spiral sector cyclotrons culminating in TRIUMF. And, of course, it has driven the development of long-pulse (even cw) facilities for electron-nucleus studies, first at MIT–Bates and then at JLab.

The interest in "discovering" new elements, i.e. elements beyond uranium, spurred activity in Berkeley (12 elements) and also in Dubna (6 elements). More recently GSI has taken up this challenge with considerable, and still ongoing, success.

This desire in nuclear physics to study ever-more-energetic collisions of heavy ions with heavy ions began with synchrotrons (Bevalac, AGS, SPS), but eventually led to the RHIC collider. In the future there will be heavy ion collisions at the LHC. The construction and operation of the RHIC and the future heavy ion capability at the LHC are motivated by the desire to study matter under extreme conditions, such as those that are believed to have existed early in the big bang. A possible future development might be the addition of electron-ion collisions to the RHIC (e-RHIC).

There is interest in nuclear physics, not only in studying matter under extreme conditions — even creating and studying the quark–gluon plasma — but also in studying rare (radioactive) species. To this end, existing facilities, such as those at TRIUMF and Michigan State University, have been extended to study radioactive species, while new facilities devoted to this same purpose are being built. A series of nested cyclotrons has been constructed at RIKEN, the FAIR facility is being built at GSI, and in the US the FRIB (previously called the RIA) is being designed.

2.2. *High energy physics*

Simply put, the particle physicists desired ever-higher energy machines while at the same time having a reaction rate adequate to see ever-smaller cross section events. The accelerator builders provided them with synchrotrons one after the other: the Cosmotron, the Bevatron (CEA), the Princeton–Penn Machine, the Dubna Synchro-phasotron, the AGS, the ZGS, the 70 GeV proton synchrotron near Serpukhov (south of Moscow), the KEK Proton Synchrotron, the Fermilab 400 GeV machine, and more. The later machines took advantage of the concept of strong focusing (Ernest Courant, M. Stanley Livingston, and Hartland Snyder; independently, also Nicholas Christofilos) as well as the ever-improving technology of vacuum systems, long straight sections, external beams magnets, and rf.

Starting in the 1950's, colliders were developed. The driving motivation was HEP and avoiding the square root increase of effective energy in fixed target experiments. Making practical colliders — hadron–hadron, electron–electron, and electron–positron, required producing intense beams (rf stacking for hadrons), maintaining the beams (an understanding of instabilities and learning how to handle them with nonlinearities and feedback), and focusing them in low-beta sections to a tiny transverse size (powerful magnets and single-particle beam dynamics). The first low-beta insertion was done at the CEA. These efforts took place in many laboratories, with the original instigators being Donald Kerst, Bruno Touschek, and Gersh Budker (Fig. 3). Experimental work was, at first, at Stanford, CERN, Novosibirsk, Frascati, and Orsay. Soon, many colliders were

Fig. 3. Gersh Budker (1918–1977). (*Courtesy of Budker Institute for Nuclear Physics*)

constructed: CEA, SPEAR, ISR, SppS, LEP, PEP I, PEP II, Adone, the VEP series, HERA, the Fermilab Tevatron and, soon to begin operation, the LHC at CERN. The ISR pioneered the science and the technology in ways that no other machine did in the late 1960's through the early 1970's.

Robert R. Wilson's vision of the Tevatron, the first superconducting synchrotron, was made possible by the ability of Helen Edwards and Alvin Tollestrup. It became the workhorse of the US HEP program, starting as a fixed target machine and then as a proton–antiproton collider in 1986. The Tevatron was a major triumph in accelerator technology and developed most of the technology that made HERA, RHIC, and LHC possible. In order to raise luminosity the counterrotating beams were placed on helical orbits, arranged so that collisions occurred only at the locations of the large detectors.

The original 400 GeV Main Ring was the first large-scale machine to use a separated function lattice. Another innovation at Fermilab was the construction of the world's first large permanent magnet synchrotron.

Driven by the needs of HEP, many accelerator advances were made. Superconductivity became widely used both in rf cavities and in magnets. At the same time, the art of detectors was advanced as new concepts (such as 4π detectors, the TPC, silicon detectors) were incorporated into ever-larger colliding beam detectors.

Colliders were not limited to being electron–positron and proton–proton colliders. Carlo Rubbia realized that protons and antiprotons could be collided in the SPS. This was accomplished using the concept of stochastic cooling, invented by Simon van der Meer. The SPS which was converted into a proton–antiproton collider had sufficient collision energy to produce and discover the W and Z bosons. For proton–antiproton colliders one needs to produce the antiprotons, cool them (stochastic cooling and later electron cooling), and incorporate many other technological advances and inventions that allowed a collider to accumulate particles and store them for a day or longer. All this was accomplished, first in the SppS collider and later in the Tevatron.

The study of proton–electron collisions was in the initial design of the PEP, but support was not forthcoming. The first machine to have ep collisions was HERA in Germany. Looking to the

Fig. 4.   An overview of the European high energy physics laboratory CERN. Superimposed on the photograph is an outline of the ring (deep underground) of the Large Hadron Collider (LHC). (*Courtesy of CERN*)

future, circular electron–positron colliders reached their ultimate size in the LEP, and linear colliders, pioneered by the SLC at SLAC, will be the method of choice in any future machine, such as the International Linear Collider (ILC). This year, a very large (27 km in circumference) proton–proton collider, the LHC (Fig. 4), will come on line. Improvements in the performance of this machine (a proposed factor of 10 in luminosity), and then the possibility of replacing the guide magnets with more powerful magnets (such as those using a Nb_3Sn superconductor) to increase its energy, will challenge many accelerator scientists in the years to come.

Each of these major facilities required a team of hundreds of skilled and dedicated physicists and engineers, each led by an accelerator physicist who was both a good scientist and an accomplished project manager.

Important science has emerged from the nonaccelerator work at the Kamioka detector, the Super K detector, and the SNO detector: the experimental demonstration that neutrino oscillations were the source of the solar neutrino problem. Detailed studies of neutrino oscillations and neutrino mass are being performed through the development of major accelerator-based neutrino experiments. The first was K2K at KEK, and then Super-K at KEK, and at Fermilab both the MINOS and MiniBooNE experiments, and, most recently, by CERN to Gran Sasso. At J-PARC a very intense neutrino beam will be created and sent to Kamiokande. Neutrino beams will play a very important role in the future.

The first accelerator neutrino experiments were proposed independently by Mel Schwartz and Bruno Pontecorvo around 1958, when they realized that the pion beams that would be produced by the AGS and the CERN PS, then under construction, could be used to produce reasonably intense neutrino beams. Neutrino experiments were a central part of the fixed target programs at the Fermilab Main Ring, its successor — the Tevatron — and the CERN SPS.

With the work on the various accelerators listed, and much theoretical work, weak interactions were understood and subsequently the standard model was established. Thus, accelerators were at the heart of perhaps the greatest accomplishment of physics in the latter half of the 20th century.

2.3. *Condensed matter physics, chemistry, and biology*

During the last few decades the scientific disciplines playing a very large — perhaps the largest — role in accelerator development have not been nuclear or high energy physics, but rather condensed matter physics, chemistry, and biology. Sure, there have been the LEP, LHC, RHIC, the nested cyclotrons at RIKEN, and FAIR, but the major recent development has been in synchrotron radiation sources (Fig. 5) and, most recently, in spallation neutron sources.

Fig. 5. The first observation of synchrotron radiation was made from this 300 MeV electron synchrotron at the General Electric Co. at Schenectady, built in the 1940's. (*Courtesy of the US Government*)

These machines that produce intense beams of synchrotron radiation or neutrons are used in condensed matter physics for the study of the magnetic properties of matter (such as is used in computer hard drives), the origin of high temperature superconductivity, wear in turbines, the catalysis in industrial chemical processes and many other studies. In chemistry, synchrotron radiation has been used to study the structure of molecules, and is now allowing the study of the dynamics of chemical reactions. Some have said the last few hundred years of chemistry were devoted to the study of equilibrium states, while the next few hundred will be devoted to dynamical studies.

In biology and medicine, synchrotron radiation is used to elucidate protein structure and cell structure. A very large activity is in the design of pharmaceuticals. Synchrotron radiation is also employed in environmental sciences, geosciences, and even art and archeology.

Most often neutron spallation studies are combined with synchrotron x-ray studies, for the two are complementary in their sensitivity as a function of the atomic number.

The idea of using a storage ring to produce synchrotron radiation came from the MURA Group. The very first such rings were TANTALUS in Madison and the ring at NIST. In the early 1970's some rings developed and operated for high energy physics were used as synchrotron radiation sources. In the US, the CEA was turned off for high energy physics. SPEAR, after several years of parasitic operation, became a dedicated synchrotron user facility. It was appreciated that storage rings with insertion devices, undulators, and wigglers, in long straight sections gave far superior x-ray beams. The first permanent magnet wiggler was built in Berkeley and employed at SPEAR. These machines were called the Second Generation, in contrast with the First Generation, which simply used bending magnets to generate x-rays. These x-ray sources became so valuable that "user demand" resulted in something like 70 of them having been built in many countries.

The next step in this evolution was the construction of very energetic rings specially designed for their resulting x-ray beams. This Third Generation includes the 6 GeV ESR in Grenoble, the 7 GeV APS at Argonne, and SPring-8 (8 GeV) in Japan, as well as the lower energy (but intense) ALS in Berkeley.

At present the NSLS at Brookhaven is being augmented with a whole new facility in the US$800 million range. And a number of lower energy (a few GeV) Third Generation facilities are now being constructed in various places around the world.

The free electron laser (FEL) has been under development since it was invented and demonstrated by John Madey *et al.*, in the late 1970's. Since that time the radiation wavelength has become shorter and shorter: the latest is the FLASH facility at DESY operating in the VUV. A major Fourth Generation machine, built with a linac (1/3 of the two-mile linac) and an FEL, is under construction at SLAC. This device, called the LCLS, will produce coherent radiation of 1.5 angstroms (0.15 nm). A similar device is under construction at SPring-8 and an XFEL using superconducting rf is under construction at DESY. Other groups are considering recirculating linac FEL's, while some groups (like LBNL) are considering whole complexes of FEL's. The aim is to produce many pulses per second (MHz, as contrasted with 100 Hz at SLAC) and beams of very short temporal extent (100 attoseconds). The requirements on emittance, timing, stability, and pulse manipulation are challenging indeed.

While x-ray sources have undergone the development described above, neutron sources — for many of the same applications as x-ray beams — have also undergone major development. The SNS at Oak Ridge was the major accelerator project in the US for the last decade (just as LCLS is for this decade). The superconducting linac is the highest energy sc hadron linac ever built, while the storage ring was a major challenge. A similar facility is being built in Japan (J-PARC) and planning is underway to build spallation facilities in Europe and China. The development of pulsed neutron sources also has an interesting history, involving contributions from many laboratories and people. The first one was built at Argonne using the ring components of the old 1.2 GeV Cornell machine. Its successor, the IPNS, provided the best and most reliable source of pulsed neutrons for the US spallation neutron source community for many years. After NIMROD was turned off around 1977, the ISIS was built at Rutherford. For the past 25 years it has been the most productive neutron spallation source in the world; only now is the SNS overtaking it. Ultimately LANSCE (LAMPF renamed), the pulsed storage ring (PSR), and the Lujan center

overtook the Argonne IPNS in the US around 2004. SNS construction at Oak Ridge began in 2000. These earlier machines pioneered the beam and instrumentation technology.

2.4. *National defense*

The first employment of accelerators in national defense was when Lawrence used the not-yet-completed 184-inch cyclotron to separate U^{235} from U^{238}. This resulted in the construction of over a thousand Calutrons through which passed all of the material of the Hiroshima bomb. These were more like spectrometers than accelerators, but the operation involved dealing with space charge effects very similar to those in accelerators.

During WWII, the Germans, the British, and the Americans employed betatrons. The Germans hoped to be able to blind the pilots of Allied bombers, the British used them to x-ray unexploded bombs and therefore help in the defusing process, and the Americans employed a betatron at Los Alamos to help in the development of the atomic bomb. It was one of these British betatrons that, after WWII, was converted into one of the very first synchrotrons.

Shortly after WWII, the American AEC became concerned that there was not enough readily available uranium to allow the buildup of what would become a very large nuclear weapon inventory. Thus the Material Testing Facility (MTA), which was to be essentially a spallation source of neutrons, was authorized. This machine was a very large linear accelerator (the accelerating tank was 90 feet long and 30 feet in radius) that was promoted, designed, and built by Lawrence and coworkers, but in fact never worked.

The desire to protect the American nation from incoming missiles was responsible for the development of induction accelerators.

Long before the SDI proposal ("Star Wars"), a conference was held in 1952 at which particle beam weapons were discussed. The actual SDI program, whose purpose was missile defense, was initiated in 1958, under which the Experimental Test Accelerator (ETA) and the Advanced Test Accelerator (ATA) — (both linear induction machines) — were constructed at Livermore (Fig. 6). During "Star Wars" times these machines were used to develop ground-based FEL's. These ideas never resulted in antimissile weapons.

Fig. 6.   Induction linac FXR at Lawrence Livermore National Laboratory, completed in 1982 to study the implosion process in nuclear weapons. (*Courtesy of Lawrence Livermore National Laboratory*)

Induction accelerators have been effective in producing x-ray pulses to study the hydrodynamics of nuclear weapon implosions. The FXR was built at Livermore, and just being commissioned is the DARHT Facility at Los Alamos.

During the period when SDI concepts were being explored, Los Alamos built, and then operated, an accelerator in space. This project was called BEAR (Beam Experiment Aboard a Rocket). More importantly, WNRF, the Weapons Neutron Research Facility storage ring system at Los Alamos (it operates in two different pulsed modes using protons from LAMPF), was successful in its mission and has led to the discovery of the electron cloud instability, a matter of importance in many machines.

The US Navy has an intense interest in defending its large ships. One possibility is an on-board FEL, and in this regard it has supported the development of ever-higher average power FEL's. This activity has not yet resulted in an anticruise missile, or an antishell weapon, but the necessary power to achieve that end is within sight.

2.5. *Medicine*

John Lawrence, a doctor and brother of Ernest Lawrence, working in Berkeley with the cyclotrons, started the field of nuclear medicine in 1936. Subsequently there has been a large need to produce radioactive nuclei both for diagnostic purposes

(PET) and for cancer therapy purposes (an example is the treatment of thyroid tumors). This has driven the development of small cyclotrons (18 MeV, 40 MeV, etc.). These machines are robust, reliable, and relatively inexpensive, and are usually located in major city hospitals. One large commercial firm has produced over 1000 of them.

The medical advantage of high energy x-rays as an external beam for cancer therapy has driven the development of x-ray-producing linacs. At first these were rather large spatially fixed machines operating below the energy for producing radioactivity, i.e. below about 10 MeV. Now there are compact linacs (still operating below 10 MeV), which are reliable and mounted directly on a gantry, so they may be rotated about the patient. A great many of these machines are in hospitals around the world. One large commercial supplier is manufacturing two or three such machines every day!

It has been appreciated, in the medical community, that external hadron beam therapy for certain types of cancer is advantageous. This realization has driven — and right now is intensively driving — the development of a variety of machines. These include cyclotrons and synchrotrons, with cyclotron-fed linacs and non-scaling fixed field alternate gradient (NS-FFAG) accelerators on the horizon. Also, a good number of very old facilities, often no longer supported for nuclear physics, are being converted to use for medical applications (examples are in Indiana and Catania).

2.6. *Industry*

Industry is a prime user of accelerators. The market, consisting of constructing accelerators, with their associated peripheral equipment, is estimated to be more than US $1000 million a year. Unlike the disciplines described above, industrial use of accelerators has not led to the development of new types of accelerators or even to the many developments of accelerator technology that have allowed advances. However, industrial use has stimulated nontrivial, ever-more-sophisticated forms of "conventional machines" and, of course, has depended on the community of accelerator scientists that are needed to make this happen. We describe present use below, but in the future one can expect that industrial use of accelerators will become even larger.

The primary use of accelerators is in the semi-conductor industry, where doping silicon with boron or phosphorus (forming p or n junctions) requires a range of energies from 100 keV to 1.5 MeV. In the high energy portion of this range, linacs and tandem Van de Graaffs are used; electrostatic machines cover the mid-energy range (tens of keV) and the low energy range (hundreds of eV). These machines are commercially produced and sold primarily in the US, Europe, Malaysia, Singapore, China, South Korea, Japan, and Taiwan.

The wafers used in satellites (SOI — silicon on insulator), and also in HEP detectors, are radiation-hard. The insulating layer of oxide is deposited with oxygen ion beams in the 100 keV range. These types of wafers are used by companies such as IBM and Honeywell in the US to manufacture special types of chips. A new technique "Smart Cut," which requires energetic hydrogen ion beams, is now being employed for the manufacture of SOI wafers.

At present, chips are made with photoresist and masks and VUV light from a KrF laser. In order to make ever-smaller objects, a different lithography technology is required and that may well be the use of electron or ion beams, probably in the range of 70 keV and produced electrostatically.

Turning from semiconductors to other types of materials, we find a wide use of accelerators in machining, cutting, surface modification, and surface analysis. For example, there is secondary ion mass spectroscopy (SIMS), which requires a tightly focused ion beam. Currently a number of Japanese companies are producing tens of electrostatic machines per year. These give 35 keV ions and a spot size diameter of 2 nm.

Another use of accelerators is in the sterilization of spices, food, medical instrumentation, and even the US mail addressed to sensitive places (White House, Pentagon, Congress, etc.). X-rays, or even direct electron beams, are employed in these systems. Most of these applications require very high power machines, typically with electrons at 6 MeV. These are either voltage-multiplying devices (Novosibirsk) or microtron-like devices called Rhodotrons (manufactured by IBA). Tens of units are produced each year.

There are many other industrial applications of accelerators. In common use is enhancement of the brilliance of gemstones. A special application is in the treatment of sugar being transformed into ethanol (such as in Brazil). Subjecting the sugar solution to 1.5 MeV, electrostatically generated electrons kills bacteria in the solution and increases the conversion efficiency by 4–5%. Accelerators make x-rays for studying the continued integrity of airplane wings, bridges, and other structures. For this application, at least one company has designed a compact, portable betatron, which it is producing and selling commercially.

3. The Accelerator Community

In this section we touch upon the community of accelerator scientists. It is interesting to see how the community developed, how the laboratories in which many of them worked operated, and how these scientists communicate and interact with one another.

3.1. The people

In the early 1930's, there were no accelerator scientists. Accelerators were designed (minimally) and built by physicists/engineers who "moved over" from other branches of physics. For example, Cockcroft was an atomic physicist, Odd Dahl an oceanographer, Van de Graaff a mechanical engineer, Lawrence a cosmic ray specialist, McMillan mostly a chemist but also a nuclear physicist, Alvarez a nuclear physicist, Hansen a physicist interested in electromagnetism, Ginzton an electrical engineer (and also a physicist), and Panofsky a nuclear physicist.

Wideröe's proposal for resonant acceleration in a linac-like device preceded Lawrence's principle for a cyclotron and influenced Lawrence. Lawrence and Livingston built the first operating cyclotron that spawned a long line of cyclotrons and later synchrocyclotrons and synchrotrons. Lawrence's cyclotrons created big science, and extraordinary science was done with them. John Adams built two machines (the PS and SPS) and provided the vision for the VBA presented at a meeting in New Orleans in 1975. The LHC is the incarnation of the VBA. Burton Richter and others developed collider technology with the Stanford electron collider. SPEAR reshaped particle physics and helped to shape synchrotron light sources. Richter was also a force behind the LEP and the SLC.

At Cornell, before becoming the Founding Director of Fermilab, Robert Wilson (Fig. 7) built a

Fig. 7. Robert Wilson breaks ground on October 3, 1969, for the 200 GeV proton accelerator, which came into operation in 1972. (*Courtesy of Fermilab*)

Fig. 8. Rolf Widerøe (1902–1996). (*Courtesy of Pedro Waloschek*)

series of electron synchrotrons and fostered a whole school of accelerator builders from among his colleagues and students. The first person to seriously propose electron–proton collisions was Bjorn Wiik for DORIS, important for not only physics but also accelerator science.

Although much activity in the field consists of building and improving accelerators following well-established and proven ideas, frequently new innovations with great promise are proposed. John Dawson and Toshi Tajima were the first to call attention to the possibility of using plasmas and lasers for particle acceleration. Robert Palmer at Brookhaven has been spearheading activity, and making important contributions to the fields of neutrino factories and muon colliders. Nicholas Christofilos, while working at Livermore, invented induction accelerators as part of the fusion energy program.

The very first training of students for a Ph.D. in accelerator physics was by E. O. Lawrence. Amongst his early students were M. Stanley Livingston ("Lawrence was my teacher when I built the first cyclotron — he got a Nobel Prize for it — I got a Ph.D."), David Sloan, and later Jackson Laslett. Widerøe (Fig. 8) was self-motivated, but he was very

much the exception. After WWII, many students were trained, in many different countries, to obtain a Ph.D., in accelerator science. It is impossible to create a comprehensive list of Ph.D. supervisors who have been active in training students, but a few institutions in which they have taught come to mind. These include, just to name a few in the US, the University of California in Berkeley, UCLA, Caltech, the University of Chicago, Cornell University, the University of Illinois, Indiana University, the University of Southern California, Stanford, Stony Brook University (SUNY), the University of Maryland, MIT, Princeton, and the University of Wisconsin.

Melvin Month was instrumental in creating the APS topical group on beam physics, which later became the APS Division of Physics of Beams.

If one looks beyond the US, one sees institutions training accelerator scientists located in most of the advanced countries of the world. One notes Aarhus University, Beijing University, the University of Bonn, Cambridge University, the ETH, Hiroshima University, the University of Kyoto, Moscow University, the University of Naples, Novosibirsk University, Oxford, the University of Pohang, the University of Rome, the University of Paris, the Technical University of Berlin, Tel Aviv University, Tokyo University, Uppsala University, and very many other places.

The reader will recognize those institutions that were more active in the past than now, but one will also note, with satisfaction, the significant number of institutions that are currently very active in accelerator science and the training of accelerator scientists. Activity often comes down to a single person. There are large centers where there is more than one person, but that tends to be the exception. It is frequently the case that an institution offers instruction in accelerator and beam physics simply because one professor is active in the field, and when that professor leaves the institution it sometimes takes many years before another person with similar interests is hired, or the institution decides to put its emphasis in another direction. Conversely, institutions not previously teaching beam physics, with one hire, became active centers. We have seen that many times over and, most importantly, in developing universities, and in developing countries.

3.2. *The laboratories*

The first accelerators were, most naturally, built in laboratories designed and operated for other kinds of physics. One thinks of the work of Cockcroft and Walton in the Cavendish Laboratory (Fig. 9), of Lawrence in Le Conte Hall on the Berkeley campus, of Gregory Breit, Odd Dahl and Merle Tuve in the Department of Terrestrial Magnetism of the Carnegie Institute, and of Van de Graaff in the Palmer Physics Laboratory in Princeton.

Soon, that was to change. Van de Graaff moved to MIT and set up his large machines in an old aircraft hanger at the Round Hill Experimental Station. Lawrence moved to a special building on the Berkeley campus, and so started, in 1931, the Radiation Laboratory (now LBNL). This growth was not without pains. Birge, then Chair of the UC Berkeley physics department, remarked, "Berkeley has become less a university with a cyclotron than a cyclotron with a university attached." In 1940 the Rad Lab was moved to its present 200-acre location high in the Berkeley hills, with a spectacular view of the San Francisco Bay Area (Fig. 10).

In the 1930s a number of universities — in many locations — were constructing cyclotrons and electrostatic machines and even betatrons. Often these machines required a special building, but were under the auspices of local physics departments. After WWII, driven by nuclear and then particle physics, a number of large specialized laboratories were built. In the US some of these laboratories had been built during the war, and they were used to house accelerators and do experiments with them.

Fig. 9.   The original Cockcroft–Walton installation at the Cavendish Laboratory. Walton is sitting in the cubicle. (*Courtesy of the University of Cambridge Cavendish Laboratory*)

Fig. 10.   Lawrence Berkeley National Laboratory, overlooking the San Francisco Bay Area. (*Courtesy of Lawrence Berkeley National Laboratory*)

The first laboratory devoted to accelerator science was Brookhaven, where first the Cosmotron was constructed, and then the AGS. At the same time the Rad Lab commissioned the 184-inch cyclotron, and then linacs, synchrotrons, and the Bevatron. At the same time also, SLAC, building on the initial, Nobel Prize–winning work at HEPL, was initiated, but with some pains that remind one of Birge's remark. A series of electron synchrotrons was constructed at Cornell University, including the first synchrotron to incorporate strong focusing.

Major laboratories were also built in the Soviet Union. Some, like the JINR at Dubna, were specially established for the construction and utilization of accelerators; others, such as Kurchatov (a laboratory built for another purpose), were converted into accelerator facilities. One of the most important — now *the* most important — was CERN. The creation of CERN was an effort to do what no European nation could do alone, namely build a large synchrotron (the PS). It was also important politically, as well as scientifically, as that was the first time that the former enemies of WWII worked together on a joint project. The formation of CERN had the support and help of the UN, just as the UN today is helping in the current activity to bring a Third Generation synchrotron radiation facility, SESAME, to the Middle East, where both Israel and Arab nations are participating.

Soon, many other accelerator laboratories were created: Frascati (Italy), DESY (Germany), GSI (Germany), Serpukhov (Soviet Union), KEK (Japan), IHEP (China), INP (Soviet Union), SIN/PSI (Switzerland), Calcutta (India), RIKEN (Japan), TRIUMF (Canada), GANIL (France), RAL (England), Fermilab, JLab, and Bates (USA). There are now 70 or so synchrotron laboratories in many countries. Simultaneously, increased accelerator construction activity occurred at the already mentioned laboratories, such as the ZGS and the superconducting ion linac at Argonne, the RHIC at Brookhaven, LAMPF at Los Alamos, and the SNS at Oak Ridge.

Although the primary purpose of these laboratories was science, they actually serve many purposes. In other cases, such as with CERN and SESAME, there is a positive political benefit to having rival countries work together. In other cases there was international rivalry, such as between Berkeley and Dubna over the discovery of transuranic elements,

or the competition between Argonne and Dubna to build the highest energy proton accelerator (the ZGS and the Synchro-phasotron). Sometimes it was simply national prestige, and sometimes it was an effort to keep scientists at home and to stimulate and encourage young scientists to stay and work in their home countries. National prestige is important, for it has proved effective in getting countries, which at first had few or no accelerators, to the very forefront of our science.

3.3. *Communication and interaction*

When we trace the development of communication and interaction between accelerator physicists, we find that it is really no different than the development of communication between all kinds of physicists (and even amongst the general population). Nevertheless, it is useful — so as to give us perspective — to look back just a bit.

Before WWII, although there were letters, and even visits, the primary method of communication — and, of course, interaction — was at a conference, often an APS meeting. Ernest O. Lawrence, for example, would travel east once a year, and it was not easy, as he would travel by train each way. In 1933 he was the only American to attend the Solvay Conference in Brussels. That required, both ways, a long train trip and an ocean voyage. And all this to report, unfortunately incorrectly, that the breakup of deuterium would be a power source.

Turning to the present, there is very extensive communication and interaction. Besides snail mail and the telephone there is e-mail, which is very widely used. (It is proper to note, if only parenthetically, that the World Wide Web, perhaps the most important development in communication and interaction in the late 20th century, came from CERN.) There are international general conferences — PAC, EPAC, and APAC — operating on a rotating basis between regions, so there is one conference each year. In addition, there are national conferences in the US (PAC in alternate years) and in Russia. There are specialized conferences such as COOL, CYCLOTRONS, DIPAC, FEL, ICALEPCS, ICAP, ICFA ABDW, and LINAC. Proceedings of these, as well as the general conferences, are available on the Joint Accelerator Conferences Website (JACoW).

In fact, JACoW has many proceedings from earlier years, now scanned and available on the site. The American Physical Society's DPB (Division of Physics of Beams) organizes accelerator physics sessions at APS meetings, and, with other APS divisions, jointly sponsors sessions in which the scientists using accelerator beams report their results. DPB is a joint sponsor with IEEE of the PAC's.

Of course, there are the many brief visits of accelerator scientists (now easy by plane) as well as longer-term visits. There are few accelerator scientists who have not spent some time at another institution, thus learning different approaches and making long-term contacts and friendships.

Finally, one notes the USPAS (US Particle Accelerator School), started by Melvin Month, and now a solid institution. The USPAS provides educational programs in the field of beams and their associated accelerator technologies not otherwise available to the community, and conducts graduate and undergraduate level courses at US universities, running two such programs per year, one in January and one in June. Students are welcomed from all corners of the world, from universities, laboratories, private companies, the government, and the military. Some students have been in the field for many years and are interested in a "refresher" course, while others are full-time students looking for classes to add to their education. Qualified teachers are chosen from national laboratories, universities, and private industry. To carry out its educational mission, the USPAS develops programs of courses suitable for universities. Major universities, in partnership with the national laboratories, underwrite the offerings and provide the necessary quality control. Through this administrative framework, universities across the nation can offer high-quality advanced technology courses. Similar schools are held in Europe, and now in Asia. They have had an important role in developing expertise amongst scientists in the accelerator community.

4. The Future

It can be dangerous to try to predict the future. But, perhaps, it is not so dangerous in accelerator science, for the subject is rather mature and, at least for major machines, the time scale is long. Some would say "too long." For example, the Next Linear Collider was first seriously studied in 1984 and will be initiated — at best — in the middle of the next decade, some 30 years later.

The long time scale for conceiving, designing, obtaining funding, choosing the site, constructing and commissioning a new big machine has a very large negative impact on careers and the training of students, not to mention its effect upon the science. Too many of our colleagues are spending their whole scientifically productive lives making proposals and doing calculations on a machine that they may never see or even actually construct.

So, throwing caution to the wind, let us try to look into the future.

4.1. *Drivers*

Clearly, all of the "drivers" mentioned in Sec. 2 will continue to drive accelerator science for the next few decades. Nuclear physicists, as already mentioned, will want to have electron-ion collisions, and new accelerators built where they can study radioactive ion collisions.

High energy physics will not end, nor can it manage without accelerators. Of course its practitioners desire to discover the Higgs, see whether supersymmetry is a correct description of nature, solve the hierarchy problem, etc., and, perhaps most importantly, contribute to our understanding of dark matter and dark energy. Improvements on the LHC are in our future. Also, surely, we can look forward to the need for some smaller machines designed to study phi mesons, tau leptons, and various rare decays. And, of course, some day, there will be a very-high-energy electron linear collider: the International Linear Collider (ILC).

Just what form the ILC will take is uncertain. The decision has been made to use the low frequency superconducting rf (TESLA) technology. But funding is clearly coming much more slowly than the community expected. Since international agreements must also be made before construction can begin, it seems unlikely that we will have the ILC before the 2020 decade. In this situation, it is not obvious that a machine limited to low energy, as is the case for the TESLA approach, is what the high energy physics community will desire. If the ILC had been built rapidly and then followed with something else, like the Two-Beam Accelerator (TBA) approach of the CLIC or a klystron-powered x-band device, it would be a different story. But if there will be only one ILC for as far into the future as one can envision (perhaps until 2050, or beyond), the technology choice may be very different. In any case, we must wait for LHC data. It will not be a surprise if the decision as to what form the ILC will take is revisited. But, in any case, sooner or later, there will be a great need for accelerator expertise.

In addition to the above, there is a compelling physics case to be made for the creation of super neutrino beams and, some day, a neutrino factory based upon the capture, cooling, and acceleration of muons, whose subsequent decay produces neutrinos. Also, of course, in our future may be a muon collider.

Condensed matter physics, chemistry, and biology have uses for synchrotron radiation that will stimulate and justify the construction of ever more Third Generation facilities. One was just completed in Australia, and surely that model will be followed in many other countries. The desire to obtain brief, coherent, high repetition rate x-ray pulses will drive the development of Fourth Generation light sources. Perhaps more than any other need, the development of Fourth Generation facilities will be the "driver" of future accelerator science. It seems clear that one can look forward to a great deal of accelerator and beam science, including many advanced technology developments in creating the Fourth Generation facilities desired by our scientific colleagues.

Also, of course, there will be complementary neutron spallation facilities. In the US and Japan there are already major facilities, but surely similar devices will be built in other places. China has already begun and Europe has had plans for a rather long time to build a neutron spallation source.

National defense, cancer therapy and industry can be counted upon to have needs for accelerators as far into the future as one can imagine. Surely, these applications alone will keep the field healthy and vibrant.

4.2. *New accelerator science*

The development of accelerator science has been, and will continue to be, driven by the physics and application needs. But many advances in accelerator science are independent of these drivers. In the past, it has been realized by the funding agencies that a certain amount of money should be spent in such an undirected manner. We can hope — and should do what we can to ensure that this will continue into the future.

Looking back, we can see so many wonderful advances, most of which, at least initally, came out of undirected research. One thinks of phase focusing, strong focusing, lattice design methods, nonlinear dynamics studies, electron cooling, stochastic cooling, laser cooling, RFQ's, FFAG, colliding beams, low beta insertions, instability theory, numerical computation of magnets and rf cavities, superconducting magnets and superconducting rf, vacuum technology, linac design, injection and extraction kickers, FELs, and so on and so forth.

And it would be wrong to think we are at the end of making further discoveries and developments. Besides the incremental improvement of what we know, there will surely be developments we cannot even envision. One has only to attend one of the specialty conferences — say, a COOL conference or a Neutrino Factory and Muon Collider Collaboration meeting or an FEL conference — to witness the excitement and innovation of accelerator science.

As we mentioned above, the CLIC scheme is a possibility for the ILC and, most people agree, a very serious possibility for the linear collider beyond the ILC. At CERN a significant effort of study has been under way and will continue into the future. There have been major proof-of-principle and prototyping of parts of the two-beam accelerator approach. We can be confident of a continuation of this effort.

Motivated by the possibility of making a neutrino factory, and also a muon collider, a great deal of effort has gone into the study of accelerating a muon beam. Since muons are produced in the nuclear decay of pions, the beam emittance is large and cooling is necessary. Only ionization cooling is fast enough compared to the muon decay time of a few microseconds (in the muon rest frame), and much thought has gone into the development of configurations that will give the necessary 3D damping. An experiment to demonstrate ionization cooling (MICE) is underway at RAL. The acceleration of muons could be done in a recirculating linac or, possibly, in an FFAG structure. An experiment to study nonscaling FFAG acceleration is now underway (EMMA) at Daresbury. Also, this activity has stimulated the experimental realization of optical stochastic cooling, which is, of course, interesting in its own right.

It is possible to drive a subcritical reactor with an incoming neutron beam. That would increase the safety of a reactor (since the driving beam could simply be switched off). Experimental demonstration of this idea is being done at KURI (part of Kyoto University). Of course, there are many other possible sources of accidents, like a fire or sabotage, that this does not address. Perhaps more interesting is the use of an accelerator to breed thorium into a fuel. And, finally, accelerator beams can be used to burn up the actinides and thus reduce the time that nuclear waste must be isolated from the environment from hundreds of thousands of years to hundreds of years. Designs that do all three things have been studied in Europe, Japan, and the US at LANL. Burning thorium is of interest to India, for they do not have uranium but do have thorium. Burning actinides is of considerable interest and is part of the Global Nuclear Energy Partnership (GNEP). However, the GNEP envisions advanced breeder reactors rather than accelerators for this purpose. The situation is in flux and it is not clear whether accelerators will ever be used for these purposes, but maybe they will be.

A second energy production use of accelerators is in fusion. Effort on heavy ion inertial fusion has been going on for the last 30 years. The Europeans at GSI have followed an approach with rf accelerators, while the Americans (mostly at LBNL, but with Princeton and LLNL also involved) have studied an induction accelerator approach. Both programs have suffered by not having sufficient support. It is not clear that heavy ion inertial fusion will be the answer to mankind's need for energy, nor is it even clear that it will play a significant role. But the energy problem is so pressing that it is disappointing, almost unbelievable, that even the modest financial support needed to significantly increase the rate of progress in this work is not forthcoming.

In the last few years interest has grown in high energy density physics. The applications are very broad and range from astrophysics to national defense. Often the method to obtain high energy density is with accelerators. Of course, lasers — such as in NIF — also provide an excellent source. To the dismay of many, the heavy ion fusion program has been redirected to high energy density studies. Yes, the two are related, but pursuit of high energy density, which is an important subject, should not have

been used to slow down the fusion effort. The world needs both.

We want to note the very interesting beam physics that is being driven by some of the new projects, like FAIR at GSI and ELENA at CERN. The cooling of stored beams is an exciting science and relevant to many applications (such as has already been seen at the Tevatron and at RHIC). The production of cold beams is of much interest for FELs and also, for example, making the very cold, and almost stopped, antiproton beams. These are used for the study of antihydrogen, a challenge undertaken at CERN and soon to be undertaken at FAIR.

Finally, we note, with enthusiasm, the very extensive effort, and considerable success, in the development of wholly new methods of acceleration. The work on laser and laser/plasma acceleration has been going on since 1978, with a great deal of progress, but no practical accelerator has yet been built. The development of accelerating plasma channels, so that a laser beam remains confined beyond the Rayleigh length, and the subsequent acceleration of electrons by a powerful laser to high energy (more than a GeV) and with a very small energy spread, has already been achieved. In another approach, the wake field from an intense, and longitudinally compact, bunch has been used to accelerate a second group of particles. In an experiment at SLAC a beam of 45 GeV was used to more than double the energy of another beam of electrons.

The present laser-driven accelerator experiments employ rather expensive and inefficient lasers. But laser technology has, ever since the invention of the laser, made continual and significant progress. It is reasonable to assume that further advances will continue to be made and that efficient and inexpensive lasers, suitable for laser/plasma acceleration, are in our future. If that is to be the case, then laser/plasma accelerators will become available for various applications, but probably not for high energy physics, at least not for very many years.

4.3. *Final words*

Having gotten this far, the reader will have seen that we have tried to answer the questions listed in the Introduction. We have delineated the many very powerful "drivers" of accelerator science and have commented upon the many interesting and innovative aspects of accelerator science. It is clear that the field of accelerator and beam physics will remain vibrant and interesting for many more years. Given the necessary long-term investment in education and research and development, it is important to see far enough into the future to be confident that young scientists and students specializing in accelerator physics are not being led down "the primrose path." We are sure it is quite the contrary, and hopefully this short article has convinced the reader of the bright future of our science.

Thus, it appears proper to attract and train young people in this field, all the while being confident that they will have both future employment and an exciting scientific career. In short, that they will have as much fun as we have had.

Acknowledgments

We thank Ka-Ngo Leung for assistance with Subsec. 2.6. This work was supported by the US Department of Energy, Office of Basic Energy Sciences, under Contract No. DE-AC02-05CH11231.

Andrew Sessler, author of more than 400 scientific papers, was the first chair of the APS Division of Physics of Beams, a former director of the Lawrence Berkeley National Laboratory, and a former president of The American Physical Society. He has been, and is, active in human rights matters. He is a member of the National Academy of Sciences.

Ernest Malamud began his accelerator career at Cornell in the 1950s where he worked on the first strong focusing synchrotron. During a 3-decade career at Fermilab he participated in the design, construction and commissioning of the Main Ring Accelerator, the anti-proton source, and modification of the Tevatron to accommodate two high luminosity interaction regions as well as the electrostatic helical orbit system. Malamud is currently Chair of the APS Forum on Education, an Adjunct Professor at the University of Nevada (Reno), lives in the Sierra Nevada foothills and enjoys the nearby skiing and hiking.

Reviews of Accelerator Science and Technology
Vol. 1 (2008) 319–329
© World Scientific Publishing Company

World Scientific
www.worldscientific.com

Book Review

Panofsky on Physics, Politics, and Peace: Pief Remembers
(by Wolfgang K. H. Panofsky and Sean Marie Deken, Springer, 2007)

Gregory Loew

Stanford Linear Accelerator Center,
2575 Sand Hill Road, Menlo Park, CA 94025, USA
galoew@slac.stanford.edu

Introduction: Genesis of the Book

When the publisher Springer approached Wolfgang K. H. Panofsky (hereafter called "Pief," as he was affectionately known to almost everybody) in February 2006 to ask him to write an autobiography, he was almost 87 years old. This was a very fortunate initiative, because Pief had always been too modest and reluctant to write an autobiography, and without an invitation from a respected science publisher, the wonderful book reviewed here would not exist. The broad community of scientists, politicians and friends who came into contact with this amazing man would have been deprived of this treasure of memories of Pief's life experiences and observations. What is also extraordinary about the book is that its contents were almost entirely dictated from memory to his assistant, Ms. Ellie Lwin, in part from a hospital bed when Pief was treated for congestive heart and lung failure in late 2006. Putting mind over matter, however, as he had done many times before, Pief subsequently returned to his office at SLAC on a regular basis. Even on the last day of his life, September 24, 2007, he came in and held several meetings with administrative and scientific colleagues before driving himself home. Had Pief not agreed to write this book, somebody else would probably have been invited to do so, but nobody could have done such a beautiful and comprehensive job. We must also thank Ms. Jean M. Deken and her archives staff at SLAC for their great editorial assistance in helping Pief with gathering missing dates and facts relevant to the material.

Nature and Nurture: Pief's Early Life

When one looks at the life and accomplishments of an individual, it is always interesting to consider how they were affected by both nature and nurture. In Pief's case, there is no doubt both played equal roles. Pief had a prodigious intelligence and memory and he came from an illustrious family. As he mentions in the first chapter of the book, he was born in Berlin on April 24, 1919, the second offspring of Erwin and Dorothea Panofsky, who met at an art history seminar. Erwin Panofsky was a world-renowned art historian; Pief's maternal grandfather, Albert Mosse, was a famous jurist assigned to assist the Japanese government in drafting a constitution during the Meiji Restoration; his aunt, Martha Mosse, was the first woman to serve as police commissioner of the city of Berlin; and his uncle, Rudolf Mosse, was the publisher of the *Berliner Tageblatt*.

One year after Pief's birth, Erwin Panofsky accepted a faculty position at the University of Hamburg, where the family lived from 1920 to 1934. Pief's elder brother (by two years), Hans, was equally bright, and both spontaneously developed an early interest in science and technology, unlike their parents, who jokingly called them "*Klempners*" (plumbers) as they were growing up. When the reviewer, as a graduate student, first heard of Pief at Stanford in 1954, there was this legend that one of the brothers was called the "smart Panofsky" and the other the "dumb Panofsky" (both were supposed to have inordinately high IQs, differing by only two digits). I never did know which Panofsky was which,

Pief at the age of about four years, playing chess with his cousin Ruth Mosse, while brother Hans looks on. (*Credit: Panofsky Family Collection.*)

and asked Pief a few months before his death if he could clarify the issue, to which he answered: "No comment."

Flashing back to "nurture", soon after the Nazis came to power in 1933, most professional German Jews like Erwin Panofsky lost their jobs and other civil rights, and Pief's father had to leave Germany, even though he lived in one of the more liberal cities in the country. He eventually secured a double appointment at New York University as well as at Princeton, where the family of four settled in 1934. One might guess that without the Nazis, Pief would have become a successful and well-known scientist in Europe anyway, but these developments and WWII certainly propelled Pief's life into the much broader intellectual and political orbit that is the subject of this book.

This Review

Pief points out modestly in the preface that his book is an "unsystematic account" of his life and work. This is true only to the extent that he does not cover in detail all that other biographers might include, namely family life and children, all his numerous friends and acquaintances, and his vast accomplishments and professional contacts. The book, however, benefits from describing all his major activities while not being too long, so that one does not get lost in the trees of the forest. Pief's "essence" is all there, and the fact that he sometimes departs from chronological order by compartmentalizing his accounts by topic in separate chapters,

makes the book that much more readable. The three major topics covered in the book are his own scientific experience, science advising and international science, and arms control. These will be reviewed here in sequence, with major emphasis on the first topic.

High School in Hamburg; University at Princeton and Caltech

Pief started his high school education in Hamburg at the Johanneum Gymnasium with essentially no science training. When he arrived in Princeton at age 15, his parents were able to enroll him and his brother directly into the university, temporarily on probation. As the reader may guess, probation was soon lifted since they were both excellent students. To their classmates who considered them somewhat as "oddballs," they were Piefke and Paffke (from two German cartoons); and the name of Piefke, "Pief" for short, stuck with him for the rest of his life. The book contains many details about his Princeton education, perhaps the most relevant to this review being his senior thesis on radiation measurements with a high-pressure ionization chamber which used isotopes produced at the small Princeton cyclotron. Familiarity with this accelerator perhaps had some influence on Pief's future career. Other experiences at Princeton which enabled him to learn about American society are well worth reading about in the book. His father befriended Einstein at the Institute for Advanced Study — and since neither could drive, Pief became their occasional chauffeur at age 16!

After graduating from Princeton in 1938, Pief received a personal letter from Robert A. Millikan to join Caltech as a graduate student, with a teaching assistantship. He accepted this offer, which turned out to be a seminal decision. Pief eventually went to do his PhD thesis under Prof. Jesse DuMond, performing a precision measurement of the ratio of Planck's constant to the charge of the electron, and also getting acquainted with the boss' eldest daughter, Adele. Pearl Harbor happened in the middle of this, and by 1942 Pief got his degree, was teaching US generals classes in electromagnetic theory, started defense work on an acoustic device called a firing error indicator (FEI), got his US citizenship, and married Adele. What a year!

Pief and future wife Adele on the steps of the Athenaeum at Caltech on the day of his PhD ceremony, 1942. (Credit: Panofsky Family Collection.)

Pief and the Bomb

Two years later, Pief's work on the FEI which measured shock waves from supersonic bullets attracted the attention of Luis Alvarez and J. Robert Oppenheimer, who were interested in measuring the yield of nuclear detonations for the Manhattan Project. As a result, Pief was invited to work at Los Alamos as a consultant, and a year later a shock wave detection device he developed was supposed to be tested by him with others on July 16, 1945, from a B-29 airplane over the Trinity plutonium bomb test in Nevada. Although the test did not take place as planned for last minute reasons of weather and safety, similar gauges were later used over Hiroshima and Nagasaki. Pief discusses these events in some detail. It is clear that his awareness of the enormity and gravity of their long term consequences shaped many of his actions for the rest of his life.

Accelerators and Physics at UCRL

After WWII ended, Pief agreed to join Alvarez at the University of California Radiation Laboratory (UCRL), directed by E. O. Lawrence, to work on proton linear accelerators, even though at that time he had no experience in nuclear physics or accelerator design. Ironically, his first two days at UCRL were spent inside the 184-inch cyclotron magnet as people noticed upon his arrival that he was the only person short enough to stand inside to make magnetic measurements. He then went on to lead the group, which successfully designed and built the 32 MeV, 40 ft drift tube proton linac. Several breakthroughs,

such as beam phase stability (discovered by McMillan and Vecksler), cavity mode-mixing remediation due to the variable length of the drift tubes (calculated by Pief) and multipacting avoidance, led to the successful completion of this enterprise. However, while the linac was used to do physics, it did not lead to the construction of higher energy proton linacs at the time because of the parallel success of the less costly proton synchrotron. Note that while all these machines were already heralding the era of "big science," the culture of the time was much less specialized than today, in that the physicists who built these machines considered them to be the necessary initial stages of their particle physics experiments and were the same people who then went on to perform these experiments. Pief was very much a beneficiary of this culture, and later it made him an example of somebody who was equally knowledgeable about both fields, and could lead both enterprises from personal knowledge.

At the time Pief began to participate in particle physics experiments, it is noteworthy that he also accepted a heavy teaching load at UC Berkeley and decided to write his textbook *Classical Electricity and Magnetism* with Melba Phillips. While he started to study proton–proton scattering at 32 MeV with the linac, his most exciting experiments turned out to involve pi mesons from the 184-inch cyclotron. From pi minus mesons impinging on protons and deuterons at rest, Pief and graduate students were able to identify the existence of the pi zero meson, and measure the masses of the pi minus and pi zero (as well as their mass difference) to about 1% accuracy by looking at the gamma rays emerging from the reactions (and their decays into electron–positron pairs). From these measurements, it was also inferred that the pions are pseudoscalar particles, namely that they have spin zero and negative intrinsic parity. Pief often indicated that this work might have been his best. Later on, using McMillan's synchrotron, the measurements were confirmed by experiments done by him in collaboration with Jack Steinberger.

Events Leading up to the Loyalty Oath

In late 1949, Pief was perhaps at the most pivotal point of his life. Following the first Soviet nuclear test on August 29 and Truman's decision to proceed with the hydrogen bomb, Lawrence and Alvarez decided that they wanted UCRL to contribute to the project and find ways to produce tritium or breed plutonium with large quantities of neutrons. After considering various methods, they converged on a proton and deuterium linear accelerator modeled after the earlier 32 MeV machine but at the much lower frequency of 12 MHz, with a diameter of 60 ft. Code-named the Materials Test Accelerator (MTA), its first stage (87 ft long) was eventually built at an abandoned naval air station near Livermore (which later became LLNL). Somewhat reluctantly, because he was already having strong second thoughts about further nuclear weapons, Pief worked on the microwave cavities for the project. But then other political events caught up with him at UCRL. These are too lengthy to recount here, but eventually they led to the Loyalty Oath, whereby the university, in Pief's words, would require its employees to "affirm their lack of Communist contamination." Pief signed the oath but became very upset when others who had refused to sign it were threatened with dismissal, and as a result he decided to resign from the lab. Alvarez tried to dissuade him but Leonard Schiff and Felix Bloch at Stanford got word of his resignation and successfully enticed him to come to Stanford. Would SLAC be here if all this had not happened? We will never know. What we do learn from this incident and many subsequent ones is that Pief never seemed to hold any grudges. Despite his fundamental disagreement with Alvarez, he maintained his friendship with him after he left Berkeley.

Stanford, the Microwave Lab and HEPL

When Pief and his growing family arrived at Stanford in early July of 1951, they moved into a big 1907 house in Los Altos, where they lived ever after. At that time the university, in particular the physics and electronics departments, was going through an unprecedented period of expansion. Pief joined both the physics department and the Microwave Laboratory. At the latter, he immediately got involved with the MARK III linear accelerator, whose conception and early construction had started under the remarkable leadership of physicist William W. Hansen, with help from the inventors of the klystron, Russell and Sigurd Varian. Hansen died prematurely of lung disease in May 1949, but the construction of MARK III continued successfully under the

leadership of Ed Ginzton with a gifted team consisting of Marvin Chodorow, R. L. Kyhl, Richard Post, Richard Neal and many others. Actually, when Pief arrived, there were a number of problems with the traveling wave accelerator sections (which arced at the design gradient) and with the reliability of the klystrons and modulators. The arcing problem forced the designers to add more sections, all running at more modest gradients, thereby lengthening the entire accelerator to the point where there was no room left at the end of the building for experiments. Robert Hofstadter, who had come from Princeton in 1950, had to begin his physics research program with a spectrometer located at the halfway point of the accelerator. This problem led to Pief's first challenge: extend the building so that it could accommodate an appropriate beam switchyard and end-station. The job required working with university architects, getting financial support from Ginzton, having the experimental area designed, etc., which taxed all his physics and administrative skills. A large amount of earth dug out for the job was piled up at the end of the new extension and served as a beam stopper. When the writer arrived at Stanford, it was known as Mount Panofsky! Emerging from the beam switchyard were two separate beam lines: one for Hofstadter's electron scattering experiments (for which he was awarded the Nobel Prize in 1961), and a second one for general use. This arrangement much later led to similar designs, albeit much larger, for SLAC.

Pief relates in considerable detail all his early experiences with his colleagues; the Office of Naval Research (ONR), which funded most of the work; the Physics Department, where he soon carried a heavy teaching load; and his interactions with Ed Ginzton. Eventually, Pief and Ed decided to split their labor and responsibilities: the laboratory was divided into two parts — the High Energy Physics Laboratory (HEPL), of which Pief became the director; and the Microwave Laboratory, for microwave tube research, headed by Ginzton. The two together were named the W. W. Hansen Laboratories.

By then, Pief had a large family with five children, was teaching, carrying out and directing research, traveling a lot and getting increasingly involved in arms control work on a national scale. Pief himself wonders (and the reader does too) how he was able to keep so many activities going simultaneously. The main explanations one can find are that he worked extremely hard, that he was an extraordinary planner, and that he was amazingly quick.

The "general use" second beam line was exploited mostly by Pief, his colleague from Berkeley, Robert Mozley (and Mozley's student Richard Taylor), and eventually a sequence of 13 other graduate students. A number of experiments were devoted to pions produced directly by electrons, in contrast to those using gamma rays at Berkeley. Various people, like Karl Brown, George Masek, Daryl Reagan, Peter Phillips and Lou Hand, came to work under Pief's supervision. With Karl Brown they designed a double-focusing spectrometer, with George Masek they electromagnetically produced muon pairs, and with Peter Phillips they designed the first single-cavity radiofrequency deflector. This experiment was also connected with what is known as the Panofsky–Wenzel theorem, which specifies the properties of electromagnetic modes capable of deflecting charged particles transversely (interestingly, Pief forgot to mention this theorem in his book!). By 1956, Pief was joined at HEPL by Research Associate Burton Richter, who, sensing the energy limitations of MARK III, together with Princeton's Gerard O'Neill and Carl Barber, designed and then built the pioneering electron–electron storage ring collider. This work was eventually superseded by Richter's and others' much more successful electron–positron colliders.

The Rise of SLAC

The success of the MARK III electron accelerator, both as a machine eventually reaching over 1 GeV in energy and as a rich source of particle physics research, inevitably led to the "next step" question. Speculations were started by Robert Hofstadter and were followed by numerous conversations involving him, Pief, Ginzton, Leonard Schiff, Richard Neal and others. A first report exploring the possibilities of a machine much larger than MARK III was presented at the 1956 CERN Symposium on High Energy Accelerators by Pief and Neal. Meanwhile, a series of meetings was organized to come up with a proposal, the first of which was held in the evening of April 10, 1956 at Pief's home. (Note in passing that quite by coincidence, April 10, 2008 was chosen as the date for an international symposium at Stanford to celebrate Pief's life.)

The formal "Proposal for a Two-Mile Electron Accelerator," which was written with the help of young English major Bill Kirk, came out in April 1957, about 51 years ago. It was a relatively short report of 64 pages plus appendices, which was submitted simultaneously to the Office of Naval Research (ONR), the Atomic Energy Commission (AEC) and the National Science Foundation (NSF). In retrospect, Pief considered the technical part of the proposal a little naïve but thought that the cost estimate was realistic. After submission of the proposal for what was originally called Project M (for "Multi-BeV" or "Monster,") there was a protracted period of ups and downs. Several controversies arose in the university, in the scientific community, the AEC (which was eventually chosen to be the funding agency), in the executive branch of the government and in Congress. The description of these anxiety-producing but fascinating (in retrospect) events occupies over ten pages of the book and is much too long to repeat here.

In the end, SLAC was created as an entity separate from the Stanford Physics Department (in Pief's words, "academically joint, administratively separate from the university") and as a national facility. Upon the death of the two Varian brothers, Ed Ginzton, who had directed the Project M research phase, resigned from Stanford to assume the leadership of Varian Associates, leaving Pief as director of SLAC. Pief used all his persuasive powers to overcome a number of the Joint Committee on Atomic Energy's positions, such as the AEC wanting to run the A&E firm, not wanting to let Stanford's H&R policies prevail over the government's, and insisting on allowing classified work on the site. Pief won out on all these points. President Eisenhower endorsed the project in 1959 and the Democratic Congress finally approved its construction on September 15, 1961 with a budget of US$114 million. The contract and a separate land lease of the Sand Hill site for 50 years were signed in April 1962, and groundbreaking started in July 1962.

Building SLAC

Of all his accomplishments, building SLAC was probably Pief's "finest hour." Again, the description of this period is much too long to repeat here in detail, but a few salient topics and incidents should be mentioned.

Central Control Room at SLAC, on April 21, 1966, the night the first electron beam was observed at the end of the first 20 sectors of the 3-km-long accelerator. This photograph, taken when it reached sector 13 (of 30), caused exhilaration among the participants. Clockwise from left to right: Pief, Gary Warren, Don Robbins, Matt Sands, Dick Neal, Ned Farinholt, Ken Crook, Dieter Walz, Ed Seppi and Greg Loew. (*Credit: SLAC Archives and History Office.*)

What made Pief such an effective director and project leader were his total commitment, his incredible intellect, his ability to delegate, his technical insights to jump in when a difficult scientific problem arose, the trust he created with his staff through his willingness to listen, and his humility. Of course Pief does not advertise these qualities in the book, but the reader may discern them intuitively.

The line organization Pief chose for the lab was nimble and efficient, and it survived for almost 40 years. People such as business manager Fred Pindar, his head of administrative services Robert Moulton, his first deputy director Matt Sands, and research division associate director Joe Ballam are often mentioned for their contributions. One of the people Pief held in enormous regard was Richard (Dick) Neal, head of the technical division, who was in charge of the construction of the entire accelerator and who carried an enormous load. The busy reader may want to look up some of the unusual stories regarding the engineering of the disk-loaded waveguide, the alignment system, the colemanite for accelerator shielding, the discovery of the 14-million-year-old

Paleoparadoxia fossil during the beam switchyard excavation, and the controversy which arose with the neighboring city of Woodside over our 220 KV power line. The last turned out to generate a schedule cliffhanger which was finally resolved amicably, in large part because of Pief's negotiating skills.

Two specific areas where Pief made personal technical contributions to the accelerator are noteworthy. One was his idea to enable the linac operators to instantly detect where the beam was poorly steered in the 3 km accelerator, producing radiation along the way. For this he proposed an argon-filled, 3-km-long coax cable along the machine which would get ionized and break down at the location where the radiation was produced. The breakdown pulse profile was constantly displayed on a scope in the control room. The device functioned very well and was called PLIC ("Panofsky's long ion chamber"). Pief's second personal contribution took place immediately after the first 15 GeV beam was steered down the accelerator in April and May 1966 and exhibited a detrimental behavior called "beam breakup" which shortened the pulse as the current was increased.

SLAC dedication, September 1967. From left to right: Glenn Seaborg (Chair, AEC), Pief, Wallace Sterling (Stanford University President), Don Hornig (President's Science Advisor), Ed Ginzton (Varian Associates CEO). (*Credit: Stanford University News Service.*)

Several explanations were offered, mostly having to do with a transverse microwave beam-induced instability, but Pief together with then postdoc Myron Bander was the first to propose a correct analytic solution for this cumulative instability growth with length, current and time (in the book, the reviewer believes, Pief somewhat erroneously labels the instability as regenerative, an adjective which is usually reserved for the beam breakup seen at much higher currents in a single section). More theoretical work on the cumulative instability was then done for several years by Richard Helm, in collaboration with many experimentalists, including the author, until mitigation to full beam specifications was attained about four years later.

The construction of the accelerator was completed in the summer of 1966, within schedule and within budget. The official dedication took place the next year in September 1967.

Physics Research at SLAC in the First Ten Years

Another of Pief's visionary contributions to the success of SLAC was his realization that if physics research was going to start promptly upon completion of the accelerator, the instruments had to be developed in parallel with the machine, and strong groups of physicists had to be hired early on to design and build them. (Many of these physicists, together with a strong particle theory group, made up the original SLAC faculty.) Hence, by 1967, the Research Yard already consisted of three fully equipped end-stations: three spectrometers in End-Station A, a 1 m bubble chamber and spark chamber in End-Station B, and a 2 m bubble chamber (from LBL) and eventually a streamer chamber in the central C-Beam.

Pief devotes an entire chapter of the book to a description of the particle physics developments preceding the inception of SLAC: pions, muons, kaons, neutrinos, the eightfold way, P and CP violation and so on, and then goes over all the early SLAC experiments. The one which he probably "enabled" the most through his contributions to the spectrometer designs was the deep inelastic scattering experiment in End-Station A, which led to the discovery of the quarks and the Nobel Prize (in 1990) for Taylor, Kendall and Friedman. Other experiments, described in less detail, include the use of photoproduction, polarized electron beams with the Yale group, muons, positrons, and of course the enormous production of bubble chamber pictures.

Paralleling these experiments, Burton Richter and David Ritson from the Physics Department proposed that an electron–positron colliding beam facility be built at SLAC, storing both 3 GeV (maximum) electrons and positrons emerging from the linac at about 1.5 GeV. A committee headed by Jackson Laslett recommended that the project be approved by the AEC but it took five years, starting in 1965, before Pief was able to convince the government to authorize construction of the SPEAR facility, eventually out of equipment funds. This project with two large detectors turned out to be a huge success, leading to the so-called November 1974 revolution, the discovery of the J/psi resonance at 3.1 GeV, and gradually, over the next four years, of the tau lepton by Martin Perl. Both these discoveries led to Nobel Prizes for Richter together with Sam Ting (MIT), and Perl, and were an immense success for the lab.

Other Accelerator Activities under Pief

The success of SPEAR as a particle physics research tool also prompted its first use as a source of synchrotron radiation, around 1974. Pief was instrumental in authorizing early "symbiotic" runs for this purpose, establishing the Stanford Synchrotron Radiation Project (SSRP) under separate university management. About 15 years later, when SPEAR particle physics ended, these runs became fully dedicated and the activity was renamed the Stanford Synchrotron Radiation Laboratory (SSRL). A separate electron injector was built and eventually SSRL was incorporated into SLAC under Burton Richter's directorship. It is an enormously successful enterprise, now in its third incarnation (SPEAR 3), attracting photon physics scientists from all over the world, and it has also led to the Linac Coherent Light Source (LCLS) currently under construction, using the last third of the SLAC linac.

Strange as it may seem, as early as 1968, the premature (in retrospect) promise of rf superconductivity gave rise to the possibility of converting the room temperature pulsed 20 GeV (7 MV/m), 0.1% duty cycle linac to a 100 GeV (33 MV/m), 6% duty cycle linac running at 2°K. Pief authorized the R&D needed for this conversion, which led to a rather optimistic but costly project. After two years of intense

research, when the maximum gradient in S-band niobium cavities barely reached 2 MV/m, Pief called a meeting and, to the dismay of the participants, put an end to the project. Despite their disappointment, it turned out that his "executive" decision was correct at the time: the 33 MV/m gradient is just becoming possible 40 years later!

Another design project was then encouraged by Pief to double the energy of the linac to 40 GeV by recirculating the beam through the machine a second time (Project RLA — Recirculating Linear Accelerator). This research led to a full proposal and budget estimate but it was turned down by the AEC in 1973 as being premature and too costly. The pressure then to "find another way" resulted in the SLED invention (SLAC Energy Doubler), using resonant cavities downstream of the klystrons and pulse compression techniques. Its gradual implementation over several years, together with higher power, longer pulse klystrons, led to a relatively inexpensive method to more than double the energy of the linac to 50 GeV.

Meanwhile, in 1976, the success of SPEAR prompted the proposal of the much larger colliding beam machine called PEP, which was the last new accelerator to come into successful operation under Pief's management, in 1980. A year before Pief's retirement in 1984, he still managed to strongly support the SLAC Linear Collider (SLC) project, proposed by Burt Richter. Its construction was started in 1983 and provided the first successful collisions at the Z boson resonance in 1989, the same year as the Loma Prieta earthquake.

All of these developments are vividly described in the penultimate chapter of Pief's book. In this chapter he also gives a fairly detailed account of his personal involvement in the ill-fated history of the SSC (Superconducting Supercollider). The author of this review would add to this controversial story a point which Pief does not mention, but seemed to the author like the "original sin," i.e. not to insist that the SSC be sited at Fermilab, where much of the infrastructure already existed.

Pief ends this chapter with some general observations on how the international landscape of high energy physics has changed over his lifetime and how these changes and increased costs are affecting the approval of the International Linear Collider (ILC).

Science Advising and International Science

Pief's familiarity with the Manhattan Project, his involvements at UCRL, his arrival at Stanford, his drafting of the "Screwdriver Report" on fissile materials detection with Hofstadter, and his overall eclectic scientific expertise propelled him as early as 1954 into a very long series of activities and panels having to do with science advising, first at NSF and later with the Air Force, culminating with the President's Science Advisory Committee (PSAC) under George Kistiakowski during the Eisenhower Administration. As a result, in 1959, while he was taking a sabbatical at CERN, he got involved in his first negotiations with the Soviets. These negotiations culminated in 1963 with the adoption of the Limited Test Ban Treaty with the USSR.

Pief stayed on the PSAC until 1964. His experience and thoughts on the roles, responsibilities, conflicts of interest and accountability of science advisors are discussed in great detail in the book and should be read by anybody who decides to accept such a position in any government agency.

In addition to his service to various US government panels, and to Stanford University in Prof. Franklin's dismissal controversy, Pief began to play a major role in international scientific organizations such as the International Union of Pure and Applied Physics (IUPAP), and was invited to many conferences to give talks and reports.

In the late 1970s, partially because of his efforts, various government-to-government collaborative science agreements were signed by the US with the Soviets, the Japanese and the People's Republic of China (PRC), and annual cooperative meetings were held on high energy physics. Pief attended many of these and played major roles in the collaborations. With regard to the PRC, he and T. D. Lee deserve personal credit for having encouraged and helped the Chinese to build an electron–positron collider in Beijing (BEPC). To get them started in this direction, in the summer of 1982 Pief invited a

PSAC meeting in Newport News, July 12, 1960. Pief is seen presenting the nuclear test ban report to President Eisenhower across the table. Clockwise from left to right: Mannie Piore, Don Hornig, George Kistiakowski, President Eisenhower, George Beadle, unknown, John Tukey, unknown, John Bardeen, Jim Killian, Al Weinberg, W. Panofsky, Jerry Wiesner, Wally Z, Detlev Bronk. (*Credit: SLAC Archives and History Office, Panofsky Collection.*)

Pief among "giants" Tom Kirk and Chen Hesheng, at the 25th Anniversary US–PRC collaboration meeting in Beijing, October 2004. (*Credit: Fred Harris, University of Hawaii.*)

delegation of about 30 Chinese physicists and engineers to SLAC to produce a preliminary design of the machine. This relationship has survived for over 25 years despite occasional problems, including the 1989 Tiananmen events, which Pief deplored. Pief, together with others, was responsible during the October 14, 1998 US–PRC meeting in Beijing for briefing Premier Zhu Rongji to approve the BEPC II upgrade.

Arms Control (1981–2007): The Unfinished Business

The control and drastic reduction of nuclear weapons was a challenge Pief confronted daily — to the last day of his life. He involved himself in each and every controversy in this area, and even though he sometimes seemed discouraged, he never gave up. What struck the reviewer was that, no matter how passionately Pief felt about a particular aspect of this subject, he never departed from rational arguments and would not allow himself to become a polemicist. In a world that is often irrational, and where an uninformed public can be driven by very superficial political arguments, being totally fair does not always produce fast results. But, by consistently behaving himself in this manner, Pief always retained the respect of his friends and adversaries.

As early as 1965, Pief was recruited to serve as a member of JASON, a group of academics who get together every summer to advise the US government on matters of general and national security interest. In 1981, he joined the Committee on

International Security and Arms Control (CISAC) of the National Academy of Sciences, a committee he chaired from 1985 to 1993. CISAC started out by holding bilateral discussions with the Russians but these were later extended to very productive contacts with China, and then with allies such as France, Italy, Germany and the UK. In Italy these contacts developed into the multinational Amaldi conferences, which Pief attended.

What positions did Pief take? He fundamentally believed that after 1945 and certainly during the Cold War, nuclear war was no longer a possible strategy for any nation, that nuclear weapons could not serve any military function for any nation, except to deter another nation from attacking it with nuclear weapons. For this, the reader is referred to his article written with his friend Spurgeon Keeney on "MAD vs. NUTS," (Mutual Assured Destruction vs. Nuclear Utilization Target Selection).

In summary, Pief supported the Limited Test Ban Treaty (LTBT), the Threshold Test Ban Treaty (TTBT), the Ban on Peaceful Nuclear Explosions (PNE), the Nuclear Nonproliferation Treaty (NPT), the Anti-Ballistic Missile Treaty (ABM), and the Comprehensive Test Ban Treaty (CTBT), which unfortunately has not yet been ratified by the US Congress. When President Reagan proposed his Star Wars project, Pief opposed it on technical grounds, arguing scientifically that it would not really work and that the "offense would always outstrip the defense" because it would be less expensive and more effective. During his CISAC chairmanship, Pief argued eloquently against lumping nuclear, chemical and biological weapons under the single WMD label, and against threatening to use nuclear weapons to deter the use of the other two weapons. He also conducted the study for the management and disposition of excess plutonium, which came up with recommendations to fabricate mixed oxide fuel (MOX) combining plutonium and uranium oxides for use in reactors, or to mix the plutonium with highly radioactive fission products which would then be disposed of in a geological repository. Presidents George W. Bush and Vladimir Putin signed the Plutonium Management and Disposition Agreement (PMDA) to dispose of 34 tons of plutonium via the MOX process, but because of mutual disagreements over bureaucratic issues, no disposition has yet taken place, seven years later.

Pief ends his book with a strong admonition to the world. There are still close to 30,000 nuclear weapons on the planet today. Such a number is far in excess of any security need. The risks of inadvertent launches due to faulty communications, desperate regional conflicts, proliferation and theft are enormous and could be totally devastating for humanity. As long as the US relies on these weapons or continues to invent new missions for them (like bunker busters) or new designs (like the Reliable Replacement Warhead, RRW), other countries will see these weapons as symbols of national power and will be tempted to acquire them. In this connection, in 2006 Pief had met with the Iranian Ambassador at the UN, and until the last day of his life he was upset that the US was not negotiating directly with that country on the nuclear problem.

Pief notes that a declaration of "no first use" of nuclear weapons has so far been embraced only by China and by none of the other nuclear weapons states. He believes that if adopted by all of them, it would at last motivate and enable them to strive for drastic reductions, revitalize the entire nuclear weapons arms control drive, and eventually lead to a worldwide prohibition against possessing nuclear weapons. He writes, "The United States, as the unquestioned leader — measured by non-nuclear armaments and economic strength — should have the strongest possible interest in leading the reining-in of nuclear weapons on an irreversible basis."

The world could not honor this wonderful scientist and human being more than by heeding his advice on dealing with this ominous threat to humanity.

Gregory Loew recently became Professor Emeritus after working at SLAC (originally Project M) for exactly fifty years. He was involved in the design and construction of the SLAC 3-km linac, and thereafter contributed to all the subsequent upgrades and projects that were added at SLAC over the years, including SPEAR, PEP-I, SLC, and PEP-II (the B-Factory). He also spent many years working on the design of future electron–positron linear colliders. He is still very interested in what happens at SLAC and plans to remain active as long as he can.